Contents

Contents

List of Boxes

Acknowledgements

The Publishers are grateful to the following for permission to reproduce copyright material:

Blackwell Scientific Publications for Figs 20.1 and 20.2 (Golterman *et al.*, 1978); Fisher Scientific for Fig. 23.1; Pharmacia Biotech for Figs 26.4a and 27.3; Microsoft Corporation for Fig. 50.1.

We would especially like to thank the following people whose feedback on *Practical Skills in Biomolecular Sciences* has been valuable in the planning of this text:

Dr M. Miller (University of Odense, Denmark)
Dr A. Radford (University of Leeds, UK)
Dr C. Smith (Manchester Metropolitan University, UK)

Preface

Practical work forms the cornerstone of scientific knowledge: our current understanding of the natural world has been shaped by the observations and experiments of earlier scientists from a broad range of disciplines. In recent years, our knowledge of the operation of biological systems at the cellular and molecular level has increased as a result of spectacular advances in techniques and methodology, for example, in nucleic acid analysis and genetic engineering. At the same time, improved methods of analysis have enabled scientists to monitor and study the smallest traces of individual biomolecules. As a result, the training required for practical work in cell and molecular bioscience is diverse, from the skills involved in the design of practical investigations and the appropriate use of a range of analytical equipment, to the presentation of results in written and oral form.

The concept for the present book arose as a direct result of feedback from the earlier text *Practical Skills in Biology*. A number of reviewers, colleagues and students commented that a text with a greater emphasis on molecular and cellular aspects of the life sciences would provide practical support to students taking courses where these aspects form the major component of the syllabus, e.g. microbiology, biochemistry, genetics, molecular biology, biomedical sciences, physiology and paramedical subjects. Alongside the production of a second edition of *Practical Skills in Biology*, we have taken the opportunity to update and modify the relevant material from the earlier text (e.g. recording, interpreting and reporting results), combining it with a number of new sections focused on analytical and molecular aspects (e.g. separation methods, cell culture, microbial genetics), to produce *Practical Skills in Biomolecular Sciences*. This title was chosen to reflect the breadth of coverage – it was never our intention to produce a detailed technical manual dealing with bio-analytical methodology or biochemical techniques for the specialist reader, but to provide students with a 'user-friendly' guide to those aspects of practical work that newcomers find most difficult. The new edition of *Practical Skills in Biology* differs from this text in its greater coverage of whole organisms and ecology, providing students with a clear choice of texts, depending on their interests.

We have selected material for inclusion in *Practical Skills in Biomolecular Sciences* based on our own teaching experience, highlighting those areas where our students have needed further guidance. It would be easy to criticize the lack of detail within a particular section from the narrow viewpoint of an individual specialist discipline. However, since most students take a number of courses in their early years, we have tried to provide a broader range of help and assistance, also dealing with those aspects that may not be covered in detail within the formal academic programme (e.g. using logarithms and exponential numbers, in Chapter 45, or study skills, in Chapter 59). We have also included specific details for a number of practical procedures and experimental protocols, so that the text can serve as a source of useful information on topics as varied as the preparation of solutions (Chapter 5), the polymerase chain reaction (Chapter 41), obtaining information via the Internet (Chapter 49) and the interpretation of statistical tests (Chapter 47).

A comparison between the second edition of *Practical Skills in Biology* and *Practical Skills in Biomolecular Sciences* shows that the latter text contains 18 new chapters (Chapters 17, 22, 25-34, 36, 37, 39–41, 49), together with a number of others that have undergone major modification (e.g. Chapters 9, 14, 21, 24, 35, 42, 46), with new sections dealing with practical aspects of biomolecules, metabolism and genetics. In other sections, the relevant material from *Practical Skills in Biology* (2nd Edn.) has been retained, e.g. in the first few chapters, covering fundamental principles, and in the final chapters, dealing with communicating information. This text is over 10 per cent longer than *Practical Skills in Biology* (2nd Edn.), reflecting the greater number of chapters and the increased coverage of analytical techniques and methodology.

Most students will have access to specialist textbooks giving in-depth coverage of the theoretical principles and knowledge covered in lectures (e.g. details of structural biochemistry and intermediary metabolism). This book aims to supplement – rather than replace – such textbooks, and covers the skills required in laboratory classes, together with practical advice, tips and hints, worked examples, definitions, key points, 'how to' boxes and checklists. The text provides an outline of the underlying theoretical principles where necessary, but emphasis throughout is on the practical applications of this information. While most of the material is aimed at students in the early years of their undergraduate courses, we are confident that the book will remain useful as students progress to honours degree level and beyond, and we have included a number of chapters dealing with more advanced aspects (e.g. spectroscopy and spectrometry, Chapter 22, electrophoresis, Chapter 27, project work, Chapter 12). More detailed coverage of specific techniques, methods, apparatus and protocols is available in the references quoted within individual chapters, and in texts such as the Practical Approach Series (IRL Press, Oxford), the Introduction to Biotechniques Series (Bios, Oxford), Methods in Enzymology (Academic Press, London) and Analytical Chemistry by Open Learning, ACOL (Wiley, Chichester).

To students who buy this book, we hope that you will find it useful in the laboratory, during your practical classes and in your project work – this is not a book to be left on the bookshelf. Lecturers should find that the text provides an effective means of supplementing the information given in practical classes, where constraints on time and resources can lead to the under-performance of students.

We would like to acknowledge the support of our families and the help provided by colleagues who read early drafts of the revised material and new chapters. Special thanks are due to: James Abbott, Jon Bookham, Brian Eddy, Mark Daniels, Martin Davies, John Dean, Jackie Eager, Howard Griffiths, Claire Halpin, Steve Hitchin, Derek Holmes, Ed Ludkin, Ian Kill, Pete Maskrey, Tom Marshall, Kate Maxwell, Steve Millam, Rachel Morris, John Raven, Zoë Reed, Pete Rowell, Bill Tomlinson, Will Whitfield, Katherine Williams, Ian Winship, Peter Wright, Bob Young and Hilary-Kay Young. Despite this help, the responsibility for any errors rests with us and we would be grateful if readers could alert us to any errors or problems, so that we can make amends as soon as possible. Please write to us at Northumbria or Dundee, or by e-mail at the following addresses: rob.reed@unn.ac.uk (RHR); david.holmes@unn.ac.uk (DH); j.d.b.weyers@dundee.ac.uk (JDBW); a.m.jones@dundee.ac.uk (AMJ).

ROB REED, DAVID HOLMES, JONATHAN WEYERS
AND ALLAN JONES

List of abbreviations

AC	affinity chromatography
ACDP	Advisory Committee on Dangerous Pathogens
ADP	adenosine diphosphate
ANOVA	analysis of variance
ATP	adenosine triphosphate
BSA	bovine serum albumin
CCCP	carbonylcyanide *m*-chlorophenylhydrazine
CE	capillary electrophoresis
CFU	colony-forming unit
CGE	capillary gel electrophoresis
COSHH	Control of Substances Hazardous to Health
CoV	coefficient of variance
CTP	cytosine triphosphate
CZE	capillary zone electrophoresis
ddNTP	dideoxyribonucleotide triphosphate
DMSO	dimethyl sulphoxide
DNA	deoxyribonucleic acid
dsDNA	double stranded DNA
dNTP	deoxyribonucleoside triphosphate
ECD	electron capture detector
EDTA	ethylenediaminetetraacetic acid
EI	electron impact ionization
EIA	enzyme immunoassay
ELISA	enzyme-linked immunosorbent assay
EMR	electromagnetic radiation
EOF	electro-osmotic flow
ESR	electron spin resonance
F	Faraday constant
FAB-MS	fast atom bombardment-mass spectrometry
FIA	fluorescence immunoassay
FID	flame ionization detector
FPLC®	fast protein liquid chromatography
FT	Fourier transformation
FT-IR	Fourier transform infra-red spectroscopy
GC	gas chromatography
GPC	gel permeation chromatography
h	Planck constant
HIC	hydrophobic interaction chromatography
HPLC	high performance liquid chromatography
IEC	ion-exchange chromatography
IEF	isoelectric focusing
Ig	immunoglobulin
IMAC	immobilized metal affinity chromatography
IR	infra-red (radiation)
IRGA	infra-red gas analyser
IRMA	immunoradiometric assay
IRMS	isotope ratio mass spectroscopy
ISE	ion selective electrode
K_m	Michaelis constant

LDH	lactate dehydrogenase
MEKC	micellar electrokinetic chromatography
MPN	most probable number
M_r	relative molecular mass
MRI	magnetic resonance imaging
MS	mass spectrometry
NAD^+	nicotinamide adenine dinucleotide (oxidized form)
NADH	nicotinamide adenine dinucleotide (reduced form)
$NADP^+$	nicotinamide adenine dinucleotide phosphate (oxidized form)
NADPH	nicotinamide adenine dinucleotide phosphate (reduced form)
NH	null hypothesis
NMR	nuclear magnetic resonance
PAGE	polyacrylamide gel electrophoresis
PAR	photosynthetically active radiation
PCR	polymerase chain reaction
PEG	polyethylene glycol
PFD	photon flux density
PFU	plaque-forming unit
PGFE	pulsed f ield gel electrophoresis
PI	photosynthetic irradiance
PPFD	photosynthetic photon flux density
PPi	pyrophosphate
PVA	polyvinyl alcohol
PY-MS	pyrolysis-mass spectroscopy
R	universal gas constant
RCF	relative centrifugal field
R_F	relative frontal mobility
RIA	radioimmunoassay
RID	radioimmunodiffusion
RNA	ribonucleic acid
RP-HPLC	reverse phase-high performance liquid chromatography
rpm	revolutions per minute
RT	reverse transcriptase
SCOT	support-coated open tubular (capillary column)
SDS	sodium dodecyl sulphate
SE	standard error (of the sample mean)
SEM	scanning electron microscopy
SI	Système Internationale D'Unités
ssRNA	single stranded RNA
STP	standard temperature and pressure
TCA	trichloracetic acid
TCD	thermal conductivity detector
TEM	transmission electron microscopy
TEMED	N,N,N′,N′-tetramethylethylenediamine
TLC	thin layer chromatography
TPMD	tetramethylphenyldiamine
TRIS	tris(hydroxymethyl)aminomethane *or* 2-amino-2-hydroxymethyl-1,3-propanediol
TTP	thymidine triphosphate
UNG	uracil-N-glycosylase
URL	uniform resource locator
UV	ultraviolet (radiation)
V_{max}	maximum velocity

Fundamental laboratory techniques

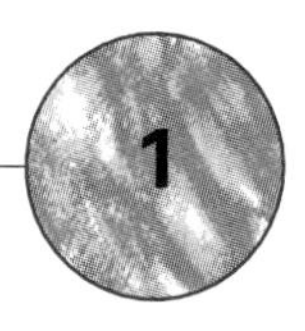

1 Basic principles

Developing practical skills – these will include:

- observing and measuring
- recording data
- designing experiments
- analysing and interpreting data
- reporting/presenting.

All knowledge and theory in science has originated from practical observation and experimentation: this is equally true for disciplines as diverse as microscopy and molecular genetics. Practical work is an important part of most courses and often accounts for a significant proportion of the assessment marks. This book aims to provide an easy-to-use reference source dealing with the basic practical techniques and skills of relevance to biomolecular sciences. The abilities developed in practical classes will continue to be useful throughout your course and beyond, some within science and others in any career you choose.

Being prepared

KEY POINT You will get the most out of practicals if you prepare well in advance. Do not go into a practical session assuming that everything will be provided, without any input on your part.

The main points to remember are:

Using textbooks in the lab – take this book along to the relevant classes, so that you can make full use of the information during the practical sessions.

- Read any handouts in advance: make sure you understand the purpose of the practical and the particular skills involved. Does the practical relate to, or expand upon, a current topic in your lectures? Is there any additional preparatory reading that will help?
- Take along appropriate textbooks, to explain aspects in the practical.
- Consider what safety hazards might be involved, and any precautions you might need to take, before you begin (p. 6).
- Listen carefully to any introductory guidance and note any important points: adjust your schedule/handout as necessary.
- During the practical session, organize your bench space – make sure your lab book is adjacent to, but not within, your working area. You will often find it easiest to keep clean items of glassware, etc. on one side of your working space, with used equipment on the other side.
- Write up your work as soon as possible, and submit it on time, or you may lose marks.
- Catch up on any work you have missed as soon as possible – preferably, before the next practical session.

Basic requirements

Recording practical results

An A4 loose-leaf ring binder offers flexibility, since you can insert laboratory handouts and lined and graph paper at appropriate points. The danger of losing one or more pages from a loose-leaf system is the main drawback. Bound books avoid this problem, although those containing alternating lined/graph or lined/blank pages tend to be wasteful – it is often better to paste sheets of graph paper into a bound book, as required.

Presenting results – while you don't need to be a graphic designer to produce work of a satisfactory standard, presentation and layout are important and you will lose marks for poorly presented work.

A good quality HB pencil or propelling pencil is recommended for recording your raw data, making diagrams, etc. as mistakes are easily corrected. Buy a black, spirit-based (permanent) marker for labelling experimental glassware, Petri plates, etc. Fibre-tipped fine line drawing/lettering pens are useful for preparing final versions of graphs and diagrams for assessment purposes. Use a clear ruler (with an undamaged edge) for graph drawing, so that you can see data points/information below the ruler as you draw.

Calculators

These range from basic machines with no pre-programmed functions and only one memory, to sophisticated programmable minicomputers with many memories. The following may be helpful when using a calculator:

- Power sources. Choose a battery-powered machine, rather than a mains-operated or solar-powered type. You will need one with basic mathematical/scientific operations including powers, logarithms (p. 260), roots and parentheses (brackets), together with statistical functions such as sample means and standard deviations (Chapter 46).
- Mode of operation. The older operating system used by e.g. Hewlett Packard calculators is known as the reverse Polish notation: to calculate the sum of two numbers, the sequence is 2 [enter] 4 + and the answer 6 is displayed. The more usual method of calculating this equation is as $2 + 4 =$, which is the system used by the majority of modern calculators. Most newcomers find the latter approach to be more straightforward. Spend some time finding out how a calculator operates, e.g. does it have true algebraic logic ($\surd$ then number, rather than number then $\surd$)? How does it deal with scientific notation (p. 260) ?
- Display. Some calculators will display an entire mathematical operation (e.g. '$2 + 4 = 6$'), while others simply display the last number/operation. The former type may offer advantages in tracing errors.
- Complexity. In the early stages, it is usually better to avoid the more complex machines, full of impressive-looking, but often unused pre-programmed functions – go for more memory, parentheses, or statistical functions rather than engineering or mathematical constants. Programmable calculators may be worth considering for more advanced studies. However, it is important to note that such calculators are often unacceptable for exams.

Presenting more advanced practical work

Presenting graphs and diagrams – ensure these are large enough to be easily read: a common error is to present graphs or diagrams that are too small, with poorly chosen scales.

In some practical reports and in project work, you may need to use more sophisticated presentation equipment. Computer-based graphics packages can be useful – choose easily-read fonts such as Arial or Helvetica for project work and posters and consider the layout and content carefully (p. 316). Alternatively, you could use fine line drawing pens and dry-transfer lettering/symbols, such as those made by Letraset®, although this approach can be more time-consuming than computer-based systems.

Printing on acetates – standard overhead transparencies are not suitable for use in laser printers or photocopiers: you need to make sure that you use the correct type.

To prepare overhead transparencies for oral presentations, you can use spirit-based markers and acetate sheets. An alternative approach is to print directly from a computer-based package, using a laser printer and special acetates, or directly onto 35 mm slides. You can also photocopy onto special acetates. Advice on content and presentation is given in Chapter 58.

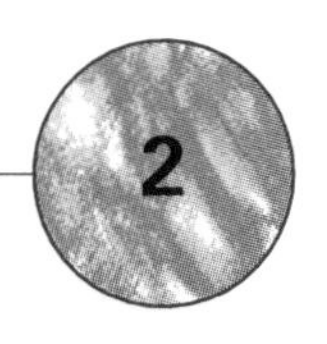

2 Health and safety

In the UK, the **Health & Safety at Work, etc. Act 1974** provides the main legal framework for health and safety. The **Control of Substances Hazardous to Health (COSHH) Regulations 1994 and 1996** impose specific legal requirements for risk assessment wherever hazardous chemicals or biological agents are used, with Approved Codes of Practice for the control of hazardous substances, carcinogens and biological agents, including pathogenic microbes.

Health and safety law requires institutions to provide a working environment that is safe and without risk to health. Where appropriate, training and information on safe working practices must be provided. Students and staff must take reasonable care to ensure the health and safety of themselves and of others, and must not misuse any safety equipment.

KEY POINT **All practical work must be carried out with safety in mind, to minimize the risk of harm to yourself and to others – safety is everyone's responsibility.**

Risk assessment

Definitions

Hazard – the ability of a substance or biological agent to cause harm.

Risk – the likelihood that a substance or biological agent might be harmful under specific circumstances.

The most widespread approach to safe working practice involves the use of risk assessment, which aims to establish:

1. The intrinsic chemical, biological and physical hazards, together with any maximum exposure limits (MELs) or occupational exposure standards (OESs), where appropriate. Chemical manufacturers provide data sheets listing the hazards associated with particular chemical compounds, while pathogenic (disease-causing) microbes are categorized according to their ability to cause illness (p. 61).
2. The risks involved, by taking into account the amount of substance to be used, the way in which it will be used and the possible routes of entry into the body (Fig. 2.1). In this regard, it is important to distinguish between the intrinsic hazards of a particular substance and the risks involved in its use in a particular exercise.
3. The persons at risk, and the ways in which they might be exposed to hazardous substances, including accidental exposure (spillage).
4. The steps required to prevent or control exposure. Ideally, a non-hazardous or less hazardous alternative should be used. If this is not feasible, adequate control measures must be used, e.g. a fume cupboard or other containment system. Personal protective equipment (e.g. lab coats, safety glasses) must be used in addition to such containment measures. A safe means of disposal will be required.

Distinguishing between hazard and risk – one of the *hazards* associated with water is drowning. However, the *risk* of drowning in a few drops of water is minimal!

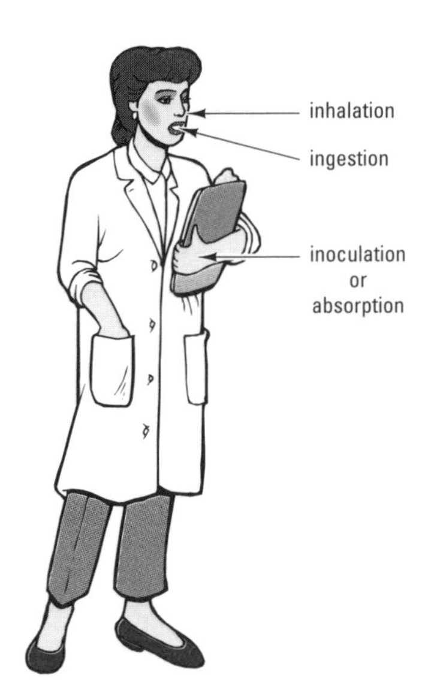

Fig. 2.1 Major routes of entry of harmful substances into the body.

The outcome of the risk assessment process must be recorded and appropriate safety information must be passed on to those at risk. For most practical classes, risk assessments will have been carried out in advance by the person in charge: the information necessary to minimize the risks to students may be given in the practical schedule. Make sure you know how your Department provides such information and that you have read the appropriate material before you begin your practical work. You should also pay close attention to the person in charge at the beginning of the practical session, as they may emphasize the major hazards and risks. In project work, you will need to be involved in the risk assessment process along with your supervisor, before you carry out any practical work.

In addition to specific risk assessments, most institutions will have a safety handbook, giving general details of safe working practices, together with the names and telephone numbers of safety personnel, first aiders, hospitals, etc. Make sure you read this and abide by any instructions.

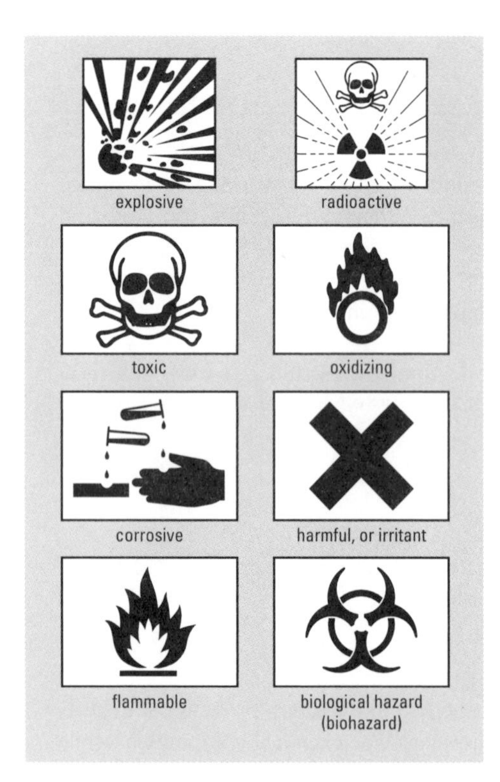

Fig. 2.2 Warning labels for specific chemical hazards.

In the UK, the *Genetic Manipulation Regulations 1989* define the legal requirements for risk assessment and for the notification of work involving genetic manipulation.

The Advisory Committee on Genetic Manipulation provides guidance on the use of genetically manipulated organisms.

The Health and Safety Executive (HSE) has specific responsibility for the operation of these regulations and is the regulatory authority for genetic manipulation in the UK.

Basic rules for laboratory work

- Make sure you know what to do in case of fire, including exit routes, how to raise the alarm, and where to gather on leaving the building. Remember that the most important consideration at all times is human safety: do not attempt to fight a fire unless it is safe to do so.
- All laboratories display notices telling you where to find the first aid kit and who to contact in case of accident/emergency. Report all accidents, even those appearing insignificant – your Department will have a formal recording procedure to comply with safety legislation.
- Wear appropriate protective clothing at all times – a clean lab coat (buttoned up), plus safety glasses if there is any risk to the eyes.
- Never smoke, eat or drink in any laboratory, because of the risks of contamination by inhalation or ingestion (Fig. 2.1).
- Never mouth pipette any liquid. Use a pipette filler (see p. 7).
- Take care when handling glassware – see p. 11 for details.
- Know the warning symbols for specific chemical hazards (see Fig. 2.2).
- Use a fume cupboard for hazardous chemicals. Make sure that it is working and then open the front only as far as is necessary: many fume cupboards are marked with a maximum opening.
- Always use the minimum quantity of any hazardous materials.
- Work in a logical, tidy manner and minimize risks by thinking ahead.
- Always clear up at the end of each session. This is an important aspect of safety, encouraging a responsible attitude towards laboratory work.

Genetic engineering and molecular genetics

Additional legal constraints apply to practical work involving genetic manipulation. A specific risk assessment must be carried out for any experiment where a cell or organism is modified by genetic engineering techniques (Chapter 42) involving the insertion of DNA into a cell or organism in which it does not normally occur. Before any practical work can be carried out, it must be authorized by the establishment's genetic manipulation safety committee and notified to the relevant authority. Such work must be carried out with appropriate containment, to prevent the accidental release of genetically modified organisms into the environment.

Practicals in molecular genetics will involve some of the techniques of genetic manipulation. Typically, these will be examples of 'self-cloning', where recombinant DNA molecules are constructed from fragments of DNA which naturally occur in that organism, for example, the transformation of laboratory strains of *Escherichia coli* using a pUC plasmid (p. 246). Such examples, using microbes unlikely to cause human disease, are usually excluded from the requirements for notification and containment.

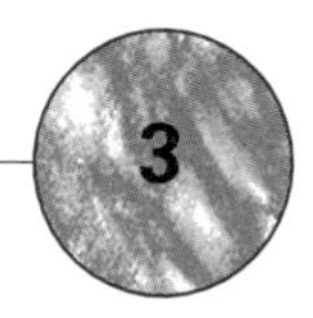

3 Working with liquids

Measuring and dispensing liquids

The equipment you should choose to measure out liquids depends on the volumes being dispensed, the accuracy required and the number of times the job must be done (Table 3.1).

Table 3.1 Criteria for choosing a method for measuring out a liquid

Method	Best volume range	Accuracy	Usefulness for repetitive measurement
Pasteur pipette	30 μl to 2 ml	Low/medium	Very good
Measuring cylinder	5–2000 ml	Medium	Good
Volumetric flask	5–2000 ml	High	Good
Burette	1–100 ml	High	Very good
Pipette/pipettor	5 μl to 25 ml	High*	Very good
Microsyringe	0.5–50 μl	High	Good
Weighing	Any (depends on accuracy of balance)	Very high	Poor
Conical flask/beaker	25–5000 ml	Very low	Good

*If correctly calibrated and used properly (see p. 9).

Reading any volumetric scale – make sure your eye is level with the bottom of the liquid's meniscus and take the reading from this point.

Certain liquids may cause problems:

- High-viscosity liquids are difficult to dispense: allow time for all the liquid to transfer.
- Organic solvents may evaporate rapidly, making measurements inaccurate: work quickly; seal containers quickly.
- Solutions prone to frothing (e.g. protein and detergent solutions) are difficult to measure and dispense: avoid forming bubbles; do not transfer quickly.
- Suspensions (e.g. cell cultures) may sediment: thoroughly mix them before dispensing.

Pasteur pipettes

Hold correctly during use (Fig. 3.1) – keep the pipette vertical, with the middle fingers gripping the barrel while the thumb and index finger provide controlled pressure on the bulb. Squeeze gently to dispense individual drops.

Pasteur pipettes should be used with care for hazardous solutions: remove the tip from the solution before fully releasing pressure on the bulb – the air taken up helps prevent spillage. To avoid the risk of cross-contamination, take care not to draw up solution into the bulb or to lie the pipette on its side. Plastic disposable 'Pastettes®' are safer and avoid contamination.

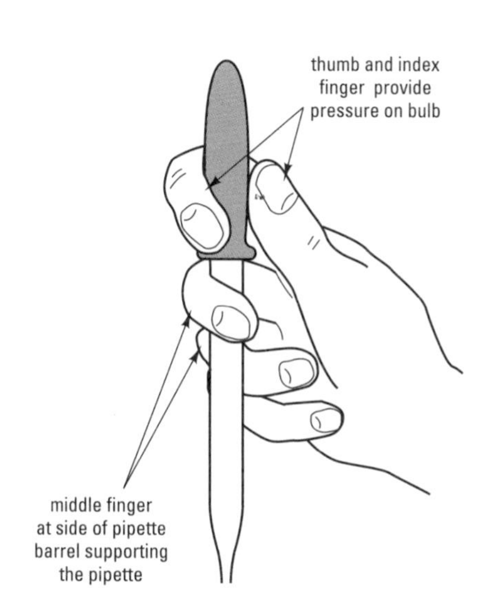

Fig. 3.1 How to hold a Pasteur pipette.

Measuring cylinders and volumetric flasks

These must be used on a level surface so that the scale is horizontal; you should first fill with solution until just below the desired mark; then fill slowly (e.g. using a Pasteur pipette) until the meniscus is level with the mark. Allow time for the solution to run down the walls of the vessel.

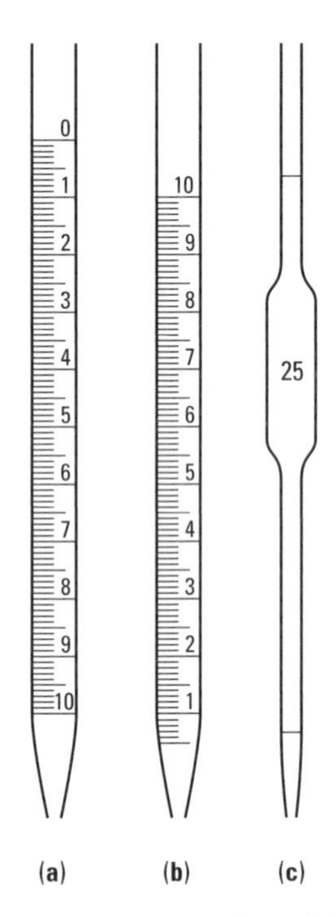

Fig. 3.2 Glass pipettes – graduated pipette, reading from zero to shoulder (a); graduated pipette, reading from maximum to tip, by gravity (b); bulb (volumetric) pipette, showing volume (calibration mark to tip, by gravity) on bulb (c).

Using a pipettor – check your technique (accuracy) by dispensing volumes of distilled water and weighing on a balance, assuming $1\,mg = 1\,\mu l = 1\,mm^3$. For small volumes, measure several 'squirts' together, e.g. $10 \times 15\,\mu l = 150\,mg$. Aim for accuracy of $\pm 1\%$.

Burettes

Burettes should be mounted vertically on a clamp stand – don't overtighten the clamp. First ensure the tap is closed and fill the body with solution using a funnel. Open the tap and allow some liquid to fill the tubing below the tap before first use. Take a meniscus reading, noting the value in your notebook. Dispense the solution via the tap and measure the new meniscus reading. The volume dispensed is the difference between the two readings. Titrations are usually performed using a magnetic stirrer to ensure thorough mixing.

Pipettes

These come in various designs, including graduated and bulb (volumetric) pipettes (Fig. 3.2). Take care to look at the volume scale before use: some pipettes empty from full volume to zero, others from zero to full volume; some scales refer to the shoulder of the tip, others to the tip by gravity or to the tip after blowing out.

KEY POINT **For safety reasons, it is no longer permissible to mouth pipette – various aids are available such as the Pi-pump®.**

Pipettors (Autopipettors)

These come in two basic types:

- Air displacement pipettors. For routine work with dilute aqueous solutions. One of the most widely used examples is the Gilson Pipetman® (Fig. 3.3). Box 3.1 gives details on its use.
- Positive displacement pipettors. For non-standard applications, including dispensing viscous, dense or volatile liquids, or certain procedures in molecular genetics, e.g. the PCR (p. 237), where an air displacement pipettor might create aerosols, leading to errors.

Air displacement and positive displacement pipettors may be:

- Fixed volume: capable of delivering a single factory-set volume.
- Adjustable: where the volume is determined by the operator across a particular range of values.
- Pre-set: movable between a limited number of values.
- Multi-channel: able to deliver several replicate volumes at the same time.

Whichever type you use, you must ensure that you understand the operating principles of the volume scale and the method for changing the volume delivered – some pipettors are easily misread.

A pipettor must be fitted with the correct disposable tip before use: each manufacturer produces different tips to fit particular models. Specialized tips are available for particular applications, e.g. fine tips for loading gels (p. 230).

Syringes

Syringes should be used by placing the tip of the needle in the solution and drawing the plunger up slowly to the required point on the scale. Check the barrel to make sure no air bubbles have been drawn up. Expel slowly and touch the syringe on the edge of the vessel to remove any liquid adhering to the end of the needle. Microsyringes should always be cleaned before and after use by repeatedly drawing up and expelling pure solvent. The dead space in the syringe needle can occupy up to 4% of the nominal syringe volume. A way of avoiding such problems is to fill the dead space with an inert substance (e.g. silicone oil) after sampling. An alternative is to use a syringe where the plunger occupies the needle space (available for small volumes only).

Box 3.1 Using a pipettor to deliver accurate, reproducible volumes of liquid

A pipettor can be used to dispense volumes with accuracy and precision, by following this step-wise procedure:

1. **Select a pipettor that operates over the appropriate range.** Most adjustable pipettors are accurate only over a particular working range and should not be used to deliver volumes below the manufacturer's specifications. Do not attempt to set the volume above the maximum limit, or the pipettor may be damaged.
2. **Set the volume to be delivered.** In some pipettors, like the Finnpipette®, you 'dial up' the required volume, using a twisting motion to set the scale (take great care not to overtighten). Other types, like the Gilson Pipetman®, have a more complex system where the scale (or 'volumeter') consists of three numbers, read from top to bottom of the barrel, and adjusted using the black knurled adjustment ring (Fig. 3.3). This number gives the first three digits of the volume scale and thus can only be understood by establishing the maximum volume of the Pipetman®, as shown on the push-button on the end of the plunger (Fig. 3.3). The following examples illustrate the principle for two common sizes of Pipetman®:

P1000 Pipetman® (maximum volume 1000 µl) if you dial up

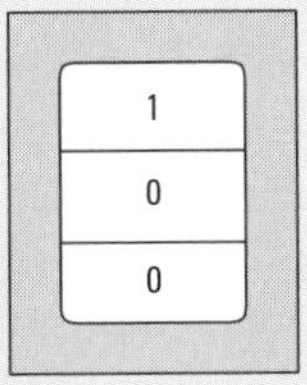

1
0
0

the volume is set at 1000 µl

P20 Pipetman® (maximum volume 20 µl) if you dial up

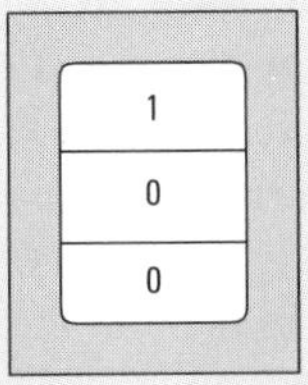

1
0
0

the volume is set at 10.0 µl

3. **Fit a new disposable tip to the end of the barrel.** Make sure that it is the appropriate type for your pipettor and that it is correctly fitted. Press the tip on firmly using a slight twisting motion – if not, you will take up less than the set volume and liquid will drip from the tip during use. Tips are often supplied in boxes, for ease of use: if sterility is important, make sure you use appropriate sterile technique at all times (p. 59). *Never, ever, try to use a pipettor without its disposable tip.*
4. **Check your delivery.** Confirm that the pipettor delivers the correct volume by dispensing volumes of distilled water and weighing on a balance, assuming 1 mg = 1 µl = 1 mm^3. The value should be within 1% of the selected volume. For small volumes, measure several 'squirts' together, e.g. 20 'squirts' of 5 µl = 100 mg. If the pipettor is inaccurate (p. 42) giving a biased result (e.g. delivering significantly more or less than the volume set), you can make a temporary correction by adjusting the volumeter scale down or up accordingly (the volume *delivered* is more important than the value *displayed* on the volumeter), or have the pipettor recalibrated. If the pipettor is imprecise (p. 42), delivering a variable amount of liquid each time, it may need to be serviced. After calibration, fit a clean (sterile) tip if necessary.
5. **Draw up the appropriate volume.** Holding the pipettor vertically, press down on the plunger/push-button until a resistance (spring-loaded stop) is met. Then place the end of the tip in the liquid. Keeping your thumb on the plunger/push-button, release the pressure slowly and evenly: watch the liquid being drawn up into the tip, to confirm that no air bubbles are present. Wait a second or so, to confirm that the liquid has been taken up, then withdraw the end of the tip from the liquid. Inexperienced users often have problems caused by drawing up the liquid too quickly/carelessly. If you accidentally draw liquid into the barrel, seek assistance from your demonstrator or supervisor as the barrel will need to be cleaned before further use.
6. **Make a quick visual check on the liquid in the tip.** Does the volume seem reasonable? (e.g. a 100 µl volume should occupy approximately half the volume of a P200 tip). The liquid will remain in the tip, without dripping, as long as the tip is fitted correctly and the pipettor is not tilted too far from a vertical position.
7. **Deliver the liquid.** Place the end of the tip against the wall of the vessel at a slight angle (10–20° from vertical) and press the plunger/push-button slowly and smoothly to the first (spring-loaded) stop. Wait a second or two, to allow any residual liquid to run down the inside of the tip, then press again to the final stop, dispensing any remaining liquid. Remove from the vessel with the plunger/push-button still depressed.
8. **Eject the tip.** Press the tip ejector button if present (Fig. 3.3). If the tip is contaminated, eject directly into an appropriate container, e.g. a beaker of disinfectant, for microbiological work, or a labelled container for hazardous solutions (p. 6). For repeat delivery, fit a new tip if necessary and begin again at step 5 above. Always make sure that the tip is ejected before putting a pipettor on the bench, as any fluid remaining will run back into the barrel of the pipettor.

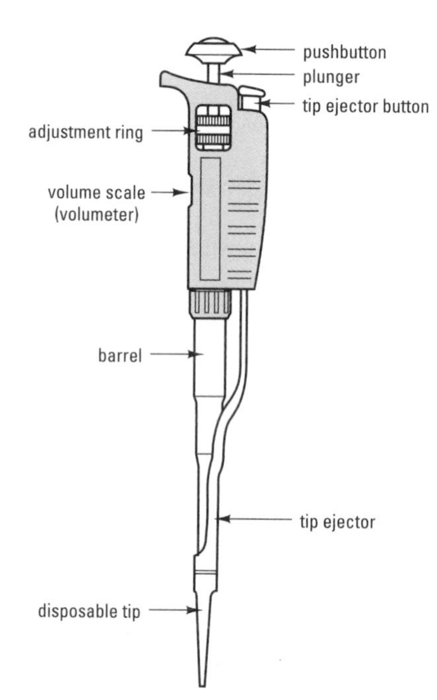

Fig. 3.3 A pipettor – the Gilson Pipetman®.

Storing light-sensitive chemicals – use a coloured vessel or wrap aluminium foil around a clear vessel

Balances

These can be used to weigh accurately (p. 17) how much liquid you have dispensed. Convert mass to volume using the equation:

$$\text{mass}/\text{density} = \text{volume} \quad [5.1]$$

e.g. 9 g of a liquid with a density of $1.2\,\text{g ml}^{-1}$ = 7.5 ml. Densities of common solvents can be found in Lide and Frederikse (1996). You will also need to know the liquid's temperature, as density is temperature dependent.

Holding and storing liquids

Test tubes

These are used for colour tests, small-scale reactions, holding cultures, etc. The tube can be sterilized by heating (p. 59) and maintained in this state with a cap or cotton wool plug.

Beakers

Beakers are used for general purposes, e.g. heating a solvent while the solute dissolves, carrying out a titration, etc. They may have volume gradations on the side: these are often inaccurate and should only be used where approximations will suffice.

Conical (Erlenmeyer) flasks

These are used for storage of solutions: their wide base makes them stable, while their small mouth reduces evaporation and makes them easier to seal. Volume gradations, where present, are usually inaccurate.

Bottles and vials

These are used when the solution needs to be sealed for safety, sterility or to prevent evaporation or oxidation. They usually have a screw top or ground glass stopper to prevent evaporation and contamination. Many types are available, including 'bijou', 'McCartney', 'universal', and 'Winkler'.

You should clearly label all stored solutions (see p. 17), including relevant hazard information, preferably marking with orange hazard warning tape. Seal vessels in an appropriate way, e.g. using a stopper or a sealing film such as Parafilm® or Nescofilm® to prevent evaporation. To avoid degradation store your solution in a fridge, but allow it to reach room temperature before use. Unless a solution containing organic constituents has been sterilized or is toxic, microbes will start growing, so older solutions may not give reliable results.

Creating specialized apparatus

Glassware systems incorporating ground glass connections such as Quickfit® are useful for setting up combinations of standard glass components, e.g. for chemical reactions. In project work, you may need to adapt standard forms of glassware for a special need. A glassblowing service (often available in chemistry departments) can make special items to order.

Choosing between glass and plastic

Bear in mind the following points:

- Reactivity. Plastic vessels often distort at relatively low temperatures; they may be inflammable, may dissolve in certain organic solvents and may be affected by prolonged exposure to ultraviolet (UV) light. Some plasticizers may leach from vessels and have been shown to have biological activity. Glass may adsorb ions and other molecules and then leach them into solutions, especially in alkaline conditions. Pyrex® glass is stronger than ordinary soda glass and can withstand temperatures up to 500 °C.

- Rigidity and resilience. Plastic vessels are not recommended where volume is critical as they may distort through time. Glass vessels are more easily broken than plastic, which is particularly important for centrifugation (see p. 126).
- Opacity. Both glass and plastic absorb light in the UV range of the EMR spectrum (Table 3.2). Quartz should be used where this is important, e.g. in cuvettes for UV spectrophotometry (see p. 112).
- Disposability. Plastic items may be cheap enough to make them disposable, an advantage where there is a risk of chemical or microbial contamination.

Table 3.2 Spectral cutoff values for glass and plastics (λ_{50} = wavelength at which transmission of EMR is reduced to 50%)

Material	λ_{50} (nm)
Routine glassware	340
Pyrex® glass	292
Polycarbonate	396
Acrylic	342
Polyester	318
Quartz	220

Cleaning glass and plastic

Beware the possibility of contamination arising from prior use of chemicals or inadequate rinsing following washing. A thorough rinse with distilled or deionized water immediately before use will remove dust and other deposits and is good practice in quantitative work, but ensure that the rinsing solution is not left in the vessel. 'Strong' basic detergents (e.g. Pyroneg®) are good for solubilizing acidic deposits. If there is a risk of basic deposits remaining, use an acid wash. If there is a risk of contamination from organic deposits, a rinse with Analar® grade ethanol is recommended. Glassware can be sterilized by washing with a sodium hypochlorite bleach such as Chloros® or with sodium metabisulphite – dilute as recommended before use and rinse thoroughly with sterile water after use. Alternatively, heat glassware to at least 121 °C for 15 min in an autoclave or 160 °C for 3 h in an oven.

Special cleaning of glass – for an acid wash use dilute acid, e.g. 100 mmol l^{-1} (100 mol m^{-3}) HCl. Rinse thoroughly at least three times with distilled or deionized water. Glassware that must be exceptionally clean (e.g. for a micronutrient study) should be washed in a chromic acid bath, but this involves toxic and corrosive chemicals and should only be used under supervision.

Safety with glass

Many minor accidents in the laboratory are due to lack of care with glassware. You should follow these general precautions:

- Wear safety glasses when there is *any* risk of glass breakage, e.g. when using low pressures or heating solutions.
- If heating glassware, use a 'soft' Bunsen flame – this avoids creating a hot spot where cracks may start. Always use tongs or special heat-resistant gloves when handling hot glassware.
- Don't use chipped or cracked glassware – it may break under very slight strains and should be disposed of in the broken glassware bin.
- Never carry large bottles by their necks – support them with a hand underneath or, better still, carry them in a basket.
- Take care when attaching tubing to glass tubes and when putting glass tubes into bungs – wear a pair of thick gloves.
- Don't force bungs too firmly into bottles – they can be very difficult to remove. If you need a tight seal, use a screw top bottle with a rubber or plastic seal or Parafilm®.
- Dispose of broken glass thoroughly and with great care – use disposable paper towels and wear thick gloves. Always put pieces of broken glass into the correct bin.

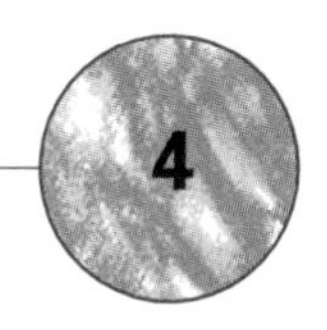

4 Basic laboratory procedures

The *Merck Index* (Budavari *et al.*, 1996) and the *CRC Handbook of Chemistry and Physics* (Lide, 1996) are useful sources of information on the physical and biological properties of chemicals, including melting and boiling points, solubility, toxicity, etc. (see Fig. 4.1).

Table 4.1 Representative risk assessment information for a practical exercise in molecular biology, involving the isolation of DNA

Substance	Hazards/comments
Sodium dodecyl sulphate (SDS)	Irritant, toxic. Wear gloves.
Sodium hydroxide (NaOH)	Highly corrosive, severe irritant. Wear gloves.
Isopropanol	Highly flammable, irritant/ corrosive, potential carcinogen. No naked flames, wear gloves.
Phenol	Highly toxic, causes skin burns, potential carcinogen. Use in fume hood, wear gloves.
Chloroform	Volatile and toxic, irritant/ corrosive, potential carcinogen. Use in fume hood, wear gloves.

Using chemicals – be considerate to others: always return store room chemicals promptly to the correct place. Report when supplies are getting low to the person responsible for looking after the store. If you empty an aspirator or wash bottle, fill it up from the appropriate source!

Using chemicals

Safety aspects

In practical classes, the person in charge has a responsibility to inform you of any hazards associated with the use of chemicals. In project work, your first duty when using an unfamiliar chemical is to find out about its properties, especially those relating to safety. For routine practical procedures, a risk assessment will have been carried out by a member of staff and relevant safety information will be included in the practical schedule: an example is shown in Table 4.1. Before you use any chemical you must find out whether safety precautions need to be taken and complete the appropriate forms confirming that you appreciate the risks involved. Your Department must provide the relevant information to allow you to do this. If your supervisor has filled out the form, read it carefully before signing.

Key safety points when handling chemicals are:

- Treat all chemicals as potentially dangerous.
- Wear a laboratory coat, with buttons fastened, at all times.
- Make sure you know where safety devices such as eye bath, fire extinguisher, first aid kit, are kept before you begin work in the lab.
- Wear gloves and safety glasses for toxic, irritant or corrosive chemicals and carry out procedures with them in a fume cupboard.
- Use aids such as pipette fillers to minimize risk of contact.
- Extinguish all naked flames when working with flammable substances.
- Never smoke, eat or drink where chemicals are handled.
- Label solutions appropriately (see pp. 6, 17, 107).
- Report all spillages and clean them up properly.
- Dispose of chemicals in the correct manner.

Selection

Chemicals are supplied in various degrees of purity and this is always stated on the manufacturer's containers. Suppliers differ in the names given to the grades and there is no conformity in purity standards. Very pure chemicals cost more, sometimes a lot more, and should only be used if the situation demands. If you need to order a chemical, your Department will have a defined procedure for doing this.

Preparing solutions

Solutions are usually prepared with respect to their molar concentrations (e.g. mmol l^{-1}, or mol m^{-3}), or mass concentrations (e.g. g l^{-1}, or kg m^{-3}): both can be regarded as an amount per unit volume, in accordance with the relationship:

$$\text{Concentration} = \frac{\text{amount}}{\text{volume}} \qquad [4.1]$$

The most important aspect of Eqn. 4.1 is to recognize clearly the units involved, and to prepare the solution accordingly: for molar concentrations, you will need the relative molecular mass of the compound, so that you can determine the mass of substance required. Further advice on concentrations and interconversion of units is given on p. 21.

Box 4.1 shows the steps involved in making up a solution. The concentration you require is likely to be defined by a protocol you are following and the grade of chemical and supplier may also be specified. Success may depend on using the same source and quality, e.g. with enzyme work. To avoid waste, think carefully about the volume of solution you require, though it is always a good idea to err on the high side because you

Box 4.1 How to make up an aqueous solution of known concentration from solid material

1. **Find out or decide the concentration of chemical required** and the degree of purity necessary.
2. **Decide on the volume of solution required.**
3. **Find out the relative molecular mass of the chemical (M_r).** This is the sum of the atomic (elemental) masses of the component element(s) and can be found on the container. If the chemical is hydrated, i.e. has water molecules associated with it, these must be included when calculating the mass required.
4. **Work out the mass of chemical that will give the concentration desired in the volume required.**
   Suppose your procedure requires you to prepare 250 ml of $0.1\,\text{mol}\,\text{l}^{-1}$ NaCl.
   (a) Begin by expressing all volumes in the same units, either millilitres or litres (e.g. 250 ml as 0.25 litres).
   (b) Calculate the number of moles required from Eqn. 4.1: $0.1 = \text{amount (mol)} \div 0.25$.
   By rearrangement, the required number of moles is thus $0.1 \times 0.25 = 0.025$ mol.
   (c) Convert from mol to g by multiplying by the relative molecular mass (M_r for NaCl $= 58.44\,\text{g}\,\text{mol}^{-1}$)
   (d) Therefore, you need to make up $0.025 \times 58.44 = 1.461$ g to 250 ml of solution, using distilled water.

   In some instances, it may be easier to work in SI units, though you must be careful when using exponential numbers (p. 260).
   Suppose your protocol states that you need 100 ml of $10\,\text{mmol}\,\text{l}^{-1}$ KCl.
   (a) Start by converting this to $100 \times 10^{-6}\,\text{m}^3$ of $10\,\text{mol}\,\text{m}^{-3}$ KCl.
   (b) The required number of mol is thus $(100 \times 10^{-6}) \times (10)$.
   (c) Each mol of KCl weighs 72.56 g (M_r, the relative molecular mass).
   (d) Therefore you need to make up $72.56 \times 10^{-3}\,\text{g} = 72.56$ mg KCl to $100 \times 10^{-6}\,\text{m}^3$ (100 ml) with distilled water.

   See Box 5.1 for additional information.
5. **Weigh out the required mass of chemical to an appropriate accuracy.** If the mass is too small to weigh to the desired degree of accuracy, consider the following options:
   (a) Make up a greater volume of solution.
   (b) Make up a stock solution which can be diluted at a later stage.
   (c) Weigh the mass first, and calculate what volume to make the solution up to afterwards using equation 4.1.
6. **Add the chemical to a beaker or conical flask then add a little less water than the final amount required.** If some of the chemical sticks to the paper, foil or weighing boat, use some of the water to wash it off.
7. **Stir and, if necessary, heat the solution to ensure all the chemical dissolves.** You can determine when this has happened visually by observing the disappearance of the crystals or powder.
8. **If required, check and adjust the pH of the solution when cool** (see p. 28).
9. **Make up the solution to the desired volume.** If the concentration needs to be accurate, use a volumetric flask; if a high degree of accuracy is not required, use a measuring cylinder.
   (a) Pour the solution from the beaker into the measuring vessel using a funnel to avoid spillage.
   (b) Make up the volume so that the meniscus comes up to the appropriate measurement line (p. 7). For accurate work, rinse out the original vessel and use this liquid to make up the volume.
10. **Transfer the solution to a reagent bottle or a conical flask and label the vessel clearly.**

may spill some or make a mistake when dispensing it. Try to choose one of the standard volumes for vessels, as this will make measuring-out easier.

Use distilled or deionized water to make up aqueous solutions and stir to make sure all the chemical is dissolved. Magnetic stirrers are the most convenient means of doing this: carefully drop a clean magnetic stirrer bar ('flea') in the beaker, avoiding splashing; place the beaker centrally on the stirrer plate, switch on the stirrer and gradually increase the speed of stirring. When the crystals or powder have completely dissolved, switch off and retrieve the flea with a magnet or another flea. Take care not to contaminate your solution when you do this and rinse the flea with distilled water.

Solubility problems – if your chemical does not dissolve after a reasonable time:

- check the limits of solubility for your compound (see *Merck Index*, Budavari *et al.* 1996),
- check the pH of the solution – solubility is often low at pH extremes.

'Obstinate' solutions may require heating but do this only if you know that the chemical will not be damaged at the temperature used. Use a stirrer-heater to keep the solution mixed as you heat it. Allow the solution to cool down before you measure its volume or pH as these are affected by temperature.

Stock solutions

Stock solutions are valuable when making up a range of solutions containing different concentrations of a reagent or if the solutions have some common ingredients. They also save work if the same solution is used over a prolonged period (e.g. a nutrient solution). The stock solution is more concentrated than the final requirement and is diluted as appropriate when the final solutions are made up. The principle is best illustrated with an example (Table 4.2).

Preparing dilutions

Making a single dilution

Making a dilution – use the relationship $[C_1]V_1 = [C_2]V_2$ to determine volume or concentration (see p. 21).

In analytical work, you may need to dilute a stock solution to give a particular mass concentration, or molar concentration. Use the following procedure:

1. Transfer an accurate volume of stock solution to a volumetric flask, using appropriate equipment (Table 3.1).
2. Make up to the calibration mark with solvent – add the last few drops from a pipette or solvent bottle, until the meniscus is level with the calibration mark.
3. Mix thoroughly, either by repeated inversion (holding the stopper firmly) or by prolonged stirring, using a magnetic stirrer. Make sure you add the magnetic flea *after* the volume adjustment step.

For routine work using dilute aqueous solutions where the highest degree of accuracy is not required, it may be acceptable to substitute test tubes or conical flasks for volumetric flasks. In such cases, you would calculate the volumes of stock solution and diluent required, with the assumption that the final volume is determined by the sum of the individual volumes of stock and diluent used (e.g. Table 4.2). Thus, a two-fold dilution would be prepared using 1 volume of stock solution and 1 volume of diluent. The dilution factor is obtained from the ratio of the initial concentration of the stock solution and the final concentration of the diluted solution. The dilution factor can be used to determine the volumes of stock and diluent required in a particular instance. For example, suppose you wanted to prepare 100 ml of a solution of NaCl at $0.2\,mol\,l^{-1}$. Using a stock solution containing $4.0\,mol\,l^{-1}$ NaCl, the

Table 4.2 Use of stock solutions. Suppose you need a set of solutions 10 ml in volume containing differing concentrations of KCl, with and without reagent Q. You decide to make up a stock of KCl at twice the maximum required concentration (50 mmol l^{-1} = 50 mol m^{-3}) and a stock of reagent Q at twice its required concentration. The table shows how you might use these stocks to make up the media you require. Note that the total volumes of stock you require can be calculated from the table (end column).

Stock solutions	Volume of stock required to make required solutions (ml)						Total volume of stock required (ml)
	No KCl plus Q	No KCl minus Q	15 mmol l^{-1} KCl plus Q	15 mmol l^{-1} KCl minus Q	25 mmol l^{-1} KCl plus Q	25 mmol l^{-1} KCl minus Q	
50 mmol l^{-1} KCl	0	0	3	3	5	5	16
[reagent Q] × 2	5	0	5	0	5	0	15
Water	5	10	2	7	0	5	29
Total	10	10	10	10	10	10	60

Using the correct volumes – it is important to distinguish between the volumes of the various liquids: a one-in-ten dilution is obtained using 1 volume of stock solution plus 9 volumes of diluent (1 + 9 = 10).

Using diluents – various liquids are used, including distilled or deionized water, salt solutions, buffers, Ringer's solution (p. 73), etc., according to the specific requirements of the procedure.

dilution factor is $0.2 \div 4.0 = 0.05 = 1/20$ (a twenty-fold dilution). Therefore, the amount of stock solution required is 1/20th of 100 ml = 5 ml and the amount of diluent needed is 19/20th of 100 ml = 95 ml.

Preparing a dilution series

Dilution series are used in a wide range of procedures, including the preparation of standard curves for calibration of analytical instruments (p. 166), and in microbiology and immunoassay, where a range of dilutions of a particular sample is often required (pp. 62, 97). A variety of different approaches can be used:

Linear dilution series

Here, the concentrations are separated by an equal amount, e.g. a series containing protein at 0, 0.2, 0.4, 0.6, 0.8, 1.0 $\mu g\,ml^{-1}$. Such a dilution series might be used to prepare a calibration curve for spectrophotometric assay of protein concentration (p. 168), or an enzyme assay (p. 194). Use $[C_1]V_1 = [C_2]V_2$ to determine the amount of stock solution required for each member of the series, with the volume of diluent being determined by subtraction.

Logarithmic dilution series

Here, the concentrations are separated by a constant proportion, often referred to as the step interval. This type of serial dilution is useful when a broad range of concentrations is required, e.g. for titration of biologically active substances (p. 84), making a plate count of a suspension of microbes (p. 84), or when a process is logarithmically related to concentration.

The most common examples are:

- Doubling dilutions – where each concentration is half that of the previous one (two-fold step interval, $\log_2$ dilution series). First, make up the most concentrated solution at twice the volume required. Measure out half of this volume into a vessel containing the same volume of diluent, mix thoroughly and repeat, for as many doubling dilutions as are required. The concentrations obtained will be 1/2, 1/4, 1/8, 1/16 etc., times the original (i.e. the dilutions will be two, four, eight and sixteen-fold, etc.).
- Decimal dilutions – where each concentration is one-tenth that of the previous one (ten-fold step interval, $\log_{10}$ dilution series). First, make up

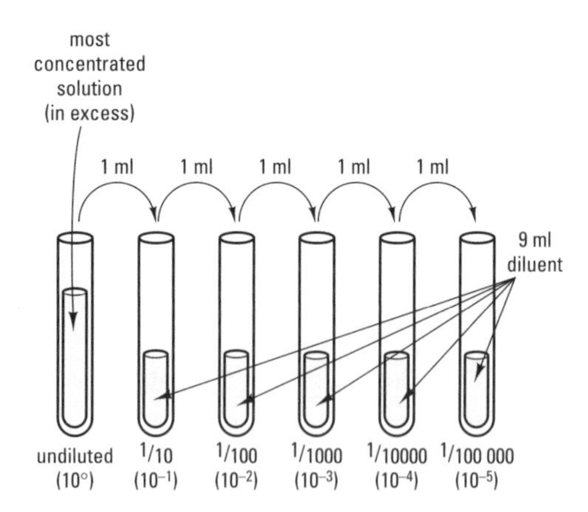

Fig. 4.1 Preparation of a dilution series. The example shown is a decimal dilution series, down to 1/100 000 (10^{-5}) of the solution in the first (left-hand) tube. Note that all solutions must be mixed thoroughly before transferring the volume to the next in the series. In microbiology and cell culture, sterile solutions and appropriate aseptic technique will be required (p. 59).

the most concentrated solution required, with at least a 10% excess. Measure out one-tenth of the volume required into a vessel containing nine times as much diluent, mix thoroughly and repeat. The concentrations obtained will be 1/10, 1/100, 1/1000 etc., times the original (i.e. dilutions of 10^{-1}, 10^{-2}, 10^{-3}, etc.). To calculate the actual concentration of solute, multiply by the appropriate dilution factor.

When preparing serial doubling or decimal dilutions, it is often easiest to add the appropriate amount of diluent to several vessels beforehand, as shown in the worked example in Fig. 4.1. When preparing a dilution series, it is essential that all volumes are dispensed accurately, e.g. using calibrated pipettors (p. 8), otherwise any inaccuracies will be compounded, leading to gross errors in the most dilute solutions.

Harmonic dilution series

Here, the concentrations in the series take the values of the reciprocals of successive whole numbers, e.g. 1, 1/2, 1/3, 1/4, 1/5, etc. The individual dilutions are simply achieved by a stepwise increase in the volume of diluent in successive vessels, e.g. by adding 0, 1, 2, 3, 4 and 5 times the volume of diluent to a set of test tubes, then adding a constant unit volume of stock solution to each vessel. Although there is no dilution transfer error between individual dilutions, the main disadvantage is that the series is non-linear, with a step interval that becomes progressively smaller as the series is extended.

Preparing a dilution series using pipettes or pipettors – use a fresh pipette or disposable tip for each dilution, to prevent carry-over of solutions.

Solutions must be thoroughly mixed before measuring out volumes for the next dilution. Use a fresh measuring vessel for each dilution to avoid contamination, or wash your vessel thoroughly between dilutions. Clearly label the vessel containing each dilution when it is made: it is easy to get confused! When deciding on the volumes required, allow for the aliquot removed when making up the next member in the series. Remember to discard any excess from the last in the series if volumes are critical.

Mixing solutions and suspensions

Various devices may be used, including:

- Magnetic stirrers and fleas. Magnetic fleas come in a range of shapes and sizes, and some stirrers have integral heaters. During use, stirrer speed may increase as the instrument warms up.
- Vortex mixers. For vigorous mixing of small volumes of solution, e.g. when preparing a dilution series in test tubes. Take care when adjusting the mixing speed – if the setting is too low, the test tube will vibrate rather than creating a vortex, giving inadequate mixing. If the setting is too high, the test tube may slip from your hand.
- Orbital shakers and shaking water baths. These are used to provide controlled mixing at a particular temperature, e.g. for long-term incubation and cell growth studies (p. 75).
- Bottle rollers. For cell culture work, ensuring gentle, continuous mixing.

Storing chemicals and solutions

Labile chemicals may be stored in a fridge or freezer. Take special care when using chemicals that have been stored at low temperature: the container and its contents must be warmed up to room temperature before use, otherwise water vapour will condense on the chemical. This may render any weighing you do meaningless and it could ruin the chemical. Other chemicals may need to be kept in a desiccator, especially if they are deliquescent.

KEY POINT Always label all stored chemicals clearly with the following information: the chemical name (if a solution, state solute(s), concentration(s) and pH if measured), plus any relevant hazard warning information, the date made up, and your name.

Separation of components of mixtures and solutions

Particulate solids (e.g. soils) can be separated on the basis of size using sieves. These are available in stacking forms which fit on automatic shakers. Sieves with the largest pores are placed at the top and the assembly is shaken for a fixed time until the sample separates. Suspensions of solids in liquids may be separated out by centrifugation (see p. 123) or filtration. Various forms of filter paper are available having different porosities and purities. Vacuum-assisted filtration speeds up the process and is best carried out with a filter funnel attached to a filter flask. Filtration through pre-sterilized membranes with very small pores (e.g. the Millipore® type) is an excellent method of sterilizing small volumes of solution. Solvents can be removed from solutes by heating, by rotary film evaporation under low pressure and, for water, by freeze drying. The last two are especially useful for heat-labile solutes – refer to the manufacturers' specific instructions for use.

Using balances

Weighing – never weigh anything directly onto a balance's pan: you may contaminate it for other users. Use a weighing boat or a slip of aluminium foil. Otherwise, choose a suitable vessel like a beaker, conical flask or aluminium tray.

Electronic balances with digital readouts are now favoured over mechanical types: they are easy to read and their self-taring feature means the mass of the weighing boat or container can be subtracted automatically before weighing an object. The most common type offers accuracy down to 1 mg over the range 1 mg to 160 g, which is suitable for most biological applications.

To operate a standard self-taring balance:

1. Check that it is level, using the adjustable feet to centre the bubble in the spirit level (usually at the back of the machine). For accurate work, make sure a draught shield is on the balance.
2. Place an empty vessel on the balance pan and allow the reading to stabilize. *If the object is larger than the pan, take care that no part rests on the body of the balance or the draught shield as this will invalidate the reading.* Press the tare bar to bring the reading to zero.
3. Place the chemical or object carefully in the vessel (powdered chemicals should be dispensed with a suitably sized clean spatula).
4. Allow the reading to stabilize and make a note of the value.
5. If you add excess chemical, take great care when removing it. Switch off if you need to clean any deposit accidentally left on or around the balance.

Larger masses should be weighed on a top-loading balance to an appropriate degree of accuracy. Take care to note the limits for the balance: while most have devices to protect against overloading, you may damage the mechanism. In the field, spring or battery-operated balances may be preferred. Try to find a place out of the wind to use them. For extremely small masses, there are electrical balances that can weigh down to 1 μg, but these are very delicate and must be used under supervision.

Measuring and controlling temperature

Heating samples

Care is required when heating samples – there is a danger of fire whenever organic material is heated and a danger of scalding from heated liquids. Safety glasses should always be worn. Use a thermostatically controlled electric stirrer-heater if possible. If using a Bunsen burner, keep the flame well away from yourself and your clothing (tie back long hair). Use a non-flammable mat beneath a Bunsen to protect the bench. Switch off when no longer required. To light a Bunsen, close the air hole first, then apply a lit match or lighter. Open the air hole if you need a hotter, more concentrated flame: the hottest part of the flame is just above the apex of the blue cone in its centre.

Ovens and drying cabinets may be used to dry specimens or glassware. They are normally thermostatically controlled. If drying organic material for dry weight measurement, do so at about 80 °C to avoid caramelizing the specimen. Always state the actual temperature used as this affects results. Check that all water has been driven off by weighing until a constant mass is reached.

Cooling samples and specimens

Heating/cooling glass vessels – take care if heating or cooling glass vessels rapidly as they may break when heat stressed. Freezing aqueous solutions in thin-walled glass vessels is risky because ice expansion may break the glass.

Fridges and freezers are used for storing stock solutions and chemicals that would either break down or become contaminated at room temperature. Normal fridge and freezer temperatures are about 4 °C and −15 °C respectively, while −70 °C freezers may be needed for long-term storage. Ice baths can be used when reactants must be kept at or close to 0 °C. Many departments will have a machine which provides flaked ice for use in these baths. If common salt is mixed with ice, temperatures below 0 °C can be achieved. A mixture of ethanol and solid CO_2 will provide a temperature of −72 °C if required. To freeze a specimen quickly, immerse in liquid N_2 (−196 °C) using tongs and wearing an apron and thick gloves, as splashes will damage your skin. Always work in a well-ventilated room.

Maintaining constant temperature

Using thermometers – some are calibrated for use in air, others require partial immersion in liquid and others total immersion – check before use. If a mercury thermometer is broken, report the spillage as mercury is a poison.

Thermostatically controlled temperature rooms and incubators can be used to keep the temperature at a desired level. Always check with a thermometer or thermograph that the thermostat is accurate enough for your study. To achieve a controlled temperature on a smaller scale, e.g. for an oxygen electrode (p. 159), use a water bath. These usually incorporate heating elements, a circulating mechanism and a thermostat. Baths for sub-ambient temperatures have a cooling element.

Controlling atmospheric conditions

Gas composition

Example Water vapour can be removed by passing gas over dehydrated $CaCO_3$ and CO_2 may be removed by bubbling through KOH solution.

The atmosphere may be 'scrubbed' of certain gases by passing through a U-tube or Dreschel bottle containing an appropriate chemical or solution.

For accurate control of gas concentrations, use cylinders of pure gas; the contents can be mixed to give specified concentrations by controlling individual flow rates. The cylinder head regulator (Fig. 4.2) allows you to control the pressure (and hence flow rate) of gas; adjust using the controls on the regulator or with spanners of appropriate size. Before use, ensure the regulator outlet tap is off (turn anticlockwise), then switch on at the

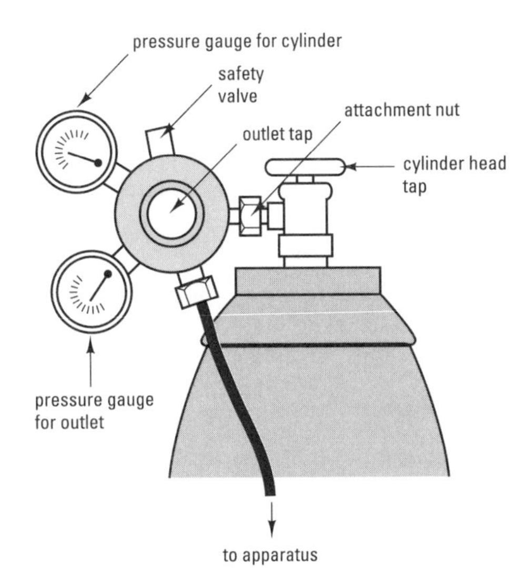

Fig. 4.2 Parts of a cylinder head regulator. The regulator is normally attached by tightening the attachment nut clockwise; the exception is with cylinders of hydrogen, where the special regulator is tightened *anticlockwise* to avoid the chance of this potentially explosive gas being incorrectly used.

Using a timer – always set the alarm before the critical time, so that you have adequate time to react.

cylinder (turn clockwise) – the cylinder dial will give you the pressure reading for the cylinder contents. Now switch on at the regulator outlet (turn slowly clockwise) and adjust to desired pressure/flow setting. To switch off, carry out the above directions in reverse order. To control dissolved gas composition in liquids, either 'degas' under vacuum or bubble with another gas, e.g. when preparing oxygen-free liquids for HPLC (p. 129).

Pressure

Many forms of pump are used to pressurize or provide partial vacuum, usually to force gas or liquid movement. Each has specific instructions for use. Many laboratories are supplied with 'vacuum' (suction) and pressurized air lines that are useful for procedures such as vacuum-assisted filtration. Make sure you switch off the taps after use. Take special care with glass items kept at very low or high pressures. These should be contained within a metal cage to minimize the risk of injury. Wear safety glasses at all times.

Measuring time

Many experiments and observations need to be carefully timed. Large-faced stopclocks allow you to set and follow 'experimental time' and remove the potential difficulties in calculating this from 'real time' on a watch or clock. Some timers incorporate an alarm which you can set to warn when readings or operations must be carried out; 24-h timers are available for controlling light and temperature regimes. Some timers count elapsed time while others count down towards zero – the latter type is less useful.

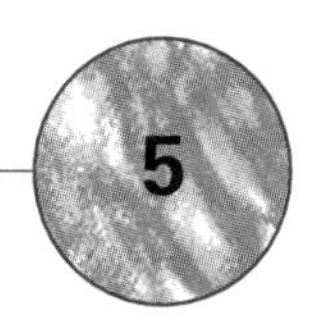

5 Principles of solution chemistry

Preparing solutions – practical advice is given on p. 13.

A solution is a homogeneous liquid, formed by the addition of solutes to a solvent (usually water in biological systems). The behaviour of solutions is determined by the type of solutes involved and by their proportions, relative to the solvent. Many laboratory exercises involve calculation of concentrations, e.g. when preparing an experimental solution at a particular concentration, or when expressing data in terms of solute concentration. Make sure that you understand the basic principles set out in this chapter before you tackle such exercises.

Solutes can affect the properties of solutions in several ways, including:

Electrolytic dissociation

Definition

Electrolyte – a substance that dissociates, either fully or partially, in water to give two or more ions.

This occurs where individual molecules of an electrolyte dissociate to give charged particles (ions). For a strong electrolyte, e.g. NaCl, dissociation is essentially complete. In contrast, a weak electrolyte, e.g. acetic acid, will be only partly dissociated, depending upon the pH and temperature of the solution (p. 26).

Osmotic effects

These are the result of solute particles lowering the effective concentration of the solvent (water). These effects are particularly relevant to biological systems since membranes are far more permeable to water than to most solutes. Water moves across biological membranes from the solution with the higher effective water concentration to that with the lower effective water concentration (osmosis).

Ideal/non-ideal behaviour

This occurs because solutions of real substances do not necessarily conform to the theoretical relationships predicted for dilute solutions of so-called ideal solutes. It is often necessary to take account of the non-ideal behaviour of real solutions, especially at high solute concentrations (see Lide and Frederikse, 1996, for appropriate data).

Concentration

Expressing solute concentrations – you should use SI units wherever possible. However, you are likely to meet non-SI concentrations and you must be able to deal with these units too.

In SI units (p. 45), the concentration of a solute is expressed in mol m^{-3}, which is convenient for most biological purposes. The concentration of a solute is usually symbolized by square brackets, e.g. [NaCl]. Details of how to prepare a solution using SI and non-SI units are given on p. 13.

A number of alternative ways of expressing the relative amounts of solute and solvent are in general use, and you may come across these terms in your practical work, or in the literature:

Molarity

Example A 1.0 molar solution of NaCl would contain 58.44 g NaCl (the molecular mass) per litre of solution.

This is the term used to denote molar concentration, [C], expressed as moles of solute per litre volume of solution (mol l^{-1}). This non-SI term continues to find widespread usage, in part because of the familiarity of working scientists with the term, but also because laboratory glassware is calibrated in millilitres and litres, making the preparation of molar and millimolar solutions relatively straightforward. However, the symbols in common use for molar (M) and millimolar (mM) solutions are at odds with the SI system and many people

Box 5.1 Useful procedures for calculations involving molar concentrations

1. **Preparing a solution of defined molarity.** For a solute of known relative molecular mass (M_r), the following relationship can be applied:

$$[C] = \frac{\text{mass of solute}/M_r}{\text{volume of solution}} \quad [5.1]$$

So, if you wanted to make up 200 ml (0.2 l) of an aqueous solution of NaCl (M_r 58.44 g) at a concentration of 500 mmol l^{-1} (0.5 mol l^{-1}), you could calculate the amount of NaCl required by inserting these values into eqn [5.1]:

$$0.5 = \frac{\text{mass of solute}/58.44}{0.2}$$

which can be rearranged to

$$\text{mass of solute} = 0.5 \times 0.2 \times 58.44 = 5.844\text{ g}$$

The same relationship can be used to calculate the concentration of a solution containing a known amount of a solute, e.g. if 21.1 g of NaCl were made up to a volume of 100 ml (0.1 l), this would give

$$[\text{NaCl}] = \frac{21.1/58.44}{0.1} = 3.61\text{ mol l}^{-1}$$

2. **Dilutions and concentrations.** The following relationship is very useful if you are diluting (or concentrating) a solution:

$$[C_1]V_1 = [C_2]V_2 \quad [5.2]$$

where $[C_1]$ and $[C_2]$ are the initial and final concentrations, while V_1 and V_2 are their respective volumes: each pair must be expressed in the same units. Thus, if you wanted to dilute 200 ml of 0.5 mol l^{-1} NaCl to give a final molarity of 0.1 mol l^{-1}, then, by substitution into eqn [5.2]:

$$0.5 \times 200 = 0.1 \times V_2$$

Thus $V_2 = 1000$ ml (in other words, you would have to add water to 200 ml of 0.5 mol l^{-1} NaCl to give a final volume of 1000 ml to obtain a 0.1 mol l^{-1} solution).

3. **Interconversion.** A simple way of interconverting amounts and volumes of any particular solution is to divide the amount and volume by a factor of 10^3: thus a molar solution of a substance contains 1 mol l^{-1}, which is equivalent to 1 mmol ml^{-1}, or 1 μmol μl^{-1}, or 1 nmol nl^{-1}, etc. You may find this technique useful when calculating the amount of substance present in a small volume of solution of known concentration, e.g. to calculate the amount of NaCl present in 50 μl of a solution with a concentration (molarity) of 0.5 mol l^{-1} NaCl:

   (a) this is equivalent to 0.5 μmol μl^{-1};

   (b) therefore 50 μl will contain 50×0.5 μmol = 25 μmol.

Alternatively, you may prefer to convert to primary SI units, for ease of calculation (see Box 6.1).

The 'unitary method' (p. 261) is an alternative approach to these calculations.

now prefer to use mol l^{-1} and mmol l^{-1} respectively, to avoid confusion. Box 5.1 gives details of some useful approaches to calculations involving molarities.

Molality

This is used to express the concentration of solute relative to the *mass* of solvent, i.e. mol kg^{-1}. Molality is a temperature-independent means of expressing solute concentration, rarely used except when the osmotic properties of a solution are of interest (p. 24).

Example A 0.5 molal solution of NaCl would contain $58.44 \times 0.5 = 29.22$ g NaCl per kg of water.

Per cent composition (% w/w)

This is the solute mass (in g) per 100 g solution. The advantage of this expression is the ease with which a solution can be prepared, since it simply requires each component to be pre-weighed (for water, a volumetric measurement may be used, e.g. using a measuring cylinder) and then mixed together. Similar terms are parts per thousand (‰), i.e. mg g^{-1}, and parts per million (ppm), i.e. μg g^{-1}.

Example A 5% w/w sucrose solution contains 5 g sucrose and 95 g water (= 95 ml water, assuming a density of 1 g ml^{-1}) to give 100 g of solution.

Per cent concentration (% w/v and % v/v)

For solutes added in solid form, this is the number of grams of solute per 100 ml solution. This is more commonly used than per cent composition, since solutions can be accurately prepared by weighing out the required

Example A 5% w/v sucrose solution contains 5 g sucrose in 100 ml of solution. A 5% v/v glycerol solution would contain 5 ml glycerol in 100 ml of solution.

Note that when water is the solvent this is often not specified in the expression, e.g. a 20% v/v ethanol solution contains 20% ethanol made up to 100 ml of solution using water.

amount of solute and then making this up to a known volume using a volumetric flask. The equivalent expression for liquid solutes is % v/v.

The principal use of mass/mass or mass/volume terms (including $g\,l^{-1}$) is for solutes whose molecular mass is unknown (e.g. cellular proteins), or for mixtures of certain classes of substance (e.g. total salt in sea water). You should *never* use the per cent term without specifying how the solution was prepared, i.e. by using the qualifier w/w, w/v or v/v. For mass concentrations, it is simpler to use mass per unit volume, e.g. $mg\,l^{-1}$, $\mu g\,\mu l^{-1}$, etc.

Activity (a)

This is a term used to describe the *effective* concentration of a solute. In dilute solutions, solutes can be considered to behave according to ideal (thermodynamic) principles, i.e. they will have an effective concentration equivalent to the actual concentration. However, in concentrated solutions ($\geqslant 500\,mol\,m^{-3}$), the behaviour of solutes is often non-ideal, and their effective concentration (activity) will be less than the actual concentration [*C*]. The ratio between the effective concentration and the actual concentration is called the activity coefficient (γ) where

$$\gamma = \frac{a}{[C]} \qquad [5.3]$$

Table 5.1 Activity coefficient of NaCl solutions as a function of molality. Data from Robinson and Stokes (1970)

Molality	Activity coefficient at 25 °C
0.1	0.778
0.5	0.681
1.0	0.657
2.0	0.668
4.0	0.783
6.0	0.986

Equation [5.3] can be used for SI units ($mol\,m^{-3}$), molarity ($mol\,l^{-1}$) or molality ($mol\,kg^{-1}$). In all cases, γ is a dimensionless term, since *a* and [*C*] are expressed in the same units. The activity coefficient of a solute is effectively unity in dilute solution, decreasing as the solute concentration increases (Table 5.1). At high concentrations of certain ionic solutes, γ may increase to become greater than unity.

KEY POINT **Activity is often the correct expression for theoretical relationships involving solute concentration (e.g. where a property of the solution is dependent on concentration). However, for most practical purposes, it is possible to use the *actual* concentration of a solute rather than the activity, since the difference between the two terms can be ignored for dilute solutions.**

Example A solution of NaCl with a molality of $0.5\,mol\,kg^{-1}$ has an activity coefficient of 0.681 at 25 °C and a molal activity of $0.5 \times 0.681 = 0.340\,mol\,kg^{-1}$.

The particular use of the term 'water activity' is considered below, since it is based on the mole fraction of solvent, rather than the effective concentration of solute.

Equivalent mass (equivalent weight)

Equivalence and normality are outdated terms, although you may come across them in older texts. They apply to certain solutes whose reactions involve the transfer of charged ions, e.g. acids and alkalis (which may be involved in H^+ or OH^- transfer), and electrolytes (which form cations and anions that may take part in further reactions). These two terms take into account the valency of the charged solutes. Thus the equivalent mass of an ion is its molecular mass divided by its valency (ignoring the sign), expressed in grams per equivalent (eq) according to the relationship:

$$\text{equivalent mass} = \frac{\text{molecular mass}}{\text{valency}} \qquad [5.4]$$

Examples For carbonate ions (CO_3^{2-}), with a molecular mass of 60.00 and a valency of 2, the equivalent mass is $60.00/2 = 30.00\,g\,eq^{-1}$.

For sulphuric acid (H_2SO_4, molecular mass 98.08), where 2 hydrogen ions are available, the equivalent mass is $98.08/2 = 49.04\,g\,eq^{-1}$.

For acids and alkalis, the equivalent mass is the mass of substance that will provide 1 mol of either H^+ or OH^- ions in a reaction, obtained by dividing the molecular mass by the number of available ions (*n*), using *n* instead of valency as the denominator in eqn [5.4].

Example A 0.5 N solution of sulphuric acid would contain $0.5 \times 49.04 = 24.52\,g\,l^{-1}$.

Normality

A 1 normal solution (1 N) is one that contains one equivalent mass of a substance per litre of solution. The general formula is:

$$\text{normality} = \frac{\text{mass of substance per litre}}{\text{equivalent mass}} \qquad [5.5]$$

Example Under ideal conditions, 1 mol of NaCl dissolved in water would give 1 mol of Na^+ ions and 1 mol of Cl^- ions, equivalent to a theoretical osmolarity of 2 osmol l^{-1}

Osmolarity

This non-SI expression is used to describe the number of moles of osmotically active solute particles per litre of solution (osmol l^{-1}). The need for such a term arises because some molecules dissociate to give more than one osmotically active particle in aqueous solution.

Osmolality

This term describes the number of moles of osmotically active solute particles per unit mass of solvent (osmol kg^{-1}). For an ideal solute, the osmolality can be determined by multiplying the molality by n, the number of solute particles produced in solution (e.g. for NaCl, $n = 2$). However, for real solutes, a correction factor (the osmotic coefficient, ϕ) is used:

$$\text{osmolality} = \text{molality} \times n \times \phi \qquad [5.6]$$

Example A 1.0 mol kg^{-1} solution of NaCl has an osmotic coefficient of 0.936 at 25 °C and an osmolality of $1.0 \times 2 \times 0.936 = 1.872$ osmol kg^{-1}.

If necessary, the osmotic coefficients of a particular solute can be obtained from tables (e.g. Table 5.2): non-ideal behaviour means that ϕ may have values > 1 at high concentrations. Alternatively, the osmolality of a solution can be measured using an osmometer.

Table 5.2 Osmotic coefficients of NaCl solutions as a function of molality. Data from Robinson and Stokes (1970)

Molality	Osmotic coefficient at 25 °C
0.1	0.932
0.5	0.921
1.0	0.936
2.0	0.983
4.0	1.116
6.0	1.271

Colligative properties and their use in osmometry

Several properties vary in direct proportion to the effective number of osmotically active solute particles per unit mass of solvent and can be used to determine the osmolality of a solution. These colligative properties include freezing point, boiling point, and vapour pressure.

An osmometer is an instrument which measures the osmolality of a solution, usually by determining the freezing point depression of the solution in relation to pure water, a technique known as cryoscopic osmometry. A small amount of sample is cooled rapidly and then brought to the freezing point (Fig. 5.1), which is measured by a temperature-sensitive thermistor

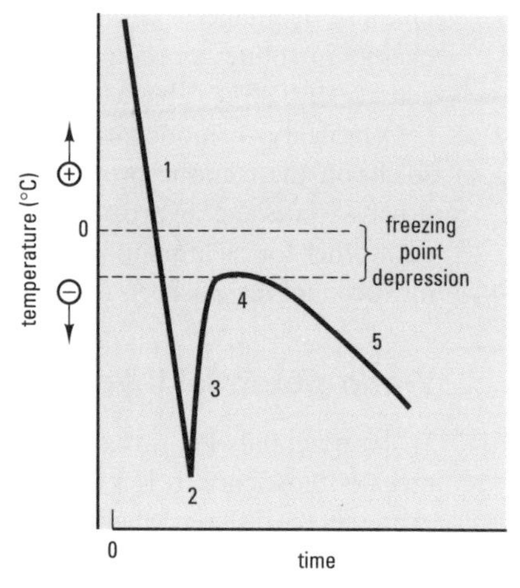

Fig. 5.1 Temperature responses of a cryoscopic osmometer. The response can be subdivided into:
1. initial supercooling
2. initiation of crystallization
3. crystallization/freezing
4. plateau, at the freezing point
5. slow temperature decrease

Using an osmometer – it is vital that the sample holder and probe are clean, otherwise small droplets of the previous sample may be carried over, leading to inaccurate measurement.

probe calibrated in mosmol kg^{-1}. An alternative method is used in vapour pressure osmometry, which measures the relative decrease in the vapour pressure produced in the gas phase when a small sample of the solution is equilibrated within a chamber.

Osmotic properties of solutions

Several inter-related terms can be used to describe the osmotic status of a solution. In addition to osmolality, you may come across the following:

Osmotic pressure

This is based on the concept of a membrane permeable to water, but not to solute molecules. For example, if a sucrose solution is placed on one side and pure water on the other, then a passive driving force will be created and water will diffuse across the membrane into the sucrose solution, since the effective water concentration in the sucrose solution will be lower. The tendency for water to diffuse into the sucrose solution could be counteracted by applying a hydrostatic pressure equivalent to the passive driving force. Thus, the osmotic pressure of a solution is the excess hydrostatic pressure required to prevent the net flow of water into a vessel containing the solution. The SI unit of osmotic pressure is the pascal, Pa ($= kg\,m^{-1}\,s^{-2}$). Older sources may use atmospheres, or bars, and conversion factors are given in Table 10.5 (p. 47). Osmotic pressure and osmolality can be interconverted using the expression 1 osmol kg^{-1} = 2.479 MPa at 25 °C.

Example A 1.0 mol kg^{-1} solution of NaCl at 25 °C has an osmolalilty of 1.872 osmol kg^{-1} and an osmotic pressure of $1.872 \times 2.479 = 4.641$ MPa.

The use of osmotic pressure has been criticized as misleading, since a solution does not exhibit an 'osmotic pressure', unless it is placed on the other side of a selectively permeable membrane from pure water!

Water activity (a_w)

This is a term often used to describe the osmotic behaviour of microbial cells. It is a measure of the relative proportion of water in a solution, expressed in terms of its mole fraction, i.e. the ratio of the number of moles of water (n_w) to the total number of moles of all substances (i.e. water and solutes) in solution (n_t), taking into account the molal activity coefficient of the solvent, water (i.e. γ_w):

$$a_w = \gamma_w \frac{n_w}{n_t} \quad [5.7]$$

The water activity of pure water is unity, decreasing as solutes are added. One disadvantage of a_w is the limited change which occurs in response to a change in solute concentration: a 1.0 mol kg^{-1} solution of NaCl has a water activity of 0.967 (Table 5.3).

Osmolality, osmotic pressure and water activity are measurements based solely on the osmotic properties of a solution, with no regard for any other driving forces, e.g. hydrostatic and gravitational forces. In circumstances where such other forces are important, you will need to measure a variable that takes into account these aspects of water status, namely water potential.

Table 5.3 Water activity (a_w) of NaCl solutions as a function of molality. Data from Robinson and Stokes (1970)

Molality	a_w
0.1	0.997
0.5	0.984
1.0	0.967
2.0	0.932
4.0	0.852
6.0	0.760

Water potential (hydraulic potential) and its applications

Water potential, Ψ_w, is the most appropriate measure of osmotic status in many areas of bioscience. It is a term derived from the chemical potential of water. It expresses the difference between the chemical potential of water in the test system and that of pure water under standard conditions and has units of pressure (i.e.

Examples A 1.0 mol kg^{-1} solution of NaCl has a (negative) water potential of −4.641 MPa.

A solution of pure water at 0.2 MPa pressure (about 0.1 MPa above atmospheric pressure) has a (positive) water potential of 0.1 MPa.

Measuring water potential – eqn [5.9] ignores the effects of gravitational forces – for systems where gravitational effects are important an additional term is required (Nobel, 1991).

Pa). It is a more appropriate term than osmotic pressure because it is based on sound theoretical principles and because it can be used to predict the direction of passive movement of water, since water will flow down a gradient of chemical potential (i.e. osmosis occurs from a solution with a higher water potential to one with a lower water potential). A solution of pure water at 20 °C and at 0.1 MPa pressure (i.e. ≈ atmospheric) has a water potential of zero. The addition of solutes will lower the water potential (i.e. make it negative), while the application of pressure, e.g. from hydrostatic or gravitational forces, will raise it (i.e. make it positive).

Often, the two principal components of water potential are referred to as the solute potential, or osmotic potential (Ψ_s, sometimes symbolized as Ψ_π or π) and the hydrostatic pressure potential (Ψ_p) respectively. For a solution at atmospheric pressure, the water potential is due solely to the presence of osmotically active solute molecules (osmotic potential) and may be calculated from the measured osmolality (osmol kg^{-1}) at 25 °C, using the relationship:

$$\Psi_w\,(\text{MPa}) = \Psi_s\,(\text{MPa}) = -2.479 \times \text{osmolality} \qquad [5.8]$$

For aquatic microbial cells, e.g. algae, fungi and bacteria, equilibrated in their growth medium at atmospheric pressure, the water potential of the external medium will be equal to the cellular water potential and the latter can be derived from the measured osmolality of the medium (eqn [5.8]) by osmometry (p. 23). The water potential of such cells can be subdivided into two major parts, the cell solute potential (Ψ_s) and the cell turgor pressure (Ψ_p) as follows:

$$\Psi_w = \Psi_s + \Psi_p \qquad [5.9]$$

To calculate the relative contribution of the osmotic and pressure terms in eqn [5.9], an estimate of the internal osmolality is required, e.g. by measuring the freezing point depression of expressed intracellular fluid. Once you have values for Ψ_w and Ψ_s, the turgor pressure can be calculated by substitution into eqn [5.9].

For terrestrial plant cells, the water potential may be determined directly using a vapour pressure osmometer, by placing a sample of the material within the osmometer chamber and allowing it to equilibrate. If Ψ_s of expressed sap is then measured, Ψ_p can be determined from eqn [5.9].

The van't Hoff relationship can be used to estimate Ψ_s, by summation of the osmotic potentials due to the major solutes, determined from their concentrations, as:

$$\Psi_s = -RTn\phi[C] \qquad [5.10]$$

where RT is the product of the universal gas constant and absolute temperature (2 479 J mol^{-1} at 25 °C), n and ϕ are as previously defined and $[C]$ is expressed in SI terms as mol m^{-3}.

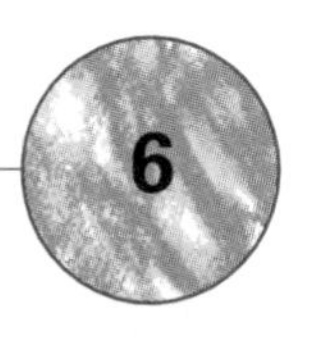

6 pH and buffer solutions

Definitions

Acid – a compound that acts as a proton donor in aqueous solution.

Base – a compound that acts as a proton acceptor in aqueous solution.

Conjugate pair – an acid together with its corresponding base.

Alkali – a compound that liberates hydroxyl ions when it dissociates. Since hydroxyl ions are strongly basic, this will reduce the proton concentration.

Ampholyte – a compound that can act as both an acid and a base. Water is an ampholyte since it may dissociate to give a proton and a hydroxyl ion (amphoteric behaviour).

pH is a measure of the amount of hydrogen ions (H^+) in a solution: this affects the solubility of many substances and the activity of most biological systems, from individual molecules to whole organisms. It is usual to think of aqueous solutions as containing H^+ ions (protons), though protons actually exist in their hydrated form, as hydronium ions (H_3O^+). The proton concentration of an aqueous solution $[H^+]$ is affected by several factors:

- Ionization (dissociation) of water, which liberates protons and hydroxyl ions in equal quantities, according to the reversible relationship:

$$H_2O \rightleftharpoons H^+ + OH^- \qquad [6.1]$$

- Dissociation of acids, according to the equation:

$$H\text{–}A \rightleftharpoons H^+ + A^- \qquad [6.2]$$

  where H–A represents the acid and A^- is the corresponding conjugate base. The dissociation of an acid in water will increase the amount of protons, reducing the amount of hydroxyl ions as water molecules are formed (eqn [6.1]). The addition of a base (usually, as its salt) to water will decrease the amount of H^+, due to the formation of the conjugate acid (eqn [6.2]).
- Dissociation of alkalis, according to the relationship:

$$X\text{–}OH \rightleftharpoons X^+ + OH^- \qquad [6.3]$$

  where X–OH represents the undissociated alkali. Since the dissociation of water is reversible (eqn [6.1]), in an aqueous solution the production of hydroxyl ions will effectively act to 'mop up' protons, lowering the proton concentration.

Many compounds act as acids, bases or alkalis: those which are almost completely ionized in solution are usually called strong acids or bases, while weak acids or bases are only slightly ionized in solution (p. 30).

In an aqueous solution, most of the water molecules are not ionized. In fact, the extent of ionization of pure water is constant at any given temperature and is usually expressed in terms of the ion product (or ionization constant) of water, K_w:

$$K_w = [H^+][OH^-] \qquad [6.4]$$

where $[H^+]$ and $[OH^-]$ represent the molar concentration (strictly, the activity) of protons and hydroxyl ions in solution, expressed as $mol\,l^{-1}$. At 25 °C, the ion product of pure water is 10^{-14} $mol^2\,l^{-2}$ (i.e. 10^{-8} $mol^2\,m^{-6}$). This means that the concentration of protons in solution will be 10^{-7} $mol\,l^{-1}$ (10^{-4} $mol\,m^{-3}$), with an equivalent concentration of hydroxyl ions (eqn [6.1]). Since these values are very low and involve negative powers of 10, it is customary to use the pH scale, where:

$$pH = -\log_{10}[H^+] \qquad [6.5]$$

and $[H^+]$ is the proton activity (see p. 22).

Example Human blood plasma has a typical H^+ concentration of approximately 0.4×10^{-7} $mol\,l^{-1}$ ($= 10^{-7.4}$ $mol\,l^{-1}$), giving a pH of 7.4.

Table 6.1 Effects of temperature on the ion product of water (K_w), H^+ ion concentration and pH at neutrality. Values calculated from Lide and Frederikse (1996)

Temp. (°C)	K_w ($mol^2\ l^{-2}$)	[H^+] at neutrality ($nmol\ l^{-1}$)	pH at neutrality
0	0.11×10^{-4}	33.9	7.47
4	0.17×10^{-4}	40.7	7.39
10	0.29×10^{-4}	53.7	7.27
20	0.68×10^{-4}	83.2	7.08
25	1.01×10^{-4}	100.4	7.00
30	1.47×10^{-4}	120.2	6.92
37	2.39×10^{-4}	154.9	6.81
45	4.02×10^{-4}	199.5	6.70

Table 6.2 Properties of some pH indicator dyes

Dye	Acid-base colour change	Useful pH range
Thymol blue (acid)	red–yellow	1.2–6.8
Bromophenol blue	yellow–blue	1.2–6.8
Congo red	blue–red	3.0–5.2
Bromocresol green	yellow-blue	3.8–5.4
Resazurin	orange–violet	3.8–6.5
Methyl red	red–yellow	4.3–6.1
Litmus	red–blue	4.5–8.3
Bromocresol purple	yellow–purple	5.8–6.8
Bromothymol blue	yellow-blue	6.0–7.6
Neutral red	red–yellow	6.8–8.0
Phenol red	yellow-red	6.8–8.2
Thymol blue (alkaline)	yellow-blue	8.0–9.6
Phenol-phthalein	none–red	8.3–10.0

KEY POINT While pH is strictly the negative logarithm (to the base 10) of H^+ activity, in practice H^+ concentration in mol l^{-1} (equivalent to kmol m^{-3} in SI terminology) is most often used in place of activity, since the two are virtually the same, given the limited dissociation of H_2O. The pH scale is not SI: nevertheless, it continues to be used widely in biological science.

The value where an equal amount of H^+ and OH^- ions are present is termed neutrality: at 25 °C the pH of pure water at neutrality is 7.0. At this temperature, pH values below 7.0 are acidic while values above 7.0 are alkaline. However, the pH of a neutral solution changes with temperature (Table 6.1), due to the enhanced dissociation of water with increasing temperature. This must be taken into account when measuring the pH of any solution and when interpreting your results.

Always remember that the pH scale is a logarithmic one, not a linear one: a solution with a pH of 3.0 is not twice as acidic as a solution of pH 6.0, but one thousand times as acidic (i.e. contains 1000 times the amount of H^+ ions). Therefore, you may need to convert pH values into proton concentrations before you carry out mathematical manipulations (see Box 46.2). For similar reasons, it is important that pH change is expressed in terms of the original and final pH values, rather than simply quoting the difference between the values: a pH change of 0.1 has little meaning unless the initial or final pH is known.

Measuring pH

pH electrodes

Accurate pH measurements can be made using a pH electrode, coupled to a pH meter. The pH electrode is usually a combination electrode, comprising two separate systems: an H^+-sensitive glass electrode and a reference electrode which is unaffected by H^+ ion concentration (Fig. 6.1). When this is immersed in a solution, a pH-dependent voltage between the two electrodes can be measured using a potentiometer. In most cases, the pH electrode assembly (containing the glass and reference electrodes) is connected to a separate pH meter by a cable, although some hand-held instruments (pH probes) have the electrodes and meter within the same assembly, often using an H^+-sensitive field effect transistor in place of a glass electrode, to improve durability and portability.

Box 6.1 gives details of the steps involved in making a pH measurement with a glass pH electrode and meter.

pH indicator dyes

These compounds (usually weak acids) change colour in a pH-dependent manner. They may be added in small amounts to a solution, or they can be used in paper strip form. Each indicator dye usually changes colour over a restricted pH range (Table 6.2): universal indicator dyes/papers make use of a combination of individual dyes to measure a wider pH range. Dyes are not suitable for accurate pH measurement as they are affected by other components of the solution including oxidizing and reducing agents and salts. However, they are useful for:

- estimating the approximate pH of a solution;
- determining a change in pH, for example at the end-point of a titration or the production of acids during bacterial metabolism (p. 71);
- establishing the approximate pH of intracellular compartments, for example the use of neutral red as a 'vital' stain (p. 37).

Buffers

Rather than simply measuring the pH of a solution, you may wish to *control* the pH, e.g in metabolic experiments, or in a growth medium for cell culture (p. 75). In fact, you should consider whether you need to control pH in any experiment involving a biological system, whether whole organisms, isolated cells, sub-cellular components or biomolecules. One of the most effective ways to control pH is to use a buffer solution.

A buffer solution is usually a mixture of a weak acid and its conjugate base. Added protons will be neutralized by the anionic base while a reduction in protons, e.g. due to the addition of hydroxyl ions, will be counterbalanced by dissociation of the acid (eqn [6.2]); thus the conjugate pair acts as a 'buffer' to pH change. The innate resistance of most biological fluids to pH change is due to the presence of cellular constituents that act as buffers, e.g. proteins, which have a large number of weakly acidic and basic groups in their amino acid side chains.

Definition

Buffer solution – one which resists a change in H^+ concentration (pH) on addition of acid or alkali.

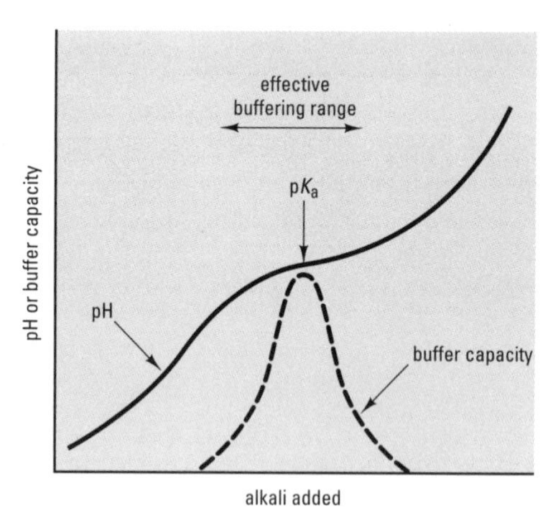

Fig. 6.1 Theoretical pH titration curve for a buffer solution. pH change is lowest and buffer capacity is greatest at the pK_a of the buffer solution.

Buffer capacity and the effects of pH

The extent of resistance to pH change is called the buffer capacity of a solution. The buffer capacity is measured experimentally at a particular pH by titration against a strong acid or alkali: the resultant curve will be strongly sigmoidal, with a plateau where the buffer capacity is greatest (Fig. 6.1). The mid-point of the plateau represents the pH where equal quantities of acid and conjugate base are present, and is given the symbol pK_a, which refers to the negative logarithm (to the base 10) of the acid dissociation constant, K_a, where

$$K_a = \frac{[H^+][A^-]}{[HA]} \qquad [6.6]$$

By rearranging eqn [6.6] and taking negative logarithms, we obtain:

$$pH = pK_a + \log_{10}\frac{[A^-]}{[HA]} \qquad [6.7]$$

This relationship is known as the Henderson–Hasselbalch equation and it shows that the pH will be equal to the pK_a when the ratio of conjugate base to acid is unity, since the final term in eqn [6.7] will be zero. Consequently, the pK_a of a buffer solution is an important factor in determining the buffer capacity at a particular pH. In practical terms, this means that a buffer solution will work most effectively at pH values about one unit either side of the pK_a.

An ideal buffer for biological purposes would possess the following characteristics:

- impermeability to biological membranes;
- biological stability and lack of interference with metabolic and biological processes;
- lack of significant absorption of ultraviolet or visible light;
- lack of formation of insoluble complexes with cations;
- minimal effect of ionic composition or salt concentration;
- limited pH change in response to temperature.

Selecting an appropriate buffer

When selecting a buffer, you should be aware of certain limitations to their use. Citric acid and phosphate buffers readily form insoluble complexes with divalent cations, while phosphate can also act as a substrate, activator or inhibitor of certain enzymes. Both of these buffers contain biologically significant quantities of cations, e.g. Na^+ or K^+. TRIS (Table 6.3) is often toxic to biological systems: due to its high lipid solubility it can penetrate membranes, uncoupling electron transport reactions in whole cells and isolated organelles. In addition, it is markedly affected by temperature, with a tenfold increase in H^+ concentration from 4 °C to 37 °C. A number of

Box 6.1 Using a glass pH electrode and meter to measure the pH of a solution

The following procedure should be used whenever you make a pH measurement: consult the manufacturer's handbook for specific information, where necessary. Do not be tempted to miss out any of the steps detailed below, particularly those relating to the effects of temperature, or your measurements are likely to be inaccurate.

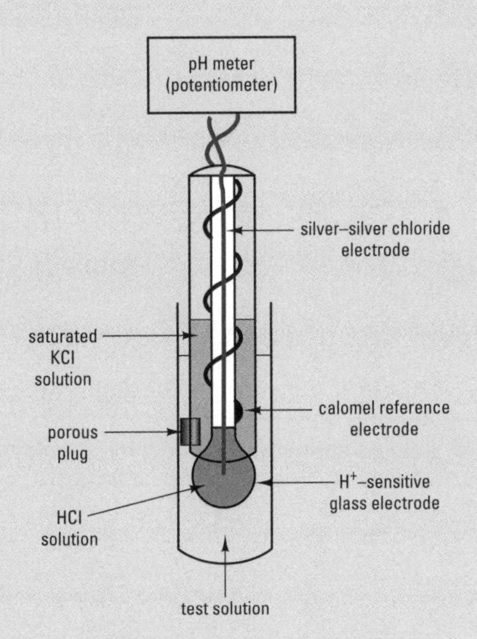

Fig. 6.2 Measurement of pH using a combination pH electrode and meter. The electrical potential difference recorded by the potentiometer is directly proportional to the pH of the test solution.

1. **Stir the test solution thoroughly before you make any measurement:** it is often best to use a magnetic stirrer. Leave the solution for sufficient time to allow equilibration at lab temperature.
2. **Record the temperature of every solution you use,** including all calibration standards and samples, since this will affect K_w, neutrality and pH.
3. **Set the temperature compensator on the meter to the appropriate value.** This control makes an allowance for the effect of temperature on the electrical potential difference recorded by the meter: it does *not* allow for the other temperature-dependent effects mentioned elsewhere. Basic instruments have no temperature compensator, and should only be used at a specified temperature, either 20 °C or 25 °C, otherwise they will not give an accurate measurement. More sophisticated systems have automatic temperature compensation.
4. **Rinse the electrode assembly with distilled water** and gently dab off the excess water onto a clean tissue: check for visible damage or contamination of the glass electrode (consult a member of staff if the glass is broken or dirty). Also check that the solution within the glass assembly is covering the metal electrode.
5. **Calibrate the instrument:** set the meter to 'pH' mode, if appropriate, and then place the electrode assembly in a standard solution of known pH, usually pH 7.00. This solution may be supplied as a liquid, or may be prepared by dissolving a measured amount of a calibration standard in water: calibration standards are often provided in tablet form, to be dissolved in water to give a particular volume of solution. Adjust the calibration control to give the correct reading. Remember that your calibration standards will only give the specified pH at a particular temperature, usually either 20 °C or 25 °C. If you are working at a different temperature, you must establish the actual pH of your calibration standards, either from the supplier, or from literature information.
6. **Remove the electrode assembly from the calibration solution and rinse again with distilled water:** dab off the excess water. Basic instruments have no further calibration steps (single-point calibration), while the more refined pH meters have additional calibration procedures.

   If you are using a basic instrument, you should check that your apparatus is accurate over the appropriate pH range by measuring the pH of another standard whose pH is close to that expected for the test solution. If the standard does not give the expected reading, the instrument is not functioning correctly: consult a member of staff.

   If you are using an instrument with a slope control function, this will allow you to correct for any deviation in electrical potential from that predicted by the theoretical relationship (at 25 °C, a change in pH of 1.00 unit should result in a change in electrical potential of 59.16 mV) by performing a two-point calibration. Having calibrated the instrument at pH 7.00, immerse in a second standard at the same temperature as that of the first standard, usually buffered to either pH 4.00 or pH 9.00, depending upon the expected pH of your samples. Adjust the slope control until the exact value of the second standard is achieved (Fig. 6.3). A pH electrode and meter calibrated using the two-point method will give accurate readings over the pH range from 3 to 11: laboratory pH electrodes are not accurate outside this range, since the

Box 6.1 (continued)

theoretical relationship between electrical potential and pH is no longer valid.

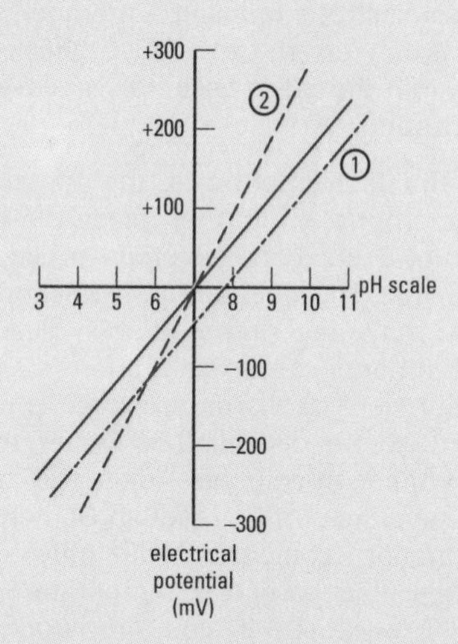

Fig. 6.3 The relationship between electrical potential and pH. The solid line shows the response of a calibrated electrode while the other plots are for instruments requiring calibration: 1 has the correct slope but incorrect isopotential point (calibration control adjustment is needed); 2 has the correct isopotential point but incorrect slope (slope control adjustment is needed).

7. **Once the instrument is calibrated, measure the pH of your solution(s)**, making sure that the electrode assembly is rinsed thoroughly between measurements. You should be particularly aware of this requirement if your solutions contain organic biological material, e.g. soil, tissue fluids, protein solutions, etc., since these may adhere to the glass electrode and affect the calibration of your instrument. If your electrode becomes contaminated during use, check with a member of staff before cleaning: avoid touching the surface of the glass electrode with abrasive material. Allow sufficient time for the pH reading to stabilize in each solution before taking a measurement: for unbuffered solutions, this may take several minutes, so do not take inaccurate pH readings due to impatience!

8. **After use, the electrode assembly must not be allowed to dry out.** Most pH electrodes should be stored in a neutral solution of KCl, either by suspending the assembly in a small beaker, or by using an electrode cap filled with the appropriate solution (typically 1.0 mol l^{-1} KCl buffered at pH 7.0). However, many labs simply use distilled water as a storage solution, leading to loss of ions from the interior of the electrode assembly. In practice, this means that pH electrodes stored in distilled water will take far longer to give a stable reading than those stored in KCl.

9. **Switch the meter to zero (where appropriate), but do not turn off the power:** pH meters give more stable readings if they are left on during normal working hours.

Problems (and solutions) include: inaccurate and/or unstable pH readings caused by cross-contamination (rinse electrode assembly with distilled water and blot dry between measurements); development of a protein film on the surface of the electrode (soak in 1% w/v pepsin in 0.1 mol l^{-1} HCl for at least an hour); deposition of organic or inorganic contaminants on the glass bulb (use an organic solvent, such as acetone, or a solution of 0.1 mol l^{-1} disodium ethylenediamine-tetraacetic acid, respectively); drying out of the internal reference solutions (drain, flush and refill with fresh solution, then allow to equilibrate in 0.1 mol l^{-1} HCl for at least an hour); cracks or chips to the surface of the glass bulb (use a replacement electrode).

zwitterionic molecules (possessing both positive and negative groups) have been introduced to overcome some of the disadvantages of the more traditional buffers. These newer compounds are often referred to as 'Good buffers', to acknowledge the early work of Dr N.E. Good and co-workers: HEPES is one of the most useful zwitterionic buffers, with a pK_a of 7.5 at 25 °C.

These zwitterionic substances are usually added to water as the free acid: the solution must then be adjusted to the correct pH with a strong alkali, usually NaOH or KOH. Alternatively, they may be used as their sodium or potassium salts, adjusted to the correct pH with a strong acid, e.g. HCl. Consequently, you may need to consider what effects such changes in ion concentration may have in a solution where zwitterions are used as buffers. In addition, zwitterionic buffers can interfere with protein determinations (e.g. Lowry method, p. 169).

Table 6.3 pK_a values at 25 °C of some acids and bases (upper section) and some large organic zwitterions (lower section) commonly used in buffer solutions. For polyprotic acids, where more than one proton my dissociate, the pK_a values are given for each ionisation step. Only the trivial acronyms of the larger molecules are provided: their full names can be obtained from the catalogues of most chemical suppliers.

Acid or base	pK_a value(s)
Acetic acid	4.8
Carbonic acid	6.1, 10.2
Citric acid	3.1, 4.8, 5.4
Glycylglycine	3.1, 8.2
Phthalic acid	2.9, 5.5
Phosphoric acid	2.1, 7.1, 12.3
Succinic acid	4.2, 5.6
TRIS*	8.3
Boric acid	9.2
MES	6.1
PIPES	6.8
MOPS	7.2
HEPES	7.5
TRICINE	8.1
TAPS	8.4
CHES	9.3
CAPS	10.4

*Note that this compound is hygroscopic and should be stored in a dessicator; also see text regarding its potential toxicity.

Table 6.4 Preparation of sodium phosphate buffer solutions for use at 25 °C. Prepare separate stock solutions of (a) disodium hydrogen phosphate and (b) sodium dihydrogen phosphate, both at 200 mol m^{-3}. Buffer solutions (at 100 mol m^{-3}) are then prepared at the required pH by mixing together the volume of each stock solution shown in the table, then diluting to a final volume of 100 ml using distilled or deionized water

Required pH (at 25 °C)	Volume of stock (a) Na_2HPO_4 (ml)	Volume of stock (b) NaH_2PO_4 (ml)
6.0	6.2	43.8
6.2	9.3	40.7
6.4	13.3	36.7
6.6	18.8	31.2
6.8	24.5	25.5
7.0	30.5	19.5
7.2	36.0	14.0
7.4	40.5	9.5
7.6	43.5	6.5
7.8	45.8	4.2
8.0	47.4	2.6

Fig. 6.4 shows a number of traditional and zwitterionic buffers and their effective pH ranges. When selecting one of these buffers, aim for a pK_a which is in the direction of the expected pH change (Tables 6.2, 6.3). For example, HEPES buffer would be a better choice of buffer than PIPES for use at pH 7.2 for experimental systems where a pH increase is anticipated, while PIPES would be a better choice for where acidification is expected.

Preparation of buffer solutions

Having selected an appropriate buffer, you will need to make up your solution to give the desired pH. You will need to consider two factors:

- The ratio of acid and conjugate base required to give the correct pH.
- The amount of buffering required; buffer capacity depends upon the absolute quantities of acid and base, as well as their relative proportions.

In most instances, buffer solutions are prepared to contain between 10 mmol l^{-1} and 200 mmol l^{-1} of the conjugate pair. While it is possible to calculate the quantities required from first principles using the Henderson–Hasselbalch equation, there are several sources which tabulate the amount of substance required to give a particular volume of solution with a specific pH value for a wide range of traditional buffers (e.g. Perrin and Dempsey, 1974). For traditional buffers, it is customary to mix stock solutions of acidic and basic components in the correct proportions to give the required pH (Table 6.4). For zwitterionic acids, the usual procedure is to add the compound to water, then bring the solution to the required pH by adding a specific amount of strong alkali or acid (obtained from tables). Alternatively, the required pH can be obtained by dropwise addition of alkali or acid, using a meter to check the pH, until the correct value is reached. When preparing solutions of zwitterionic buffers, the acid may be relatively insoluble. Do not wait for it to dissolve fully before adding alkali to change the pH – the addition of alkali will help bring the acid into solution (but make sure it has all dissolved before the desired pH is reached).

Remember that buffer solutions will only work effectively if they have sufficient buffering capacity to resist the change in pH expected during the course of the experiment. Thus a weak solution of HEPES (e.g. 10 mmol l^{-1}, adjusted to pH 7.0 with NaOH) will not be able to buffer the growth medium of a dense suspension of cells for more than a few minutes.

Finally, when preparing a buffer solution based on tabulated information, always confirm the pH with a pH meter before use.

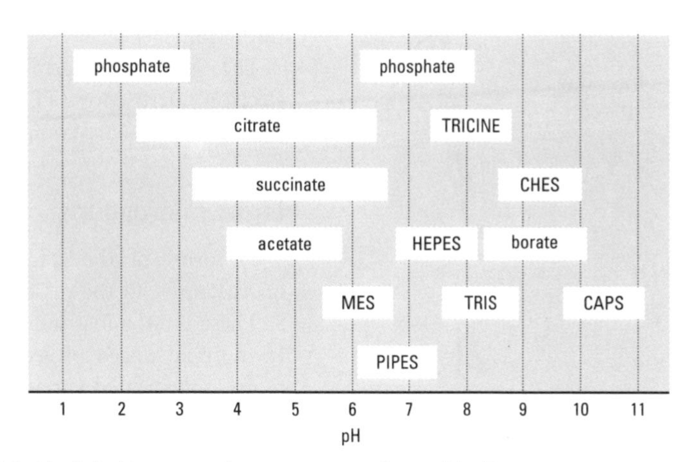

Fig. 6.4 Useful pH ranges of some commonly used buffers.

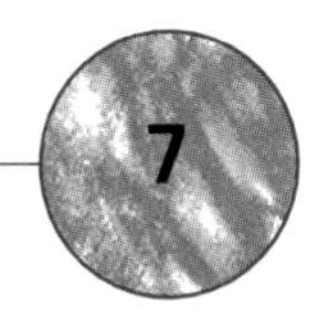

7 Introduction to microscopy

Many features of interest in biological systems are too small to be seen by the naked eye and can only be observed with a microscope. All microscopes consist of a coordinated system of lenses arranged so that a magnified image of a specimen is seen by the viewer (Fig. 7.1). The main differences are the wavelengths of electromagnetic radiation used to produce the image, the nature and arrangement of the lens systems and the methods used to view the image.

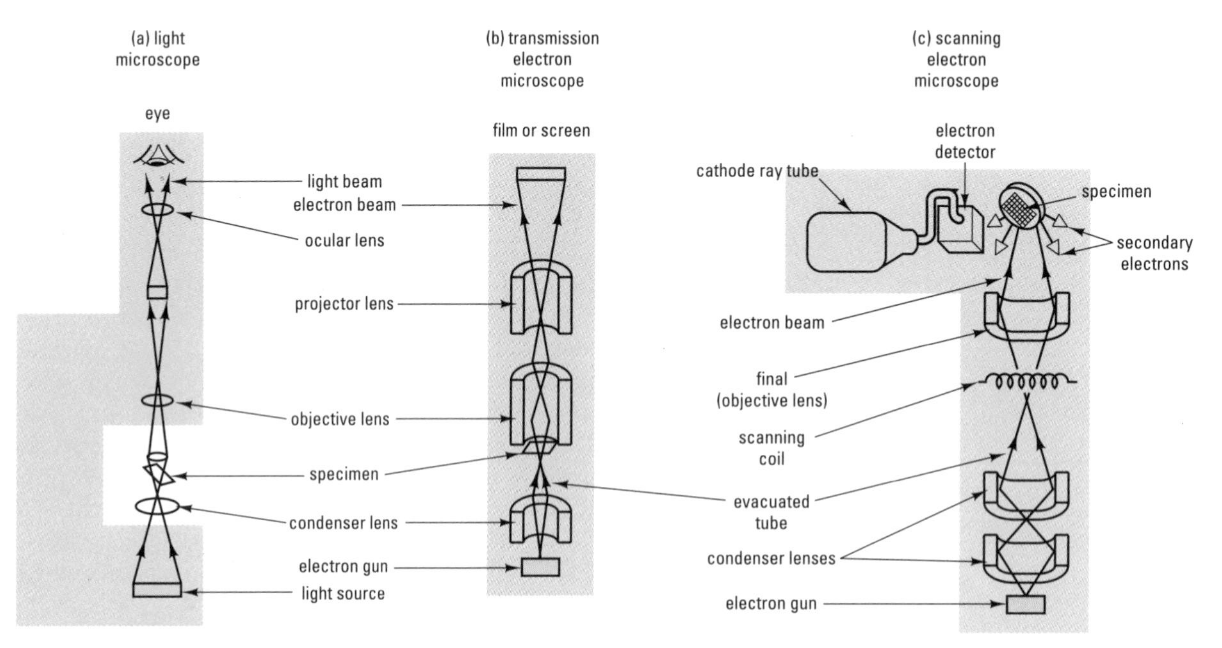

Fig. 7.1 Simplified diagrams of light and electron microscopes. Note that the electron microscopes are drawn upside-down to aid comparison with the light microscope.

Definitions

Resolution – the ability to distinguish between two points on the specimen – the better the resolution, the 'sharper' the image.

Contrast – the difference in intensity perceived between different parts of an image.

Microscopes allow objects to be viewed with increased resolution and contrast. Resolution is affected by lens design and inversely related to the wavelength of radiation used. Contrast can be enhanced (a) by the use of stains, and (b) by adjusting microscope settings, usually at the expense of resolution.

The three main forms of microscopy are light microscopy, transmission electron microscopy (TEM) and scanning electron microscopy (SEM). Their main properties are compared in Table 7.1.

Light microscopy

Two forms of the standard light microscope, the binocular (compound) microscope and the dissecting microscope, are described in detail in Chapter 8. These are the instruments most likely to be used in routine practical work. In more advanced project work, you may use one or more of the following more sophisticated variants of light microscopy to improve image quality:

- Dark field illumination involves a special condenser which causes reflected and diffracted light from the specimen to be seen against a dark

Table 7.1 Comparison of microscope types. Resolution is that obtained by a skilled user. LM, light microscope; SEM, scanning electron microscope; TEM, transmission electron microscope

	Type of microscope		
Property	LM	SEM	TEM
Resolution	200 nm	10 nm	1 nm
Depth of focus	Low	High	Medium
Field of view	Good	Good	Limited
Specimen preparation (ease)	Easy	Easy	Skilled
Specimen preparation (speed)	Rapid	Quite rapid	Slow
Relative cost of instrument	Low	High	High

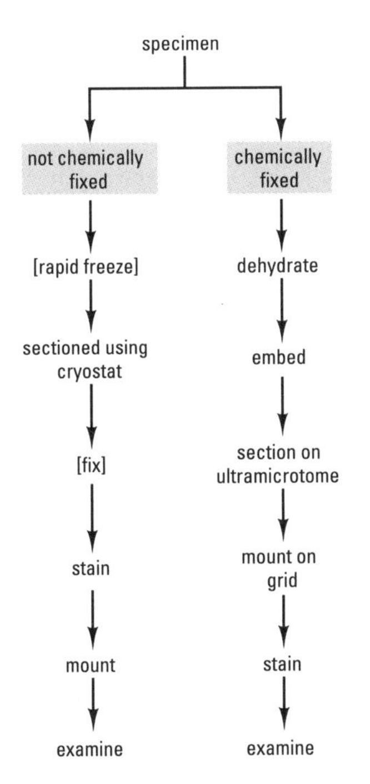

Fig. 7.2 Flowchart of procedures necessary to prepare specimens for transmission electron microscopy (TEM). Steps in brackets are optional.

background. The method is particularly useful for near-transparent specimens and for delicate structures like flagella. Care must be taken with the thickness of slides used – air bubbles and dust must be avoided and immersion oil must be used between the dark field condenser and the underside of the slide.

- Ultraviolet microscopy uses short-wavelength UV light to increase resolution. Fluorescence microscopy uses radiation at UV wavelengths to make certain naturally fluorescent substances (e.g. chlorophyll or fluorescent dyes which bind to specific cell components) emit light of visible wavelengths. Special light sources, lenses and mountants are required for UV and fluorescence microscopy and filters must be used to prevent damage to users' eyes.
- Phase contrast microscopy is useful for increasing contrast when viewing transparent specimens. It is superior to dark field microscopy because a better image of the interior of specimens is obtained. Phase contrast operates by causing constructive and destructive interference effects in the image, visible as increased contrast. Adjustments must be made for each objective lens and the microscope must be set up carefully to give optimal results.
- Nomarski or Differential Interference Contrast (DIC) microscopy gives an image with a three-dimensional quality. However, the relief seen is optical rather than morphological, and care should be taken in interpreting the result. One of the advantages of the technique is the extremely limited depth of focus which results: this allows 'optical sectioning' of a specimen.
- Polarized light microscopy can be used to reveal the presence and orientation of optically active components within specimens (e.g. starch grains, cellulose fibres), showing them brightly against a dark background.
- Confocal microscopy allows 3-dimensional views of cells or thick sections. A finely focused laser is used to create electronic images of layered horizontal 'slices', usually after fluorescent staining. Images can be viewed individually or reconstructed to provide a 3-D computer-generated image of the whole specimen.

Electron microscopes

Electron microscopes offer an image resolution up to 200 times better than light microscopes (Table 7.1) because they utilize radiation of shorter wavelength in the form of an electron beam. The electrons are produced by a tungsten filament operating in a vacuum and are focused by electromagnets. TEM and SEM differ in the way in which the electron beam interacts with the specimen: in TEM, the beam passes through the specimen (Fig. 7.1b), while in SEM the beam is scanned across the specimen and is reflected from the surface (Fig. 7.1c). In both cases, the beam must fall on a fluorescent screen before the image can be seen. Permanent images ('electron micrographs') are produced after focusing the beam on photographic film.

You are unlikely to use either type of electron microscope as part of undergraduate practical work because of the time required for specimen preparation and the need for detailed training before these complex machines can be operated correctly. However, electron microscopy is extremely important in understanding cellular and sub-cellular structures and you may be shown electron micrographs, either to demonstrate cell ultrastructure (TEM) or surface features (SEM), or to investigate changes in the number, size, shape or condition of cells and organelles (TEM). Figure 7.2 shows the main stages in preparing material for TEM.

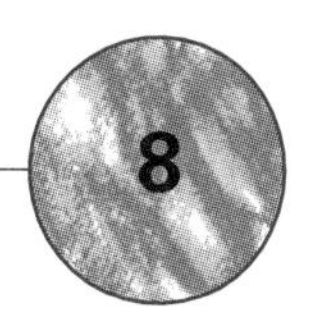

Setting up and using a light microscope

Correct use of the light microscope is a basic laboratory skill. A standard undergraduate binocular microscope (Fig. 8.1) consists of three main types of optical unit: eyepiece, objective and condenser. These are attached to a stand which holds the specimen on a stage.

Setting up a binocular light microscope

Before using any microscope, familiarize yourself with its component parts.

KEY POINT Never assume that the previous person to use your microscope has left it set up correctly: apart from differences in users' eyes, the microscope needs to be properly set up for each lens combination used.

The procedures outlined below are simplified to allow you to set up microscopes like those of the Olympus CH series (Fig. 8.1). For monocular microscopes, disregard instructions for eyepiece adjustment.

1. Place the microscope at a convenient position on the bench. Adjust your seating so that you are comfortable operating the focus and stage controls. Unwind the power cable, plug in and switch on after first ensuring that the lamp setting is at a minimum. Adjust the lamp setting to about two-thirds of the maximum.
2. Select a low-power (e.g. $\times 10$) objective. Ensure that the lens clicks home.
3. Set the eyepiece lenses to your interpupillary distance; this can usually be read off a scale on the turret. You should now see a single circular field of vision. If you do not, try adjusting in either direction.
4. Put a prepared slide on the stage. Examine it first against a light source and note the position, colour and rough size of the specimen. Place the slide on the stage (correct way up!) and, viewing from the side, position it with the stage adjustment controls so that the specimen is illuminated.
5. Focus the image of the specimen using first the coarse and then the fine focusing controls. The image will be reversed and upside-down compared to that seen by viewing the slide directly.
   (a) If both eyepiece lenses are adjustable, set your interpupillary distance on the scale on each lens. Close your left eye, look through the right eyepiece with your right eye and focus the image with the normal controls. Now close your right eye, look through the left eyepiece with your left eye and focus the image by rotating the eyepiece holder. Take a note of the setting for future use.
   (b) If only the left eyepiece is adjustable, close your left eye, look with the right eye through the static right eyepiece and focus the image with the normal controls. Now close your right eye, look through the left eyepiece with your left eye and focus the image by rotating the eyepiece holder. Take a note of the setting for future use.
6. Close the condenser–iris diaphragm (aperture–iris diaphragm), then open it to a position such that further opening has no effect on the brightness of the image (the 'threshold of darkening'). The edge of the diaphragm should not be in view. Turn down the lamp if it is too bright.

Using binocular eyepieces – if you do not know your interpupillary distance, ask someone to measure it with a ruler. You should stare at a fixed point in the distance while the measurement is taken. Take a note of the value for future use.

Problems with spectacles – those who wear glasses can remove them for viewing as microscope adjustments will accommodate most deficiencies in eyesight (except astigmatism). This is more comfortable and stops the spectacle lenses being scratched by the eyepiece holders. However, it creates difficulties in focusing when drawing diagrams.

Fig. 8.1 Diagram of the Olympus binocular microscope model CH.

- The lamp (**1**) in the base of the stand (**2**) supplies light; its brightness is controlled by an on–off switch and voltage control (**3**). Never use maximum voltage or the life of the bulb will be reduced – a setting two-thirds to three-quarters of maximum should be adequate for most specimens. A field–iris diaphragm may be fitted close to the lamp to control the area of illumination (not present on this model).
- The condenser control (**4**) focuses light from the condenser lens system (**5**) onto the specimen and projects the specimen's image onto the front lens of the objective. Correctly used, it ensures optimal resolution.
- The condenser–iris diaphragm (**6**) controls the amount of light entering and leaving the condenser; its aperture can be adjusted using the condenser–iris diaphragm lever (**7**). Use this to reduce glare and enhance image contrast by cutting down the amount of stray light reaching the objective lens.
- The specimen (normally mounted on a slide) is fitted to a mechanical stage or slide holder (**8**) using a spring mechanism. Two controls allow you to move the slide in *x* and *y* planes. Vernier scales on the slide holder can be used to return to the same place on a slide. The fine and coarse focus controls (**9**) adjust the height of the stage relative to the lens systems. Take care when adjusting the focus controls to avoid hitting the lenses with the stage or slide.
- The objective lens (**10**) supplies the initial magnified image; it is the most important component of any microscope because its qualities determine resolution, depth of field and optical aberrations. The objective lenses are attached to a revolving nosepiece (**11**). Take care not to jam the longer lenses onto the stage or slide as you rotate the nosepiece. You should feel a distinct click as each lens is moved into position. The magnification of each objective is written on its side; a normal complement would be ×4, ×10, ×40 and ×100 (oil immersion).
- The eyepiece lens (**12**) is used to further magnify the image from the objective and to put it in a form and position suitable for viewing. Its magnification is written on the holder (normally ×10). By twisting the holder for one or both of the eyepiece lenses you can adjust their relative heights to take account of optical differences between your eyes. The interpupillary distance scale (**13**) and adjustment knob allow compensation to be made for differences in the distance between users' pupils.
- The turret clamping screw (**14**) allows the eyepiece turret (**15**) to be rotated so a demonstrator can view your specimen without exchanging position with you. If loosened too much, the turret can come off, so take care and always re-tighten after use.

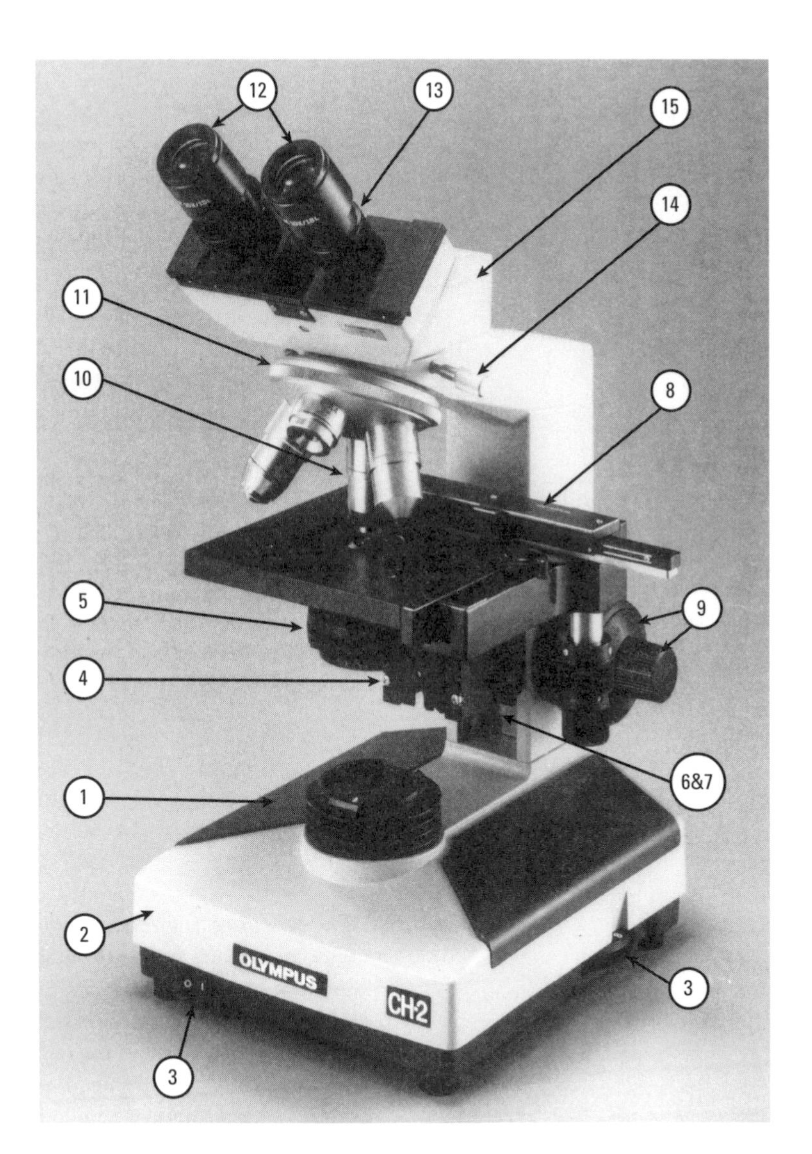

Adjusting a microscope with a field–iris diaphragm – adjust this *before* the condenser–iris diaphragm: close it until its image appears in view, if necessary focusing its image with the condenser controls and centring it. Now open it so the whole field is just illuminated.

High-power objectives – *never remove a slide while a high power objective lens (i.e. ×40 or ×100) is in position.* Always turn back to the ×10 first. Having done this, lower the stage and remove the slide.

7. Focus the condenser. Place an opaque pointed object (e.g. a sharp pencil point) on the centre of the light source. Adjust the condenser setting until both the specimen and needle tip/pencil point are in focus together. Check that the condenser–iris diaphragm is just outside the field of view.
8. For higher magnifications, swing in the relevant objective (e.g. ×40), carefully checking that there is space for it. Adjust the focus using the fine control only. If the object you wish to view is in the centre of the field with the ×10 objective, it should remain in view (magnified, of course) with the ×40. Adjust the condenser–iris diaphragm and focus the condenser as before – the correct setting for each lens will be different.
9. When you have finished using the microscope, remove the last slide and clean the stage if necessary. Turn down the lamp setting to its minimum, then switch off. Clean the eyepiece lenses with lens tissue. Check that the objectives are clean. Unplug the microscope from the mains and wind the cable round the stand and under the stage. Replace the dust cover.

If you have problems in obtaining a satisfactory image, refer to Box 8.1; if this doesn't help, refer the problem to the class supervisor.

Box 8.1 Problems in light microscopy and possible solutions

- No image; very dark image; image dark and illuminated irregularly – microscope not switched on (check plug and base); illumination control at low setting or off; objective nosepiece not clicked into place over a lens; diaphragm closed down too much or off centre; lamp failure
- Image visible and focused but pale and indistinct – diaphragm needs to be closed down further; condenser requires adjustment
- Image blurred and cannot be focused – dirty objective; dirty slide; slide in upside-down; slide not completely flat on stage; eyepiece lenses not set up properly for user's eyes; fine focus at end of travel
- Dust and dirt in field of view – eyepiece lenses dirty; slide dirty; dirt on lamp glass or upper condenser lens

Procedure for observing transparent specimens

Some stained preparations and all colourless objects are difficult to see when the microscope is adjusted as above. Contrast can be improved by closing down the condenser–iris diaphragm. Note that when you do this, diffraction haloes appear round the edges of objects. These obscure the image of the true structure of the specimen and may result in loss of resolution. Nevertheless, an image with increased contrast may be easier to interpret.

Stains and staining procedures

Table 8.1 gives a number of widely used stains and their uses on plant, animal and microbial cells and tissues.

Procedure for oil immersion objectives

These provide the highest resolution of which the light microscope is capable. They must be used with immersion oil filling the space between the objective lens and the top of the slide. The oil has the same refractive index as the glass lenses, so loss of light by reflection and refraction at the glass/air interface is

Table 8.1 A selection of stains for light microscopy of sections

Stain	What it stains	Comments
Plant tissue stains		
Chlorazol black	Cell walls: black Nuclei: black, yellow or green Suberin: amber	The solvent used (70% ethanol in water or water alone) affects colours developed
Neutral Red	Living cells: pink (pH <7)	A 'vital' stain used to determine cell viability or to visualize plant protoplasts in plasmolysis experiments; best used at neutral external pH
Phloroglucinol/HCl	Lignified cell walls: red	Care is required because the acid may damage microscope lenses
Ruthenium red	Pectins: red	Shows up the middle lamella
Safranin + Fast green	Nuclei, chromosomes, cuticle and lignin: red Other components: green	Stain in safranin first, then counterstain with fast green (light green will substitute). A differentiation step is required
Toluidine blue	Lignified cell walls: blue Cellulose cell walls: purple	Best to apply dilute and allow progressive staining to occur
Animal tissue stains		
Azure A/eosin B	Nuclei, RNA: blue Basophilic cells: blue–violet Most other cells: pale blue Muscle cells: pink Necrosing cells: pink Cartilage matrix: red–violet Bone: pink Red blood cells: orange–red Mucins: green–blue/blue–violet	Used in pathology – shows up bacteria as blue; must be fresh; care required over pH: Mann's methyl blue/eosin gives similar results
Chlorazol black	Chitin: greenish-black Nuclei: black, yellow or green Glycogen: pink or red	Solvent (70% v/v ethanol in water or water alone) affects colours formed
Iron haematoxylin	Nuclei, chromosomes and red blood cells: black Other structures: grey or blue–black	Good for resolving fine detail; iron alum used as mordant before haematoxylin to differentiate
Mallory	Nuclei: red Nucleoli: yellow Collagen, mucus:; blue Red blood cells: yellow Cytoplasm: pink or yellow	Simple, one-stage stain; fades within a year; not to be used with osmium-containing fixatives. Heldenhain's azan gives similar results but does not fade. Cason's one-step Mallory is a rapidly applied stain which is particularly good for connective tissue
Masson's trichrome	Collagen, mucus: green Cytoplasm: pink	Used as a counterstain after, for example, iron haematoxylin, which will have stained nuclei black. Not to be used after osmium fixation
Microbial stains		
Giemsa	Bacterial chromosome: purple Bacterial cytoplasm: colourless	Also used in zoology to stain protozoa
Gram	Gram-positive bacteria: violet/purple Gram-negative bacteria: red/pink yeasts: violet/purple	See p. 70 for procedure
Gray	Bacterial flagella: red	Uses toxic chemicals: mercuric chloride and formaldehyde. Leifson's stain is an alternative
Lactophenol cotton blue	Fungal cytoplasm: blue (hyphal wall unstained)	Shrinkage may occur
Nigrosin or India ink	Background: grey–black	Negative stains for visualization of capsules: requires a very thin film
Proca–Kayser	Viable bacteria: purple Dead bacteria: pink/red Viable endospores: pink Dead endospores: blue	Recent method involves acridine orange and UV microscopy
Shaeffer and Fulton	Bacterial endospores: green Vegetative cells: pink/red	Malachite green is primary stain, heated for 5 min. Counterstained with safranin
Ziehl–Neelsen	Actinomycetes: red Bacterial endospores: red Other microbes: blue	Requires heat treatment of fuchsin primary stain, decolorization with ethanol–HCl and a methylene blue counterstain (acid-fast structures remain red)

reduced. This increases the resolution, brightness and clarity of the image and reduces aberration. Use oil immersion objective(s) as follows:

1. Check that the object of interest is in view using the ×40 objective.
2. Apply a single small droplet of immersion oil to the illuminated spot on the top of the slide, having first swung the ×40 objective away. Never use too much oil: it can run off the slide and mess up the microscope.
3. Move the high power objective into position carefully, checking first that there is space for it. Focus on the specimen using the fine control only.
4. Perform condenser–iris diaphragm, condenser focusing and lamp brightness adjustments as for the other lenses.
5. When finished, clean the oil immersion lens by gently wiping it with clean lens tissue. If the slide is a prepared one, wipe the oil off with lens tissue.

Establishing scale and measuring objects

Spurious accuracy – avoid putting too many significant figures in any estimates of dimensions: there may be quite large errors in estimating print magnifications which could make the implied accuracy meaningless (p. 42).

The magnification of a light microscope image is calculated by multiplying the objective magnification by the eyepiece magnification. However, the magnification of the image bears no certain relation to the magnification of any drawing of the image – you may equally well choose to draw the same image 10 mm or 10 cm long. For this reason, *it is essential to add a scale to any diagram.* You can provide either a bar giving the estimated size of an object of interest, or a bar of defined length (e.g. 100 μm).

The simplest method of estimating linear dimensions is to compare the size of the image to the diameter of the field of view. You can make a rough estimate of the field diameter by focusing on the millimetre scale of a transparent ruler using the lowest power objective. Estimate the diameter of this field directly, then use the information to work out the field diameters at the higher powers *pro rata.* For example, if the field at an overall magnification of ×40 is 4 mm, at an overall magnification of ×100 it will be: $40/100 \times 4\,\text{mm} = 1.6\,\text{mm}$.

Using an eyepiece graticule – choose the eyepiece lens corresponding to your stronger eye (usually the same side as the hand you write with) and check that you have made the correct adjustments to the eyepiece lenses as detailed on p. 34.

Greater accuracy can be obtained if an eyepiece micrometer (graticule) is used. This carries a fine scale and fits inside an eyepiece lens. The eyepiece micrometer is calibrated using a stage micrometer, basically a slide with a fine scale on it. Say you find at ×400 overall magnification that 1 stage micrometer unit of 0.1 mm is equal to 39 eyepiece micrometer units. Each eyepiece micrometer unit = 0.1/39 mm = 2.56 μm. The eyepiece scale can now be used to measure objects: in the above example, the scale reading is multiplied by 2.56 to give a value in micrometres. Alternatively, you could put a 100 μm bar on your diagram corresponding to the length of 39 eyepiece micrometer units on your specimen.

Care and maintenance of your microscope

Microscopes are delicate precision instruments. Handle them with care and never force any of the controls. Never touch any of the glass surfaces with anything other than clean, dry lens tissue. Bear in mind that a replacement would be very expensive.

If moving a microscope, hold the stand above the stage with one hand and rest the base of the stand on your other hand. Always keep the microscope vertical (or the eyepieces may fall out). Put the microscope down gently.

Clean lenses by gently wiping with clean, dry lens tissue. Use each piece of tissue once only. Try not to touch lenses with your fingers as oily fingerprints are difficult to clean off. Do not allow any solvent (including water) to come into contact with a lens; salt solutions are particularly damaging.

The investigative approach

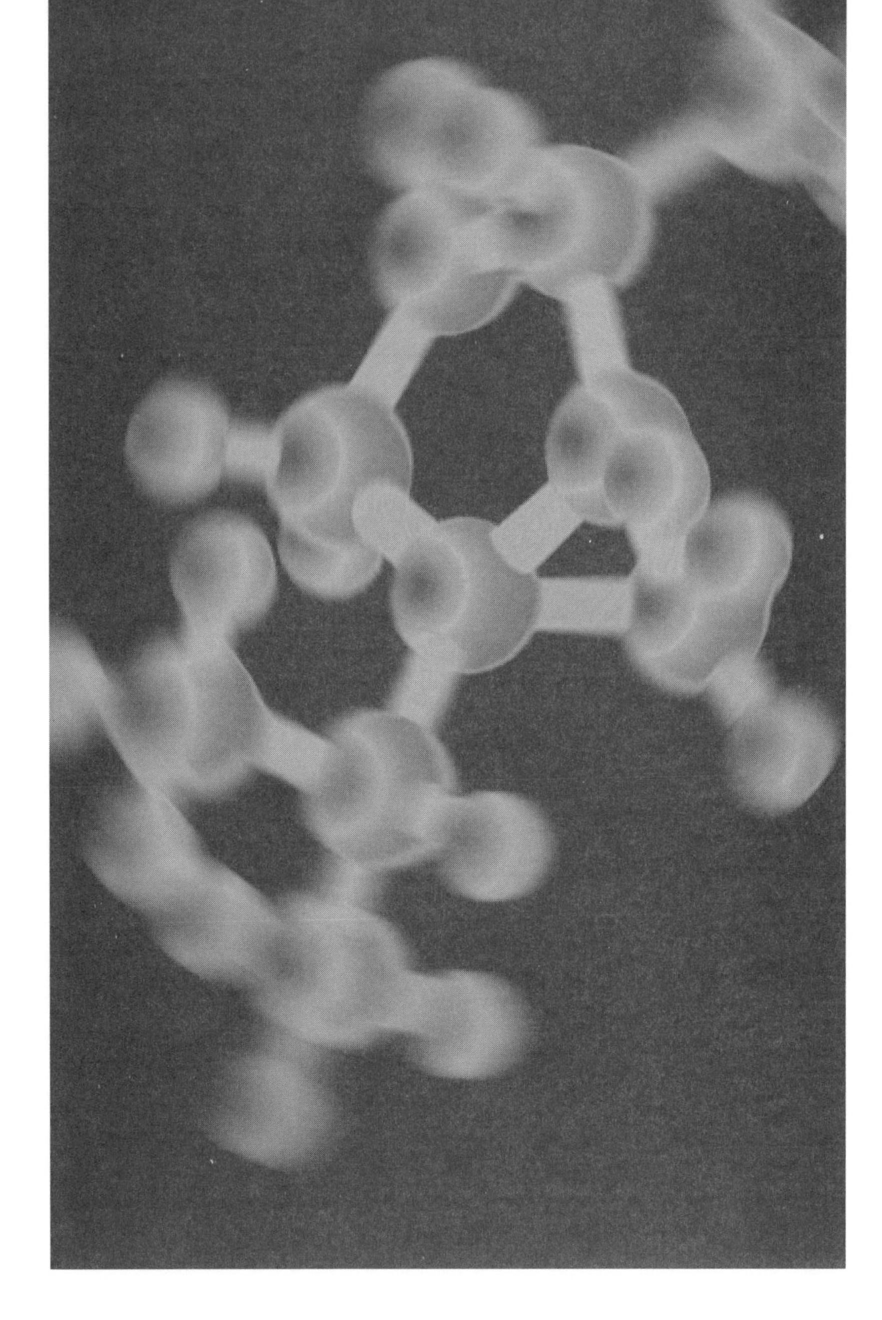

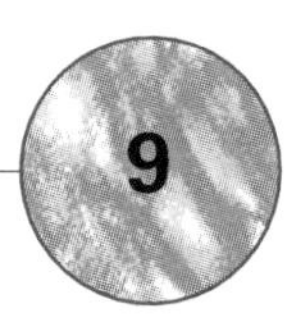

9 Making and recording measurements

Definition

Variable – any characteristic or property which can take one of a range of values

Parameter – a numerical constant or mathematical function used to describe a particular population (e.g. the mean height of 18-year-old females).

Statistic – an estimate of a parameter obtained from a sample (e.g. the height of 18-year-old females based on those in your class).

The term data (singular = datum, or data value or variate) refers to measurements of a particular characteristic, or variable, classified as:

- Quantitative: where the individual values are described on a numerical scale which may be either (i) continuous, taking any value on the measurement scale, or (ii) discontinuous (or discrete), where only integer values are possible. Many of the variables measured in biomolecular science are continuous and quantitative, e.g. weight, temperature, time, amount of product formed in an enzyme reaction.
- Ranked: where the data values can be listed in order of magnitude. Where such data are given numbered ranks, they are sometimes called 'semi-quantitative data'. Note that such ranks cannot be treated as 'real' numbers and they should not be added, averaged, etc.
- Qualitative: where individual values are assigned to a descriptive category, e.g. morphological characteristics in a genetics cross (p. 218), or presence/absence of a particular biomolecule in a sample (p. 165).

Variables may be independent or dependent. Usually, the variable under the control of the experimenter (e.g. time, substrate concentration, pH, etc.) is the independent variable, while the variable being measured (biological response) is the dependent variable (p. 50). Sometimes, it is inappropriate to describe variables in this way, and they are often referred to as interdependent. Another group of values, often termed derived (or computed) data are calculated from two or more individual measurements, and these include ratios, percentages and rates.

Measurement scales

Examples A nominal scale for temperature is not feasible, since the relevant descriptive terms can be ranked in order of magnitude.

An ordinal scale for temperature measurement might use descriptive terms, ranked in ascending order e.g. cold = 1, cool = 2, warm = 3, hot = 4.

The Celsius scale is an interval scale for temperature measurement, since the arbitrary zero corresponds to the freezing point of water (0 °C).

The Kelvin scale is a ratio scale for temperature measurement since 0 K represents a temperature of absolute zero (for information, the freezing point of water is 273.15 K on this scale).

Variables may be measured on different types of scale:

- Nominal – where classification is based on a descriptive characteristic (e.g. colour). This is the only scale for qualitative data.
- Ordinal – this classifies by numerical rank, from smallest to greatest, but with no assumption of equal spacing between ranks.
- Interval – for certain quantitative variables, where numbers on an equal unit scale are related to an arbitrary zero, e.g. temperature in °C.
- Ratio – similar to the interval scale, except that the zero point represents an absence of that character, i.e. it is an absolute zero.

Note that you may be able to make measurements of a particular variable in more than one way: for example, you could measure light on a nominal scale based on colour ('blue', 'red', etc.), an ordinal scale based on visual intensity (where 1 = dull, 2 = moderate, 3 = bright, etc.), or a ratio scale based on irradiance ($W\,m^{-2}$, between specified wavelengths of the electromagnetic radiation spectrum, p. 108). In general, you should aim to make quantitative measurements using a ratio scale, to allow you to use the broadest range of mathematical operations and statistical procedures.

Accuracy and precision

Accuracy is the closeness of a measured or derived data value to its true value, while precision is the closeness of repeated measurements to each other (Fig. 9.1). A balance with a fault in it (i.e. a bias, see below) could give

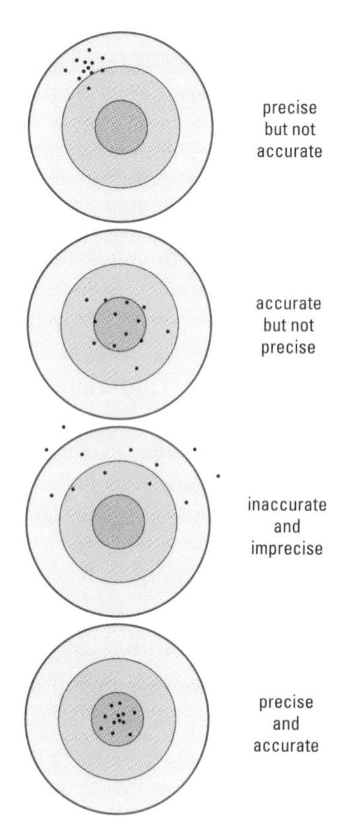

Fig. 9.1 'Target' diagrams illustrating precision and accuracy.

precise (i.e. very repeatable) but inaccurate (i.e. untrue) results. Unless there is bias in a measuring system, precision will lead to accuracy and it is precision that is generally the most important practical consideration, if there is no reason to suspect bias. You can investigate the precision of any measuring system by repeated measurements of the same sample: the nearer the replicate values are to each other, the more precise the measurement.

Absolute accuracy and precision are impossible to achieve, due to the limitations of measuring systems. It is particularly important to avoid spurious accuracy in the presentation of results; include only those digits which the accuracy of the measuring system implies. This type of error is common when changing units (e.g. inches to metres) and in derived data, especially when calculators give results to a large number of decimal places.

Bias (systematic error)

Bias is a systematic or non-random inaccuracy and is one of the most troublesome difficulties in using numerical data. Biases may be associated with incorrectly calibrated instruments, e.g. a faulty pipettor, or with experimental manipulations, e.g. loss of enzyme activity during storage. Bias in measurement can also be subjective, or personal, e.g. an experimenter's pre-conceived ideas about an 'expected' result.

Bias can be minimized by using a carefully standardized procedure, with fully calibrated instruments. Investigate bias in 'trial runs' by measuring a variable in several different ways, to see if the same result is obtained.

To avoid personal bias, 'blind' measurements should be made where the identity of each sample is unknown to the operator, e.g. use a coding system.

Minimizing errors – determine early in your study what the dominant errors are likely to be and concentrate your time and effort on reducing these.

Working with derived data – special effort should be made to reduce measurement errors because their effects can be magnified when differences, ratios, indices or rates are calculated.

Measurement error

All measurements are subject to error, but the dangers of misinterpretation are reduced by recognizing and understanding the likely sources of error and by adopting appropriate protocols and calculation procedures.

A common source of measurement error is carelessness, e.g. reading a scale in the wrong direction or parallax errors. This can be reduced greatly by careful recording and may be detected by repeating the measurement. Other errors arise from faulty or inaccurate equipment, but even a perfectly functioning machine has distinct limits to the accuracy and precision of its measurements. These limits are often quoted in manufacturers' specifications and are applicable when an instrument is new; however, you should allow for some deterioration with age.

One major influence virtually impossible to eliminate is the effect of the investigation itself: even putting a thermometer in a liquid may change the temperature of the liquid. The very act of measurement may give rise to a confounding variable (p. 50) as discussed in Chapter 11. You should include descriptions of the possible sources of error(s) and estimates of their importance in any report. However, do not use 'biological variability' as a catch-all excuse for poor technique or inadequacies in your experimental design.

Recording primary data – never be tempted to jot down data on scraps of paper: you are likely to lose them, or to forget what individual values mean.

Collecting and recording primary data

When carrying out lab work or research projects, you will need to master the important skills of recording and managing data. Individual observations (e.g. laboratory temperature) can be recorded in the text of your notes, but tables are the most convenient way to collect large amounts of information.

Designing a table for data collection – make sure there is sufficient space in each column for the values – if in doubt, err on the generous side.

Recording numerical data – write down only those numbers that can be justified by your measurement technique (significant figures).

Choosing a lab notebook – a spiral-bound notebook is good for making a primary record – it lies conveniently open on the bench and provides a simple method of dealing with major mistakes!

KEY POINT **A good set of lab notes should:**

- **outline the purpose of your experiment or observation;**
- **set down all the information required to describe your materials and methods;**
- **record all relevant information about your results or observations and provide a visual representation of the data;**
- **note your immediate conclusions and suggestions for further experiments.**

When preparing a table for data collection, you should:

1. Use a concise title or a numbered code for cross-referencing.
2. Decide on the number of variables to be measured and their relationship with each other and lay out the table appropriately:
   (a) The first column of your table should show values of the independent (controlled) variable, with subsequent columns for the individual (measured) values for each replicate or sample.
   (b) If several variables are measured for the same organism or sample, each should be given a row.
   (c) In time-course studies, put the replicates as columns grouped according to treatment, with the rows relating to different times.
3. Make sure the arrangement reflects the order in which the values will be collected. Your table should be designed to make the recording process as straightforward as possible, to minimize the possibility of mistakes. For final presentation, a different arrangement may be best (Chapter 44).
4. Consider whether additional columns are required for subsequent calculations. Create a separate column for each mathematical manipulation, so the step-by-step calculations are clearly visible. Use a computer spreadsheet (p. 287) if you are manipulating lots of data.
5. Use a pencil to record data so that mistakes can be easily corrected.
6. Take sufficient time to record quantitative data unambiguously – use large clear numbers, making sure that individual numerals cannot be confused.
7. Record numerical data to an appropriate number of significant figures, reflecting the accuracy and precision of your measurement (p. 42). Do not round off data values, as this might affect the subsequent analysis.
8. Record the actual observations, not your interpretation, e.g. the colour of a particular biochemical test, rather than whether the test was positive or negative. Take care not to lose any of the information content of the data: for instance, if you only write down means and not individual values, this will affect your ability to carry out subsequent statistical analyses.
9. Prepare duplicated recording tables/checklists for repeated experiments.
10. Explain any unusual results in a footnote. Don't rely on memory.

Recording details of project work

The recommended system is one where you make a dual record.

Primary record

The primary record is made at the bench and you must concentrate on the detail of materials, methods and results. Include information that would not be used elsewhere, but which might prove useful in error tracing: for example, if you note how a solution was made up (exact volumes and weights used rather than concentration alone), this could reveal whether a miscalculation had been the cause of a rogue result. Note the origin, type and state of the chemicals and organism(s) used. In Materials and Methods, the basic rule is to record enough information to allow a reasonably competent scientist to repeat your work exactly. You must tread a line between the extremes of pedantic, irrelevant detail and the omission of information essential for proper interpretation – better perhaps to err on the side of extra detail to begin with. An experienced worker can tell you which subtle shifts in technique are important (e.g. batch numbers for an important chemical, or when a new stock solution is prepared). Many important scientific advances have been made because of careful observation and record taking and because coincident data were recorded that did not seem of immediate value. Make rough diagrams to show the arrangement of replicates, equipment, etc. If forced to use loose paper to record data, make sure each sheet is dated and taped to your lab book, collected in a ring binder, or attached with a treasury tag. The same applies to traces, printouts and graphs.

The basic order of the primary record should mirror that of a research report (see p. 311), including: the title and date, brief introduction, comprehensive materials and methods, the data and short conclusions.

Secondary record

You should make a secondary record concurrently or later in a bound book and it ought to be neater, in both organization and presentation. This book will be used when discussing results with your supervisor, and when writing up a report or thesis, and may be part of your course assessment. Writing a second, neater version forces you to consider again details that might have been overlooked in the primary record and provides a duplicate in case of loss or damage. While these notes should retain the essential features of the primary record, they should be more concise and the emphasis should move towards analysis of the experiment. Don't repeat Materials and Methods for a series of similar experiments; use devices such as 'method as for Expt. B4'. A photocopy may be sufficient if the method is derived from a text or article (check with your supervisor). Outline the aims more carefully at the start and link the experiment to others in a series (e.g. 'Following the results of Expt. D24, I decided to test whether...'). You should present data in an easily digested form, e.g. as tables of means or as summary graphs. Use appropriate statistical tests (p. 268) to support your analysis of the results. Always analyse and think about data immediately after collecting them as this may influence your subsequent activities. Write down any conclusions: sometimes those which seem obvious at the time of doing the work are forgotten when the time comes to write up a report or thesis. Likewise, ideas for further studies may prove valuable later. Even if your experiment appears to be a failure, suggestions as to the likely causes might prove useful.

Using communal records

If working with a research team, you may need to use their communal databases. These avoid duplication of effort and ensure uniformity in techniques. They may also form part of the legal safety requirements for lab work. They might include:

- a shared notebook of common techniques (e.g. media or solutions);
- a set of simplified step-by-step instructions for use of equipment;
- an alphabetical list of suppliers of equipment and consumables;
- a list of chemicals required by the group and where they are stored;
- the risk assessment sheets for dangerous procedures (p. 5);
- a record of the use and disposal of radioisotopes or clinical samples.

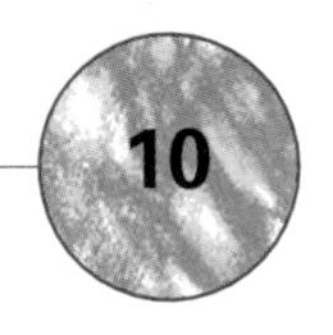

SI units and their use

Dimensionless measurements – some quantities can be expressed as dimensionless ratios or logarithms (e.g. absorbance and pH), and in these cases you do not need to use a qualifying unit.

When describing a measurement, you normally state both a number and a unit (e.g. 'the length is 1.85 metres'). The number expresses the ratio of the measured quantity to a fixed standard, while the unit identifies that standard measure or dimension. Clearly, a single unified system of units is essential for efficient communication of such data within the scientific community. The Système International D'Unités (SI) is the internationally ratified form of the metre-kilogram-second system of measurement and represents the accepted scientific convention for measurements of physical quantities.

Another important reason for adopting consistent units is to simplify complex calculations where you may be dealing with several measured quantities (see p. 258). Although the rules of the SI are complex and the scale of the base units is sometimes inconvenient, to gain the full benefits of the system you should observe its conventions strictly.

The description of measurements in SI involves:

- seven base units and two supplementary units, each having a specified abbreviation or symbol (Table 10.1);
- derived units, obtained from combinations of base and supplementary units, which may also be given special symbols (Table 10.2);
- a set of prefixes to denote multiplication factors of 10^3, used for convenience to express multiples or fractions of units (Table 10.3).

Table 10.1 The base and supplementary SI units

Measured quantity	Name of SI unit	Symbol
Base units		
Length	metre	m
Mass	kilogram	kg
Amount of substance	mole	mol
Time	second	s
Electric current	ampere	A
Temperature	kelvin	K
Luminous intensity	candela	cd
Supplementary units		
Plane angle	radian	rad
Solid angle	steradian	sr

Table 10.2 Some important derived SI units

Measured quantity	Name of unit	Symbol	Definition in base units	Alternative in derived units
Energy	joule	J	$m^2\,kg\,s^{-2}$	N m
Force	newton	N	$m\,kg\,s^{-2}$	$J\ m^{-1}$
Pressure	pascal	Pa	$kg\,m^{-1}\,s^{-2}$	$N\ m^{-2}$
Power	watt	W	$m^2\,kg\,s^{-3}$	$J\ s^{-1}$
Electric charge	coulomb	C	A s	$J\ V^{-1}$
Electric potential difference	volt	V	$m^2\,kg\,A^{-1}\,s^{-3}$	$J\ C^{-1}$
Electric resistance	ohm	Ω	$m^2\,kg\,A^{-2}\,s^{-3}$	$V\ A^{-1}$
Electric conductance	siemens	S	$s^3\,A^2\,kg^{-1}\,m^{-2}$	$A\ V^{-1}$ or Ω^{-1}
Electric capacitance	farad	F	$s^4\,A^2\,kg^{-1}\,m^{-2}$	$C\ V^{-1}$
Luminous flux	lumen	lm	cd sr	
Illumination	lux	lx	$cd\,sr\,m^{-2}$	$lm\ m^{-2}$
Frequency	hertz	Hz	s^{-1}	
Radioactivity	becquerel	Bq	s^{-1}	
Enzyme activity	katal	kat	mol substrate s^{-1}	

Table 10.3 Prefixes used in the SI

Multiple	Prefix	Symbol	Multiple	Prefix	Symbol
10^{-3}	milli	m	10^3	kilo	k
10^{-6}	micro	μ	10^6	mega	M
10^{-9}	nano	n	10^9	giga	G
10^{-12}	pico	p	10^{12}	tera	T
10^{-15}	femto	f	10^{15}	peta	P
10^{-18}	atto	a	10^{18}	exa	E

Recommendations for describing measurements in SI units

Basic format

- Express each measurement as a number separated from its units by a space. If a prefix is required, no space is left between the prefix and the unit it refers to. Symbols for units are only written in their singular form and do not require full stops to show that they are abbreviated or that they are being multiplied together.

Example 10 μg is correct, while 10μg, 10 μg. and 10μ g are incorrect. 2.6 mol is right, but 2.6 mols is wrong.

Example n stands for nano and N for newtons.

- Give symbols and prefixes appropriate upper or lower case initial letters as this may define their meaning. Upper case symbols are named after persons but when written out in full they are not given initial capital letters.
- Show the decimal sign as a full point on the line. Some metric countries continue to use the comma for this purpose and you may come across this in the literature: commas should not therefore be used to separate groups of thousands. In numbers that contain many significant figures, you should separate multiples of 10^3 by spaces rather than commas.

Example 1 982 963.192 309 kg (perhaps better expressed as 1.982 963 192 309 Gg).

Compound expressions for derived units

- Take care to separate symbols in compound expressions by a space to avoid the potential for confusion with prefixes. Note, for example, that 200 m s (metre-seconds) is different from 200 ms (milliseconds).
- Express compound units using negative powers rather than a solidus (/): for example, write mol m^{-3} rather than mol/m^3. The solidus is reserved for separating a descriptive label from its units (see p. 251).
- Use parentheses to enclose expressions being raised to a power if this avoids confusion: for example, a photosynthetic rate might be given in mol CO_2 (mol photons)$^{-1}$ s^{-1}.
- Where there is a choice, select relevant (natural) combinations of derived and base units: e.g. you might choose units of Pa m^{-1} to describe a hydrostatic pressure gradient rather than kg m^{-2} s^{-1}, even though these units are equivalent and the measurements are numerically the same.

Examples 10 μm is preferred to 0.000 01 m or 0.010 mm.

1 mm^2 = 10^{-6} m^2 (not one-thousandth of a square metre).

1 dm^3 (1 litre) is more properly expressed as 1×10^{-3} m^3.

Avogadro's constant is $6.022\,174 \times 10^{23}$ mol^{-1}.

State as MW m^{-2} rather than W mm^{-2}.

Use of prefixes

- Use prefixes to denote multiples of 10^3 (Table 10.3) so that numbers are kept between 0.1 and 1000.
- Treat a combination of a prefix and a symbol as a single symbol. Thus, when a modified unit is raised to a power, this refers to the whole unit including the prefix.
- Avoid the prefixes deci (d) for 10^{-1} and centi (c) for 10^{-2} as they are not strictly SI.
- Express very large or small numbers as a number between 1 and 10 multiplied by a power of 10 if they are outside the range of prefixes shown in Table 10.3.
- Do not use prefixes in the middle of derived units: they should be attached only to a unit in the numerator (the exception is in the unit for mass, kg).

KEY POINT **For the foreseeable future, you will need to make conversions from other units to SI units, as much of the literature quotes data using imperial, c.g.s. or other systems. You will need to recognize these units and find the conversion factors required. Examples relevant to biology are given in Box 10.1. Table 10.4 provides values of some important physical constants in SI units.**

Table 10.4 Some physical constants in SI terms

Physical constant	Symbol	Value and units
Avogadro's constant	N_A	$6.022\,174 \times 10^{23}$ mol^{-1}
Boltzmann's constant	k	$1.380\,626$ J K^{-1}
Charge of electron	e	$1.602\,192 \times 10^{-19}$ C
Gas constant	R	$8.314\,43$ J K^{-1} mol^{-1}
Faraday's constant	F	$9.648\,675 \times 10^{4}$ C mol^{-1}
Molar volume of ideal gas at STP	V_0	$0.022\,414$ m^3 mol^{-1}
Speed of light *in vacuo*	c	$2.997\,924 \times 10^{8}$ m s^{-1}
Planck constant	h	$6.626\,205 \times 10^{-34}$ J s

Box 10.1 Conversion factors between some redundant units and the SI

Quantity	SI unit/symbol	Old unit/symbol	Multiply number in old unit by this factor for equivalent in SI unit*	Multiply number in SI unit by this factor for equivalent in old unit*
Area	square metre/m^2	acre	$4.046\,86 \times 10^3$	$0.247\,105 \times 10^{-3}$
		hectare/ha	10×10^3	0.1×10^{-3}
		square foot/ft^2	0.092 903	10.763 9
		square inch/in^2	645.16×10^{-9}	$1.550\,00 \times 10^6$
		square yard/yd^2	0.836 127	1.195 99
Angle	radian/rad	degree/°	$17.453\,2 \times 10^{-3}$	57.295 8
Energy	joule/J	erg	0.1×10^{-6}	10×10^6
		kilowatt hour/kWh	3.6×10^6	$0.277\,778 \times 10^{-6}$
Length	metre/m	Ångstrom/Å	0.1×10^{-9}	10×10^9
		foot/ft	0.304 8	3.280 84
		inch/in	25.4×10^{-3}	39.370 1
		mile	$1.609\,34 \times 10^3$	$0.621\,373 \times 10^{-3}$
		yard/yd	0.914 4	1.093 61
Mass	kilogram/kg	ounce/oz	$28.349\,5 \times 10^{-3}$	35.274 0
		pound/lb	0.453 592	2.204 62
		stone	6.350 29	0.157 473
		hundredweight/cwt	50.802 4	$19.684\,1 \times 10^{-3}$
		ton (UK)	$1.016\,05 \times 10^3$	$0.984\,203 \times 10^{-3}$
Pressure	pascal/Pa	atmosphere/atm	101 325	$9.869\,23 \times 10^{-6}$
		bar/b	100 000	10×10^{-6}
		millimetre of mercury/mmHg	133.322	$7.500\,64 \times 10^{-3}$
		torr/Torr	133.322	$7.500\,64 \times 10^{-3}$
Radioactivity	becquerel/Bq	curie/Ci	37×10^9	$27.027\,0 \times 10^{-12}$
Temperature	kelvin/K	centigrade (Celsius) degree/°C	°C + 273.15	K − 273.15
		Fahrenheit degree/°F	(°F + 459.67) × 5/9	(K × 9/5) − 459.67
Volume	cubic metre/m^3	cubic foot/ft^3	0.028 316 8	35.314 7
		cubic inch/in^3	$16.387\,1 \times 10^{-6}$	$61.023\,6 \times 10^3$
		cubic yard/yd^3	0.764 555	1.307 95
		UK pint/pt	$0.568\,261 \times 10^{-3}$	1 759.75
		US pint/liq pt	$0.473\,176 \times 10^{-3}$	2 113.38
		UK gallon/gal	$4.546\,09 \times 10^{-3}$	219.969
		US gallon/gal	$3.785\,41 \times 10^{-3}$	264.172

*In the case of temperature measurements, use formulae shown

In this book, we use l and ml where you would normally find equipment calibrated in that way, but use SI units where this simplifies calculations. In formal scientific writing, constructions such as $1 \times 10^{-6}\,m^3$ (= 1 ml) and $1\,mm^3$ (= 1 μl) may be used.

Expressing enzyme activity – the derived SI unit is the katal (kat) which is the amount of enzyme that will transform 1 mol of substrate in 1 s (see Chapter 35).

Definition

STP – Standard Temperature and Pressure = 293.15 K and 0.101 325 MPa.

Some implications of SI in bioscience

Volume

The SI unit of volume is the cubic metre, m^3, which is rather large for practical purposes. The litre (l) and the millilitre (ml) are technically obsolete, but are widely used and glassware is still calibrated using them.

Mass

The SI unit for mass is the kilogram (kg) rather than the gram (g): this is unusual because the base unit has a prefix applied.

Amount of substance

You should use the mole (mol, i.e. Avogadro's constant, see Table 10.4) to express very large numbers. The mole gives the number of atoms in the atomic mass, a convenient constant. Always specify the elementary unit referred to in other situations (e.g. mol photons $m^{-2}\,s^{-1}$).

Concentration

The SI unit of concentration, $mol\,m^{-3}$, is quite convenient for biological systems. It is equivalent to the non-SI term 'millimolar' (mM) while 'molar' (M) becomes $kmol\,m^{-3}$. Note that the symbol M in the SI is reserved for mega and hence should not be used for concentrations. If the solvent is not specified, then it is assumed to be water (see Chapter 5).

Time

In general, use the second (s) when reporting physical quantities having a time element (e.g. give photosynthetic rates in mol $CO_2\,m^{-2}\,s^{-1}$). Hours (h), days (d) and years should be used if seconds are clearly absurd (e.g., samples were taken over a 5-year period). Note, however, that you may have to convert these units to seconds when doing calculations.

Temperature

The SI unit is the kelvin, K. The degree Celsius scale has units of the same magnitude, °C, but starts at 273.15 K, the melting point of ice at STP. Temperature is similar to time in that the Celsius scale is in widespread use, but note that conversions to K may be required for calculations. Note also that you must not use the degree sign (°) with K and that this symbol must be in upper case to avoid confusion with k for kilo; however, you *should* retain the degree sign with °C to avoid confusion with the coulomb, C.

Light

While the first six base units in Table 10.1 have standards of high precision, the SI base unit for luminous intensity, the candela (cd) and the derived units lm and lx (Table 10.2), are defined in 'human' terms. They are, in fact, based on the spectral responses of the eyes of 52 American GIs measured in 1923! Clearly, few organisms 'see' light in the same way as this sample of humans. Also, light sources differ in their spectral quality. For these reasons, it is better to use expressions based on energy or photon content (e.g. $W\,m^{-2}$ or mol photons $m^{-2}\,s^{-1}$), in studies other than those on human vision. Ideally you should specify the photon wavelength spectrum involved (see Chapter 20).

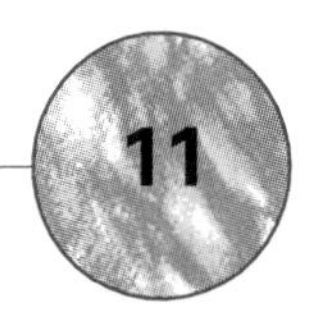

11 Scientific method and design of experiments

Definitions

Hypothesis – One possible explanation for an observed event. A mechanistic hypothesis is one based on some intuition about the mechanism underlying a phenomenon.

Practical life science – note that many practical procedures used in bioscience laboratories are not 'experiments', even though they may be called so!

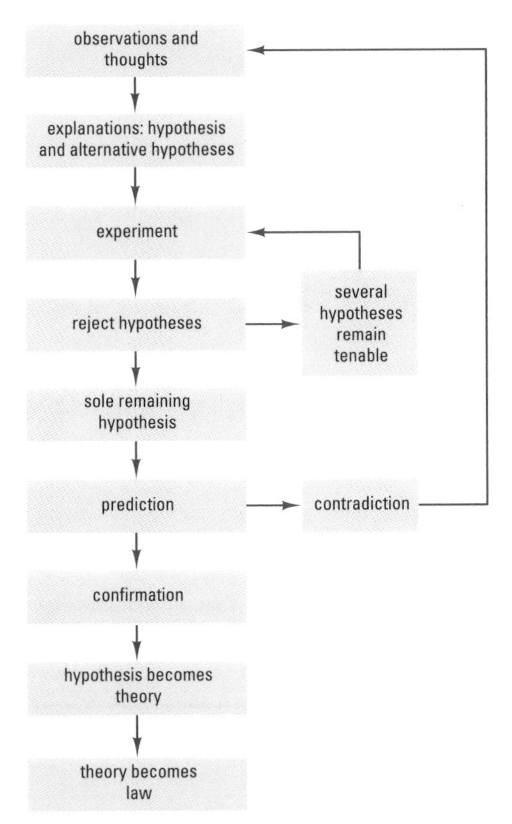

Fig. 11.1 How scientific investigations proceed.

Definition

Mathematical model – an algebraic summary of the relationship between the variables in a system.

Science is a systematized body of knowledge derived from observation and experiment. Scientists carry out experiments, make observations and attempt to explain the results; these tentative explanations are called hypotheses and their validity is tested by systematically forming and rejecting alternative explanations.

Some branches of life science are designed to provide fundamental information, rather than to test a particular hypothesis: for example, the purification and characterization of a newly discovered enzyme. In contrast, an *experiment* is a contrived situation designed to test one or more hypotheses under conditions controlled by the investigator. Any hypothesis that cannot be rejected from the results of an experiment is provisionally accepted. This 'sieve' effect leaves us with a set of current explanations for our observations. These explanations are not permanent and may be rejected on the basis of a future investigation. A hypothesis that has withstood many such tests and has been shown to allow predictions to be made is known as a theory, and a theory may generate such confidence through its predictive abilities to be known as a law (Fig. 11.1).

Observations are a prelude to experimentation, but they are preconditioned by a framework of peripheral knowledge. While there is an element of luck in being at the right place and time to make important observations, as Pasteur stated, 'chance favours only the prepared mind'. A fault in scientific method is that the design of the experiment and choice of method may influence the outcome – the decisions involved may not be as objective as some scientists assume. Another flaw is that radical alternative hypotheses may be overlooked in favour of a modification to the original hypothesis, and yet just such leaps in thinking have frequently been required before great scientific advances.

No hypothesis can ever be rejected with certainty. Statistics allow us to quantify as vanishingly small the probability of an erroneous conclusion, but we are nevertheless left in the position of never being 100% certain that we have rejected all relevant alternative hypotheses, nor 100% certain that our decision to reject some alternative hypotheses was correct! However, despite these problems, experimental science has yielded and continues to yield many important findings.

KEY POINT **The fallibility of scientific 'facts' is essential to grasp. No explanation can ever be 100% certain as it is always possible for a new alternative hypothesis to be generated. Our understanding of biology changes all the time as new observations and methods force old hypotheses to be retested.**

Quantitative hypotheses involve a mathematical description of the system. They can be formulated concisely by mathematical models. Models are often useful because they force deeper thought about mechanisms and encourage simplification of the system. A mathematical model:

- is inherently testable through experiment;
- identifies areas where information is lacking or uncertain;
- encapsulates many observations;
- allows you to predict the behaviour of the system.

Remember, however, that assumptions and simplifications required to create a model may result in it being unrealistic. Further, the results obtained from any model are only as good as the information put into it.

Experimentation and variables

Definitions

Treatment – a particular set of conditions applied to one or more experimental subjects.

Block – a group of replicates: a sub-division of the entire set of experimental subjects (the field). This terminology originates from agricultural experiments.

In many experiments, the aim is to provide evidence for causality. If x causes y, we expect, repeatably, to find that a change in x results in a change in y. Hence, the ideal experiment of this kind involves measurement of y, the dependent (measured) variable, at one or more values of x, the independent variable, and subsequent demonstration of some relationship between them. Experiments therefore involve comparisons of the results of treatments – changes in the independent variable as applied to an experimental subject. The change is engineered by the experimenter under controlled conditions. Experimental subjects given the same treatment are known as replicates.

Interpretation of experiments is seldom clear-cut because uncontrolled variables always change when treatments are given.

Confounding variables

Example In an experiment using a metabolic inhibitor prepared in an organic solvent, the concentration of solvent will vary alongside the concentration of inhibitor. A control, using solvent alone, will allow its effects to be determined.

These increase or decrease systematically as the independent variable increases or decreases. Their effects are known as systematic variation. This form of variation can be disentangled from that caused directly by treatments by incorporating appropriate controls in the experiment. A control is really just another treatment where a potentially confounding variable is adjusted so that its effects, if any, can be taken into account. The results from a control may therefore allow an alternative hypothesis to be rejected. There are often many potential controls for any experiment.

The consequence of systematic variation is that you can never be certain that the treatment, and the treatment alone, has caused an observed result. By careful design, you can, however, 'minimize the uncertainty' involved in your conclusion. Methods available include:

- Ensuring, through experimental design, that the independent variable is the only major factor that changes in any treatment.
- Incorporating appropriate controls to show that potential confounding variables have little or no effect.
- Selecting experimental subjects randomly to cancel out systematic variation arising from biased selection.
- Matching or pairing individuals among treatments so that differences in response due to their initial status are eliminated.
- Arranging subjects and treatments randomly so that responses to systematic differences in conditions do not influence the results.
- Ensuring that experimental conditions are uniform so that responses to systematic differences in conditions are minimized.

Avoiding personal bias – it may be necessary to encode the subjects, so that the investigator is 'blind' as to which subject is in which treatment regime.

Nuisance variables

These are uncontrolled variables which cause differences in the value of y independently of the value of x, resulting in random variation. Experimental bioscience is characterized by the high number of nuisance variables that are found and their relatively great influence on results: biological data tend to have large errors! To reduce and assess the consequences of nuisance variables:

- incorporate replicates to allow random variation to be quantified;
- choose experimental subjects that are as similar as possible;
- control random fluctuations in environmental conditions.

Evaluating design constraints – a good way to do this is by processing an individual subject through the experimental procedures – a 'preliminary run' can help to identify potential difficulties.

Constraints on experimental design

Box 11.1 outlines the important stages in designing an experiment. At an early stage, you should find out how your resources may constrain the design. For example, limits may be set by availability of subjects, cost of treatment, availability of a chemical or bench space. Logistics may be a factor (e.g. time taken to record or analyse data), or your equipment and facilities may affect design because you cannot regulate conditions as well as you might desire. You should also consider what statistical tests you intend to make (p. 268), as this is an important part of experimental design.

Using independent replicates – remember that the degree of independence of replicates is important: sub-samples cannot act as replicate samples – they tell you about variability in the measurement method but not in the quantity being measured.

Using replicates

Replicate results show how variable the response is within treatments. They allow you to compare the differences among treatments in the context of the variability within treatments – you can do this via statistical tests such as analysis of variance (Chapter 47). Larger sample sizes tend to increase the precision of estimates of statistical parameters and increase the chances of showing a significant difference between treatments if one exists. For statistical reasons (weighting, ease of calculation, fitting data to certain tests), it is best to keep the number of replicates even.

Box 11.1 Checklist for designing and executing an experiment

1. **Preliminaries**
   (a) **Read background material** and decide on a subject area to investigate.
   (b) **Formulate a simple hypothesis to test.** It is preferable to have a clear answer to one question than to be uncertain about several questions.
   (c) **Decide which dependent variable you are going to measure and how:** is it relevant to the problem? Can you measure it accurately, precisely and without bias?
   (d) **Think about and plan the statistical analysis of your results.** Will this affect your design?
2. **Designing**
   (a) **Find out the limitations on your resources.**
   (b) **Choose treatments which alter the minimum of confounding variables.**
   (c) **Incorporate as many effective controls as possible.**
   (d) **Keep the number of replicates as high as is feasible.**
   (e) **Ensure that the same number of replicates is present in each treatment with random allocation to individual treatments.**
3. **Planning**
   (a) **List all the materials you will need.** Order any chemicals and make up solutions; grow, collect or prepare the experimental material you require; check equipment is available.
   (b) **Organize space and/or time** in which to do the experiment.
   (c) **Account for the time taken to apply treatments and record results.** Make out a timesheet if things will be hectic.
4. **Carrying out the experiment**
   (a) **Record the results and make careful notes of everything you do.** Make additional observations to those planned if interesting things happen.
   (b) **Repeat experiment** if time and resources allow.
5. **Analysing**
   (a) **Graph data as soon as possible** (during the experiment if you can). This will allow you to visualize what has happened and make adjustments to the design (e.g. timing of measurements).
   (b) **Carry out any planned statistical analysis.**
   (c) **Jot down conclusions and new hypotheses** arising from the experiment.

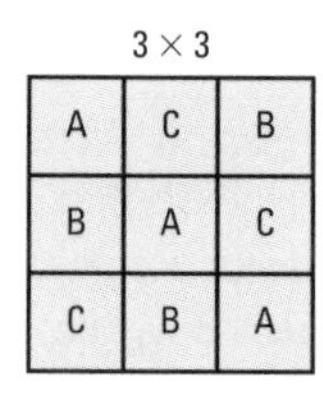

4 × 4

A	C	B	D
D	B	C	A
C	D	A	B
B	A	D	C

Fig. 11.2 Examples of Latin square arrangements for 3 and 4 treatments. Letters indicate treatments; the number of possible arrangements increases greatly as the number of treatments increases.

If the total number of replicates available for an experiment is limited by resources, you may need to compromise between the number of treatments and the number of replicates per treatment. Statistics can help here, as it is possible to work out the minimum number of replicates you would need to show a certain difference between pairs of means (say 10%) at a specified level of significance (say $P = 0.05$). For this, you need to obtain a prior estimate of variability within treatments (see Sokal and Rohlf, 1994).

Randomization of treatments

For relatively simple experiments, you can adopt a completely randomized design; here, the position and treatment assigned to any subject is defined randomly. You can draw lots, use a random number generator on a calculator, or use the random number tables which can be found in most books of statistical tables.

A completely randomized layout has the advantage of simplicity but cannot show how confounding variables alter in space or time. This information can be obtained if you use a blocked design in which the degree of randomization is restricted. Here, the experimental space or time is divided into blocks, each of which accommodates the complete set of treatments. When analysed appropriately, the results for the blocks can be compared to test for differences in the confounding variables and these effects can be separated out from the effects of the treatments. The size and shape (or timing) of the block you choose is important: besides being able to accommodate the number of replicates desired, the suspected confounding variable should be relatively uniform within the block.

Example If you knew that soil type varied in a graded fashion across a field, you might arrange blocks to be long thin rectangles at right angles to the gradient to ensure conditions within the block were as even as possible.

A Latin square is a method of placing treatments so that they appear in a balanced fashion within the experimental area. Treatments appear once in each column and row (see Fig. 11.2), so the effects of confounding variables can be 'cancelled out' in two directions at right angles to each other. This is effective if there is a smooth gradient in some confounding variable over the experimental area. It is less useful if the variable has a patchy distribution, where a randomized design might be better.

Latin square designs are useful in serial experiments where different treatments are given to the same subjects in a sequence (e.g. Fig. 11.3). A disadvantage of Latin squares is the fact that the number of columns and rows is equal to the number of replicates, so increases in the number of replicates can only be made by the use of further Latin squares.

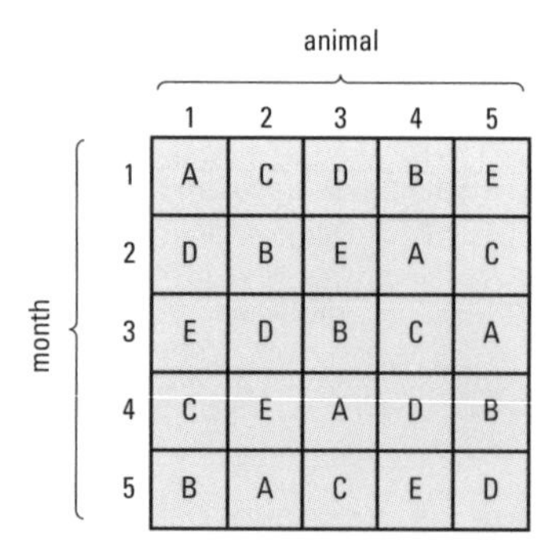

Fig. 11.3 Example of how to use a Latin square design to arrange sequential treatments. The experimenter wishes to test the effect of drugs A–E on weight gain, but only has five animals available. Each animal is fed on control diet for the first 3 weeks of each month, then on control diet plus drug for the last week. Weights are taken at start and finish of each treatment. Each animal receives all treatments.

Pairing and matching subjects

The paired comparison is a special case of blocking used to reduce systematic variation when there are two treatments. Examples of its use are:

- 'Before and after' comparison. Here, the pairing removes variability arising from the initial state of the subjects, e.g. weight gain of mice on a diet, where the weight gain may depend on the initial weight.
- Application of a treatment and control to parts of the same subject or to closely related subjects. This allows comparison without complications arising from different origin of subjects, e.g. drug or placebo given to sibling rats, virus-containing or control solution swabbed on left or right halves of a leaf.
- Application of treatment and control under shared conditions. This allows comparison without complications arising from different environment of subjects, e.g. rats in a cage, plants in a pot.

Example A Latin square format is used for agar bioassays of antibodies and vitamins. Replicates are arranged so that gradients across the plate are cancelled out.

Matched samples represent a restriction on randomization where you make a balanced selection of subjects for treatments on the basis of some attribute or attributes that may influence results, e.g. age, gender, prior history. The effect of matching should be to 'cancel out' the unwanted source(s) of variation. Disadvantages include the subjective element in choice of character(s) to be balanced, inexact matching of quantitative characteristics, the time matching takes and possible wastage of unmatched subjects.

When analysed statistically, both paired comparisons and matched samples can show up differences between treatments that might otherwise be rejected on the basis of a fully randomized design. Note that the statistical analysis of paired samples is slightly different from that of fully random or matched samples (see Sokal and Rohlf, 1994).

Definition

Interaction – where the effects of treatments given together is greater or less than the sum of their individual effects.

Multifactorial experiments

The simplest experiments are those in which one treatment (factor) is applied at a time to the subjects. This approach is likely to give clear-cut answers, but it could be criticized for lacking realism. In particular, it cannot take account of interactions among two or more conditions that are likely to occur in real life. A multifactorial experiment (Fig. 11.4) is an attempt to do this; the interactions among treatments can be analysed by specialized forms of analysis of variance.

Multifactorial experiments are economical on resources because of 'hidden replication'. This arises when two or more treatments are given to a subject because the result acts statistically as a replicate for each treatment. Choice of relevant treatments to combine is important in multifactorial experiments; for instance, an interaction may be present at certain concentrations of a chemical but not at others (perhaps because the response is saturated). It is also important that the measurement scale for the response is consistent, otherwise spurious interactions may occur. Beware when planning a multifactorial experiment that the numbers of replicates do not get out of hand: you may have to restrict the treatments to 'plus' or 'minus' the factor of interest (as in Fig. 11.4).

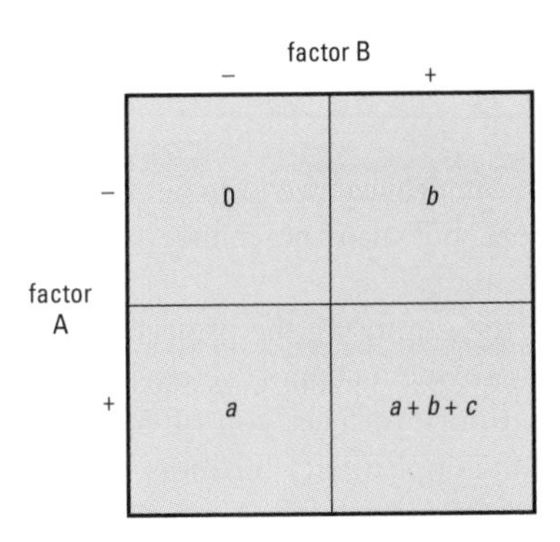

Fig. 11.4 Design of a simple multifactorial experiment. Factors A and B have effects *a* and *b* when applied alone. When both are applied together, the effect is denoted by $a + b + c$.

- If $c = 0$, there is no interaction (e.g. $2 + 2 + c = 4$).
- If c is positive, there is a positive interaction (synergism) between A and B (e.g. $2 + 2 + c = 5$).
- If c is negative, there is a negative interaction (antagonism) between A and B (e.g. $2 + 2 + c = 3$).

Repetition of experiments

Even if you have taken great care to ensure that your experiment is well designed and statistically analysed, you are limited in the conclusions that can be made. Firstly, what you can say is valid for a particular place and time, with a particular investigator, experimental subject and method of applying treatments. Secondly, if your results were significant at the 5% level of probability, there is still an approximately one-in-twenty chance that the results did arise by chance. To guard against these possibilities, it is important that experiments are repeated. Ideally, this would be done by an independent scientist with independent materials. However, it makes sense to repeat work yourself so that you can have full confidence in your conclusions. Many scientists recommend that all experiments are carried out three times in total. This may not be possible in undergraduate practical classes or project work!

Reporting results – it is good practice to report how many times your experiments were repeated (in Materials and Methods); in the Results section, you need either a statement saying that the illustrated experiment is representative or one explaining the differences between results obtained.

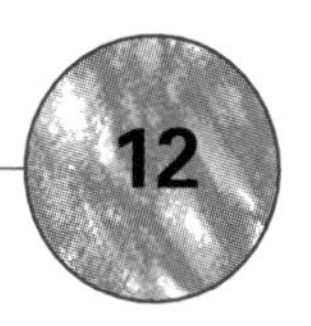

Project work

Research projects are an important component of the final year syllabus for most degree programmes in the life sciences, while shorter projects may also be carried out during courses in earlier years. Project work presents difficulties at many stages but can be extremely rewarding. The assessment of your project is likely to contribute significantly to your degree grade, so all aspects of this work should be approached in a thorough manner.

The Internet as an information source – since many university departments have home pages on the World Wide Web, searches using relevant key words may indicate where research in your area is currently being carried out. Academics usually respond positively to e-mailed questions about their area of expertise.

Deciding on a topic to study

Assuming you have a choice, this important decision should be researched carefully. Make appointments to visit possible supervisors and ask them for advice on topics that you find interesting. Use library texts and research papers to obtain further background information. Perhaps the most important criterion is whether the topic will sustain your interest over the whole period of the project. Other things to look for include:

- Opportunities to learn new skills. Ideally, you should attempt to gain experience and skills that you might be able to 'sell' to a potential employer.
- Ease of obtaining valid results. An ideal project provides a means to obtain 'guaranteed' data for your report, but also the chance to extend knowledge by doing genuinely novel research.
- Assistance. What help will be available to you during the project? A busy lab with many research students might provide a supportive environment should your potential supervisor be too busy to meet you often; on the other hand, a smaller lab may provide the opportunity for more personal interaction with your supervisor.
- Impact. It is not outside the bounds of possibility for undergraduate work to contribute to research papers. Your prospective supervisor can alert you to such opportunities.

Asking around – one of the best sources of information about supervisors, laboratories and projects is past students. Some of the postgraduates in your department may be products of your own system and they could provide an alternative source of advice.

Planning your work

As with any lengthy exercise, planning is required to make the best use of the time allocated. This is true on a daily basis as well as over the entire period of the project. It is especially important not to underestimate the time it will take to write and produce your thesis (see below). If you wish to benefit from feedback given by your supervisor, you should aim to have drafts in his/her hands in good time. Since a large proportion of marks will be allocated to the report, you should not rush its production.

If your Department requires you to write an interim report, look on this as an opportunity to clarify your thoughts and get some of the time-consuming preparative work out of the way. If not, you should set your own deadlines for producing drafts of the introduction, materials and methods section, etc.

Project work can be very time-consuming at times. Try not to neglect other aspects of your course – make sure your lecture notes are up to date and collect relevant supporting information as you go along.

Liaising with your supervisor(s) – this is essential if your work is to proceed efficiently. Specific meetings may be timetabled e.g. to discuss a term's progress, review your work plan or consider a draft introduction. Most supervisors also have an 'open-door' policy, allowing you to air current problems. Prepare well for all meetings: have a list of questions ready before the meeting; provide results in an easily digestible form (but take your lab. notebook along); be clear about your future plans for work.

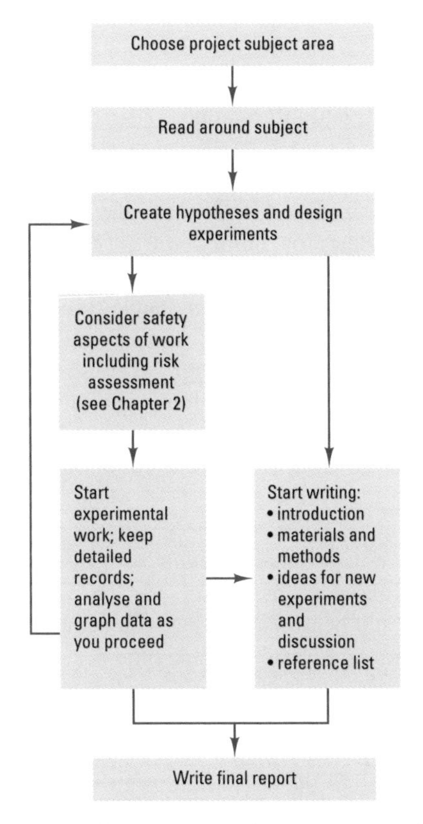

Fig. 12.1 Flowchart showing a recommended sequence of events in carrying out an undergraduate research project.

Getting started

Fig. 12.1 is a flowchart illustrating how a project might proceed; at the start, don't spend too long reading the literature and working out a lengthy programme of research. Get stuck in and do an experiment. There's no substitute for 'getting your hands dirty' for stimulating new ideas:

- even a 'failed' experiment will provide some useful information which may allow you to create a new or modified hypothesis;
- pilot experiments may point out deficiencies in experimental technique that will need to be rectified;
- the experience will help you create a realistic plan of work.

Designing experiments or sampling procedures

Design of experiments is covered in Chapter 11. Avoid being too ambitious at the start of your work! It is generally best to work with a simple hypothesis and design your experiments or sampling around this. A small pilot experiment or test sample will highlight potential stumbling blocks including resource limitations, whether in materials or time or both.

Working in a laboratory environment

During your time as a project student, you are effectively a guest in your supervisor's laboratory.

- Be considerate – keep your 'area' tidy and offer to do your share of lab duties such as calibrating the pH meter, replenishing stock solutions, distilled water, etc., maintaining cultures, tending plants or animals.
- Use instruments carefully – they could be worth more than you'd think. Careless use may invalidate calibration settings and ruin other people's work as well as your own.
- Do your homework on techniques you intend to use – there's less chance of making costly mistakes if you have a good background understanding of the methods you will be using.
- Always seek advice if you are unsure of what you are doing.

KEY POINT **It is essential that you follow all the safety rules applying to the laboratory or field site. Make sure you are acquainted with all relevant procedures – normally there will be prominent warnings about these. If in doubt, ask!**

Keeping notes and analysing your results

Tidy record keeping is often associated with good research, and you should follow the advice and hints given in Chapter 9. Try to keep copies of all files relating to your project. As you obtain results, you should always calculate, analyse and graph data as soon as you can (see Fig. 12.1). This can reveal aspects that may not be obvious in numerical or readout form. Don't be worried by negative results – these can sometimes be as useful as positive results if they allow you to eliminate hypotheses – and don't be dispirited if things do not work first time. Thomas Edison's maxim 'Genius is one per cent inspiration and ninety-nine per cent perspiration' certainly applies to research work!

Writing the report

The structure of scientific reports is dealt with in Chapter 55. The following advice concerns methods of accumulating relevant information.

Brushing up on IT skills – word processors and spreadsheets are extremely useful when producing a thesis. Chapters 50 and 51 detail key features of these programs. You might benefit from attending courses on the relevant programs or studying manuals or texts so that you can use them more efficiently.

Introduction This is a big piece of writing that can be very time-consuming. Therefore, the more work you can do on it early on, the better. You should allocate some time at the start for library work (without neglecting benchwork), so that you can build up a database of references (p. 297). While photocopying can be expensive, you will find it valuable to have copies of key reviews and references handy when writing away from the library. Discuss proposals for content and structure with your supervisor to make sure your effort is relevant. Leave space at the end for a section on aims and objectives. This is important to orientate readers (including assessors), but you may prefer to finalize the content after the results have been analysed!

Materials and methods You should note as many details as possible when doing the experiment or making observations. Don't rely on your memory or hope that the information will still be available when you come to write up. Even if it is, chasing these details might waste valuable time.

Using drawings or photographs – these can provide valuable records of sampling sites or experimental set-ups and could be useful in your report. Plan ahead and do the relevant work at the time of carrying out your research rather than afterwards.

Results Show your supervisor graphed and tabulated versions of your data promptly. These can easily be produced using a spreadsheet (p. 287), but you should seek your supervisor's advice on whether the design and print quality is appropriate to be included in your thesis. You may wish to access a specialist graphics program to produce publishable-quality graphs and charts: allow some time for learning its idiosyncrasies! If you are producing a poster for assessment (Chapter 57), be sure to mock up the design well in advance. Similarly, think ahead about your needs for any seminar or poster you will present.

Discussion Because this comes at the end of your thesis, and some parts can only be written after you have all the results in place, the temptation is to leave the discussion to last. This means that it can be rushed – not a good idea because of the weight attached by assessors to your analysis of data and thoughts about future experiments. It will help greatly if you keep notes of aims, conclusions and ideas for future work *as you go along* (Fig. 12.1). Another useful tip is to make notes of comparable data and conclusions from the literature as you read papers and reviews.

Acknowledgements Make a special place in your notebook for noting all those who have helped you carry out the work, for use when writing this section of the report.

References Because of the complex formats involved (p. 300), these can be tricky to type. To save time, process them in batches as you go along.

KEY POINT **Make sure you are absolutely certain about the deadline for submitting your report and try to submit a few days before it. If you leave things until the last moment, you may find access to printers, photocopiers and binding machines is difficult.**

Handling cells and tissues

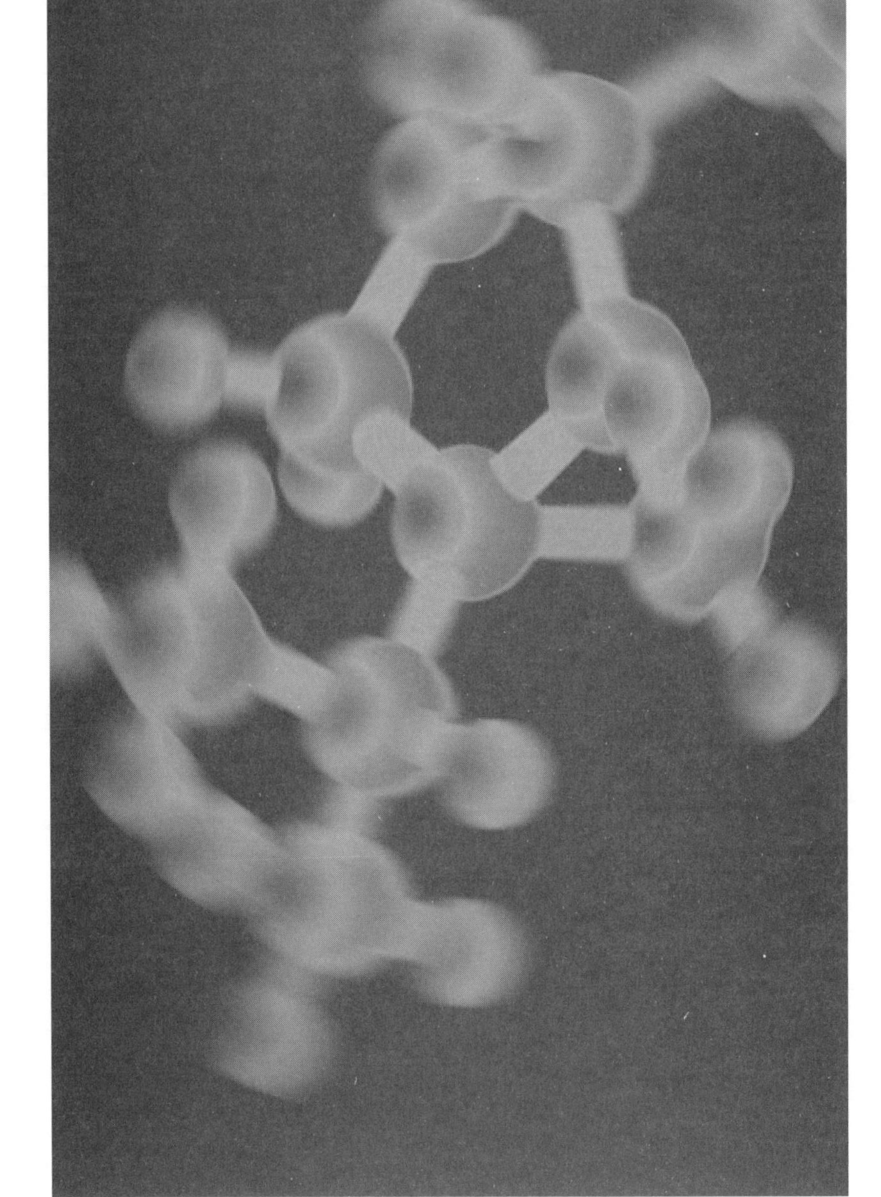

13 Sterile technique and microbial culture

Sterile technique (aseptic technique) is the name given to the procedures used in cell culture. While the same general principles apply to all cell types, you are most likely to learn the basic procedures using bacteria and most of the examples given in this section refer to bacterial culture.

Sterile technique serves two main purposes:

1. To prevent accidental contamination of laboratory cultures due to microbes from external sources, e.g. skin, clothing or the surrounding environment.
2. To prevent microbial contamination of laboratory workers, in this instance you and your fellow students.

KEY POINT ***All* microbial cell cultures should be treated as if they contained potentially harmful organisms. Sterile technique forms an important part of safety procedures, and must be followed whenever cell cultures are handled in the laboratory.**

Care is required because:

- You may accidentally isolate a harmful microbe as a contaminant when culturing a relatively harmless strain.
- Some individuals are more susceptible to infection and disease than others – not everyone exposed to a particular microbe will become ill.
- Laboratory culture involves purifying and growing large numbers of microbial cells – this represents a greater risk than small numbers of the original microbe.
- A microbe may change its characteristics, perhaps as a result of gene exchange or mutation.

Achieving a sterile state – you should assume that *all* items of laboratory equipment have contaminating microbes on their surfaces, unless they have been destroyed by some form of sterilization. Such items will only remain sterile if they do not come into contact with the non-sterile environment.

Sterilization procedures

Given the ubiquity of microbes, the only way to achieve a sterile state is by their destruction or removal. Several methods can be used to achieve this objective:

Heat treatment

This is the most widespread form of sterilization employed in several basic laboratory procedures including the following:

- Red heat sterilization. Achieved by heating metal inoculating loops, forceps, needles, etc., in a Bunsen flame. This is a simple and effective form of sterilization as no microbe will survive even a brief exposure to a naked flame. Flame sterilization using alcohol is used for glass rods and spreaders (see below).
- Dry heat sterilization. Here, a hot air oven is used at a temperature of at least 160 °C for at least 2 h. This method is used for the routine sterilization of laboratory glassware. Dry heat procedures are of little value for items requiring repeated sterilization during use.
- Moist heat sterilization. This is the method of choice for many laboratory items, including most fluids, apart from heat-sensitive media. It is also used to decontaminate liquid media and glassware after use. The laboratory autoclave is used for these purposes. Typically, most items will be sterile after 15 min at 121 °C, although large items may require a

longer time. The rapid killing action results from the latent heat of condensation of the pressurized steam, released on contact with cool materials in the autoclave.

Radiation

Many disposable plastic items used in microbiology and cell biology are sterilized by exposure to UV or ionizing radiation. They are supplied commercially in sterile packages, ready for use. Ultraviolet radiation has limited use in the laboratory, while ionizing radiation (e.g. γ-rays) requires industrial facilities and cannot be operated on a laboratory scale.

Filtration

Using a sterile filter – most filters are supplied as pre-sterilized items. Make sure you follow a procedure that does not contaminate the filter on removal from its protective wrapping.

Heat-labile solutions (e.g. complex macromolecules, including proteins, antibiotics, serum) are particularly suited to this form of sterilization. The filters come in a variety of shapes, sizes and materials, usually with a pore size of either 0.2 μm or 0.45 μm. The filtration apparatus and associated equipment is usually sterilized by autoclaving, or by dry heat. Passage of liquid through a sterile filter of pore size 0.2 μm into a sterile vessel is usually sufficient to remove bacteria but not viruses, so filtered liquids are not necessarily virus-free.

Chemical agents

These are known as disinfectants, or biocides, and are most often used for the disposal of contaminated items following laboratory use, e.g. glass slides and pipettes. They are also used to treat spillages. The term 'disinfection' implies destruction of disease-causing bacterial cells, although spores and viruses may not always be destroyed. Remember that disinfectants require time to exert their killing effect – any spillage should be covered with an appropriate disinfectant and left for at least 10 min before mopping up.

Use of laboratory equipment

Using molten agar – a water bath (at 45–50 °C) can be used to keep an agar-based medium in its molten state after autoclaving. Always dry the outside of the container on removal from the water bath, to reduce the risk of contamination from microbes in the water, e.g. during pour plating (p. 63).

Working area

One of the most important aspects of good sterile technique is to keep your working area as clean and tidy as possible. Start by clearing all items from your working surface, wipe the bench down with disinfectant and then arrange the items you need for a particular procedure so that they are close at hand, leaving a clear working space in the centre of your bench.

Media

Cells may be cultured in either a liquid medium (broth), or a solidified medium (p. 79). The gelling agent used in most solidified media is agar, a complex polysaccharide from red algae that produces a stiff transparent gel when used at 1–2% (w/v). Agar is used because it is relatively resistant to degradation by most bacteria and because of its rheological properties – an agar medium melts at 98 °C, remaining solid at all temperatures used for routine laboratory culture. However, once melted, it does not solidify until the temperature falls to about 44 °C. This means that heat-sensitive constituents (e.g. vitamins, blood, cells, etc.) can be added aseptically to the medium after autoclaving.

Plastic disposable loops – these are used in many research laboratories: pre-sterilized and suitable for single use, they avoid the hazards of naked flames and the risk of aerosol formation during heating. Discard into a disinfectant solution after use.

Inoculating loops

The initial isolation and subsequent transfer of microbes between containers can be achieved using a sterile inoculating loop. Most teaching laboratories use nichrome wire loops in a metal handle. A wire loop can be repeatedly sterilized by heating the wire, loop downwards and almost vertical, in the hottest part of a Bunsen flame until the whole wire becomes red hot. Then the loop is removed from the flame to minimize heat transfer to the handle. After cooling for 8–10 s (without touching any other object), it is ready for use.

When re-sterilizing a contaminated wire loop in a Bunsen flame after use, do not heat the loop too rapidly, as the sample may spatter, creating an aerosol: it is better to soak the loop for a few minutes in disinfectant than to risk heating a fully charged (contaminated) inoculating loop.

Containers

Using glass pipettes – these are plugged with cotton wool at the top before being autoclaved inside a metal can. Flame the open end of the can on removal of a pipette, to prevent contamination of the remaining pipettes. Autopipettors and sterile disposable tips (p. 8) offer an alternative approach.

There is a risk of contamination whenever a sterile bottle, flask or test tube is opened. One method that reduces the chance of airborne contamination is quickly to pass the open mouth of the glass vessel through a flame. This destroys any microbes on the outer surfaces nearest to the mouth of the vessel. In addition, by heating the air within the neck of the vessel, an outwardly-directed air flow is established, reducing the likelihood of microbial contamination.

It is general practice to flame the mouth of each vessel immediately after opening and then repeat the procedure just before replacing the top. Caps, lids and cotton wool plugs must not be placed on the bench during flaming and sampling: they should be removed and held using the smallest finger of one hand, to minimize the risk of contamination. This also leaves the remaining fingers free to carry out other manipulations. With practice, it is possible to remove the tops from two tubes, flame each tube and transfer material from one to the other while holding one top in each hand.

Laminar flow cabinets

Using a Bunsen burner to reduce airborne contamination – working close to the updraught created by a Bunsen flame reduces the likelihood of particles falling from the air into an open vessel.

These are designed to prevent airborne contamination, e.g. when preparing media or subculturing microbes or tissue cultures. Sterile air is produced by passage through a high efficiency particulate air (HEPA) filter: this is then directed over the working area, either horizontally (towards the operator) or downwards. The operator handles specimens, media, etc., through an opening at the front of the cabinet. Standard laminar flow cabinets do *not* protect the worker from contamination and must not be used with pathogenic microbes: special safety cabinets are used for work with ACDP hazard group 3 and 4 microbes (Table 13.1) and for samples that might contain such pathogens.

Table 13.1 Classification of microbes on the basis of hazard. The following categories are recommended by the UK Advisory Committee on Dangerous Pathogens (ACDP)

Hazard group	Comments
1	Unlikely to cause human disease
2	May cause disease: possible hazard to laboratory workers, minimal hazard to community
3	May cause severe disease: may be a serious hazard to laboratory workers, may spread to community
4	Causes severe disease: is a serious hazard to laboratory workers, high risk to community

Microbiological hazards

KEY POINT The most obvious risks when handling microbial cultures are those due to ingestion or entry via a cut in the skin – all cuts should be covered with a plaster or disposable plastic gloves. A less obvious source of hazard is the formation of aerosols of liquid droplets from microbial suspensions, with the risk of inhalation, or surface contamination of other objects.

The following steps will minimize the risk of aerosol formation:

- Use stoppered tubes when shaking, centrifuging or mixing microbial suspensions.
- Pour solutions gently, keeping the difference in height to a minimum.
- Discharge pipettes onto the side of the container.

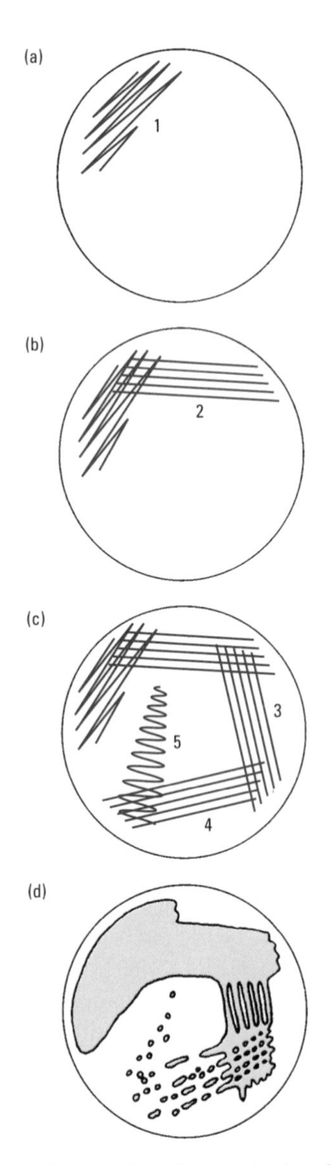

Fig. 13.1 Preparation of a streak plate for single colonies. (a) Using a sterile metal loop, take a small sample of the material to be streaked. Distribute the sample over a small sector of the plate (area 1), then flame the loop and allow to cool (approximately 8–10 s). (b) Make several small streaks from the initial sector into the adjacent sector (area 2), taking care not to allow the streaks to overlap. Flame the loop and allow to cool. (c) Repeat the procedure for areas 3 and 4, re-sterilizing the loop between each step. Finally, make a single, long streak, as shown for area 5. (d) The expected result after incubation at the appropriate temperature (e.g. 37 °C for 24 h): each step should have diluted the inoculum, giving individual colonies within one or more sectors on the plate. Further sub-culture of an individual colony should give a pure (clonal) culture.

Other general rules which apply in all laboratories include:

- Take care with sharp instruments, including needles and glass Pasteur pipettes.
- Do not pour waste cultures down the sink – they must be autoclaved.
- Put other contaminated items (e.g. slides, pipettes) into disinfectant after use.
- Wipe down your bench with disinfectant when practical work is complete.
- Always wash your hands before leaving the laboratory.

Plating methods

Many culture methods make use of a solidified medium within a Petri plate. A variety of techniques can be used to transfer and distribute the organisms prior to incubation. The three most important procedures are described below.

Streak dilution plate

Streaking a plate for single colonies is one of the most important basic skills in microbiology, since it is used in the initial isolation of a cell culture and in maintaining stock cultures, where a streak dilution plate with single colonies all of the same type confirms the purity of the strain. A sterile inoculating loop is used to streak the organisms over the surface of the medium, thereby diluting the sample (Fig. 13.1). The aim is to achieve single colonies at some point on the plate: ideally, such colonies are derived from single cells (e.g. in the case of unicellular bacteria, animal and plant cell lines) or from groups of cells of the same species (in filamentous or colonial forms). Single colonies, containing cells of a single species and derived from a single parental cell, form the basis of all pure culture methods (p. 65).

Note the following:

- Keep the lid of the Petri plate as close to the base as possible to reduce the risk of aerial contamination.
- Allow the loop to glide over the surface of the medium. Hold the handle at the balance point (near the centre) and use light, sweeping movements, as the agar surface is easily damaged and torn.
- Work quickly, but carefully. Do not breathe directly onto the exposed agar surface and replace the lid as soon as possible.

Spread plate

This method is used with cells in suspension, either in a liquid growth medium or in an appropriate sterile diluent. It is one method of quantifying the number of viable cells (or colony forming units) in a sample, after appropriate dilution (p. 84).

An L-shaped glass spreader is sterilized by dipping the end of the spreader in a beaker containing a small amount of 70% v/v alcohol, allowing the excess to drain from the spreader and then igniting the remainder in a Bunsen flame. After cooling, the spreader is used to distribute a known volume of cell suspension across the plate (Fig. 13.2). *There is a significant fire risk associated with this technique*, so take care not to ignite the alcohol in the beaker, e.g. by returning an overheated glass rod to the beaker. The alcohol will burn with a pale blue flame that may be difficult to see, but will readily ignite other materials (e.g. a laboratory coat). Another source of risk

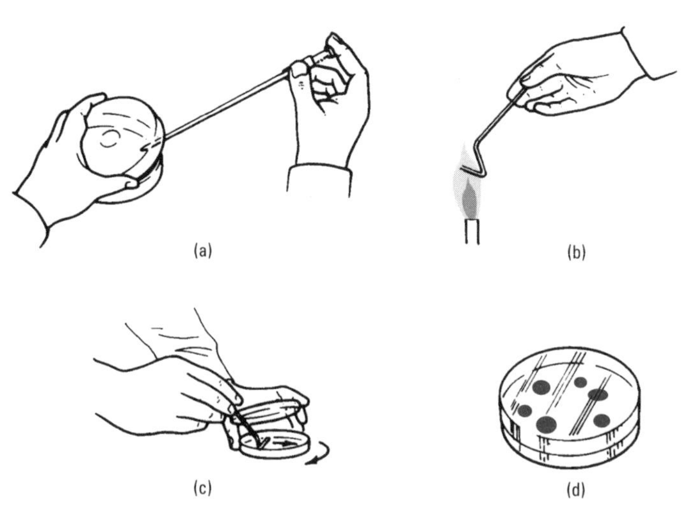

Fig. 13.2 Preparation of a spread plate. (a) Transfer a small volume of cell suspension (0.05–0.5 ml) to the surface of a solidified medium in a Petri plate. (b) Flame sterilize a glass spreader and allow to cool (8–10 s). (c) Distribute the liquid over the surface of the plate using the sterile spreader. Make sure of an even coverage by rotating the plate as you spread: allow the liquid to be absorbed into the agar medium. Incubate under suitable conditions. (d) After incubation, the microbial colonies should be distributed over the surface of the plate.

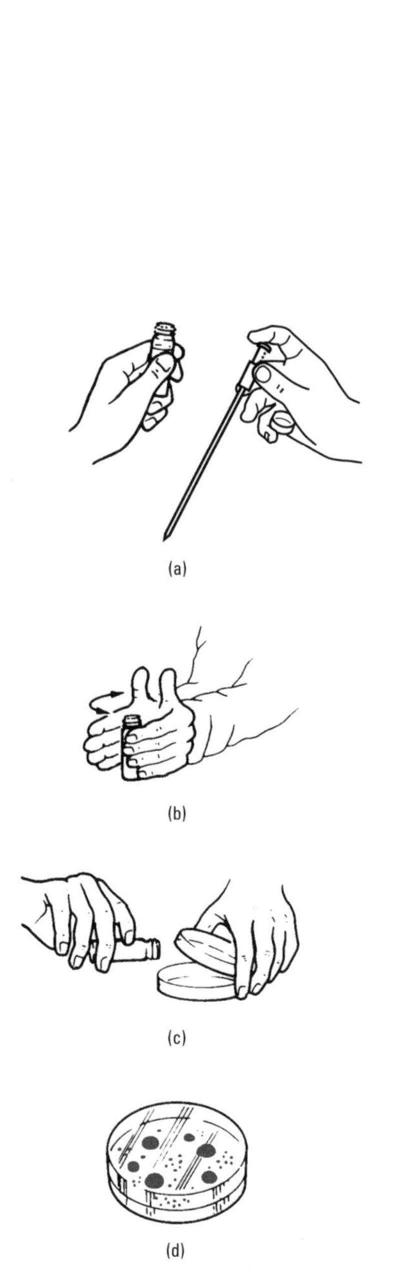

Fig. 13.3 Preparation of a pour plate. (a) Add a known volume of cell suspension (0.05–1.0 ml) to a small bottle of molten agar medium from a 45 °C water bath. (b) Mix thoroughly, by rotating between the palms of the hands: do not shake or this will cause frothing of the medium. (c) Pour the mixture into an empty, sterile Petri plate and allow to set. Incubate under suitable conditions. (d) After incubation, the microbial colonies will be distributed throughout the medium: any cells deposited at the surface will give larger, spreading colonies.

comes from small droplets of flaming alcohol shed by an overloaded spreader onto the bench and this is why you *must* drain excess alcohol from the spreader *before* flaming. Some laboratories now provide plastic disposable spreaders for student use, to avoid the risk of fire.

Pour plate

This procedure also uses cells in suspension, but requires molten agar medium, usually in screw-capped bottles containing sufficient medium to prepare a single Petri plate (i.e. 15–20 ml), maintained in a water bath at 45–50 °C. A known volume of cell suspension is mixed with this molten agar, distributing the cells throughout the medium. This is then poured without delay into an empty sterile Petri plate and incubated, giving widely spaced colonies (Fig. 13.3). Furthermore, as most of the colonies are formed within the medium, they are far smaller than those of the surface streak method, allowing higher cell numbers to be counted (e.g. up to 1 000 colonies per plate): some workers pour a thin layer of molten agar onto the surface of a pour plate after it has set, to ensure that no surface colonies are produced.

Most bacteria and fungi are not killed by brief exposure to temperatures of 45–50 °C, though the procedure may be more damaging to microbes from low temperature conditions, e.g. psychrophilic bacteria.

One disadvantage of the pour plate method is that the typical colony morphology seen in surface grown cultures will not be observed for those colonies that develop within the agar medium. A further disadvantage is that some of the suspension will be left behind in the screw-capped bottle. This can be avoided by transferring the suspension to the Petri plate, adding the molten agar, then swirling the plate to mix the two liquids. However, even when the plate is swirled repeatedly and in several directions, the liquids are not mixed as evenly as in the former procedure.

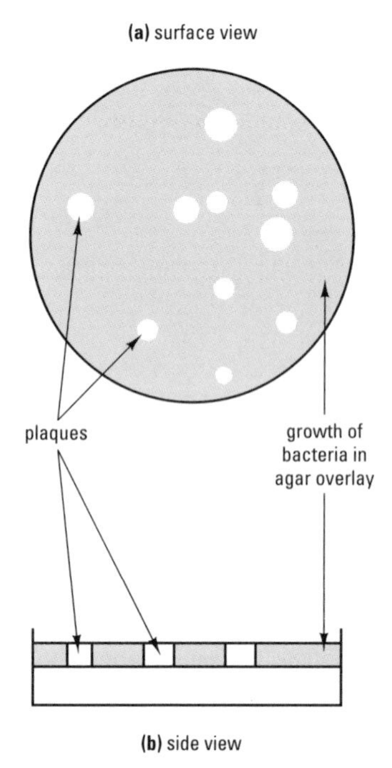

Fig. 13.4 Phage plaques in a 'lawn' of susceptible host-bacterium.

Counting plaques in phage assays – view plaques against a black background to make them easier to see: mark each plaque with a spirit-based marker to ensure an accurate count.

Working with phages

Bacterial viruses ('bacteriophages', or simply 'phages') are often used to illustrate the general principles involved in the detection and enumeration of viruses. They also have a role in genome mapping in bacteria (p. 224). Individual phage particles (virions) are too small to be seen by light microscopy, but are detected by their effects on susceptible host cells:

- Virulent phages will infect and replicate within actively growing host cells, causing cell lysis and releasing new infective phages – this 'lytic cycle' takes ≈30 min for T-even phages of *E. coli*, e.g. T4.
- Temperate phages are a specialized group, capable either of lytic growth or an alternative cycle, termed lysogeny – the phage becomes latent within a host cell (lysogen), typically by insertion of its genetic information into the host cell genome, becoming a 'prophage' (p. 224). At a later stage, termed induction, the prophage may enter the lytic cycle. A widely used example is λ phage of *E. coli*.

The lytic cycle can be used to detect and quantify the number of phages in a sample. A known volume of sample is mixed with susceptible bacterial cells in molten soft agar medium (45-50 °C), then poured on top of a plate of the same medium, creating a thin layer of 'top agar'. The upper layer contains only half the normal amount of agar, to allow phages to diffuse through the medium and attach to susceptible cells. On incubation, the bacteria will grow throughout the agar to produce a homogeneous 'lawn' of cells, except in those parts of the plate where a phage particle has infected and lysed the cells to create a clear area, termed a plaque (Fig. 13.4). Each plaque is due to a single functional phage (i.e. a plaque-forming unit, or PFU). A count of the number of plaques can be used to give the number of phages in a particular sample (e.g. as PFU ml^{-1}), with appropriate correction for dilution and the volume of sample counted in an analogous manner to a bacterial plate count (p. 84). Temperate phages often produce cloudy plaques, because many of the infected cells will be lysogenized rather than lysed, creating turbidity within the plaque. Samples of material from within the plaque can be used to sub-culture the phage for further study, perhaps in a broth culture where the phages will cause widespread cell lysis and a decrease in turbidity. Alternatively, phages can be stored by adding chloroform to aqueous suspensions – this will prevent contamination by cellular micro-organisms. A similar approach can be used to detect and count animal or human viruses, using a monolayer of a susceptible animal or human cells.

Electron microscopy (EM, p. 33) provides an alternative approach to the detection of viruses, avoiding the requirement for culture of infected host cells, and giving a faster result. However, it requires access to specialized equipment and expertise. EM counts are often higher than culture-based methods, for similar reasons to those described for bacteria (p. 82).

Labelling Petri plates – the following information should be recorded:

- date
- the growth medium
- your name or initials
- brief details of the experimental treatment

Labelling your plates and cultures

Petri plates should always be labelled on the base, rather than the lid. Restrict your labelling to the outermost region of the plate, to avoid problems when counting colonies, assessing growth, etc. After labelling, Petri plates usually are incubated upside-down in a temperature-controlled incubator (often at 37 °C) for an appropriate period (usually 18–72 h).

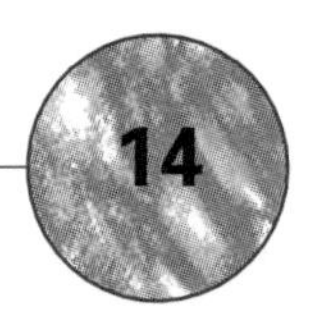

Isolating and identifying microbes

Microorganisms have a broad range of applications in the life sciences:

- Microbes are widely used as model systems, as they are often easier to study under controlled laboratory conditions than 'higher' organisms.
- Bacteria and fungi are used as sources of particular biomolecules, e.g. for the characterization of a specific enzyme *in vitro*.
- Pathogenic microbes are studied at the molecular level, to find new methods of identification using biochemical 'markers' and to investigate the molecular basis of their pathogenicity.
- Microbes are used in biotechnology and industrial microbiology as sources of particular biomolecules, e.g. production of blood clotting factors using yeast cells.

In your practical classes, you are likely to gain experience of a wide range of laboratory exercises involving microbes, particularly bacteria such as *Escherichia coli* (*E. coli*), using the procedures of sterile technique described in Chapter 13.

Isolating a particular microbe

Separation methods

Most microbial isolation procedures involve some form of separation to obtain individual microbial cells. The most common approach is to use an agar medium for primary isolation, with streak dilution, spread plating or pour plating to produce single colonies, each derived from an individual microorganism (p. 62). It is often necessary to dilute samples before isolation, so that a small number of individual microbial cells is transferred to the growth medium. Strict serial dilution (p. 15) of a known amount of sample is needed for quantitative work.

KEY POINT If your aim is to isolate a particular microbe, perhaps for further investigation, you will need to sub-culture individual colonies from the primary isolation plate to establish a pure culture, also known as an axenic culture.

Obtaining a pure culture – if a single colony from a primary isolation medium is used to prepare a streak dilution plate and all the colonies on the second plate appear identical, then a pure culture has been established. Otherwise, you cannot assume that your culture is pure.

Using a sonicator – minimize heat damage with short treatment 'bursts' (typically up to 1 min), cooling the sample between bursts.

Pure cultures of most microbes can be maintained indefinitely, using sterile technique and microbial culture methods (Chapter 13).

Other separation techniques include:

- Dilution to extinction. This involves diluting the sample to such an extent that only one or two microbes are present per millilitre: small volumes of this dilution are then transferred to a liquid growth medium. After incubation, most of the tubes will show no growth, but some tubes may show growth, having been inoculated with a single microbe at the outset. This should give a pure culture, though it is wasteful of resources.
- Sonication/homogenization (p. 90). Useful for separating individual microbial cells from each other and from inert particles, prior to isolation. However, some decrease in viability is likely.
- Filtration. This can be useful where the number of microbes is low. Samples can be passed through a sterile cellulose ester filter (pore size 0.2 μm), which is then incubated on the surface of an appropriate solidified medium.

Avoiding contamination during sampling – always remember that *you* are the most important source of contamination of field samples: components of the oral or skin microflora are the most likely contaminants.

Definitions

Psychrophile – a microbe with an optimum temperature for growth of <20 °C (Lit. 'cold-loving').

Psychrotroph – a microbe with an optimum temperature for growth of ⩾ 20 °C, but capable of growing at lower temperature, typically 0–5 °C (Lit. 'cold-feeding').

Thermophile – a microbe with an optimum growth temperature of >45 °C (Lit. 'heat-loving').

Mesophile – a microbe with an optimum growth temperature of 20–45 °C (Lit. 'middle-loving').

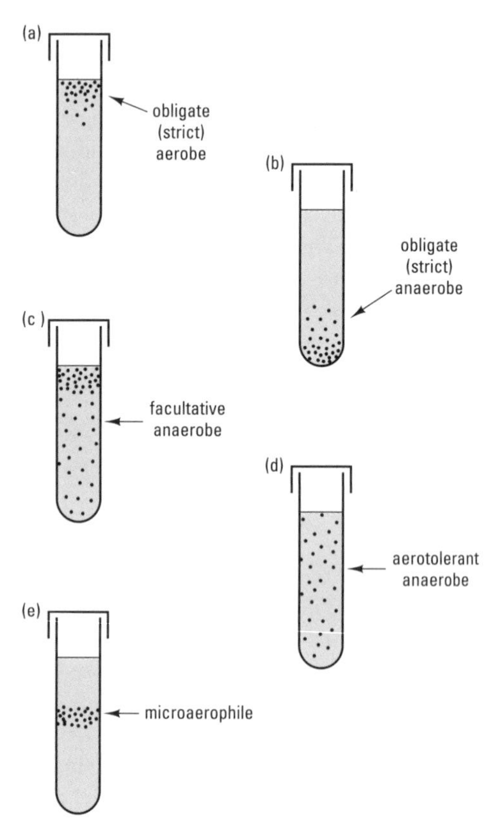

Fig. 14.1 Agar shake tubes. Bacteria are suspended in molten agar at 45–50 °C and allowed to cool. The growth pattern after incubation reflects the atmospheric (oxygen) requirements of the bacterium.

- Micromanipulation. It may be possible to separate the microbe from contaminants using a micropipette and dissecting microscope. The microbe can then be transferred to an appropriate growth medium, to give a pure culture. However, this is rarely an easy task for the novice.
- Motility. Phototactic microbes (including photosynthetic flagellates and motile cyanobacteria) will move towards a light source, while heterotrophic flagellate bacteria will move through a filter of appropriate pore size into a nutrient solution, or away from unfavourable conditions.

Selective and enrichment methods

Selective methods are based on the use of conditions that will permit the growth of a particular group of microbes while inhibiting others. Laboratory incubation under selective conditions will allow the particular microbes to be isolated in pure culture. Enrichment techniques encourage the growth of certain bacteria, usually by providing additional nutrients in the growth medium. The difference between selective and enrichment techniques is that the former use growth conditions unfavourable for competitors while the latter provide improved growth conditions for the chosen microbes.

Selective and enrichment techniques can be considered together, since they both enhance the growth of a particular microbe when compared with its competitors. Methods based on specific physical conditions include:

- Temperature. Psychrophilic and psychrotrophic microbes can be isolated by incubating the growth medium at 4 °C, while thermophilic microbes require temperatures above 45 °C for isolation. Short-term heat treatment of samples can be used to select for endospore-forming bacteria, e.g. 70–80 °C for 5–15 min, prior to isolation.
- Atmosphere. Many eukaryotic microbes are obligate aerobes, requiring an adequate supply of oxygen to grow. Bacteria vary in their responses to oxygen: obligate anaerobes are the most demanding, growing only under anaerobic conditions (e.g. in an anaerobic cabinet). The oxygen requirements of a bacterium can be determined using the agar shake tube method as part of the isolation procedure (Fig. 14.1). Some pathogenic bacteria grow best in an atmosphere with a reduced oxygen status and increased CO_2 concentration: such carboxyphilic bacteria (capnophiles) are grown in an incubator where the gas composition can be adjusted.
- Centrifugation. This can be used to separate buoyant microbes from their non-buoyant counterparts – on centrifugation, such organisms will collect at the surface while the remaining microbes will sediment. Alternatively, density gradient methods may be used (p. 124). Centrifugation can be combined with repeated washing, to separate microbes from contaminants.
- Ultraviolet irradiation. Some microbes are tolerant of UV treatment and can be selected by exposing samples to UV light. However, the survivors may show a greater rate of mutation.
- Illumination. Samples may be enriched for cyanobacteria and microalgae by incubation under a suitable light regime. For dilute samples, where the number of photosynthetic microbes is too low to give the sample any visible green coloration, there is a risk of photoinhibition and loss of viability if the irradiance is too high. Such samples need shading during initial growth.

Chemical methods form the mainstay of bacteriological isolation techniques and various media have been developed for the isolation of specific groups of bacteria (Table 14.1). The chemicals involved can be subdivided into:

Table 14.1 Selective agents in bacteriological media

Substance	Selective for
Azide salts	*Enterococcus* spp.
Bile salts	Intestinal bacteria
Brilliant green	Gram-negative bacteria
Crystal violet	*Streptococcus* spp.
Gentian violet	Gram-negative bacteria
Lauryl sulphate	Gram-negative bacteria
Methyl violet	*Vibrio* spp.
Malachite green	*Mycobacterium*
Polymyxin	*Bacillus* spp.
Sodium selenite	*Salmonella* spp.
Sodium chloride	Halotolerant bacteria *Staphylococcus aureus*
Sodium tetrathionate	*Salmonella* spp.
Trypan blue	*Streptococcus* spp.

- Selectively toxic substances: for example, salt-tolerant, Gram-positive cocci can be grown in a medium containing 7.5% w/v NaCl, which prevents the growth of most common heterotrophic bacteria. Several media include dyes as selective agents, particularly against Gram-positive bacteria.
- Antibiotics: for example, the use of antibacterial agents (penicillin, streptomycin, chloramphenicol) in media designed to isolate fungi, or the use of antifungal agents (cycloheximide, nystatin) in bacterial media. Some antibacterial agents show a narrow spectrum of toxicity and these can be incorporated into selective isolation media for resistant bacteria.
- Nutrients which encourage the growth of certain microbes: including the addition of a particular carbon source, or specific inorganic nutrients.
- Substances that affect the pH of the medium: for example, the use of alkaline peptone water at pH 8.6 for the isolation of *Vibrio* spp.

KEY POINT **Note that sub-cultures from primary isolation media must be grown in a non-selective medium, to confirm the purity of the isolate.**

Many of the selective and enrichment media used in bacteriology are able to distinguish between different types of bacteria: such media are termed differential media or diagnostic media and they are often used in the preliminary stages of an identification procedure. Box 14.1 gives details of the constituents of MacConkey medium, a selective, differential medium used in clinical microbiology (e.g. for isolation of certain faecal bacteria).

Identifying a particular microbe

Most of the methods described in this chapter were developed for the identification of bacteria, and bacterial examples are used to illustrate the principles involved. While the basic techniques are equally applicable to other types of microbe, the identification systems for some protozoa, fungi and algae rely predominantly on microscopic appearance. Identification of viruses requires electron microscopy or immunological techniques.

KEY POINT **Identification of bacteria is often based on a combination of a number of different features, including growth characteristics, microscopic examination, physiological or biochemical characterization, and, where necessary, immunological tests.**

Direct observation

Once a microbe has been isolated and cultured in the laboratory, the visual appearance of individual colonies on the surface of a solidified medium may provide useful information. Bacteria typically produce smooth, glistening colonies, varying in diameter from <1 mm to >1 cm. Actinomycete colonies are often <1 cm, with a shrivelled, powdery surface. Filamentous fungi usually grow as large, spreading colonies with a matt appearance and are identified by microscopy, using the morphological characteristics of their reproductive structures. Yeasts produce smaller, glistening colonies: identification usually involves microscopy, combined with physiological and biochemical tests similar to those used for bacteria.

Colony characteristics

The characteristics of a microbial colony on a particular medium include:

- Size: some bacteria produce punctiform colonies, with a diameter of less than 1 mm, while motile bacteria may spread over the entire plate.

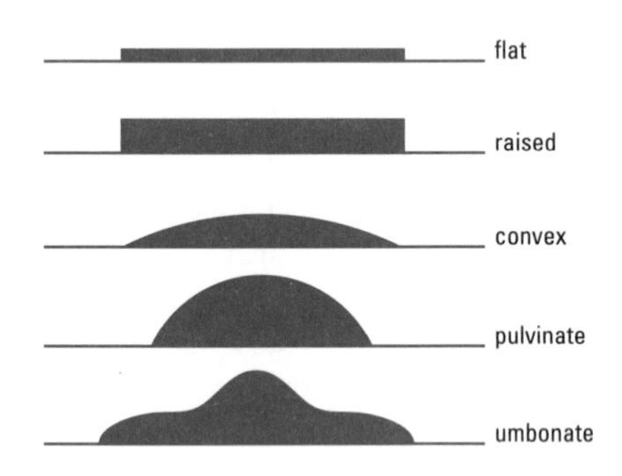

Fig. 14.2 Colony elevation (cross-sectional profile).

Box 14.1 Differential media for bacterial isolation: an example

MacConkey agar is both a selective and a differential medium, useful for the isolation and identification of intestinal Gram-negative bacteria. Each component in the medium has a particular role:

- Crystal violet: selectively inhibits the growth of Gram-positive bacteria.
- Bile salts: toxic to most microbes except those growing in the intestinal tract.
- Peptone: (a meat digest) provides a rich source of complex organic nutrients, to support the growth of non-exacting bacteria.
- Lactose: present as an additional specific carbon source.
- Neutral red: a pH indicator dye, to show the decrease in pH which accompanies the breakdown of lactose.

Any intestinal Gram-negative bacterium capable of fermenting lactose will grow on MacConkey agar to produce large purple–red colonies, the red coloration being due to the neutral red indicator under acidic conditions while the purple coloration, often accompanied by a metallic sheen, is due to the precipitation of bile salts and crystal violet at low pH.

In contrast, enteric Gram-negative bacteria unable to metabolize lactose will give colonies with no obvious pigmentation. This differential medium has been particularly useful in medical microbiology, since many enteric bacteria are unable to ferment lactose (e.g. *Salmonella, Shigella*) while others metabolize this carbohydrate (e.g. *Escherichia coli, Klebsiella* spp.). Colonial morphology on such a medium can give an experienced bacteriologist important clues to the identity of an organism, e.g. capsulate *Klebsiella* spp. characteristically produce large, convex, mucoid colonies with a weak pink coloration, due to the fermentation of lactose, while *E. coli* produces smaller, flattened pink colonies with a metallic sheen.

Measuring colony size – choose a typical colony, well spaced from any others, as colony size is affected by competition for nutrients.

- Form: colonies may be circular, irregular, lenticular (spindle-shaped) or filamentous.
- Elevation: colonies may be flat, raised, convex, etc. (see Fig. 14.2).
- Margin: the edge of a colony may be distinctive, e.g. undulate or filamentous.
- Consistency: colonies may be viscous (or mucoid), butyrous (of similar consistency to butter) or friable (dry and granular), etc.
- Colour: some bacteria produce characteristic pigments. A few pigments are fluorescent under UV light.
- Optical properties: colonies may be translucent or opaque.
- Odour: some actinomycetes and cyanobacteria produce earthy odours, while certain bacteria and yeasts produce fruity or 'off' odours. However, odour is not a reliable characteristic in bacterial identification.
- Haemolytic reactions on blood agar: many pathogenic bacteria produce characteristic zones of haemolysis. Alpha haemolysis is a partial breakdown of the haemoglobin from the erythrocytes, producing a green zone around the colony, while beta haemolysis is the complete destruction of haemoglobin, producing a clear zone.

Safe working with mould cultures – never attempt to smell mould cultures, because of the risk of inhaling large numbers of spores.

Further visual characterization requires the use of a microscope and an oil immersion objective at a total magnification of $\times 1000$ (p. 36).

Hanging drop technique – place a drop of bacterial suspension on a coverslip and invert over a cavity slide so that the drop does not make contact with the slide: motile aerobes are observed at the edge of the droplet, where oxygen is most abundant.

Motility

Wet mounts can be prepared by placing a small drop of bacterial suspension on a clean, degreased slide, adding a coverslip and examining the film without delay. For aerobes, areas near air bubbles or by the edge of the coverslip give best results, while anaerobes show greatest motility in the centre of the preparation, with rapid loss of motility due to oxygen toxicity.

Prepare wet mounts using young cultures in exponential growth in a liquid medium (p. 79). It is best to work with cultures grown at 20 or 25 °C, since those grown at 37 °C may not be actively motile on cooling to room temperature. It is essential to distinguish between the following:

- True motility, due to the presence of flagella: bacteria dart around the field of view, changing direction in zigzag, tumbling movements.
- Brownian motion: non-motile bacteria show a localized, vibratory, random motion, due to bombardment of bacterial cells by molecules in the solution.
- Passive motion, due to currents within the suspension: all cells will be swept in the same direction at a similar rate of movement.
- Gliding motility: a slower, intermittent movement, parallel to the longitudinal axis of the cell, requiring contact with a solid surface.

Assessing motility – if you have not seen bacterial motility before, it is worth comparing your unknown bacterium to a positive and a negative control.

Cell shape

Bacteria are subdivided into the following groups:

- Cocci (singular, coccus): spherical, or almost spherical, cells, sometimes growing in pairs (diplococci), chains or clumps.
- Rods: straight, cylindrical cells with flattened, tapered or rounded ends, termed bacilli. Short rods are sometimes called cocco-bacilli.
- Curved rods: the curvature varies according to the organism, from short curved rods, sometimes tapered at one end, to spiral shapes.
- Branched filaments: characteristic of actinomycete bacteria.

Using cell shape in microbial identification – many bacteria are pleomorphic, varying in size and shape according to the growth conditions and the age of the culture: thus, other characteristics are required for identification.

Gram staining

This is the most important differential staining technique in bacteriology (Box 14.2 gives details). It enables us to divide bacteria into two distinct groups, Gram-positive and Gram-negative, according to a particular staining procedure (the technique is given a capital letter, since it is named after its originator, H.C. Gram). The basis of the staining reaction is the different structure of the cell walls of Gram-positive and Gram-negative bacteria. Heat fixation of air-dried bacteria causes some shrinkage, but cells retain their shape: to measure cell dimensions use a chemical fixative.

Gram staining should be carried out using light smears of young, active cultures, since older cultures may give variable results. In particular, certain Gram-positive bacteria may stain Gram-negative if older cultures are used. This Gram-variability is due to autolytic changes in the cell wall of Gram-positive bacteria. Developing spores are often visible as unstained areas within older vegetative cells of *Bacillus* and *Clostridium*. Other stains are required to demonstrate spores, capsules or flagella (p. 37).

Assessing the Gram status of an unknown bacterium – if a pure culture gives both Gram-positive and Gram-negative cells, identical in size and shape, it can be regarded as a Gram-positive organism that is demonstrating Gram-variability.

Basic laboratory tests

At least two simple biochemical tests are usually performed:

1. Oxidase test. This identifies cytochrome *c* oxidase, an enzyme found in obligate aerobic bacteria. Soak a small piece of filter paper in a fresh solution of 1% (w/v) N-N-N′-N′-tetramethyl-p-phenylenediamine

Box 14.2 Preparation of a heat-fixed, Gram-stained smear

Preparation of a heat-fixed smear
The following procedure will provide you with a thin film of bacteria on a microscope slide, for staining.

1. **Take a clean microscope slide and pass it through a Bunsen flame twice**, to ensure it is free of grease. Allow to cool.
2. **Using a sterile inoculating loop, place a single drop of water in the centre of the slide and then mix in a small amount of sample** from a single bacterial colony with the drop, until the suspension is slightly turbid. Smear the suspension over the central area of the slide, to form a thin film. For liquid cultures, use a single drop of culture fluid, spread in a similar manner.
3. **Allow to air-dry at room temperature**, or high above a Bunsen flame: air drying must proceed gently, or the cells will shrink and become distorted.
4. **Fix the air-dried film by passage through a Bunsen flame.** Using a slide holder or forceps, pass the slide, film side up, rapidly through the hottest part of the flame (just above the blue cone). The temperature of the slide should be just too hot for comfort on the back of your hand: note that you must not overheat the slide or you may burn yourself (you will also ruin the preparation).
5. **Allow to cool**: the smear is now ready for staining.

Gram-staining procedure
The version given here is a modification of the Hucker method, since acetone is used to decolorize the smear. Note that some of the staining solutions used are flammable, especially the acetone decolorizing solvent: you must make sure that all Bunsens are turned off during staining. The procedure should be carried out with the slides suspended over a sink, using a staining rack.

1. **Flood a heat-fixed smear with 2% w/v crystal violet in 20% v/v ethanol: water** and leave for 1 min.
2. **Pour off the crystal violet and rinse briefly with tap water. Flood with Gram's iodine** (2 g KI and 1 g I_2 in 300 ml water) for 1 min.
3. **Rinse briefly with tap water** and leave the tap running gently.
4. **Tilt the slide and decolorize with acetone** for 2–3 s: acetone should be added dropwise to the slide until no colour appears in the effluent. This step is critical, since acetone is a powerful decolorizing solvent and must not be left in contact with the slide for too long.
5. **Immediately immerse the smear in a gentle stream of tap water**, to remove the acetone.
6. **Pour off the water and counterstain for 10–15 s using 2.5% w/v safranin** in 95% v/v ethanol:water.
7. **Pour off the counterstain, rinse briefly with tap water, then dry the smear** by blotting gently with absorbent paper: all traces of water must be removed before the stained smear is examined microscopically.
8. **Place a small drop of immersion oil on the stained smear: examine directly** (without a coverslip) using an oil-immersion objective (p. 36).

Gram-positive bacteria retain the crystal violet (primary stain) and appear purple while Gram-negative bacteria are decolorized by acetone and counterstained by the safranin, appearing pink or red when viewed microscopically.

Other decolorizing solvents are sometimes used, including ethanol:water, ethyl ether:acetone and acetone:alcohol mixtures. The time of decolorization must be adjusted, depending upon the strength of the solvents used, e.g. 95% v/v ethanol:water is less powerful than acetone, requiring around 30 s to decolorize a smear.

Performing the oxidase test - *never* use a nichrome wire loop, as this will react with the reagent, giving a false positive result.

dihydrochloride on a microscope slide. Rub a small amount from the surface of a young, active colony onto the filter paper using a glass rod or *plastic* loop: a purple-blue colour within 10 s is a positive result.

2. Catalase test. This identifies catalase, an enzyme found in obligate aerobes and in most facultative anaerobes, which catalyses the breakdown of hydrogen peroxide into water and oxygen. Transfer a small sample of your unknown bacterium onto a coverslip using a disposable plastic loop or glass rod. Invert onto a drop of hydrogen peroxide: the appearance of bubbles within 30 s is a positive reaction. This method minimizes the dangers from aerosols formed when gas bubbles burst.

The oxidase and catalase tests effectively allow us to sub-divide bacteria on the basis of their oxygen requirements, without using agar shake cultures

Avoiding false negatives – ensure you use sufficient material during oxidase and catalase testing, otherwise you may obtain a false negative result: a clearly visible 'clump' of bacteria should be used.

(p. 66) and overnight incubation, since, for the most part: obligate aerobes will be oxidase and catalase positive; facultative anaerobes will be oxidase negative and catalase positive; microaerophilic bacteria, aerotolerant anaerobes and strict anaerobes will be oxidase and catalase negative – the latter group will grow only under anaerobic conditions (p. 66).

Once you have reached this stage (colony characteristics, motility, shape, Gram reaction, oxidase and catalase status) it may be possible to make a tentative identification, at least for certain Gram-positive bacteria, at the generic level. To identify Gram-negative bacteria, particularly the oxidase-negative, catalase-positive rods, further tests are required.

Identification tables: further laboratory tests

Bacteria are asexual organisms and strains of the same species may give different results for individual biochemical/physiological tests. This variation is allowed for in identification tables, based on the results of a large number of tests. Identification tables are often used for particular sub-groups of bacteria, after Gram staining and basic laboratory tests have been performed: an example is shown in Table 14.2.

A large number of specific biochemical and physiological tests are used in bacterial identification including:

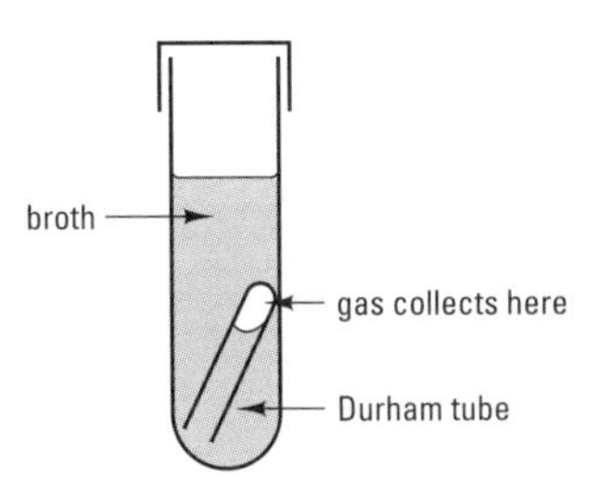

Fig. 14.3 Durham tube in carbohydrate utilization broth. Air within the Durham tube is replaced by broth during the autoclaving procedure.

- Carbohydrate utilization tests. Some bacteria can use a particular carbohydrate as a carbon and energy source. Acidic end-products can be identified using a pH indicator dye (p. 27) while CO_2 is detected in liquid culture using an inverted small test tube (Durham tube, Fig. 14.3). Aerobic breakdown is termed oxidation while anaerobic breakdown is known as fermentation. Identification tables usually incorporate tests for several different carbohydrates e.g. Table 14.2.
- Enzyme tests. Most of these incorporate a substance which changes colour if the enzyme is present, e.g. a pH indicator, or a chromogenic substrate.
- Tests for specific end-products of metabolism, e.g. the production of indole due to the metabolic breakdown of the amino acid tryptophan, or H_2S from sulphur-containing amino acids.

New approaches to microbial identification – several novel methods of detection and identification are based on nucleic acid techniques, including use of the PCR and Southern blotting to detect particular microbes (p. 233).

Table 14.2 Identification table for selected motile, oxidase-negative, catalase-positive, Gram-negative rods

	Biochemical test								
Bacterium	1	2	3	4	5	6	7	8	9
Escherichia coli	v	+	−	+	−	v	v	+	−
Proteus mirabilis	−	−	v	−	+	−	+	−	+
Morganella morganii	−	−	−	−	+	−	+	+	−
Vibrio parahaemolyticus	−	+	v	−	−	+	+	+	−
Salmonella spp.	−	+	v	−	−	+	+	−	+

Key to biochemical tests and symbols
1. sucrose utilization,
2. mannitol utilization,
3. citrate utilization,
4. β-galactosidase activity,
5. urease activity,
6. lysine decarboxylase activity,
7. ornithine decarboxylase activity,
8. indole production,
9. H_2S production,

+, >90% of strains tested positive,
−, <10% of strains tested positive,
v, 10–90% of strains tested positive.

Using identification kits – some biochemical tests are now supplied in kit form, e.g. the API® 20E system incorporates 20 tests within a sterile plastic strip: test results are converted into a 7-digit code, for comparison with known bacteria using either a reference book (Analytical Profile Index), or a computer program.

Immunological tests

Tests used in diagnostic microbiology include:

- Agglutination tests: based on the reaction between specific antibodies and a particular bacterium (p. 96). These tests are particularly useful for sub-dividing biochemically similar bacteria.
- Fluorescent antibody tests: the reaction between a labelled antibody and a particular bacterium can be visualized using UV microscopy. The direct fluorescent antibody test uses fluorescein isothiocyanate as the label.
- Enzyme-linked immunoassay tests using antibodies labelled with a particular enzyme, e.g. the double antibody sandwich ELISA test (p. 99).

While such tests can give specific and accurate confirmation of the identity of a bacterium under controlled laboratory conditions, they are too expensive and time-consuming for routine identification purposes. Antigen–antibody reactions are also used in the technique of immunomagnetic separation (IMS): antibodies attached to paramagnetic microspheres can be used to separate particular antigens from mixtures, e.g. a specific bacterium from a water sample.

Typing methods

The identification of bacteria at sub-species level is known as typing: this is usually done in a specialist laboratory, e.g. as part of an epidemiological study to establish the source of an infection. Various methods are used:

Practical applications of bacterial typing – *E. coli* O157:H7 is a serotype of this bacterium that is capable of causing severe human disease: it can be identified on the basis of an agglutination reaction with an appropriate antiserum.

- Antigen typing or serotyping is based on immunological tests.
- Phage typing is based on the susceptibility of different strains to certain bacterial viruses (phages).
- Biotyping is based on biochemical differences between different strains e.g. enzyme profiles or antibiotic resistance screening ('antibiograms').
- Bacteriocin typing: bacteriocins are proteins released by bacteria which inhibit the growth of other members of the same species.

Naming microbes

Since many microbes are asexual organisms, the traditional definition of a biological species – based on the concept of interbreeding – is not applicable and is replaced by the idea that members of the same species will share a number of morphological and biochemical characteristics, as described above. Cellular microorganisms (e.g. bacteria, archaea, fungi, protozoa, algae) are given two Latin terms to identify their genus and species (a Latin 'binomial'), for example, *Escherichia coli*, followed by a term for type or sub-species where appropriate. After first use in a text, the genus may be abbreviated to a single letter, e.g. *E. coli*, as long as this does not cause confusion with other genera. Where the species name is unknown, the (non-italicized) abbreviation 'sp.' should be used, e.g. *Enterococcus* sp.

Writing taxonomic names – always underline or italicize genus and species names to avoid confusion: thus 'bacillus' is a descriptive term for rod-shaped bacteria, while *Bacillus* is a generic name.

The classification of viruses is less advanced than for cellular microbes and the current nomenclature has been arrived at on a piecemeal, *ad hoc* basis, despite international attempts to create a unified system based on nucleic acid type and host preferences. Many viruses are referred to by their trivial names, often reflecting the diseases they cause (e.g. influenza virus), or by code names, e.g. the bacterial viruses ϕX174, T4, etc. Often, a three-letter abbreviation is used, e.g. HIV (for human immunodeficiency virus) or TMV (for tobacco mosaic virus).

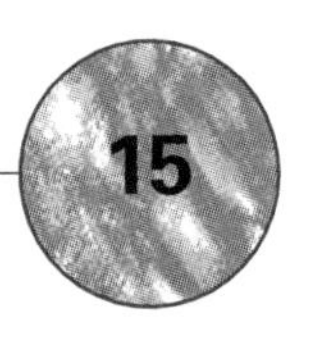

15 Working with animal and plant tissues and cells

Definitions

In vivo – occurring within a living organism (literally 'in life').

In vitro – occurring outside a living organism, in an artificial environment (literally 'in glass').

While the aim of many studies is to isolate, quantify and characterize individual molecules from a biological system, e.g. the purification of a particular enzyme (p. 185), this is not always the most appropriate course of action. Depending on the purpose of the investigation, it may be more relevant to study the functioning of biomolecules within more complex systems, in order to understand their role in a particular biological process. At one extreme this may be carried out *in vivo*, using whole multicellular organisms (e.g. individual animals or plants), while the other extreme is represented by *in vitro* studies, using sub-cellular 'cell-free' extracts. Between these two extremes, a range of tissue and cell culture techniques offers some of the biological complexity of the intact organism combined with a degree of experimental control that may not be obtainable *in vivo*. The ethics and costs of whole animal experimentation have provided an additional stimulus to the development of *in vitro* methods, e.g. toxicity tests using mammalian cell culture rather than laboratory animals, and it is likely that developments in molecular genetics will lead to further applications of cell and tissue culture.

Animal tissues and organs

Working with vertebrate animals and their organs/tissues – remember that procedures must be consistent with the law, i.e. in the UK, the Animals Scientific Procedures Act 1986.

Physiological experiments are carried out using either whole organisms, or a range of animal organs and tissues, including heart, liver, muscle, etc.

KEY POINT A major practical consideration is that the tissue should be studied as soon as possible after the death of the animal, typically under laboratory conditions that mimic the *in vivo* environment as closely as possible.

In most instances, the experiments are relatively short-term (<24 h) and the aim is to maintain the tissue in a physiological state similar to that within the living organism. For metabolic studies, the whole organ or a tissue slice (typically 1–10 mm thick) will be bathed in an appropriate perfusion fluid, supplied either by gravity or by peristaltic pump. Practical considerations include:

- Inorganic solute requirements – the chemical composition of the perfusion fluid is usually chosen to reflect the major inorganic ion requirements of the tissue. For short-term studies, a number of so-called *physiological salt solutions* may be used, e.g. Ringer's solution, one formulation of which is given in Table 15.1.
- Oxygen requirements – it may be necessary to increase the O_2 content of the perfusion fluid by bubbling with O_2, in order to meet the oxygen demand of the innermost parts of the tissue. However, this can lead to oxygen toxicity in the outermost parts and an alternative approach is simply to increase the rate of perfusion.
- Physico-chemical conditions – including temperature (usually controlled to ±1 °C of normal body temperature), water status (the perfusion fluid

Table 15.1 Composition of Ringer's solution (simplified formulation, for amphibians, etc.)

Compound	Amount per litre (g)
NaCl	6.0
KCl	0.075
$CaCl_2$	0.1
$NaHCO_3$	0.1

Definitions

Plasmodesmata – transverse connections through the cell wall, linking the cytoplasm of adjacent plant cells and creating a symplasm.

Apoplasm – that part of the plant body outside the symplasm.

Light compensation point – the amount of photosynthetically active radiation (PAR) at which photosynthetic CO_2 uptake is balanced by CO_2 production due to respiration and photorespiration (p. 207).

Table 15.2 Components of Long Ashton medium (nitrate version)

Stock solution: mass required per litre of solution (g)	Volume of stock solution to make 1 litre of medium (ml)
Major nutrients	
KNO_3: 50.60	8
$Ca(NO_3)_2$: 80.25	8
$MgSO_4.7H_2O$: 46.00	8
$NaH_2PO_4.2H_2O$: 52.00	4
Micronutrients	
FeKEDTA: 3.30	5
$MnSO_4.4H_2O$: 2.23	1
$ZnSO_4.7H_2O$: 0.29	1
$CuSO_4.5H_2O$: 0.25	1
H_3BO_3: 3.10	1
$Na_2MoO_4.2H_2O$: 0.12	1
NaCl: 5.85	1
$CoSO_4.7H_2O$: 0.056	1

Effects of humans on plants – remember that your exhaled breath will be nearly saturated with water vapour and will contain CO_2 at 3–4% v/v, some 100 times more concentrated than atmospheric CO_2: your breath can thus affect rates of transpiration and photosynthesis.

should be isotonic with the tissue), pH and buffering capacity (e.g. some perfusion fluids have elevated $NaHCO_3$, to mimic the buffering capacity of mammalian serum).

- Organic nutrient requirements – for longer-term studies, suitable organic nutrients will be required: these may be chemically defined additives, e.g. vitamins, amino acids, proteins, etc., or biological fluids such as plasma or serum. Glucose is often added as carbon and energy source.

Plant tissues and organs

Individual plant components (e.g. leaves, leaf slices and epidermal strips) can be isolated from the main plant body for study under controlled conditions. Since photosynthetic plant parts are autotrophic, they may be maintained *in vitro* for longer than animal organs, given adequate light and CO_2. However, most plant cells are joined via plasmodesmata, and the separation of such cells when the component is removed from the plant often leads to death when these connections are broken. Plants also show wound responses that may affect the metabolic processes under study. For these reasons, the most suitable systems for longer-term studies are often whole organs, e.g. whole leaves or entire root systems. Water culture (e.g. in Long Ashton medium, Table 15.2) is an alternative approach, offering greater control over the root environment (Dodds, 1995).

It is best to use vigorous, healthy stock plants and to follow a well-established procedure, taking account of the following:

- Sterility – strict attention to sterile technique (p. 59) can be essential to the success of many longer-term experiments. Decontamination of plant organs may be especially difficult where specimens are obtained from soil: to achieve this, use a surface wash with disinfectant (e.g. 10% w/v sodium hypochlorite), followed by several rinses with sterile water.
- Gaseous environment – in general, the experimental system should be well ventilated. Actively photosynthetic tissues will rapidly deplete the atmospheric CO_2 in a closed vessel: plant parts may also produce physiologically active gases, such as ethylene, especially at wound sites. Turgor loss may occur in isolated plant parts unless a high humidity is maintained.
- Nutrition – plant tissues may benefit from a supply of inorganic ions, including K^+, SO_4^{2-}, etc. and may require certain vitamins, micronutrients and plant hormones for prolonged studies.
- Physico-chemical conditions – light is the most important environmental requirement for green plant parts. At atmospheric CO_2 concentrations, the light compensation point is about 5–10 μmol photons PAR $m^{-2} s^{-1}$ (p. 109) and photosynthesis is usually saturated between 500 and 2000 μmol photons PAR $m^{-2} s^{-1}$, depending on the plant type. Light quality and photoperiod (daylength) are also important: fluorescent tubes and incandescent bulbs that mimic the photosynthetic spectrum of sunlight are available (see p. 109), while photoperiod can be controlled using an electric timer. The water potential of aqueous media can be adjusted using a membrane impermeant solute such as mannitol, and pH values are often kept close to those of the apoplasm ($\approx$pH 6) using appropriate buffers, if necessary (p. 28).

Cell and tissue cultures

KEY POINT One of the main differences between organ and tissue incubation techniques and those used in cell/tissue culture is that the former aim simply to maintain metabolic and physiological activity for a limited period, while the latter provide conditions suitable for cell growth, division and development *in vitro* over an extended time scale, from a few days to several months.

The basic principles involved in culturing animal and plant cells are broadly similar to those described for microbial cell culture (p. 59). Definitions for several key terms are given in the margin.

Definitions

Primary culture – a cell culture derived from tissue or organ fragments (explants). Primary culture ends on first sub-culture.

Passage – an alternative term for sub-culture.

Cell line – a cell culture derived by passage of a primary culture.

Finite cell line – a culture with a limited capacity for growth *in vitro* (maximum number of cell doublings).

Continuous cell line – a culture with the capacity for unlimited multiplication *in vitro*. Sometimes termed an established cell line.

Clone – a population of cells derived from a single original cell, i.e. sharing the same genotype.

Transformation – a permanent alteration in the growth characteristics of a finite cell line that may include (i) changes in morphology, (ii) an increased growth rate and/or (iii) the acquisition of an infinite lifespan, often termed immortalization. Transformation may be spontaneous or may be induced by chemical agents or viruses, and often involves a change in chromosome number.

Senescence – the end point in the limited lifespan of a finite cell line, characterized by the lack of proliferation. Conversion to an established (continuous) cell line requires 'escape' from senescence.

Applications of cell and tissue culture

The main uses of animal and plant cell culture systems include:

- Experimental model systems in biochemistry, pharmacology and physiology: cell culture offers certain advantages over whole organism studies, with greater control over environmental conditions and biological variability. The use of genetically defined clones of cells may simplify the analysis of experimental data. Conversely, results obtained with specialized cell-based systems might be unrepresentative of a broader range of cell types and may be more difficult to interpret in terms of the whole organism.
- Studies of the growth requirements of particular cells: including studies of the positive effects of growth factors or growth-promoting substances, and the negative effects of xenobiotics or cytotoxic compounds and events linked to programmed cell death (apoptosis). The use of cell culture in bioassays and mutagenicity testing is considered in Chapter 16 (p. 84).
- Studies of cell development and differentiation: including aspects of the cell cycle and gene expression. Cell cultures retaining their ability to differentiate *in vitro* are particularly interesting to researchers, while the lack of differentiation and uncontrolled growth of many animal cell lines makes them useful models of tumour development and neoplasia.
- Pathological studies: including the culture of foetal cells for karyotyping and the detection of genetic abnormality, e.g. trisomy, translocation, etc.
- Genetic manipulation: cell culture techniques have played an essential role in the development of molecular biology, including the production of transgenic animals and plants by techniques such as transfection, etc.
- Biotechnology: including the industrial production of therapeutic proteins, vaccines, monoclonal antibodies, etc. using large-scale batch and continuous culture techniques similar to those used in microbiology (p. 80).

Animal cell culture systems

These may be established either from whole organisms (e.g. chick embryo), discrete organs (e.g. rat liver) or from blood (e.g. lymphocytes), typically using *mild* enzymic tissue disruption techniques, where necessary (p. 89). While, in theory, it is possible to culture nucleated cells from virtually any source, in practice, the highest rates of success are most often achieved with young, actively growing tissues. The principal considerations in animal cell culture are:

- Safety: it is important to be aware of the potential dangers of infection from cell cultures. Although avian and rodent cells present a reduced risk of disease transmission compared to human cells, all cell cultures must be regarded as a potential source of pathogenic microbes, and appropriate sterile technique should be used at all times (see p. 59). Work involving human tissues and cell cultures must be carried out in a recirculating (Class II) safety cabinet by trained, experienced personnel – because of the possible risks, you are unlikely to gain practical experience with such cultures in the early stages of your course. For other cell cultures, you will need to follow the code of safe practice of your Department – consult your Departmental Safety Officer if you have any doubts about safe working procedures.

- Whether to use a primary culture or a continuous cell line: freshly isolated cells are more likely to reflect the biochemical activities of cells *in vivo*, though they will have a limited lifespan in culture, requiring repeated isolation for longer-term projects. Continuous cell lines are more easily cultured and offer the advantage that their growth requirements in culture may be known in some detail, especially for the more widely used cell lines (e.g. BHK, HeLa). Because continuous cell lines are clonal, their responses are more reproducible, giving less variable results than primary cultures.
- The need for a solid substratum: some cells must be attached to a solid surface in order to grow. Anchorage dependence is a typical feature of primary cultures and finite cell lines – such cultures show density-dependent growth inhibition once the cells have formed a confluent monolayer on the surface of the substratum. An alternative approach is to grow such cells on a particulate support using 'microcarrier' beads. In contrast, many continuous cell lines can be maintained in suspension culture, as individual cells or aggregates.
- The physico-chemical conditions, including pH (typically 7.2–7.5) and buffering capacity, osmolality (usually $300 \pm 20\,\text{mosmol}\,\text{kg}^{-1}$) and temperature (e.g. 35–37 °C for mammalian cells).
- The requirements of the culture medium: these will include the provision of inorganic ions (as a balanced salt solution), a carbon/energy source plus other organic nutrients, and a supplement containing antimicrobial agents to counter the risks of contamination. For example, Dulbecco's modified Eagle's medium is used for many mammalian cell types that grow as adherent monolayers, while suspension cultures of continuous cell lines can be maintained using less stringent media. To support growth, the basal medium is usually supplemented with serum (usually foetal calf serum, at up to 20% v/v), or a chemically defined serum-like supplement containing a mixture of proteins, polypeptides, hormones, lipids and trace components. A typical antimicrobial supplement might include antibacterial agents, e.g. penicillin and streptomycin, an antifungal agent, e.g. griseofulvin and an antimycoplasmal agent, e.g. gentamicin. The *in vitro* level of CO_2 and O_2 must also be considered: many cell cultures are buffered using bicarbonate, and must be maintained in an atmosphere of elevated CO_2, either in a sealed culture vessel or in a CO_2 incubator, to maintain pH balance. In some cases, a pH-sensitive dye (e.g. phenol red, p. 27) may be incorporated into the growth medium, to provide a visual check on pH status during growth.
- The equipment required: this may include a laminar flow hood or safety cabinet to reduce the possibility of microbial contamination, suitable culture vessels (typically pre-sterilized, disposable polystyrene dishes, bottles and flasks, treated to create a negatively charged, hydrophilic surface), a supply of high purity water (typically distilled, deionized and carbon filtered), a suitable incubator with temperature control of ±0.5 °C or better, often with CO_2 control and mechanical mixing, and an inverted phase contrast microscope to examine adherent cell monolayers during growth.

Example Dulbecco's modified Eagle's medium contains Ca^{2+}, Fe^{3+}, Mg^{2+}, K^{+}, Na^{+}, Cl^{-}, $SO4_2^{-}$, $PO4_2^{-}$, glucose, 20 amino acids, 10 vitamins, inositol and glutathione. Foetal calf serum is usually added at up to 20% v/v.

Using a horizontal laminar flow hood – note that this is designed to minimize contamination of the culture. It must *not* be used as a substitute for a safety cabinet.

Definitions

Explant – a fragment of tissue used to initiate a culture (the term is also used in animal culture).

Callus – an aggregation of undifferentiated plant cells in culture.

Totipotency – the ability (of any plant cell) to de-differentiate and re-differentiate into any of the cell types found in the mature plant.

Embryogenic callus – tissue with the capacity to differentiate under defined laboratory conditions, typically in response to plant growth regulators in the medium.

Protoplasts – cells lacking their cell walls.

Sphaeroplasts – cells with attached fragments of their cell walls: osmotically sensitive.

The successive stages of isolation of animal components are shown in Fig. 15.1a.

Plant tissue and cell culture systems

Plant tissue cultures can be established by growing explants of sterilized tissue on the surface of an agar-based growth medium to give a callus of undifferentiated cells. Initial sterilization is usually achieved by incubation for 15–20 min in 10% w/v sodium hypochlorite.

KEY POINT **For most plants, cell cultures can be established from a broad range of tissue types, reflecting the totipotency of many plant cells.**

Embryogenic callus may be induced to differentiate, forming tissues and organs on a medium containing appropriate plant hormones: in many cases, these cultures will develop to form plantlets that can be grown on to mature plants, or encapsulated to produce so-called 'artificial seeds'. This approach can be used to study the conditions necessary for differentiation and development, or to propagate rare plants and other valuable stock (e.g. virus-free stock, or genetically altered plants). Callus derived from anthers can be used to provide haploid cell cultures and haploid plants – these are often useful for experimental genetics and breeding purposes.

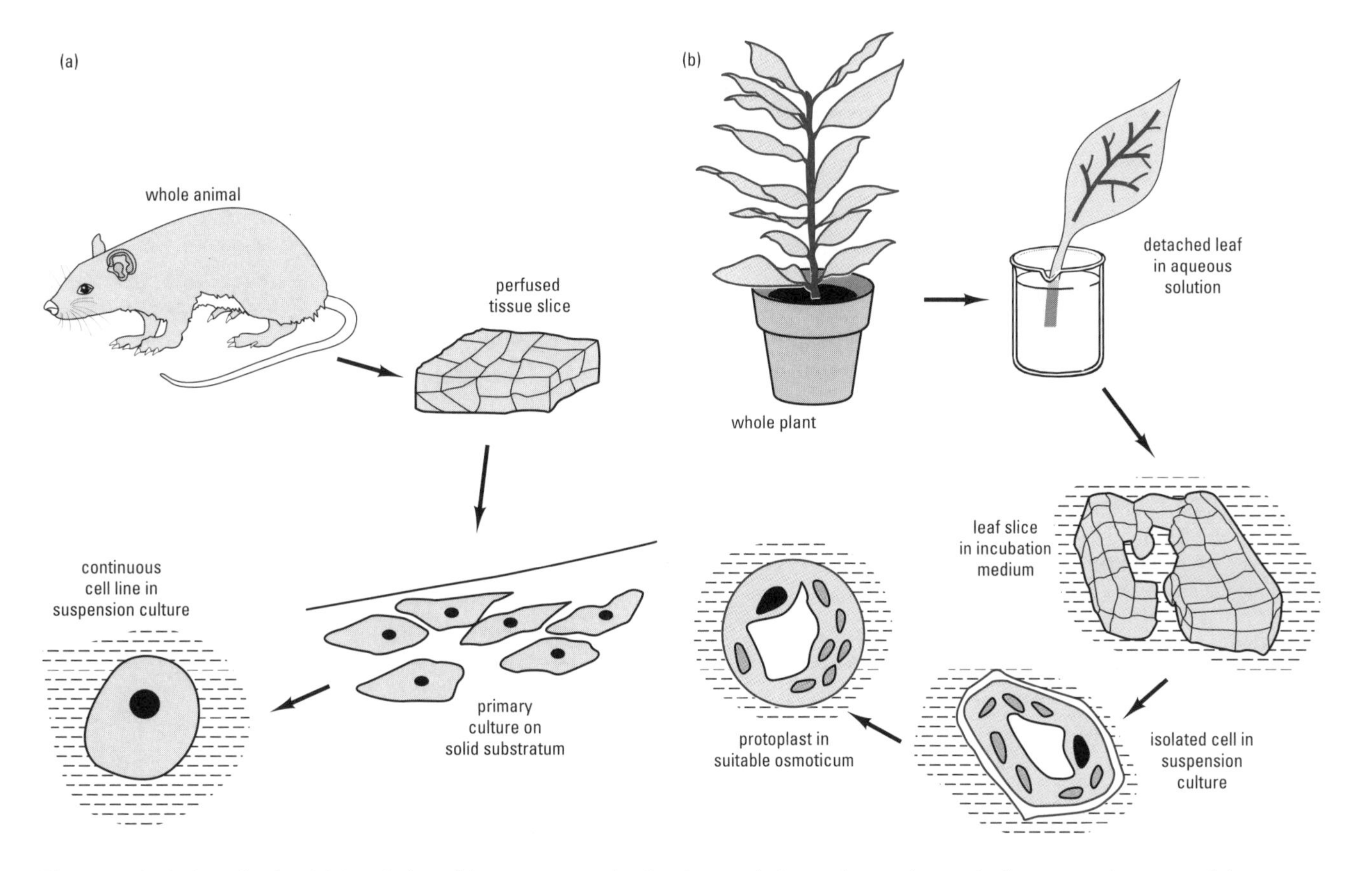

Fig. 15.1 Isolation of animal (a) and plant (b) components for *in vitro* study (note decreasing scale, from organisms to cells).

Plant cell suspension cultures are usually obtained by transferring fragments of actively growing callus to a 'shake' flask containing liquid medium in an orbital incubator: gentle agitation causes fragmentation of the callus tissue, to give a suspension culture that will contain individual plant cells and cell aggregates. This culture can be maintained by repeated sub-culturing of material from the upper layers of the liquid, encouraging the growth of small aggregates. However, in contrast to animal cell suspension cultures, it is rare for plant suspension cultures to be entirely unicellular, due to the presence of plasmodesmata and plant cell walls. Suspension cultures often require a minimum inoculum size on sub-culture – an inoculum volume of 10% v/v may be necessary to ensure successful sub-culture. It may also be helpful to add a small amount of 'conditioned' medium from a previous culture. Commercial-scale suspension cultures are used to produce certain plant pigments and secondary metabolites such as flavourings and high-value pharmaceutical compounds.

Example Murashige and Skoog's medium contains a balanced mixture of the principal inorganic ions, plus 7 trace element compounds, 3 vitamins, inositol, glycine and sucrose (at 30 g/l) as the major carbon source.

The growth media used for callus and suspension cultures are more complex than those required for intact plant organs and tissues, with organic nutrient supplements in addition to a balanced salt solution. Such organic supplements often include a major carbon and energy source, e.g. sucrose, plus various vitamins and growth regulators (typically, at least one auxin and a cytokinin), together with undefined components such as yeast extract and hydrolysed casein in some instances. Otherwise, the techniques are broadly similar to those described for microbial systems (p. 62).

Plant protoplasts

Working with protoplasts – remember that all solutions must contain a suitable osmoticum, to prevent bursting. Mannitol is widely used at $\approx 400\,\text{mmol}\,\text{l}^{-1}$.

Some experimental procedures using plant cells require the enzymatic removal of their cell walls, creating protoplasts that can be manipulated *in vitro*, then maintained under conditions that allow the regeneration of cell walls, with subsequent growth and differentiation to give genetically modified plants. Protoplast isolation often involves pre-treatment in a concentrated osmoticum (e.g. sucrose or mannitol, at $>500\,\text{mmol}\,\text{l}^{-1}$) to plasmolyse the cells, weakening the linkage between cell wall and plasma membrane. Enzymatic treatment is often prolonged, taking several hours in a suitable mixture of enzymes, e.g. cellulase and macerozyme®. The resulting material can be sieved through fine nylon mesh, centrifuged at low speed and then resuspended, to remove sub-cellular debris and cell aggregates. The protoplast preparation can be checked for fragments of cell wall using a suitable stain, e.g. 0.1% w/v calcofluor white and a fluorescence microscope.

Assessing cell and protoplast viability – cytochemical techniques are often used, e.g. exclusion of the mortal stain Trypan blue or Evan's blue. The fluorogenic vital stain fluorescein diacetate is an alternative: it is cleaved by esterases within living cells, liberating fluorescein and giving green–yellow fluorescent cells when viewed by UV fluorescence microscopy.

The fusion of protoplasts from different plants can be used to produce a somatic hybrid: this process can be used to circumvent inter-species reproductive barriers, creating novel plants. Protoplast fusion can be induced by chemical 'fusogens', e.g. using polyethylene glycol (PEG) at high concentration, or by electrofusion (incubation under low alternating current to encourage aggregation, then brief exposure to a high voltage electrical field – typically $1\,000\,\text{V}\,\text{cm}^{-1}$ for 1–2 ms – causing protoplast fusion). The successive stages of isolation of plant components are illustrated in Fig. 15.1b.

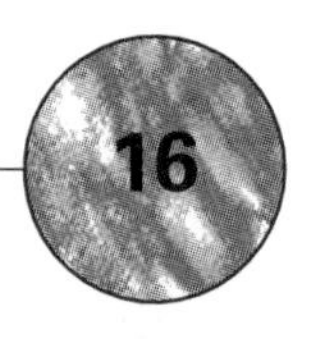

16 Culture systems and growth measurement

Microbial, animal and plant cell culture methods are based on the same general principles, requiring:

- a pure culture (also known as an axenic culture), perhaps isolated as part of an earlier procedure, or from a culture collection;
- a suitable nutrient medium to provide the necessary components for growth. This medium must be sterilized before use;
- satisfactory growth conditions including temperature, pH, atmospheric requirements, ionic and osmotic conditions;
- sterile technique (p. 59) to maintain the culture in pure form.

Heterotrophic animal cells, fungi and many bacteria require appropriate organic compounds as sources of carbon and energy. Non-exacting bacteria can utilize a wide range of compounds and they are often grown in media containing complex natural substances (including meat extract, yeast extract, soil, blood). Animal cells have more stringent growth requirements (p. 75).

Photoautotrophic bacteria, cyanobacteria and algae are grown in a mineral medium containing inorganic ions including chelated iron, with a light source and CO_2 supply. Plant cells may require additional vitamins and hormones (p. 77). For chemoautotrophic bacteria, the light source is replaced by a suitable inorganic energy source, e.g. H_2S for sulphur-oxidizing bacteria, NH_3/NH_4^+ for nitrifying bacteria, etc.

Definitions

Heterotroph – an organism that uses complex organic carbon compounds as a source of carbon and energy.

Photoautotroph – an organism that uses light as a source of energy and CO_2 as a carbon source (photosynthetic metabolism).

Chemoautotroph – an organism that acquires energy from the oxidation of simple inorganic compounds, fixing CO_2 as a source of carbon (chemosynthetic metabolism).

Growth on solidified media

Many organisms can be cultured on an agar-based medium (p. 60). An important benefit is that an individual cell inoculated onto the surface can develop to form a visible colony: this is the basis of most microbial isolation and purification methods, including the streak dilution, spread plate and pour plate procedures (p. 62).

Animal cells are often grown as an adherent monolayer on the surface of a plastic or glass culture vessel (p. 76), rather than on an agar-based medium.

Several types of culture vessel are used:

- Petri plates (Petri dishes): usually the pre-sterilized, disposable plastic type, providing a large surface area for growth.
- Glass bottles or test tubes: these provide sufficient depth of agar medium for prolonged growth of bacterial and fungal cultures, avoiding problems of dehydration and salt crystallization. Inoculate aerobes on the surface and anaerobes by stabbing down the centre, into the base (stab culture).
- Flat-sided bottles: these are used for animal cell culture, to provide an increased surface area for attachment and allow growth of cells as a surface monolayer. Usually plastic and disposable.

Harvesting bacteria from a plate – colonies can be harvested using a sterile loop, providing large numbers of cells without the need for centrifugation. The cells are relatively free from components of the growth medium; this is useful if the medium contains substances which interfere with subsequent procedures.

Sub-culturing – when sub-culturing microbes from a colony on an agar medium, take your sample from the growing edge, so that viable cells are transferred.

The dynamics of growth are usually studied in liquid culture, apart from certain rapidly growing filamentous fungi, where increases in colony diameter can be measured accurately, e.g. using Vernier calipers.

Growth in liquid media

Many cells, apart from primary cultures of animal cells (p. 75), can be grown as a homogeneous unicellular suspension in a suitable liquid medium, where growth is usually considered in terms of cell number (population growth) rather than cell size. Most liquid culture systems need agitation, to ensure adequate mixing and to keep the cells in suspension. An Erlenmeyer flask (100-2000 ml capacity) can be used to grow a batch culture on an orbital shaker, operating at 20–250 cycles per minute. For aerobic organisms, the surface area of such a culture should be as large as possible: restrict the volume of medium to not more than 20% of the flask volume. Larger cultures ('fermenters') may need to be gassed with sterile air and mixed using a magnetic stirrer rather than an orbital shaker. The simplest method of air sterilization is filtration, using glass wool, non-absorbent cotton wool or a commercial filter unit of appropriate pore size (usually 0.2 μm). Air is introduced via a sparger (a glass tube with many small holes, so that small bubbles are produced) near the bottom of the culture vessel to increase the surface area and enhance gas exchange. More complex fermenters have baffles and paddles to further improve mixing and gas exchange.

Liquid culture systems may be sub-divided under two broad headings:

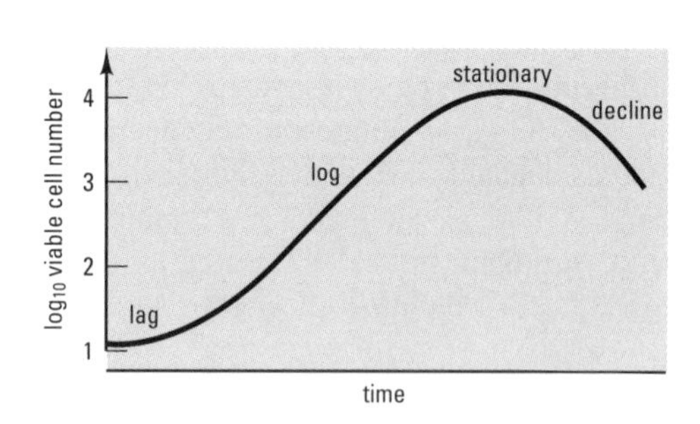

Fig. 16.1 Population growth curve for cells in batch culture (liquid medium).

Batch culture

This is the most common approach for routine liquid culture. Cells are inoculated into a sterile vessel containing a fixed amount of growth medium. Your choice of vessel will depend upon the volume of culture required: larger-scale vessels (e.g. 1 litre and above) are often called 'fermenters' or 'bioreactors', particularly in biotechnology. Growth within the vessel usually follows a predictable S-shaped (sinusoidal) curve when plotted in log-linear format (Fig. 16.1), divided into four components:

1. Lag phase: the initial period when no increase in cell number is seen. The larger the inoculum of active cells the shorter the lag phase will be, provided the cells are transferred from similar growth conditions.
2. Log phase, or exponential phase: where cells are growing at their maximum rate. This may be quantified by the growth rate constant, or specific growth rate (μ), where:

$$\mu = \frac{2.303\,(\log N_x - \log N_0)}{(t_x - t_0)} \qquad [16.1]$$

where N_0 is the initial number of cells at time t_0 and N_x is the number of cells at time t_x. For times specified in hours, μ is expressed as h^{-1}.

Prokaryotes grow by binary fission while eukaryotes grow by mitotic cell division; in both cases each cell divides to give two identical offspring. Consequently, the doubling time or generation time (g, or T_2) is:

$$g = \frac{0.301\,(t_x - t_0)}{\log N_x - \log N_0} \qquad [16.2]$$

Cells grow at different rates, with doubling times ranging from under 20 min for some bacteria to 24 h or more for animal and plant cells. Exponential phase cells are often used in laboratory experiments, since growth and metabolism are nearly uniform.

> **Example** Suppose you counted 2×10^3 cells ($\log_{10} = 3.30$) per unit volume at t_0 and 6.3×10^4 cells ($\log_{10} = 4.80$) after 2 h (t_x).
> Substitution into Eqn. 16.1 gives $[2.303(4.8 - 3.3)] \div 2 = 1.727\ h^{-1}$ (or 0.0288 min^{-1}).
> Substituting the same values into Eqn. 16.2 gives: $[0.301 \times 2] \div [4.8 - 3.3] = 0.40$ h (or 24 min).

3. Stationary phase: growth decreases as nutrients are depleted and waste products accumulate. Any increase in cell number is offset by death. This phase is usually termed the 'plateau' in animal cell culture.
4. Decline phase, or death phase: this is the result of prolonged starvation and toxicity, unless the cells are sub-cultured. Like growth, death often

shows an exponential relationship with time, which can be characterized by a rate constant (death rate constant), equivalent to that used to express growth or, more often, as the decimal reduction time (d, or T_{90}), the time required to reduce the population by 90%:

$$d = \frac{t_x - t_0}{\log N_x - \log N_0} \qquad [16.3]$$

Some cells undergo rapid autolysis at the end of the stationary period while others show a slower decline.

Batch culture methods can be used to maintain stocks of particular organisms; cells are sub-cultured onto fresh medium before they enter the decline phase. However, primary cultures of animal cells have a finite life in unless transformed to give a continuous cell line, capable of indefinite growth (p. 75).

Continuous culture

This is a method of maintaining cells in exponential growth for an extended period by continuously adding fresh growth medium to a culture vessel of fixed capacity. The new medium replaces nutrients and displaces some of the culture, diluting the remaining cells and allowing further growth.

After inoculating the vessel, the culture is allowed to grow for a short time as a batch culture, until a suitable population size is reached. Then medium is pumped into the vessel: the system is usually set up so that any increase in cell number due to growth will be offset by an equivalent loss due to dilution, i.e. the cell number within the vessel is maintained at a steady state. The cells will be growing at a particular rate (μ), counterbalanced by dilution at an equivalent rate (D):

$$D = \frac{\text{flow rate}}{\text{vessel volume}} \qquad [16.4]$$

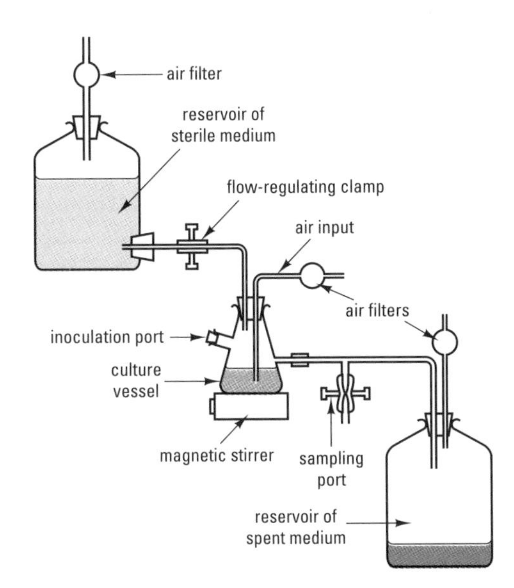

Fig. 16.2 Example of a two-dimensional lab equipment diagram of components of a chemostat.

where D is expressed per unit time (e.g. h^{-1}). In a chemostat, the growth rate is limited by the availability of some nutrient in the inflowing medium, usually either carbon or nitrogen (see Fig. 16.2). In a turbidostat, the input of medium is controlled by the turbidity of the culture, measured using a photocell. A turbidostat is more complex than a chemostat, with additional equipment and controls.

To determine the specific growth rate (μ) of a continuous culture:

1. Measure the flow of medium through the vessel over a known time interval (e.g. connect a sterile measuring cylinder or similar volumetric device to the outlet), to calculate the flow rate.
2. Divide the flow rate by the vessel volume (eqn [16.4]) to give the dilution rate (D).
3. This equals the specific growth rate, since $D = \mu$ at steady state.
4. If you want to know the doubling time (g), calculate using the relationship:

$$g = \frac{0.693}{\mu} \qquad [16.5]$$

(Note that eqn [16.5] also applies to exponential phase cells in batch culture and is useful for interconverting g and μ.)

Example Suppose you wanted to convert a doubling time of 20 min to a specific growth rate. Rearrangement of Eqn. 16.5 gives $\mu = 0.693 \div 20 = 0.03465$ min^{-1} ($= 2.08\,h^{-1}$).

Continuous culture systems are more complex to set up than batch cultures. They are prone to contamination, having additional vessels for fresh medium and waste culture: strict aseptic technique is necessary when the medium reservoirs are replaced, and during sampling and harvesting. However, they offer several advantages over batch cultures, including the following:

- The physiological state of the cells is more clearly defined, since actively growing cells at the same stage of growth are provided over an extended time period. This is useful for biochemical and physiological studies.
- Monitoring and control can be automated and computerized.
- Modelling can be carried out for biotechnology/fermentation technology.

Measuring growth in cell cultures

The most widely used methods of measuring growth are based on cell number:

Direct microscopic counts

Alternative approaches to measuring growth – these include biomass, dry weight, turbidity, absorbance or any major cellular component, e.g. protein, nucleic acid, ATP, etc.

One of the simplest methods is to count the cells in a known volume of medium using a microscope and a counting chamber or haemocytometer (Box 16.1). While this gives a rapid assessment of the total cell number, it does not discriminate between living and dead cells. It is also time-consuming as a large number of cells must be counted for accurate measurement. It may be difficult to distinguish individual cells, e.g. for cells growing as clumps.

Electronic particle counters

These instruments can be used to give a direct (total) count of a suspension of microbial cells. The Coulter® counter detects particles due to change in electrical resistance when they pass through a small aperture in a glass tube. It gives a rapid count based on a larger number of cells than direct microscopy. It is well suited for repeat measurements or large sample numbers and can be linked to a microcomputer for data processing. If correctly calibrated, the counter can also measure cell sizes. A major limitation of electronic counters is the lack of discrimination between living cells, dead cells, cell clumps and inanimate particles (e.g. dust). In addition, the instrument must be set up and calibrated by trained personnel. Flow cytometry is a more specialized alternative, since particles can be sorted as well as counted.

Culture-based counting methods

A variety of culture-based techniques can be used to determine the number of microbes in a sample. A major assumption of such methods is that, under suitable conditions, an individual viable microbial cell will be able to multiply and grow to give a visible change in the growth medium, i.e. a colony on an agar-based medium, or turbidity ('cloudiness') in a liquid medium. You are most likely to gain practical experience using bacterial cultures, counted by one or more of the following methods:

- Spread or pour plate methods ('plate counts', p. 62). The most widespread approach is to transfer a suitable amount of the sample to an agar medium, incubate under appropriate conditions and then count the resulting colonies (Box 16.2).

Box 16.1 How to use a counting chamber or haemocytometer

A counting chamber is a specially designed slide containing a chamber of known depth with a grid etched onto its lower surface. When a flat coverslip is placed over the chamber, the depth is uniform. Use as follows:

1. **Place the special coverslip over the chamber.** Press the edges firmly, to ensure that the coverslip makes contact with the surface of the slide, but take care that you do not break the slide or coverslip by using too much force. When correctly positioned, you should be able to see interference rings (Newton's rings) at the edge of the coverslip.
2. **Add a small amount of your cell suspension to fill the central space above the grid.** Place on the microscope stage and allow the cells to settle (2–3 min).
3. **Examine the grid microscopically,** using the $\times 10$ objective lens first, since the counting chamber is far thicker than a standard microscope slide. Then switch to the $\times 40$ objective: take care not to scratch the surface of the objective lens, as the special coverslip is thicker than a normal coverslip. For a dense culture, the small squares are used, while the larger squares are used for dilute suspensions. You may need to dilute your suspension if it contains more than thirty cells per small square.
4. **Count the number of cells in several squares:** at least 600 cells should be counted for accurate measurements. Include those cells that cross the upper and left-hand boundaries, but not those that cross the lower or right-hand rulings. A hand tally may be used to aid counting. Motile cells must be immobilized prior to counting (e.g. by killing with glutaraldehyde).
5. **Divide the total number of cells (C) by the number of squares counted (S),** to give the mean cell count per square.
6. **Determine the volume (in ml) of liquid corresponding to a single square (V),** e.g. a Petroff–Hausser chamber has small squares of linear dimension 0.2 mm, giving an area of $0.04\,mm^2$; since the depth of the chamber is 0.02 mm, the volume is $0.04 \times 0.02 = 0.0008\,mm^3$; as there are $1000\,mm^3$ in 1 ml, the volume of a small square is 8×10^{-7} ml; similarly, the volume of a large square (equal to 25 small squares) is 2×10^{-5} ml. Note that other types of counting chamber will have different volumes: check the manufacturer's instructions.
7. **Calculate the cell number per ml by dividing the mean cell count per square by the volume of a single square (in ml).**
8. **Remember to take account of any dilution of your original suspension** in your final calculation by multiplying by the reciprocal of the dilution (M), e.g. if you counted a one in twenty dilution of your sample, multiply by twenty, or if you diluted to 10^{-5}, multiply by 10^5.

The complete equation for calculating the total microscopic count is:

$$\text{Total cell count (per ml)} = (C / S / V) \times M \qquad [16.6]$$

e.g. if the mean cell count for a hundred-fold dilution of a cell suspension, counted using a Petroff–Hausser chamber, was 12.4 cells in ten small squares, the total count would be

$$(12.4 / 10 / 8 \times 10^{-7}) \times 10^2 = 1.55 \times 10^8\,ml^{-1}.$$

A simpler, less accurate approach is to use a known volume of sample under a coverslip of known area on a standard glass slide, counting the number of cells per field of view using a calibrated microscope of known field diameter, then multiplying up to give the cell number per ml.

Definition

CFU – colony-forming unit: a cell or group of cells giving rise to a single colony on a solidified medium.

- Membrane filtration. For bacterial samples where the expected cell number is lower than $10\,CFU\,ml^{-1}$, pass the sample through a sterile filter (pore size $0.2\,\mu m$ or $0.45\,\mu m$). The filter is then incubated on a suitable medium until colonies are produced, giving a count by dividing the mean colony count per filter by the volume of sample filtered.
- Multiple tube count, or most probable number (MPN). A bacteriological technique where the sample is diluted and known volumes are transferred to several tubes of liquid medium (typically, five tubes at three volumes), chosen so that there is a low probability of the smallest volumes containing a viable cell. After incubation, the number of tubes showing growth (turbidity) is compared to tabulated values to give the most probable number (MPN per ml).

Box 16.2 How to make a plate count of bacteria using an agar-based medium

1. **Prepare serial decimal dilutions of the sample in a sterile diluent (p. 15).** The most widely used diluents are 0.1% w/v peptone water or 0.9% w/v NaCl, buffered at pH 7.3. Take care that you mix each dilution before making the next one. For soil, food, or other solid samples, make the initial decimal dilution by taking 1 g of sample and making this up to 10 ml using a suitable diluent. Gentle shaking or homogenization may be required for organisms growing in clumps. The number of decimal dilutions required for a particular sample will be governed by your expected count: dilute until the expected number of viable cells is around 100–1 000 ml^{-1}.
2. **Transfer an appropriate volume (e.g. 0.05–0.5 ml) of the lowest dilution to an agar plate** using either the spread plate method or the pour plate procedure (p. 62). At least two, and preferably more, replicate plates should be prepared for each sample. You may also wish to prepare plates for more than one dilution, if you are unsure of the expected number of viable cells.
3. **Incubate under suitable conditions for 18-72 h, then count the number of colonies on each replicate plate at the most appropriate dilution.** The most accurate results will be obtained for plates containing 30–300 colonies. Mark the base of the plate with a spirit-based pen each time you count a colony. Determine the mean colony count per plate at this dilution (*C*).
4. **Calculate the colony count per ml of that particular dilution** by dividing by the volume (in ml) of liquid transferred to each plate (*V*).
5. **Now calculate the count per ml of the original sample** by multiplying by the reciprocal of the dilution: this is the multiplication factor (*M*); e.g. for a dilution of 10^{-3}, the multiplication factor would be 10^3. For soil, food or other solid samples, the count should be expressed per g of sample.

The complete equation for calculating the viable count is:

$$\text{Count per ml (or per g)} = (C / V) \times M \quad [16.7]$$

e.g. for a sample with a mean colony count of 5.5 colonies per plate for a volume of 0.05 ml at a dilution of 10^{-7}, the count would be:

$$(5.5 / 0.05) \times 10^7 = 1.1 \times 10^9 \text{ CFU ml}^{-1}$$

Strictly speaking, the count should be reported as colony-forming units (CFU) per ml, rather than as cells per ml, since a colony may be the product of more than one cell, particularly in filamentous microbes or in organisms with a tendency to aggregate.

Counting injured or stressed microbes – a resuscitation stage may be required, to allow cells to grow under selective conditions, p. 66.

The principal advantage of culture-based counting procedures is that dead cells will not be counted. However, for such techniques, the incubation conditions and media used may not allow growth of all cells, underestimating the true viable count. Further problems are caused by cell clumping and dilution errors. In addition, such methods require sterile apparatus and media and the incubation period is lengthy before results are obtained. An alternative approach is to use direct microscopy, combined with 'vital' or 'mortal' staining. For example, the direct epifluorescence technique (DEFT) uses acridine orange and UV epifluorescence microscopy to separate living and dead bacteria, while neutral red is a vital stain used for plant cells. Chapter 8 gives examples of vital/mortal stains for other cell types.

Bioassays and their applications

A bioassay is a method of quantifying a chemical substance (analyte) by measuring its effect on a biological system under controlled conditions. The hypothetical underlying phenomena are summarized by the relationship:

$$\text{A} + \text{Rec} \rightleftharpoons \text{ARec} \rightarrow\rightarrow \text{R} \quad [16.8]$$

where A is the analyte, Rec the receptor, ARec the analyte–receptor complex and R the response. This relation is analogous to the formation of product from an enzyme–substrate complex and, using similar mathematical arguments to those of enzyme kinetics (p. 194), it can be shown that the

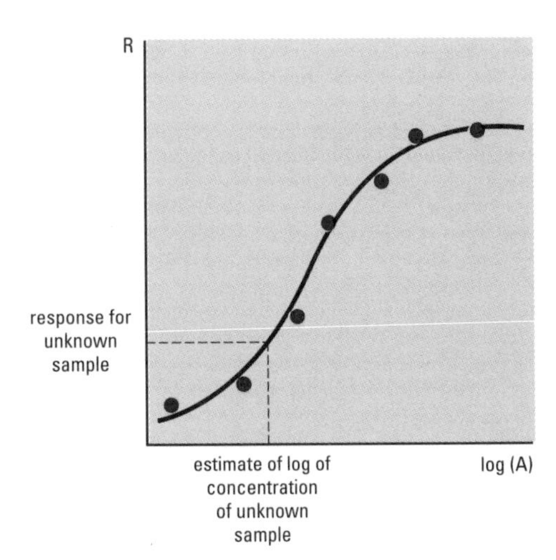

Fig. 16.3 Typical bioassay response curve, showing estimation of an unknown sample. Symbols as noted in text. Closed circles represent responses with standard samples.

Examples The following are typical bioassays:

- measuring the amount of antibiotic in a clinical sample using a sensitive bacterium;
- measuring the amount of a vitamin in a foodstuff using a microbe that requires that vitamin for growth;
- measuring the mutagenic properties of a chemical compound in the Ames test (see Box 16.3).

Legal use of bioassays – in the UK, where bioassays involving 'higher' animals are controlled by the Animals Scientific Procedures Act (1986), they can only be carried out under the direct supervision of a scientist licensed by the Home Office.

expected relationship between [A] and rate of response is hyperbolic (sigmoidal in a log–linear plot like Fig. 16.3). This pattern of response is usually observed in practice if a wide enough concentration range of analyte is tested.

To carry out the assay, the response elicited by the unknown sample is compared to the response obtained for differing concentrations of the substance, as shown in Fig. 16.3. When fitting a curve to standard points and estimating unknowns, the available methods, in order of increasing accuracy, are:

1. fitting by eye;
2. using linear regression on a restricted 'linear' portion of the assay curve;
3. linearization followed by regression (e.g. by probit transformation);
4. non-linear regression (e.g. to the Morgan–Mercer–Flodin equation).

In general, bioassay techniques have more potential faults than physico–chemical assay techniques. These may include the following:

- A greater level of variability: error in the estimate of the unknown compound will result because no two organisms will respond in exactly the same way. Assay curves vary through time, and because they are non-linear, a full standard curve is required each time the assay is carried out.
- Lack of chemical information: bioassays provide information about *biological* activity; they say little about the chemical structure of an unknown compound. The presence of a specific compound may need to be confirmed by a physico–chemical method (e.g. mass spectrometry).
- Possibility of interference: while many bioassays are very specific, it is possible that different chemicals in the extract may influence the results.

Despite these problems, bioassays are still much used. They are 'low-tech' and generally cheap to set up. They often allow detection at very low concentrations. Bioassays also provide the means to assess the biological activity of chemicals and to study changes in sensitivity to a chemical, which physico-chemical techniques cannot do (see Box 16.3). Changes in sensitivity may be evident in the shape of the dose–response curve and its position on the concentration axis.

Bioassays can involve responses of whole organisms or parts of organisms. 'Isolated' responding systems (e.g. excised tissues or cells) decrease the possibility of interference from other parts of the organism. Disadvantages include disruption of nutrition and wound damage during excision. Isolation can continue down to the molecular level, as in immunoassays (Chapter 18).

Bioassays are the basis for characterizing the effectiveness of drugs and the toxicity of chemicals. Here, response is often treated as a quantal (all-or-nothing) event. The E_{50} is defined as that concentration of a compound causing 50% of the organisms to respond. Where death is the observed response, the LD_{50} describes the concentration of a chemical that would cause 50% of the test organisms to die within a specified period under a specified set of conditions. Box 16.3 presents details of the Ames test, a widely used bioassay used to assess the mutagenicity of chemicals.

Considerations when setting up a bioassay

- The response should be easily measured and as metabolically 'close' to the initial binding event as possible.
- The experimental conditions should mimic the *in vivo* environment.
- The standards should be chemically identical to the compound being measured and spread over the expected concentration range being tested.
- The samples should be purified if interfering compounds are present and diluted so the response will be on the 'linear' portion of the assay curve.

Box 16.3 Mutagenicity testing using the Ames test – an example of a widely used bioassay

Chemical carcinogens can be identified by the formation of tumours in laboratory animals exposed to the compound under controlled conditions. However, such animal bioassays are time-consuming and expensive. Dr B.N. Ames and co-workers have shown that most carcinogens are also mutagens, i.e. they will induce mutational changes in DNA. The Ames test makes use of this correlation to provide a simple, rapid and inexpensive bioassay for the initial screening of potential carcinogens. The test makes use of particular strains of *Salmonella typhimurium* with the following characteristics:

- histidine auxotrophy – the tester strains are unable to grow on a minimal medium without added histidine: this characteristic is the result of specific mutational changes to the DNA of these strains, including base substitutions and frame shifts;
- increased cell envelope permeability, to permit access of the test compound to the cell interior;
- defects in excision repair systems and enhanced error-prone repair systems, to reduce the likelihood of DNA repair after treatment with a potential mutagen.

When grown in the presence of a chemical mutagen, the bacteria may revert to prototrophy (p. 222) as a result of back mutations that restore the wild-type phenotype: such revertants grow independently of external histidine and are able to form colonies on minimal medium, unlike the original test strains. The extent of reversion can be used to assess the mutagenic potential of a particular chemical compound. Since many chemicals must be activated *in vivo*, the test incorporates a rat liver homogenate (so-called 'S-9 activator', containing microsomal enzymes) to simulate the metabolic events within the liver. The tester strains are mixed with S-9 activator and a small amount of molten soft agar, then poured as a thin agar overlay (top agar) on a minimal medium plate. The top agar layer contains a very small amount of histidine, to allow the tester strains to divide a few times and express any mutational changes (i.e. prototrophy).

The test can be performed in one of two ways:

1. Spot test: a concentrated drop of the test compound is placed at the centre of the plate, either directly on the agar surface, or on a small filter paper disc. The test compound will diffuse into the agar and revertants appear as a ring of colonies around the site of inoculation, as shown in Fig. 16.4. The distance of the ring of colonies from this site provides a measure of the toxicity of the compound, while the number of colonies within the ring gives an indication of the relative mutagenicity of the test substance. The spot test is often carried out in a simplified form, without added S-9 activator, as a rapid preliminary test prior to quantitative analysis.

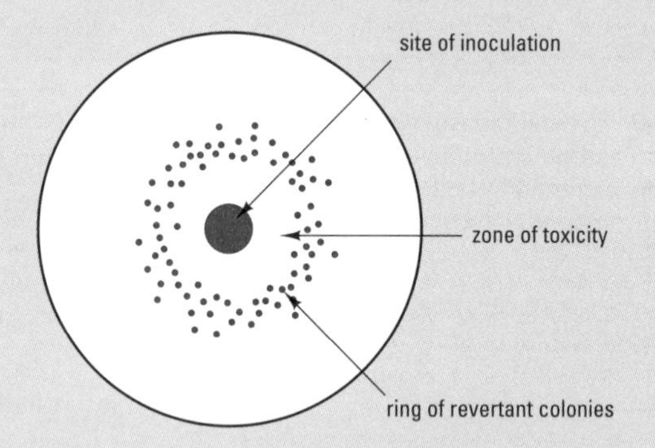

Fig. 16.4 Typical outcome of spot test (Ames test)

2. Agar incorporation test: known amounts of the test compound are mixed separately with molten top agar and the other constituents and poured onto separate minimal agar plates. After incubation for 48 h, revertants will appear as evenly dispersed colonies throughout the agar overlay. The number of colonies reflects the relative mutagenicity of the test compounds, with a direct relationship between colony count and the amount of mutagen. Agar incorporation tests can be used to generate dose–response curves similar to that shown in Fig. 16.3. The simplicity, sensitivity and reproducibility of the Ames test has resulted in its widespread use for screening potential carcinogens in many countries, though it is not an infallible test for carcinogenicity.

N.B. Safe handling procedures *must* be followed at all times, as the test substances may be carcinogenic – testers should wear gloves and avoid skin contact or ingestion.

To check for interference, the bioassay may be standardized against another method (preferably physico–chemical). Related compounds known to be present in the analyte solution should be shown to have minimal activity in the bioassay. If an interfering compound is present, this may show up if a known amount of standard is added to sample vials – the result will not be the sum of independently determined results for the standard and sample.

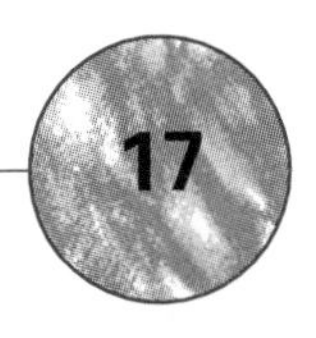

17 Homogenization and fractionation of cells and tissues

Definitions

Disruption – a process involving structural damage to cells and tissues, to an extent where the biomolecule or complex of interest is released.

Homogenization – a process where cells and tissues are broken into fragments small enough to create a uniform, stable emulsion (the homogenate).

Homogenizer – a general term for any equipment used to disrupt or homogenize cells and tissues.

Most biological molecules and sub-cellular complexes must be isolated from their source material in order to be studied in detail. Unless the biomolecule is already a component of an aqueous medium (e.g. plasma, tissue exudate), the first step will be to disrupt the structure of the appropriate cells or tissues. Following disruption and/or homogenization, the *in vitro* environment will be very different to that of the intact cell, and it is important that the integrity of the biomolecule or sub-cellular complex is preserved as far as possible during the isolation procedure.

KEY POINT **Disruption may be achieved by chemical, physical, or mechanical procedures – the rigour of the technique(s) required will depend on the intracellular location of the molecule and the nature of the source material.**

Types of cell and their susceptibility to disruption

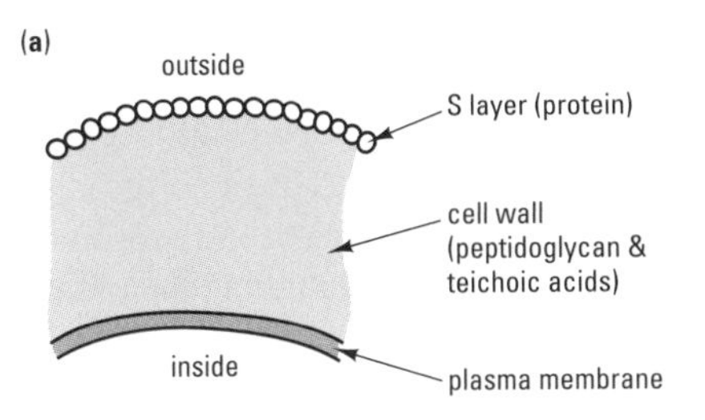

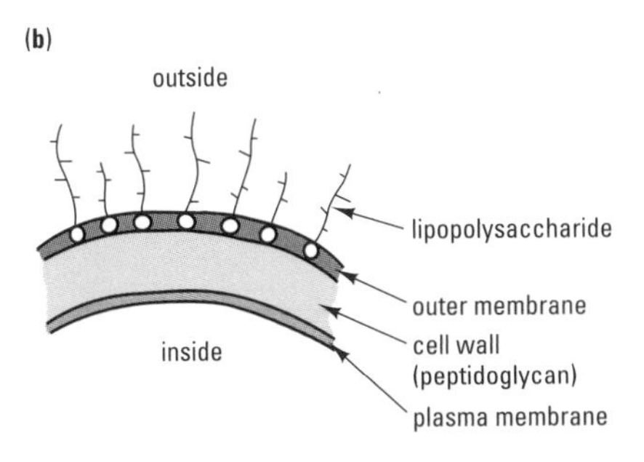

Fig. 17.1 Cell envelopes of (a) Gram-positive and (b) Gram-negative bacteria.

Cell walls are a major obstacle to disruption in many organisms and the technique used for a given application must take into account this aspect. With animal cells, there is no cell wall and the plasma membrane and cytoskeleton are relatively weak: unicellular suspensions of animal cells (e.g. blood cells) can be disrupted by gentle techniques. Animal cells within tissues are more difficult to disrupt, due to the presence of connective tissue. Muscle requires vigorous techniques, as a result of additional contractile proteins. For plant cells, the plasma membrane is surrounded by a cellulose cell wall, sometimes with additional components, e.g. lignin or waxes. As a result, large shear forces are often required to disrupt plant cells.

Most Gram-positive and Gram-negative bacteria are surrounded by a rigid, protective cell wall. Peptidoglycan is a major structural component and this may be degraded by the enzyme lysozyme, particularly in certain Gram-positive organisms. In contrast, the outer membrane of Gram-negative bacteria (Fig. 17.1b) protects against lysozyme: however, prior treatment of Gram-negative bacteria with EDTA destabilizes the outer membrane by removing Ca^{2+}, making the cell sensitive to lysozyme. Note also that some Gram-positive bacteria may be insensitive to lysozyme, due either to a modified peptidoglycan structure or to additional protective outer layers (proteinaceous S-layers: Fig. 17.1a).

The cell walls of filamentous fungi and yeasts are very robust, containing up to 90% polysaccharide (e.g. chitin, mannan, etc.), with embedded protein microfibrils – they are often difficult to disrupt, requiring enzymic digestion and/or mechanical homogenization.

Homogenizing media

The solution used for homogenization serves several purposes, since it acts as a solvent or suspension medium for the released components; it serves as a cooling medium (since many biological macromolecules are denatured by heat); and it contains various reagents that may help to preserve the biological integrity of components. A typical medium will contain:

- Buffer. This replaces the intracellular buffer systems and is needed to prevent pH changes that might denature proteins. TRIS and phosphate buffers are often used for 'physiological' pH values (7.0–8.0); alternatives include various zwitterionic buffers (p. 28).
- Inorganic salts. The intracellular ionic strength is quite high, so KCl and NaCl are often included to maintain the ionic strength of the homogenate. However, the total concentration of inorganic salts should be kept below $100\,mmol\,l^{-1}$, to avoid thickening of the homogenate due to solubilization of structural proteins.
- Sucrose. This is used to prevent osmotic lysis of organelles (e.g. mitochondria, lysosomes), and stabilizes proteins from hydrophobic intracellular environments by reducing the polarity of the aqueous medium.
- Mg^{2+}. This helps to preserve the integrity of membrane systems by counteracting the fixed negative charges of membrane phospholipids.
- EDTA. This chelates divalent cations and removes metal ions (e.g. Cu^{2+}, Pb^{2+}, Hg^{2+}) that inactivate proteins by binding to thiol groups. In addition, it removes Ca^{2+} which could activate certain proteases, nucleases and lipases in the homogenate.
- Protease inhibitors (e.g. phenylmethanesulphonylfluoride, PMSF; L-*trans*-epoxysuccinylleucylamido-(4-guanodino)-butane, E-64; leupeptin). These protect solubilized proteins from digestion by intracellular proteases, mainly released from lysosomes on disruption of the cell. Lysosomal proteases have acid pH optima, another reason for maintaining the pH of the medium close to neutrality.
- Reducing agents (e.g. 2-mercaptoethanol, dithiothreitol, cysteine at $\approx 1\,mmol\,l^{-1}$). These reagents prevent oxidation of certain proteins, particularly those with free thiol groups that may be oxidized to disulphide bonds when released from the cell under aerobic conditions.
- Detergents (e.g. Triton X-100, SDS). These cause dissociation of proteins and lipoproteins from the cell membrane, aiding the release of membrane-bound and intracellular components.

Using chelating agents – EDTA (ethylenediaminetetraacetic acid) and EGTA (ethylenebis(oxethylenenitrilo)-tetraacetic acid). If Mg^{2+} is an important component of the medium, use EGTA, which does not chelate Mg^{2+}.

Extracting proteins from plants – carry out all post-homogenization procedures as quickly as possible, because plant cell vacuoles may contain phenols that will inactivate proteins when released.

Methods of disrupting cells and tissues

Prior to disruption, animal tissues will need to be freed of any visible fat deposits and connective components, and then randomly sliced with fine scissors or a scalpel. Any fibrous and vascular tissue should be removed from plant material. Disruption can be achieved by mechanical and non-mechanical means: the principal applications of various methods are described in Table 17.1 for the major cell and tissue types.

Definitions

Isotonic – a medium with the same water potential (p. 24) as the cells or tissues.

Hypertonic – a medium with a more negative water potential, compared with the cells or medium.

Hypotonic – a medium with a less negative water potential, compared with the cells or medium.

Non-mechanical methods

- Osmotic shock: cells are first placed in a hypertonic solution of high osmolality (p. 24), e.g. 20% w/v sucrose, leading to loss of water. On dilution of this solution (e.g. by addition of water or transfer to a hypotonic solution), the cells will burst due to water influx. Osmotic shock treatment is effective for wall-less cells or protoplasts.

- Freezing and thawing: causing leakage of intracellular material, following cell wall and membrane damage and internal disruption due to ice crystal formation.
- Lytic enzymes: damaging the cell wall and/or plasma membrane. Cells can then be disrupted by osmotic shock or gentle mechanical treatment.

Avoiding protein denaturation during homogenization – excessive frothing of the homogenate indicates denaturation of proteins (think of whipping egg whites for meringue).

Mechanical methods

All mechanical procedures for cell disruption generate heat, and this may denature proteins. Therefore it is very important to cool the starting material, the homogenizing medium, and, if possible, the homogenizer itself (to $\approx 4\,^{\circ}C$). The homogenization should be carried out in short bursts, and the homogenate should be cooled in an ice bath between each burst. Cooling will also reduce the activity of any degradative enzymes in the homogenate. Ideally, carry out homogenization in a walk-in cooler ('cold room').

Equipment commonly used includes:

- Mixers and blenders. These are similar to domestic liquidizers, with a static vessel and rotating blades. The Waring blender is widely used: it has a stainless steel vessel that will stay cool if pre-chilled. The vessel and blades are designed to maximize turbulence, both disrupting and homogenizing tissues and cells.

Table 17.1 Summary of techniques for the disruption of tissues and cells – note that safety glasses should be worn for all procedures

Technique	Suitability	Comments
Non-mechanical methods		
Osmotic shock	Animal soft tissues Some plant cells	Small scale only
Freeze/thaw	Animal soft tissues Some bacteria	Time consuming; small scale; closed system – suitable for pathogens with appropriate safety measures; some enzymes are cold-labile
Lytic enzymes, e.g. lipases; proteases pectinase; cellulase	Animal cells Plant cells	Mild and selective; small scale; expensive; enzymes must be removed once lysis is complete
Lysozyme	Some bacteria	Gram-negative bacteria must be pre-treated with EDTA. Suitable for some organisms resistant to mechanical disruption.
Mechanical methods		
Pestle and mortar + abrasives	Tough tissues	Not suitable for delicate tissues
Ball mills + glass beads	Bacteria and fungi	May cause organelle damage in eukaryotes
Blenders and rotor-stators	Plant and animal tissues	Ineffective for microbes
Homogenizers (glass & Teflon)	Soft, delicate tissues e.g. white blood cells, liver	Glass may shatter – wear safety glasses during use
Solid extrusion (Hughes press)	Tough plant material; bacteria; yeasts	Small scale
Liquid extrusion (French pressure cell)	Microbial cells	Small scale
Ultrasonication	Microbial cells	Cooling required; small scale; may cause damage to organelles, especially in eukaryotic cells.

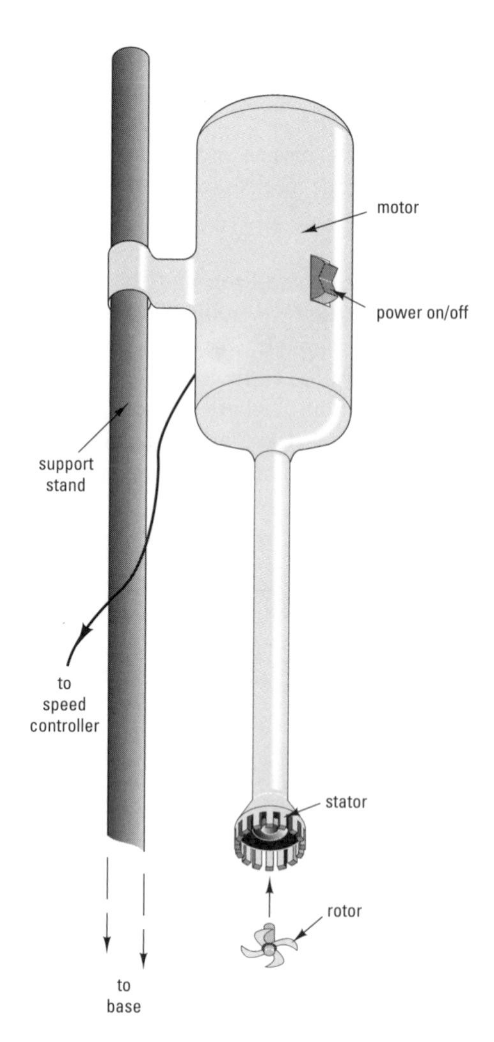

Fig. 17.2 Components of a rotor-stator homogenizer.

- Ball mills (e.g. Retch mixer mill, Mickle mill). These devices contain glass beads that vibrate and collide with each other and with tissues/cells, leading to disruption.
- Liquid extrusion devices (e.g. French pressure cell). Cells are forced from a vessel to the outside, through a very narrow orifice at high pressures ($\approx$100 MPa). The resulting pressure changes are a powerful means of disrupting cells.
- Solid extrusion (e.g. Hughes press). Here, a frozen cell paste is forced through a narrow orifice, where the shear forces and the abrasive properties of the ice crystals cause cell disruption.
- Rotor-stators (e.g. Ultra-turrax® homogenizer). These have a rotor (a set of stainless steel blades) and a stator (a slotted stainless steel cylinder) at the tip of a stainless steel shaft, immersed in the homogenizing medium: the arrangement is illustrated in Fig. 17.2. The high speed of the rotor blades causes material in the homogenizing fluid to be sucked into the dispersing head, where it is pressed radially through the slots in the stator. Along with the cutting action of the rotor blades, the material is subjected to very high shear and thrust and the resulting turbulence in the gap between rotor and stator gives effective mixing. The vigour of the homogenization process can be altered by varying the rotor speed setting. Various sizes of rotor-stator are available, with typical diameters in the range 8–65 mm: the smaller sizes are particularly useful for small-scale preparations.
- Sonicators. Ultrasonic waves are transmitted to an aqueous suspension of cells via a metal probe. The ultrasound creates bubbles within the liquid and these produce shock waves when they collapse. Successful disruption depends on the correct choice of power and incubation time, together with pH, temperature and ionic strength of the suspension medium, often obtained by trial and error. You can reduce the effects of heating during ultrasonication by using short 'bursts' of power (10–30 s), with rests of 30–60 s in between, and by keeping your cell suspension on ice during disruption. An ultrasonic water bath provides a more gentle means of disrupting certain types of cells, e.g. some bacterial and animal cells.
- Homogenizers. These involve the reciprocating movement of a ground glass or Teflon® pestle within a glass tube (Fig. 17.3). Cells are forced against the walls of the tube, releasing their contents. For glass pestles, the tubes also have ground glass homogenizing surfaces and may have an overflow chamber. The homogenizer can either be hand operated (e.g. Dounce), or motorized (e.g. Potter-Elvejham). The clearance between the pestle and the tube (range 0.05 mm to 0.5 mm) must be chosen to suit the particular application.

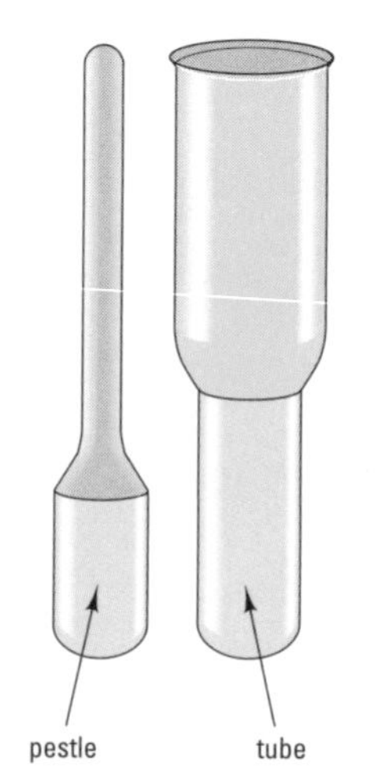

Fig. 17.3 Ground-glass homogenizer.

Cell fractionation and the isolation of organelles

The fractionation and separation of organelles from a cell homogenate by differential centrifugation is described in Chapter 23. Particular organelles can be obtained by appropriate choice of source tissue and homogenization method, as illustrated in Table 17.2 for the major types of organelle.

Table 17.2 Isolation and fractionation procedures for various organelles

Stage	Nuclei	Mitochondria	Microsomes	Chloroplasts*
Source	Thymus tissue, which has little cytoplasm, giving high yields.	Beef heart, with fat and connective tissue removed, then cubed & minced. Keep at pH 7.5 using TRIS buffer.	Rat liver, stored overnight to reduce glycogen content.	Spinach leaves, de-ribbed and cut into 1 cm strips.
Pre-treatment	Rinse with buffered physiological saline. Suspend in homogenizing medium.	Suspend in 2× volume of ice-cold homogenizing medium. Squeeze through muslin.	Chop finely with scissors and wash in 2× volume of homogenizing medium.	Rinse, then suspend in 3× volume of pre-chilled homogenizing medium.
Homogenizing medium	250 mmol l^{-1} sucrose; 10 mmol l^{-1} TRIS/HCl buffer, pH 7.6; 5 mmol l^{-1} $MgCl_2$; 0.2–0.5% v/v Triton X-100	250 mmol l^{-1} sucrose; 10 mmol l^{-1} TRIS/HCl buffer pH 7.7, containing 1 mmol l^{-1} succinic acid and 0.2 mmol l^{-1} EDTA.	250 mmol l^{-1} sucrose; 50 mmol l^{-1} TRIS/HCl buffer, pH 7.5; 25 mmol l^{-1} KCl; 5 mmol l^{-1} $MgCl_2$.	400 mmol l^{-1} sucrose; 25 mmol l^{-1} HEPES/NaOH buffer at pH 7.6; 2 mmol l^{-1} EDTA.
Homogenization	Waring blender, low speed, 3 min.	Bring to pH 7.8 using 2 mol l^{-1} TRIS base: Waring blender, high speed, 15 s: check and adjust pH to 7.8 and repeat blending step, 5 s.	Potter-Elvejham glass homogenizer with a Teflon pestle – 3 × 5 min at 800 rpm.	Pre-chilled Waring blender or rotor-stator.
Filtration/ Centrifugation	Filter through gauze. Spin at 2000 g for 10 min; discard supernatant. Repeat the homogenization, filtration and centrifugation stages to improve purity of organelles.	Spin at 1200 g for 20 min: filter supernatant through muslin: centrifuge at 26 000 g for 15 min. Remove and discard upper (lighter) layer of pellet. Resuspend lower layer and re-homogenize (2 × 5 s). Centrifuge at 26 000 g for 15 min.	Centrifuge at 680 g for 10 min – discard pellet; centrifuge at 10 000 g for 10 min – discard pellet; centrifuge at 100 000 g for 60 min – retain pellet. Resuspend in buffer, pH 8.0 and re-centrifuge at 100 000 g for 60 min.	Pass through several layers of muslin (wear gloves); centrifuge at 2500 g for 60 s; resuspend pellet in buffer, pH 7.6; re-centrifuge at 2500 g for 60 s. The colour of the supernatants gives a visual indication of chloroplast damage (e.g. if green).
Before use	Resuspend in homogenization medium without Triton X-100.	Resuspend pellet in buffer, pH 7.8 and either use immediately, or store at −20°C overnight.	Resuspend in buffer solution at pH 8.0.	Resuspend pellet in appropriate incubation medium, containing sucrose, e.g. for CO_2/O_2 studies, p. 207.

*An alternative approach is to use plant protoplasts (p. 78) as the starting material, releasing the chloroplasts by gentle lysis – diluting the medium with water.

Analytical techniques

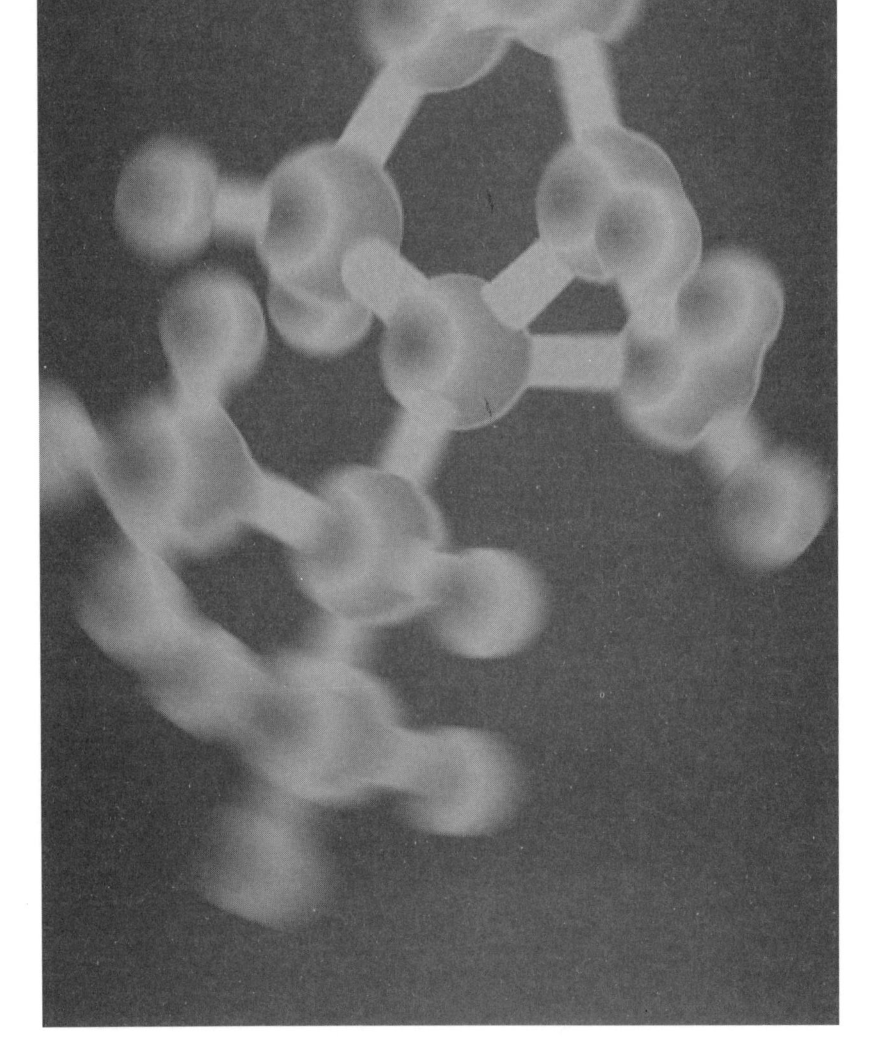

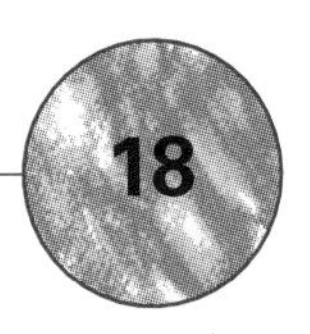

Immunological methods

Definitions

Antibody – a protein produced in response to an antigen (an antibody-generating foreign macromolecule).

Epitope – a site on the antigen that determines its interaction with a particular antibody.

Hapten – a substance which contains at least one epitope, but is too small to induce antibody formation unless it is linked to a macromolecule.

Ligand – a molecule or chemical group that binds to a particular site on another molecule.

Antibodies are an important component of the immune system, which protects animals against certain diseases (see Roitt, 1991). They are produced in response to foreign macromolecules (antigens). A particular antibody will bind to a site on a specific antigen, forming an antigen–antibody complex (immune complex). Immunological assays use the specificity of this interaction for:

- identifying macromolecules, cellular components or whole cells;
- quantifying a particular substance.

Antibody structure

An antibody is a complex globular protein, or immunoglobulin (Ig). While there are several types, IgG is the major soluble antibody in vertebrates and is used in most immunological assays. Its main features are:

- Shape: IgG is a Y-shaped molecule (Fig. 18.1), with two antigen-binding sites.
- Specificity: variation in amino acid composition at the antigen-binding sites explains the specificity of the antigen–antibody interaction.
- Flexibility: each IgG molecule can interact with epitopes which are different distances apart, including those on different antigen molecules.
- Labelling: regions other than the antigen-binding sites can be labelled, e.g. using a radioisotope or enzyme (p. 98).

KEY POINT **The presence of two antigen-binding sites on a single flexible antibody molecule is relevant to many immunological assays, especially the agglutination and precipitation reactions.**

Producing polyclonal antibodies – in the UK, this is controlled by Government regulations, since it involves vertebrate animals: personnel must be licensed by the Home Office and must operate in accordance with the Animals Scientific Procedures Act (1986).

Fig. 18.1 Diagrammatic representation of IgG (antibody).

Antibody production

Polyclonal antibodies

These are commonly used at undergraduate level. They are produced by repeated injection of antigen into a laboratory animal. After a suitable period (3–4 weeks) blood is removed and allowed to clot, leaving a liquid phase (polyclonal antiserum) containing many different IgG antibodies, resulting in:

- cross-reaction with other antigens or haptens;
- batch variation, as individual animals produce slightly different antibodies in response to the same antigen;
- non-specificity, as the antiserum will contain many other antibodies.

Standardization of polyclonal antisera therefore is difficult. You may need to assess the amount of cross-reaction, inter-batch variation or non-specific binding using appropriate controls, assayed at the same time as the test samples.

Monoclonal antibodies

These are specific to a single epitope and are produced from individual clones of cells, grown using cell culture techniques (p. 75). Such cultures provide a stable source of antibodies of known, uniform specificity. While monoclonal antibodies are likely to be used increasingly in future years, polyclonal antisera are currently employed for most routine immunological assays.

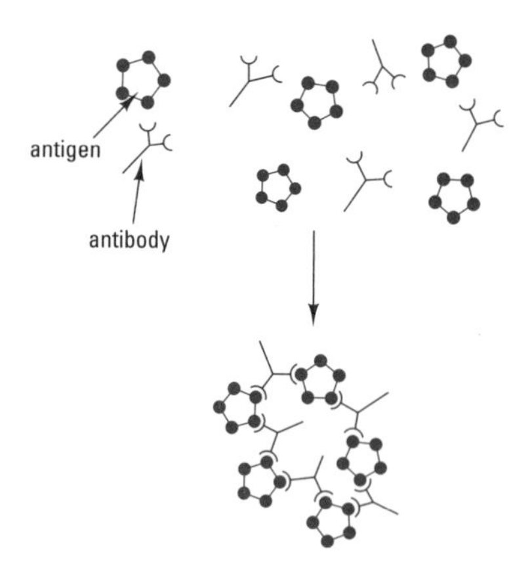

Fig. 18.2 Formation of an antigen–antibody complex.

Agglutination tests

When antibodies interact with a suspension of a particulate antigen, e.g. cells or latex particles, the formation of immune complexes (Fig. 18.2) causes visible clumping, termed agglutination. Agglutination tests are used in several ways:

- Microbial identification: at the species or subspecies level (serotyping), e.g. mixing an unknown bacterium with the appropriate antiserum will cause the cells to agglutinate.
- Latex agglutination (bound antigens): by coating soluble (non-particulate) antigens onto microscopic latex spheres, their reaction with a particular antibody can be visualized.
- Latex agglutination (bound antibodies): antibodies can be bound to latex microspheres, leaving their antigen-binding sites free to react with soluble antigen.
- Haemagglutination: red blood cells can be used as agglutinating particles. However, in some instances, such reactions do not involve antibody interactions (e.g. some animal viruses may haemagglutinate unmodified red blood cells).

Precipitin tests

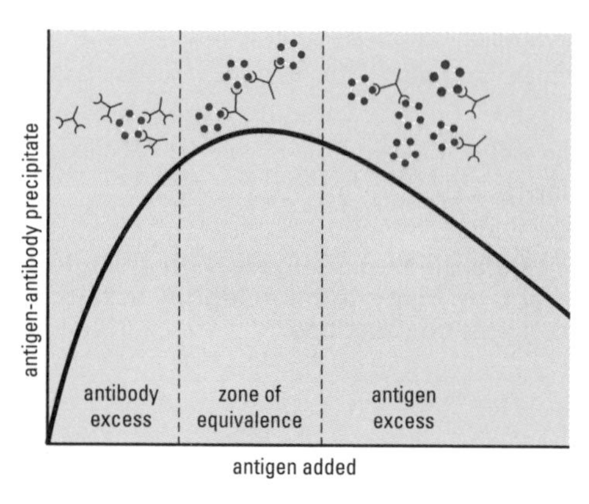

Fig. 18.3 Precipitation curve for an antigen titrated against a fixed amount of antibody.

Immune complexes of antibodies and soluble antigens (or haptens) usually settle out of solution as a visible precipitate: this is termed a precipitin test, or precipitation test. The formation of visible immune complexes in agglutination and precipitation reactions only occurs if antibody and antigen are present in an optimal ratio (Fig. 18.3). It is important to appreciate the shape of this curve: cross-linkage is maximal in the zone of equivalence, decreasing if either component is present in excess. The quantitative precipitin test can be used to measure the antibody content of a solution (Clausen (1989) gives details). Visual assessment of precipitation reactions forms the basis of several other techniques, described below.

Immunodiffusion assays

These techniques are easier to perform and interpret than the quantitative precipitin test. Precipitation of antibody and antigen occurs within an agarose gel, giving a visible line corresponding to the zone of equivalence (Fig. 18.3). The most widespread techniques are detailed below.

Single radial immunodiffusion (RID) (Mancini technique)

This is used to quantify the amount of antigen in a test solution, as follows:

1. Prepare an agarose gel (1.5% w/v), containing a fixed amount of antibody: allow to set on a glass slide or plate, on a level surface.
2. Cut several circular wells in the gel. These should be of a fixed size between 2 and 4 mm in diameter (see Fig. 18.4a).
3. Add a known amount of the antigen or test solution to each well.
4. Incubate on a level surface at room temperature in a moist chamber: diffusion of antigen into the gel produces a precipitin ring. This is usually measured after 2–7 days, depending on the molecular mass of the antigen.
5. Examine the plates against a black background (with side illumination), or stain using a protein dye (e.g. Coomassie blue).
6. Measure the diameter of the precipitin ring, e.g. using Vernier calipers.

Preparing wells for RID – cut your wells carefully. They should have straight sides and the agarose must not be torn or lifted from the glass plate. All wells should be filled to the top, with a flat meniscus, to ensure identical diffusion characteristics. Non-circular precipitin rings, resulting from poor technique, should not be included in your analysis.

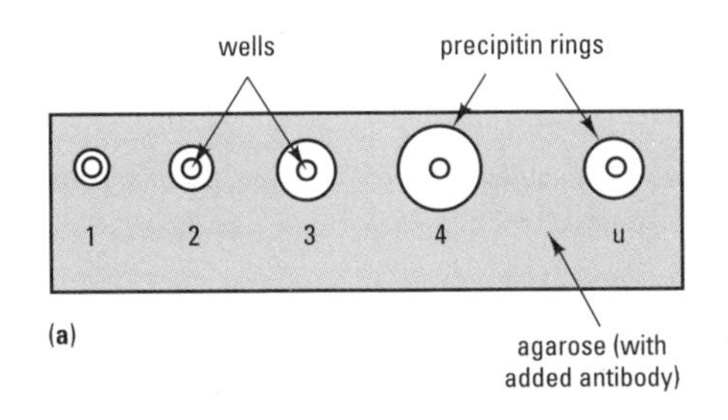

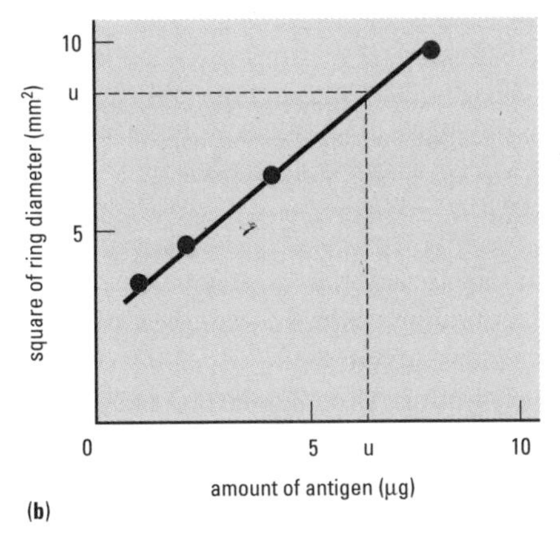

Fig. 18.4 Single radial immunodiffusion (RID). (a) Assay: four standards are shown (wells 1 to 4, each one double the strength of the previous standard), and an unknown (u), run at the same time, (b) Calibration curve. The unknown contains 6.25 μg of antigen. Note the non-zero intercept of the calibration curve, corresponding to the square of the well diameter: do not force such calibration lines through the origin.

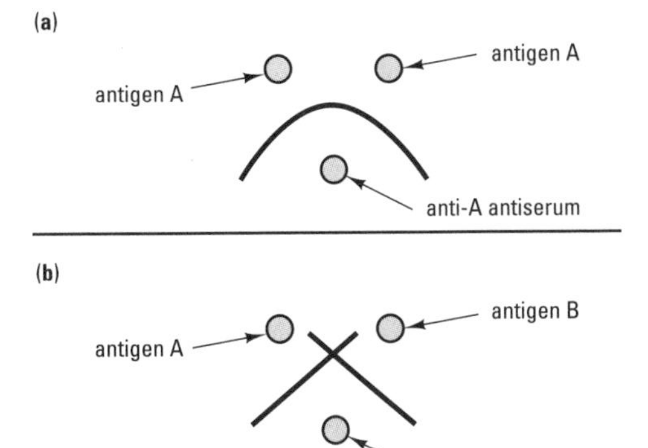

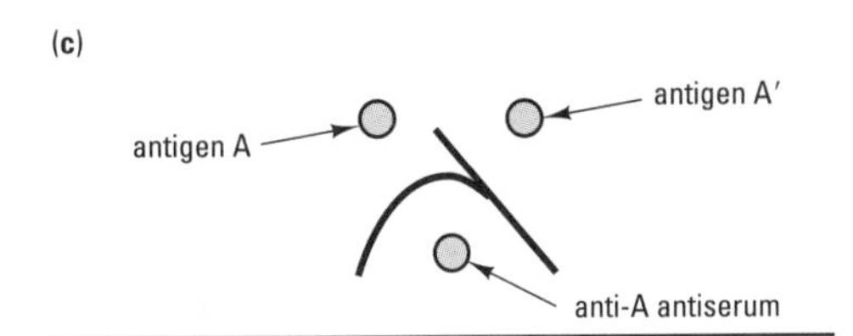

Fig. 18.5 Precipitin reactions in double diffusion immunoassay: (a) identity; (b) non-identity; (c) partial identity.

7. Prepare a calibration curve from the samples containing known amounts of antigen (Fig. 18.4b): the squared diameter of the precipitin ring is directly proportional to the amount of antigen in the well.
8. Use the calibration curve to quantify the amount of antigen in your test solutions, assayed at the same time.

Double diffusion immunoassay (Ouchterlony technique)

This technique is widely used to detect particular antigens in a test solution, or to look for cross-reaction between different antigens.

1. Prepare an agarose gel (1.5% w/v) on a level glass slide or plate: allow to set.
2. Cut several circular wells in the gel.
3. Add test solutions of antigen or polyclonal antiserum to adjacent wells. Both solutions diffuse outwards, forming visible precipitin lines where antigen and corresponding antibody are present in optimal ratio (Fig. 18.5).

The various reactions between antigen and antiserum are:

- Identity: two wells containing the same antigen, or antigens with identical epitopes, will give a fused precipitin line (identical interaction between the antiserum and the test antigens, Fig. 18.5a).
- Non-identity: where the antiserum contains antibodies to two different antigens, each with its own distinct epitopes, giving two precipitin lines which intersect without any interaction (no cross-reaction, Fig. 18.5b).
- Partial identity: where two antigens have at least one epitope in common, but where other epitopes are present, giving a fused precipitin line with a spur (cross-reaction, Fig. 18.5c).

Immunoelectrophoretic assays

These methods combine the precipitin reaction with electrophoretic migration, providing sensitive, rapid assays with increased separation and resolution. The principal techniques are:

Cross-over electrophoresis (counter-current electrophoresis)

Similar to the Ouchterlony technique, since antigen and antibody are in separate wells. However, the movement of antigen and antibody towards each other is driven by a voltage gradient (p. 142): most antigens migrate towards the anode, while IgG migrates towards the cathode. This method is faster and more sensitive than double immunodiffusion, taking 15–20 min to reach completion.

Quantitative immunoelectrophoresis (Laurell rocket immunoelectrophoresis)

Similar to RID, as the antibody is incorporated into an agarose plate while the antigen is placed in a well. However, a voltage gradient moves the antigen into the gel, usually towards the anode, while the antibody moves towards the cathode, giving a sharply peaked, rocket-shaped precipitin line, once equivalence is reached (within 2–10 h). The height of each rocket shape at equivalence is directly proportional to the amount of antigen added to each well. A calibration curve for samples containing known amount of antigen can be used to quantify the amount of antigen present in test samples (Fig. 18.6).

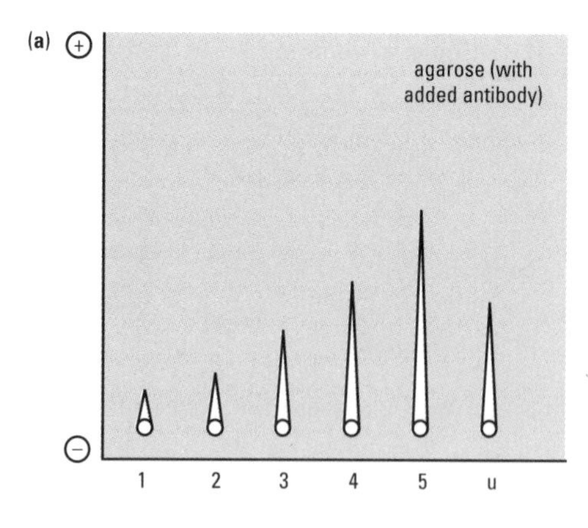

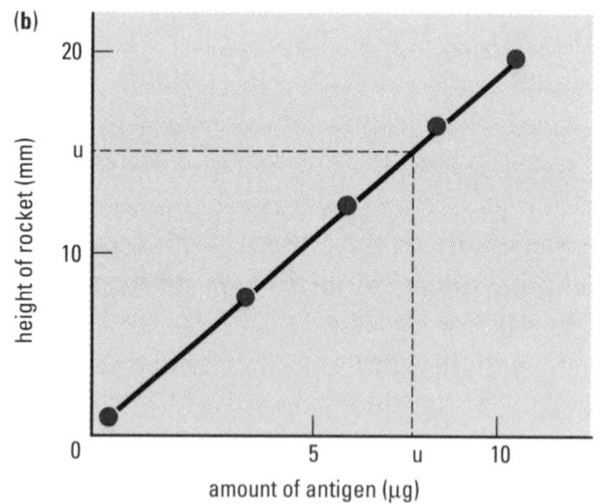

Fig. 18.6 Laurell rocket immunoelectrophoresis. (a) Assay: precipitin rockets are formed by electrophoresis of five standards of increasing concentration (wells 1 to 5) and an unknown (u). (b) Calibration curve: the unknown sample contains 7.4 μg of antigen.

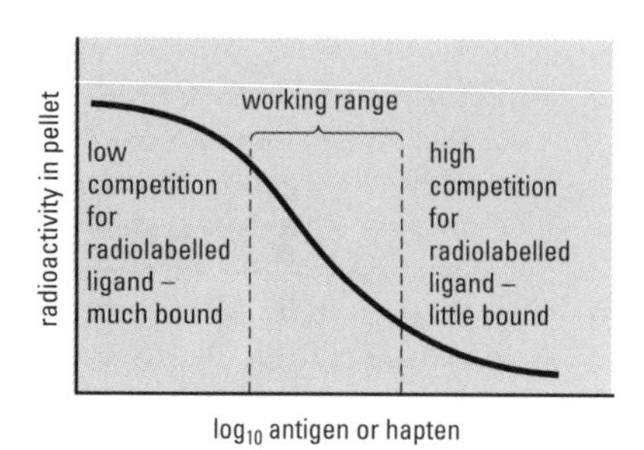

Fig. 18.7 Radioimmunoassay calibration curve. Note that the assay is insensitive at very low and very high antigen levels.

Radioimmunological methods

These methods use radioisotopes to detect and quantify the antigen–antibody interaction, giving improved sensitivity over agglutination and precipitation methods. The principal techniques are described here.

Radioimmunoassay (RIA)

This is based on competition between a radioactively labelled antigen (or hapten) and an unlabelled antigen for the binding sites on a limited amount of antibody. The quantity of antigen in a test solution can be determined using a known amount of radiolabelled antigen and a fixed amount of antibody (Fig. 18.7). As with other immunoassay methods, it is important to perform appropriate controls, to screen for potentially interfering compounds. The basic procedure for RIA is:

1. Add appropriate volumes of a sample to a series of small test tubes. Prepare a further set of tubes containing known quantities of the substance to be assayed to provide a standard curve.
2. Add a known amount of radiolabelled antigen (or hapten) to each tube (sample and standard).
3. Add a fixed amount of antibody to each tube (the antibody must be present in limited quantity).
4. Leave at constant temperature for a fixed time (usually 24 h), to allow antigen–antibody complexes to form.
5. Precipitate the antibody and bound antigen using saturated ammonium sulphate, followed by centrifugation.
6. Determine the radioactivity of the supernatant or the precipitate (p. 102).
7. Prepare a calibration curve of radioactivity against $\log_{10}$ antigen (Fig. 18.7). The curve is most accurate in the central region, so adjust the amount of antigen in your test sample to fall within this range.

Note the following:

- You must be registered to work with radioactivity: check with the Departmental Safety Officer, if necessary (p. 106).
- Measure all volumes as accurately as possible as the end result depends on the volumetric quantities of unlabelled (sample) antigen, radiolabelled antigen and antibody: an error in any of these reagents will invalidate the assay.
- Incorporate replicates, so that errors can be quantified.
- Seek your supervisor's advice about fitting a curve to your data: this can be a complex process.

Immunoradiometric assay (IRMA)

This technique uses radiolabelled antibody, rather than antigen, for direct measurement of the amount of antigen (or hapten) in a sample. Most immunoradiometric assays are similar to the double antibody sandwich method described below, except that the second antibody is labelled using a radioisotope. The important advantages over RIA are:

- linear relationship between amount of radioactivity and test antigen;
- wider working range for test substance;
- improved stability/longer shelf-life.

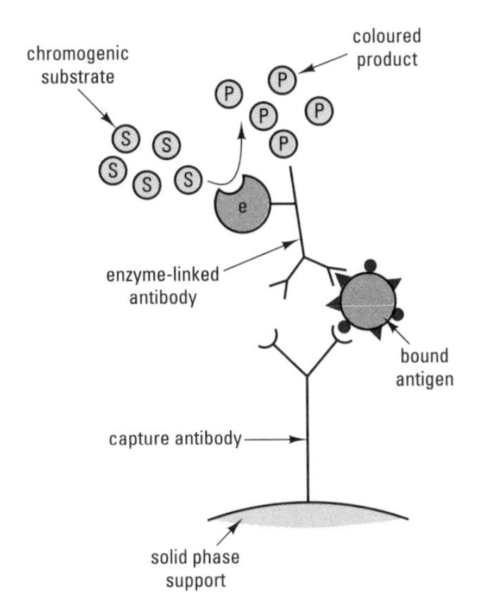

Fig. 18.8 Double antibody sandwich ELISA.

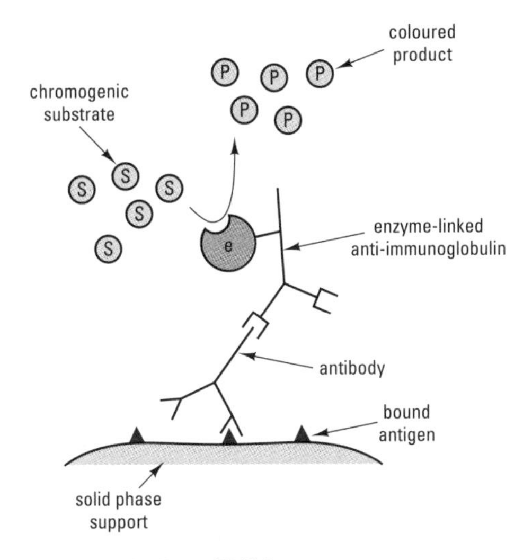

Fig. 18.9 Indirect ELISA.

Enzyme immunoassays (EIA)

These techniques are also known as enzyme-linked immunosorbent assays (ELISA). They combine the specificity of the antibody–antigen interaction with the sensitivity of enzyme assays using either an antibody or an antigen conjugated (linked) to an enzyme at a site which does not affect the activity of either component. The enzyme is measured by adding an appropriate substrate which yields a coloured product. Enzymes offer the following advantages over radioisotopic labels:

- Increased sensitivity: a single enzyme molecule can produce many product molecules, amplifying the signal.
- Simplified assay: enzyme assays are usually easier than radioisotope assays (p. 192).
- Improved stability of reagents: components are generally more stable than their radiolabelled counterparts, giving them a longer shelf-life.
- No radiological hazard: no requirement for specialized containment/disposal facilities.
- Automation is straightforward: using disposable microtitre plates and an optical scanner.

The principal techniques are:

Double antibody sandwich ELISA

This is used to detect specific antigens, involving a three-component complex between a capture antibody linked to a solid support, the antigen, and a second, enzyme-linked antibody (Fig. 18.8). This can be used to detect a particular antigen, e.g. a virus in a clinical sample, or to quantify the amount of antigen.

Indirect ELISA

This is used for antibody detection, with a specific antigen attached to a solid support. When the appropriate antibody is added, it binds to the antigen and will not be washed away during rinsing. Bound antibody is then detected using an enzyme-linked anti-immunoglobulin, e.g. a rabbit IgG antibody raised against human IgG (Fig. 18.9). One advantage of the indirect assay is that a single enzyme-linked anti-immunoglobulin can be used to detect several different antibodies, since the specificity is provided by the bound antigen.

19 Radioactive isotopes and their uses

Examples $^{12}_{6}C$, $^{13}_{6}C$ and $^{14}_{6}C$ are three of the isotopes of carbon. About 98.9% of naturally occurring carbon is in the stable $^{12}_{6}C$ form. $^{13}_{6}C$ is also a stable isotope but it only occurs at 1.1% natural abundance. Trace amounts of radioactive $^{14}_{6}C$ are found naturally; this is a negatron-emitting radioisotope (see Table 19.2).

The isotopes of a particular element have the same number of protons in the nucleus but a different number of neutrons, giving them the same proton number (atomic number) but a different nucleon number (mass number, i.e. number of protons + number of neutrons). Isotopes may be stable or radioactive. Radioactive isotopes (radioisotopes) disintegrate spontaneously at random to yield radiation and a decay product.

Radioactive decay

There are three forms of radioactivity (Table 19.1) arising from three main types of nuclear decay:

Table 19.1 Types of radioactivity and their properties

Radiation	Range of maximum energies (MeV*)	Penetration range in air (m)	Suitable shielding material
Alpha (α)	4–8	0.025–0.080	Unnecessary
Beta (β)	0.01–3	0.150–16	Plastic (e.g. Perspex®)
Gamma (γ)	0.03–3	1.3–13†	Lead

*Note that 1 MeV = 1.6×10^{-13} J
†Distance at which radiation intensity is reduced to half

Examples

^{226}Ra decays to ^{222}Rn by loss of an alpha particle, as follows:

$$^{226}_{88}Ra \rightarrow ^{222}_{86}Rn + ^{4}_{2}He^{2+}$$

^{14}C shows beta decay, as follows:

$$^{14}_{6}C \rightarrow ^{14}_{7}N + \beta^{-}$$

^{22}Na decays by positron emission, as follows:

$$^{22}_{11}Na \rightarrow ^{22}_{10}Ne + \beta^{+}$$

^{55}Fe decays by electron capture and the production of an X-ray, as follows:

$$^{55}_{26}Fe \rightarrow ^{55}_{25}Mn + X$$

The decay of ^{22}Na by positron emission (β^{+}) leads to the production of a γ ray when the positron is annihilated on collision with an electron.

- Alpha decay involves the loss of a particle equivalent to a helium nucleus. Alpha (α) particles, being large and positively charged, do not penetrate far in living tissue, but they do cause ionization damage and this makes them generally unsuitable for tracer studies.
- Beta decay involves the loss or gain of an electron or its positive counterpart, the positron. There are three sub-types:
  (a) Negatron (β^{-}) emission: loss of an electron from the nucleus when a neutron transforms into a proton. This is the most important form of decay for radioactive tracers used in biology. Negatron-emitting isotopes of biological importance include ^{3}H, ^{14}C, ^{32}P and ^{35}S.
  (b) Positron (β^{+}) emission: loss of a positron when a proton transforms into a neutron. This only occurs when sufficient energy is available from the transition and may involve the production of gamma rays when the positron is later annihilated by collision with an electron.
  (c) Electron capture (EC): when a proton 'captures' an electron and transforms into a neutron. This may involve the production of X-rays as electrons 'shuffle' about in the atom (as with ^{125}I) and it frequently involves electron emission.
- Internal transition involves the emission of electromagnetic radiation in the form of gamma (γ) rays from a nucleus in a metastable state and always follows initial alpha or beta decay. Emission of gamma radiation leads to no further change in atomic number or mass.

Note from the above that more than one type of radiation may be emitted when a radioisotope decays. The main radioisotopes used in biology and their properties are listed in Table 19.2.

Table 19.2 Properties of some isotopes used commonly in biology. Physical data obtained from Lide and Frederikse (1996)

Isotope	Emission(s)	Maximum energy (MeV)	Half-life	Main uses	Advantages	Disadvantages
^{3}H	β^-	0.018 61	12.3 years	Suitable for labelling organic molecules in wide range of positions at high specific activity	Relatively safe	Low efficiency of detection, high isotope effect, high rate of exchange with environment
^{14}C	β^-	0.156 48	5715 years	Suitable for labelling organic molecules in a wide range of positions	Relatively safe, low rate of exchange with environment	Low specific activity
^{22}Na	β^+ (90%), EC	2.842 (β^+)	2.60 years	Transport studies	High specific activity	Short half-life, hazardous
^{32}P	β^-	1.710	14.3 days	Labelling proteins and nucleotides (e.g. DNA)	High specific activity, ease of detection	Short half-life, hazardous
^{35}S	β^-	0.167	87.2 days	Labelling proteins and nucleotides	Low isotope effect	Low specific activity
^{36}Cl	β^-, β^+, EC	0.709 (β^-) 1.142 (β^+, EC)	300 000 years	Transport studies	Low isotope effect	Low specific activity, hazardous
^{125}I	EC + γ	0.178 (EC)	59.9 days	Labelling proteins and nucleotides	High specific activity	Hazardous
^{131}I	$\beta^- + \gamma$	0.971 (β^-)	8.04 days	Labelling proteins and nucleotides	High specific activity	Hazardous

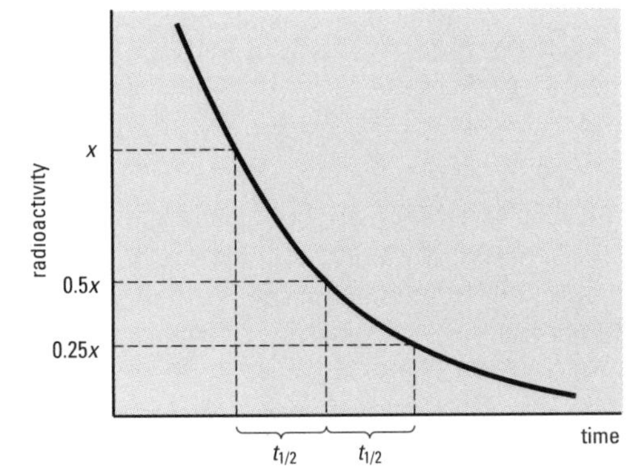

Fig. 19.1 Decay of a radioactive isotope with time. The time taken for the radioactivity to decline from x to $0.5x$ is the same as the time taken for the radioactivity to decline from $0.5x$ to $0.25x$, and so on. This time is the half-life ($t_{1/2}$) of the isotope.

Each radioactive particle or ray carries energy, usually measured in electron volts (eV). The particles or rays emitted by a particular radioisotope exhibit a range of energies, termed an energy spectrum, characterized by the maximum energy of the radiation produced, E_{max} (Table 19.2). The energy spectrum of a particular radioisotope is relevant to the following:

- Safety: isotopes with the highest maximum energies will have the greatest penetrating power, requiring appropriate shielding (Table 19.1).
- Detection: different instruments vary in their ability to detect isotopes with different energies.
- Discrimination: some instruments can distinguish between isotopes, based on the energy spectrum of the radiation produced (p. 105).

The decay of an individual atom (a 'disintegration') occurs at random, but that of a population of atoms occurs in a predictable manner. The radioactivity decays exponentially, having a characteristic half-life ($t_{1/2}$). This is the time taken for the radioactivity to fall from a given value to half that value (Fig. 19.1). The $t_{1/2}$ values of different radioisotopes range from fractions of a second to more than 10^{19} years (see also Table 19.2). If the $t_{1/2}$ is very short, as with ^{15}O ($t_{1/2} \approx 2$ min), then it is generally impractical to use the isotope in experiments because you would need to account for the decay during the experiment and counting period.

To calculate the fraction (f) of the original radioactivity left after a particular time (t), use the following relationship:

$$f = e^{x}, \text{ where } x = -0.693t/t_{1/2} \quad [19.1]$$

Note that the same units must be used for t and $t_{1/2}$ in the above equation.

Measuring radioactivity

Table 19.3 Relationships between units of radioactivity

1 Bq = 1 d.p.s.
1 Bq = 60 d.p.m.
1 Bq = 27 pCi
1 d.p.s. = 1 Bq
1 d.p.m. = 0.0167 Bq
1 Ci = 37 GBq
1 mCi = 37 MBq
1 μCi = 37 kBq
1 Sv = 100 rem
1 Gy = 100 rad
1 Gy ≈ 100 roentgen
1 rem = 0.01 Sv
1 rad = 0.01 Gy
1 roentgen ≈ 0.01 Gy

The SI unit of radioactivity is the becquerel (Bq), equivalent to one disintegration per second (d.p.s.), but disintegrations per minute (d.p.m.) are also used. The curie (Ci) is a non-SI unit equivalent to the number of disintegrations produced by 1 g of radium (37 GBq). Table 19.3 shows the relationships between these units. In practice, most instruments are not able to detect all of the disintegrations from a particular sample, i.e. their efficiency is less than 100% and the rate of decay may be presented as counts min^{-1} (c.p.m.) or counts s^{-1} (c.p.s.). Most modern instruments correct for background radiation and inefficiencies in counting, converting count data to d.p.m. Alternatively, the results may be presented as the measured count rate, although this is only valid where the efficiency of counting does not vary greatly among samples.

KEY POINT The specific activity is a measure of the quantity of radioactivity present in a known amount of the substance:

$$\text{specific activity} = \frac{\text{radioactivity (Bq, Ci, d.p.m., etc.)}}{\text{amount of substance (mol, g, etc.)}} \qquad [19.2]$$

This is an important concept in practical work involving radioisotopes, since it allows inter-conversion of disintegrations (activity) and amount of substance (see Box 19.1).

Two SI units refer to doses of radioactivity and these are used when calculating exposure levels for a particular source. The sievert (Sv) is the amount of radioactivity giving a dose in man equivalent to 1 gray (Gy) of X-rays: 1 Gy = an energy absorption of 1 J kg^{-1}. The dose received in most biological experiments is a negligible fraction of the maximum permitted exposure limit. Conversion factors from older units are given in Table 19.3.

The most important methods of measuring radioactivity for biological purposes are described below.

The Geiger–Müller (G–M) tube

This operates by detecting radiation when it ionizes gas between a pair of electrodes across which a voltage has been applied. You should use a hand-held Geiger–Müller tube for routine checking for contamination (although it will not pick up ^{3}H activity).

The scintillation counter

This operates by detecting the scintillations (fluorescence 'flashes') produced when radiation interacts with certain chemicals called fluors. In solid (or external) scintillation counters (often referred to as 'gamma counters') the radioactivity causes scintillations in a crystal of fluorescent material held close to the sample. This method is only suitable for radioisotopes producing penetrating radiation.

Liquid scintillation counters are mainly used for detecting beta decay and they are especially useful in biology. The sample is dissolved in a suitable solvent containing the fluor(s) – the 'scintillation cocktail'. The radiation first interacts with the solvent, and the energy from this interaction is passed to the fluors which produce detectable light. The scintillations are measured by photomultiplier tubes which turn the light pulses into electronic pulses, the magnitude of which is directly related to the energy of the original radioactive event. The spectrum of electronic pulses is thus related to the energy spectrum of the radioisotope.

Box 19.1 How to determine the specific activity of an experimental solution

Suppose you need to make up a certain volume of an experimental solution, to contain a particular amount of radioactivity. For example, 50 ml of a mannitol solution at a concentration of 25 mmol l^{-1}, to contain 5 Bq μl^{-1} – using a manufacturer's stock solution of ^{14}C-labelled mannitol (specific activity = 0.1 Ci/mmol^{-1}).

1. **Calculate the total amount of radioactivity in the experimental solution**, in this example $5 \times 1\,000$ (to convert μl to ml) $\times$ 50 (50 ml required) $= 2.5 \times 10^5$ Bq (i.e. 250 kBq).
2. **Establish the volume of stock radioisotope solution required**: for example, a manufacturer's stock solution of ^{14}C-labelled mannitol contains 50 μCi of radioisotope in 1 ml of 90% v/v ethanol: water. Using Table 19.3, this is equivalent to an activity of $50 \times 37 = 1\,850$ kBq. So, the volume of solution required is 250/1850 of the stock volume, i.e. 0.135 1 ml (135 μl).
3. **Calculate the amount of non-radioactive substance required** as for any calculation involving concentration (see pp. 13, 21), e.g. 50 ml (0.05 l) of a 25 mmol l^{-1} (0.025 mol l^{-1}) mannitol (relative molecular mass 182.17) will contain $0.05 \times 0.025 \times 182.17 = 0.227\,7$ g.
4. **Check the amount of radioactive isotope to be added.** In most cases, this represents a negligible amount of substance, e.g. in this instance, 250 kBq of stock solution at a specific activity of 14.8×10^6 kBq/mmol^{-1} (converted from 0.4 Ci mmol^{-1} using Table 19.3) is equal to 250/ 14 800 000 = 16.89 nmol, which is equivalent to approximately 3 μg mannitol. This can be ignored in calculating the mannitol concentration of the experimental solution.
5. **Make up the experimental solution** by adding the appropriate amount of non-radioactive substance and the correct volume of stock solution.
6. **Measure the radioactivity in a known volume of the experimental solution.** If you are using an instrument with automatic correction to Bq, your sample should contain the predicted amount of radioactivity, e.g. an accurately dispensed volume of 100 μl of the mannitol solution should give a corrected count of $100 \times 5 = 500$ Bq (or $500 \times 60 = 30\,000$ d.p.m.).
7. **Note the specific activity of the experimental solution**: in this case, 100 μl (1×10^{-4} l) of the mannitol solution at a concentration of 0.025 mol l^{-1} will contain 25×10^{-7} mol (2.5 μmol) mannitol. Dividing the radioactivity in this volume (30 000 d.p.m.) by the amount of substance (eqn [19.2]) gives a specific activity of $30\,000/2.5 = 12\,000$ d.p.m. μmol^{-1}, or 12 d.p.m. nmol^{-1}. This value can be used:
   - (a) To assess the accuracy of your protocol for preparing the experimental solution: if the measured activity is substantially different from the predicted value, you may have made an error in making up the solution.
   - (b) To determine the counting efficiency of an instrument; by comparing the measured count rate with the value predicted by your calculations.
   - (c) To interconvert activity and amount of substance: the most important practical application of specific activity is the conversion of experimental data from counts (activity) into amounts of substance. This is only possible where the substance has not been metabolized or otherwise converted into another form; e.g., a tissue sample incubated in the experimental solution described above with a measured activity of 245 d.p.m. can be converted to nmol mannitol by dividing by the specific activity, expressed in the correct form. Thus $245/12 = 20.417$ nmol mannitol.

Correcting for quenching – find out how your instrument corrects for quenching and check the quench indication parameter (QIP) on the printout, which measures the extent of quenching of each sample. Large differences in the QIP would indicate that quenching is variable among samples and might give you cause for concern.

Modern liquid scintillation counters use a series of electronic 'windows' to split the pulse spectrum into two or three components. This may allow more than one isotope to be detected in a single sample, provided their energy spectra are sufficiently different (Fig. 19.2). A complication of this approach is that the energy spectrum can be altered by pigments and chemicals in the sample, which absorb scintillations or interfere with the transfer of energy to the fluor; this is known as quenching (Fig. 19.2). Most instruments have computer-operated quench correction facilities (based on measurements of standards of known activity and energy spectrum) which correct for such changes in counting efficiency.

Box 19.2 Tips for preparing samples for liquid scintillation counting

Modern scintillation counters are very simple to operate; problems are more likely to be due to inadequate sample preparation than to incorrect operation of the machine. Common pitfalls are the following:

- Incomplete dispersal of the radioactive compound in the scintillation cocktail. This may lead to underestimation of the true amount of radioactivity present:
  - **(a)** Water-based samples may not mix with the scintillation cocktail – change to an emulsifier-based cocktail. Take care to observe the recommended limits, upper and lower, for amounts of water to be added or the cocktail may not emulsify properly.
  - **(b)** Solid specimens may absorb disintegrations or scintillations: extract radiochemicals using an intermediate solvent like ethanol (ideally within the scintillation vial) and then add the cocktail. Tissue solubilizing compounds such as Soluene® are effective, particularly for animal material, but extremely toxic, so the manufacturer's instructions must be followed closely. Radioactive compounds on slices of agarose or polyacrylamide gels may be extracted using a product such as Protosol®. Agarose gels can be dissolved in a small volume of boiling water.
  - **(c)** Particulate samples may sediment to the bottom of the scintillation vial – suspend them by forming a gel. This can be done with certain emulsifier-based cocktails by adding a specific amount of water.
- Chemiluminescence. This is where a chemical reacts with the fluors in the scintillation cocktail causing spurious scintillations, a particular risk with solutions containing strong bases or oxidizing agents. Symptoms include very high initial counts which decrease through time. Possible remedies are:
  - **(a)** leave the vials at room temperature for a time before counting. Check with a suitable blank that counts have dropped to an acceptable level.
  - **(b)** Neutralize basic samples with acid (e.g. acetic acid or HCl).
  - **(c)** Use a scintillation cocktail that resists chemiluminescence such as Hiconicfluor®.
  - **(d)** Raise the energy of the lower counts detected to about 8 keV – most chemiluminescence pulses are weak (0–7 keV). This approach is not suitable for ^{3}H.

Liquid scintillation counting of high-energy β-emitters – β-particles with energies greater than 1 MeV can be counted in water (Čerencov radiation), with no requirement for additional fluors (e.g. ^{32}P).

Many liquid scintillation counters treat the first sample as a 'background', subtracting whatever value is obtained from the subsequent measurements as part of the procedure for converting to d.p.m. If not, you will need to subtract the background count from all other samples. Make sure that you use an appropriate background sample, identical in all respects to your radioactive sample but with no added radioisotope, in the correct position within the machine. Check that the background reading is reasonable (15–30 c.p.m.). Tips for preparing samples for liquid scintillation counting are given in Box 19.2.

Autoradiography

This is a method where photographic film is exposed to the isotope. It is used mainly to locate radioactive tracers in thin sections of an organism or on chromatography papers and gels, but quantitative work is possible. The radiation interacts with the film in a similar way to light, silver grains being formed in the developed film where the particles or rays have passed through. The radiation must have enough energy to penetrate into the film, but if it has too much energy the grain formation may be too distant from the point where the isotope was located to identify precisely the point of origin (e.g. high-energy β-emitters). Autoradiography is a relatively specialized method and individual lab protocols should be followed for particular isotopes/applications.

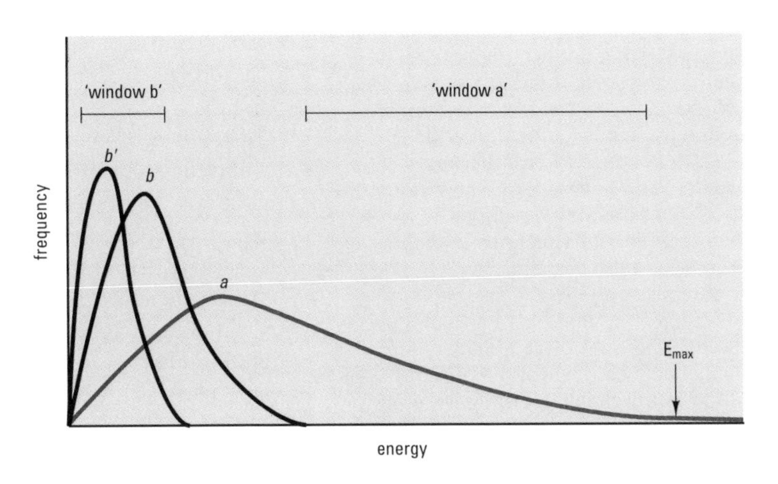

Fig. 19.2 Energy spectra for three radioactive samples, detected using a scintillation counter. Sample *a* is a high-energy β-emitter while *b* contains a low-energy β-emitter, giving a lower spectral range. Sample *b'* contains the same amount of low-energy β-emitter, but with quenching, shifting the spectral distribution to a lower energy band. The counter can be set up to record disintegrations within a selected range (a 'window'). Here, 'window a' could be used to count isotope *a* while 'window b' could give a value of isotope *b*, by applying a correction for the counts due to isotope *a*, based on the results from 'window a'. Dual counting allows experiments to be carried out using two isotopes (double labelling).

Biological applications for radioactive isotopes

The main advantages of using radioactive isotopes in biological experiments are:

- Radioactivity is readily detected. Methods of detection are sufficiently sensitive to measure extremely small amounts of radioactive substances.
- Studies can be carried out on intact, living organisms. If care is taken, minimal disruption of normal conditions will occur when radiolabelled compounds are introduced.
- Protocols are simple compared to equivalent methods for chemical analysis.

The main disadvantages are:

- The 'isotope effect'. Molecules containing different isotopes of the same atom may react at slightly different rates and behave in slightly different ways to the natural isotope. The isotope effect is more extreme the smaller the atom and is most important for ^{3}H-labelled compounds of low molecular mass.
- The possibility of mistaken identity. The presence of radioactivity does not tell you anything about the compound in which the radioactivity is present: it could be different from the one in which it was applied, due to metabolism or spontaneous breakdown of a ^{14}C-containing organic compound.

Metabolism of radiolabelled compounds – you may need to separate individual metabolites before counting, e.g. using chromatography (p. 128), or electrophoresis (p. 142).

The main types of experiments are:

- Investigations of metabolic pathways: a radioactively labelled substrate is added (often to an *in vitro* experimental 'system' rather than a whole organism) and samples taken at different time intervals. By identifying the labelled compounds and plotting their appearance through time, an indication of the pathway of metabolism can be obtained.
- Translocation studies: radioisotopes are used to follow the fate of molecules within an organism. Uptake and translocation rates can be determined with relative ease.
- Ecological studies: radioisotopic tracers provide a convenient method for determining food web interrelationships and for investigating behaviour patterns, while environmental monitoring may involve following the 'spectral signature' of isotopes deliberately or accidentally released.

Example Carbon dating – living organisms have essentially the same ratio of ^{14}C to ^{12}C as the atmosphere; however, when an organism dies, its $^{14}C/^{12}C$ falls because the radioactive ^{14}C isotope decays. Since we know the half-life of ^{14}C (5 715 years), a sample's $^{14}C/^{12}C$ ratio will allow us to estimate its age; e.g. if the ratio were exactly 1/8 of that in the atmosphere, the sample is three half-lives old and was formed 17 145 years before present. Such estimates carry an error of the order of 10% and are unreliable for samples older than 50 000 years, for which longer-lived isotopes can be used.

- Radio-dating: the age of plant or mineral samples can be determined by measuring the amount of a radioisotope in the sample. The age of the specimen can be found using the $t_{1/2}$ by assuming how much was originally incorporated.
- Mutagenesis and sterilization: radioactive sources can be used to induce mutations, particularly in microorganisms. Gamma emitters of high energy will kill microbes and are used to sterilize equipment such as disposable Petri dishes.
- Assays: radioisotopes are used in several quantitative detection methods of value to biologists. Radioimmunoassay is described on p. 98. Isotope dilution analysis works on the assumption that introduced radiolabelled molecules will equilibrate with unlabelled molecules present in the specimen. The amount of substance initially present can be worked out from the change in specific activity of the radioisotope when it is diluted by the 'cold' material. A method is required whereby the substance can be purified from the sample and sufficient substance must be present for its mass to be measured accurately.

Billington *et al.* (1992) give further details and practical advice on using radioisotopes in biological experiments.

Working practices when using radioactive isotopes

In the UK, institutions must be registered for work with specific radioisotopes under the Radioactive Substances Act (1960).

In the UK, the Ionizing Radiations Act (1985) provides details of local arrangements for the supervision of radioisotope work.

By law, undergraduate work with radioactive isotopes must be very closely supervised. In practical classes, the protocols will be clearly outlined, but in project work you may have the opportunity to plan and carry out your own experiments, albeit under supervision. Some of the factors that you should take into account, based on the assumption that your department and laboratory are registered for radioisotope use, are discussed below:

1. Must you use radioactivity? If not, it may be a legal requirement that you use the alternative method.
2. Have you registered for radioactive work? Normal practice is for all users to register with a local Radiation Protection Supervisor. Details of the project may have to be approved by the appropriate administrator(s). You may have to have a short medical examination before you can start work.
3. What labelled compound will you use? Radioactive isotopes must be ordered well in advance through your department's Radiation Protection Supervisor. Aspects that need to be considered include:
   (a) The radionuclide. With many organic compounds this will be confined to ^{3}H and ^{14}C (but see Table 19.2). The risk of a significant 'isotope effect' may influence this decision (see above).
   (b) The labelling position. This may be a crucial part of a metabolic study. Specifically labelled compounds are normally more expensive than those that are uniformly ('generally') labelled.
   (c) The specific activity. The upper limit for this is defined by the isotope's half-life, but below this the higher the specific activity, the more expensive the compound.
4. Are suitable facilities available? You'll need a suitable work area, preferably out of the way of general lab traffic and within a fume cupboard for those cases where volatile radioactive substances are used or may be produced.

Each new experiment should be planned carefully and experimental protocols laid down in advance so you work as safely as possible and do not waste expensive radioactively labelled compounds. In conjunction with your supervisor, decide whether your method of application will introduce enough radioactivity into the system, how you will account for any loss of radioactivity during recovery of the isotope and whether there will be enough activity to count at the end. You should be able to predict approximately the amount of radioactivity in your samples, based on the specific activity of the isotope used, the expected rate of uptake/exchange and the amount of sample to be counted. Use the isotope's specific activity to estimate whether the non-radioactive ('cold') compound introduced with the radiolabelled ('hot') compound may lead to excessive concentrations being administered. Advice for handling data is given in Box 19.1.

Carrying out a 'dry run' – consider doing this before working with radioactive compounds, perhaps using a dye to show the movement or dilution of introduced liquids, as this will lessen the risks of accident and improve your technique.

Safety and procedural aspects

Make sure the bench surface is one that can be easily decontaminated by washing (e.g. Formica®) and always use a disposable surfacing material such as Benchkote®. It is good practice to carry out as many operations as possible within a Benchkote®-lined plastic tray so that any spillages are contained. You will need a lab coat to be used exclusively for work with radioactivity, safety spectacles and a supply of thin latex or vinyl disposable gloves. Suitable vessels for liquid waste disposal will be required and special plastic bags for solids – make sure you know beforehand the disposal procedures for liquid and solid wastes. Wash your hands after handling a vessel containing a radioactive solution and again before removing your gloves. Gloves should be placed in the appropriate disposal bag as soon as your experimental procedures are complete.

Using Benchkote® – the correct way to use Benchkote® and similar products is with the waxed surface down (to protect the bench or tray surface) and the absorbent surface up (to absorb any spillage). Write the date in the corner when you put down a new piece. Monitor using a G–M tube and replace regularly under normal circumstances. If you are aware of spillage, replace immediately and dispose of correctly.

It is important to comply with the following guidelines:

- Read and obey the local rules for safe usage of radiochemicals.
- Maximize the distance between you and the source as much as possible.
- Minimize the duration of exposure.
- Wear protective clothing (properly fastened lab coat, safety glasses, gloves) at all times.
- Use appropriate shielding at all times (Table 19.1).
- Monitor your working area for contamination frequently.
- Mark all glassware, trays, bench work areas, etc., with tape incorporating the international symbol for radioactivity (Fig. 19.3).
- Keep adequate records of what you have done with a radioisotope – the stock remaining and that disposed of in waste form must agree.
- Store radiolabelled compounds appropriately and return them to storage areas immediately after use.
- Dispose of waste promptly and with due regard for local rules.
- Make the necessary reports about waste disposal, etc., to your departmental Radiation Protection Supervisor.
- Clear up after you have finished each experiment.
- Wash thoroughly after using radioactivity.
- Monitor the work area and your body when finished.

Fig. 19.3 Tape showing the international symbol for radioactivity.

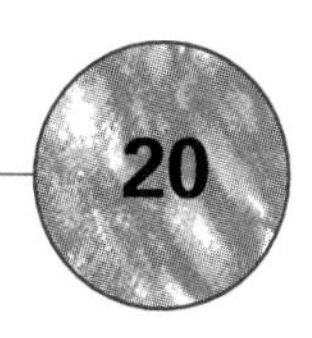

20 Light measurement

The nature of light

Instruments for producing *and* measuring light of particular energy (wavelength) – these are considered in detail in Ch. 21, p. 111.

Light is most strictly defined as that part of the spectrum of electromagnetic radiation detected by the human eye. However, the term is also applied to radiation just outside that visible range (e.g. UV and infra-red 'light'). Electromagnetic radiation is emitted by the sun and by other sources (e.g. an incandescent lamp) and the electromagnetic spectrum is a broad band of radiation, ranging from cosmic rays to radio waves (Fig. 20.1). Most biological experiments involve measurements within the UV, visible and infra-red regions (generally, within the wavelength range 200–1 000 nm).

Radiation has the characteristics of a particle and of a vibrating wave, travelling in discrete particulate units, or 'packets', termed photons. A quantum is the amount of energy contained within a single photon (it is important not to confuse these two terms, although they are sometimes used interchangeably in the literature). In some circumstances, it is appropriate to measure light in terms of the number of photons, usually expressed directly in moles (6.02×10^{23} photons = 1 mol); older textbooks may use the redundant term Einstein as the unit of measurement, where 1 Einstein = 1 mol photons). Alternatively, the energy content (power) may be measured (e.g. in $W\,m^{-2}$). Radiation also behaves as a vibrating electrical and magnetic field moving in a particular direction, with the magnetic and electrical components vibrating perpendicular to one another and perpendicular to the direction of travel. The wave nature of radiation gives rise to the concepts of wavelength (λ, usually measured in nm), frequency (ν, measured in s^{-1}), speed (c, the speed of electromagnetic radiation, which is $3 \times 10^8\,m\,s^{-1}$ in a vacuum), and direction. In other words, radiation is a vector quantity, where

$$c = \lambda\nu \qquad [20.1]$$

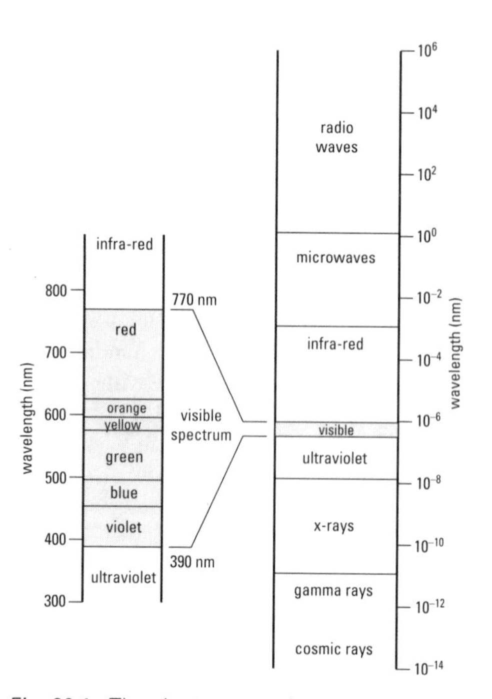

Fig. 20.1 The electromagnetic spectrum.

Photometric and radiometric measurements

Photometric measurements

These are based on the energy perceived by a 'standard' human eye, with maximum sensitivity in the yellow–green region, around 555 nm (Fig. 20.2). The unit of measurement is the candela, a base unit in the SI system, defined in terms of the visual appearance of a specific quantity of platinum at its freezing point. Derived units are used for the luminous flow (lumen) and luminous flow per unit area (lux). These units were once used in photobiology and you may come across them in older literature. However, it is now recognized that such measurements are of little direct relevance to biologists, including even those who may wish to study visual responses, because they are not based on fundamental physical principles.

Judging light quality by eye – the human eye rapidly compensates for changes in light climate by varying the size of the pupil and is a poor guide to light quantity.

Radiometric measurements

The radiometric system is based on physical properties of the electromagnetic radiation itself, expressed either as the number of photons, or their energy content. The following terms are used (units of measurement in parentheses):

- Photon flux (mol photons s^{-1}) is the number of photons arriving at an object within the specified time interval.

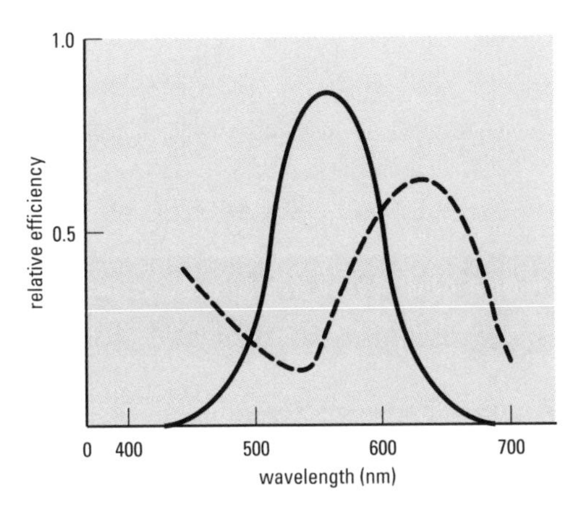

Fig. 20.2 Relative efficiency of vision (solid line) and photosynthesis (dotted line) as a function of wavelength.

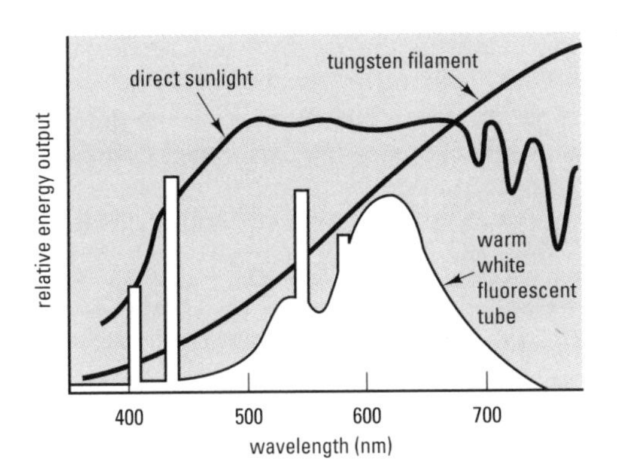

Fig. 20.3 Spectral distribution of energy output from various sources. (Adapted from Golterman *et al.* (1978).)

Table 20.1 Approximate conversion factors for a photosynthetic irradiance (PI) of 1 W m^{-2} to photosynthetic photon flux density (PPFD)

Source	PPFD (μmol photons m^{-2} s^{-1})
Sunlight	4.6
'Cool white' fluorescent tube	4.6
Osram 'daylight' fluorescent tube	4.6
Quartz-iodine lamp	5.0
Tungsten bulb	5.0

(*Source*: Lüning 1981)

- Photon exposure (mol photons m^{-2}) is the total number of photons received by an object, usually expressed per unit surface area.
- Photon flux density (mol photons m^{-2} s^{-1}) or PFD is the most commonly used term to describe the number of photons arriving at a particular surface, expressed per unit surface area and per unit time interval.
- Photosynthetically active radiation or PAR is radiation within the waveband 400–700 nm, since the photosynthetic pigments (chlorophylls, carotenoids, etc.) show maximum absorption within this band.
- Photosynthetic photon flux density (mol photons m^{-2} s^{-1}) or PPFD is the number of photons within the waveband 400–700 nm arriving at a particular surface per unit time interval (power) expressed per unit surface area. Often this term is used interchangeably with PFD.
- Irradiance (J m^{-2} s^{-1} = W m^{-2}) is the amount of energy arriving at a surface per unit time interval (power) expressed per unit surface area.
- Photosynthetic irradiance (W m^{-2}) or PI is the energy of radiation within the waveband 400–700 nm arriving at a surface, expressed per unit surface area and per unit time interval.

Choice of measurement scale

Photon flux density

This is often the most appropriate unit of measurement for biological systems where individual photons are involved in the underlying process, e.g. in photosynthetic studies, where PPFD is measured, since each photochemical reaction involves the absorption of a single photon by a pigment molecule. Most modern light-measuring instruments (radiometers) can measure this quantity, giving a reading in μmol photons m^{-2} s^{-1}.

Irradiance

This is appropriate if you are interested in the energy content of the light, e.g. if you are studying energy balance, or thermal effects. Many radiometers measure photosynthetic irradiance within the waveband 400–700 nm, giving a reading in W m^{-2}. It is possible to make an approximate conversion between PPFD and PI measurements, providing the spectral properties of the light source are known (see Table 20.1).

Spectral distribution

This can be determined using a spectroradiometer, e.g. to compare different light sources (Fig. 20.3). A spectroradiometer measures irradiance or photon flux density in specific wavebands. This instrument consists of a monochromator (p. 112) to allow separate narrow wavebands (5–25 nm bandwidth) to be measured by a detector; some instruments provide a plot of the spectral characteristics of the source.

Using a radiometer ('light meter')

The main components of a radiometer are:

- Receiver: either flat-plate, hemispherical or spherical, depending upon requirements. Most incorporate a protective diffuser, to reduce reflection.
- Detector: either thermoelectric or photoelectric. Some photoelectric detectors suffer from fatigue, with a decreasing response on prolonged exposure: check the manufacturer's handbook for exposure times.

- Processor and readout device to convert the output from the detector into a visible reading, in analogue or digital form.

Box 20.1. gives practical details of the steps involved in using a radiometer.

Box 20.1 Measuring photon flux density or irradiance using a battery-powered radiometer

1. **Check the battery.** Most instruments have a setting that gives a direct readout of battery voltage. Recharge if necessary before use.
2. **Select the appropriate type of measurement** (e.g. photon flux density or irradiance over the PAR waveband, or an alternative range): the simpler instruments have a selection dial for this purpose.
3. **Place the sensor in the correct location and position** for the measurement: it may be appropriate to make several measurements at different positions, and take an average.
4. **Choose the most appropriate scale** for the readout device: for needle-type meters, the choice of maximum reading is usually selected by a dial, within the range 0.3 to 30 000. Start at a high range and work down until the reading is on the scale. Your final scale should be chosen to provide the most accurate reading, e.g. a reading of 15 μmol photons m^{-2} s^{-1} should be made using the 0–30 scale, rather than a higher range.
5. **Read the value from the meter.** For needle-type instruments there may be two scales, the upper one marked from 0 to 10 and the lower one from 0 to 3: make sure you use the correct one, e.g. a half-scale deflection on the 0–30 scale is 15.
6. **Check that the answer is realistic,** e.g. full sunlight has a PPFD of up to 2 000 μmol photons m^{-2} s^{-1} (PI $\equiv$ 400 W m^{-2}), though the value will be far lower on a dull or cloudy day, while the PPFD at a distance of 1 m from a mercury lamp is around 150 μmol photons m^{-2} s^{-1}, and 50 μmol photons m^{-2} s^{-1} at the same distance from a fluorescent lamp.

In your write-up give full details of how the measurement was made, e.g. the type of light source, instrument used, where the sensor was placed, whether an average was calculated, etc.

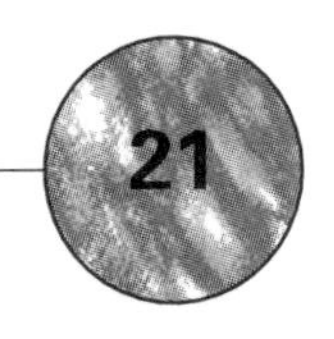

21 Basic spectroscopy

Spectroscopic techniques can be used to:

- tentatively identify compounds, by determining their absorption or emission spectra;
- quantify substances, either singly or in the presence of other compounds, by measuring the signal strength at an appropriate wavelength;
- determine molecular structure;
- follow reactions, by measuring the disappearance of a substance, or the appearance of a product as a function of time.

The absorption and emission of electromagnetic radiation of specific energy (wavelength) is a characteristic feature of many molecules, involving the movement of electrons between different energy states, in accordance with the laws of quantum mechanics. Electrons in atoms or molecules are distributed at various energy levels, but are mainly at the lowest energy level, usually termed the ground state. When exposed to energy (e.g. from electromagnetic radiation), electrons may be excited to higher energy levels (excited states), with the associated absorption of energy at specific wavelengths giving rise to an absorption spectrum. One quantum of energy is absorbed for a single electron transition from the ground state to an excited state. On the other hand, when an electron returns to its ground state, one quantum of energy is released; this may be dissipated to the surrounding molecules (as heat) or may give rise to an emission spectrum. The energy change (ΔE) for an electron moving between two energy states, E_1 and E_2, is given by the equation:

$$\Delta E = E_1 - E_2 = h\nu \qquad [21.1]$$

where h is the Planck constant (p. 47) and ν is the frequency of the electromagnetic radiation expressed in Hz or s^{-1}). Frequency is related to wavelength (λ, usually expressed in m) and the speed of electromagnetic radiation, c (p. 47) by the expression:

$$\nu = c/\lambda \qquad [21.2]$$

UV/visible spectrophotometry

This is a widely used technique for measuring the absorption of radiation in the visible and UV regions of the spectrum. A spectrophotometer is an instrument designed to allow precise measurement at a particular wavelength, while a colorimeter is a simpler instrument, using filters to measure broader wavebands (e.g. light in the green, red or blue regions of the visible spectrum).

Principles of light absorption

Two fundamental principles govern the absorption of light passing through a solution:

- The absorption of light is exponentially related to the number of molecules of the absorbing solute that are encountered, i.e. the solute concentration $[C]$.
- The absorption of light is exponentially related to the length of the light path through the absorbing solution, l.

These two principles are combined in the Beer–Lambert relationship, which is usually expressed in terms of the intensity of the incident light (I_0) and the emergent light (I):

$$\log_{10}(I_0/I) = \varepsilon l[C] \qquad [21.3]$$

where ε is a constant for the absorbing substance and the wavelength, termed the absorption coefficient or absorptivity, and $[C]$ is expressed either as mol l^{-1} or g l^{-1} (see p. 20) and l is given in cm. This relationship is extremely useful, since most spectrophotometers are constructed to give a direct measurement of $\log_{10}(I_0/I)$, termed the absorbance (A), or extinction (E), of a solution (older texts may use the outdated term optical density). Note that

Definition

Absorbance (*A*) – this is given by:

$A = \log_{10}(I_0/I)$.

Definition

Transmittance (T) – this is usually expressed as a percentage, where

$T = (I/I_0) \times 100\,(\%)$.

for substances obeying the Beer–Lambert relationship, A is linearly related to $[C]$. Absorbance at a particular wavelength is often shown as a subscript, e.g. A_{550} represents the absorbance at 550 nm. The proportion of light passing through the solution is known as the transmittance (T), and is calculated as the ratio of the emergent and incident light intensities.

Some instruments have two scales:

- an exponential scale from zero to infinity, measuring absorbance;
- a linear scale from 0 to 100, measuring (per cent) transmittance.

For most practical purposes, the Beer–Lambert relationship applies and you should use the absorbance scale.

UV/visible spectrophotometer

The principal components of a UV/visible spectrophotometer are shown in Fig. 21.1. High intensity tungsten bulbs are used as the light source in basic instruments, capable of operating in the visible region (i.e. 400–700 nm). Deuterium lamps are used for UV spectrophotometry (200–400 nm); these lamps are fitted with quartz envelopes, since glass does not transmit UV radiation.

A major improvement over the simple colorimeter is the use of a diffraction grating to produce a parallel beam of monochromatic light from the (polychromatic) light source. In practice the light emerging from such a monochromator is not of a single wavelength, but is a narrow band of wavelengths. This bandwidth is an important characteristic, since it determines the wavelengths used in absorption measurements – the bandwidth of basic spectrophotometers is around 5–10 nm while research instruments have bandwidths of less than 1 nm.

Bandwidth is affected by the width of the exit slit (the slit width), since the bandwidth will be reduced by decreasing the slit width. To obtain accurate data at a particular wavelength setting, the narrowest possible slit width should be used. However, decreasing the slit width also reduces the amount of light reaching the detector, decreasing the signal-to-noise ratio. The extent to which the slit width can be reduced depends upon the sensitivity and stability of the detection/amplification system and the presence of stray light.

Using plastic disposable cuvettes – these are adequate for work in the near-UV region, e.g. for enzyme studies using nicotinamide coenzymes, at 340 nm (p. 193), as well as the visible range.

Most UV/visible spectrophotometers are designed to take cuvettes with an optical path length of 10 mm. Disposable plastic cuvettes are suitable for routine work in the visible range using aqueous and alcohol-based solvents, while glass cuvettes are useful for other organic solvents. Glass cuvettes are manufactured to more exacting standards, so use optically matched glass cuvettes for accurate work, especially at low absorbances (< 0.1), where any

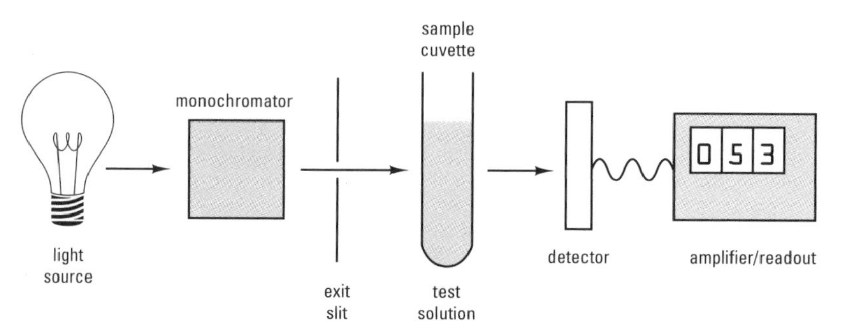

Fig. 21.1 Components of a UV/visible spectrophotometer.

Examples The molar absorptivity of NADH is 6.22×10^3 l mol^{-1} cm^{-1} at 340 nm. For a test solution giving an absorbance of 0.21 in a cuvette with a light path of 5 mm, using eqn [21.3] this is equal to a concentration of:

$0.21 = 6.22 \times 10^3 \times 0.5 \times [C]$
$[C] = 0.0000675$ mol l^{-1}
(or 67.5 μmol l^{-1}).

The specific absorptivity (10 g l^{-1}) of double-stranded DNA is 200 at 260 nm, therefore a solution containing 1 g l^{-1} will have an absorbance of 200/10 = 20. For a DNA solution, giving an absorbance of 0.35 in a cuvette with a light path of 1.0 cm, using eqn [21.3] this is equal to a concentration of:

$0.35 = 20 \times 1.0[C]$
$[C] = 0.0175$ g l^{-1}
(equivalent to 17.5 μg ml^{-1}).

Chlorophylls a and b of vascular plants and green algae can be extracted in 90% v/v acetone/water and assayed by measuring the absorbance of the mixed solution at 2 wavelengths, according to the formulae:

Chlorophyll *a* (mg l^{-1}) = $11.93\,A_{664} - 1.93\,A_{647}$

Chlorophyll *b* (mg l^{-1}) = $20.36\,A_{647} - 5.5\,A_{664}$

Note: different equations are required for other solvents.

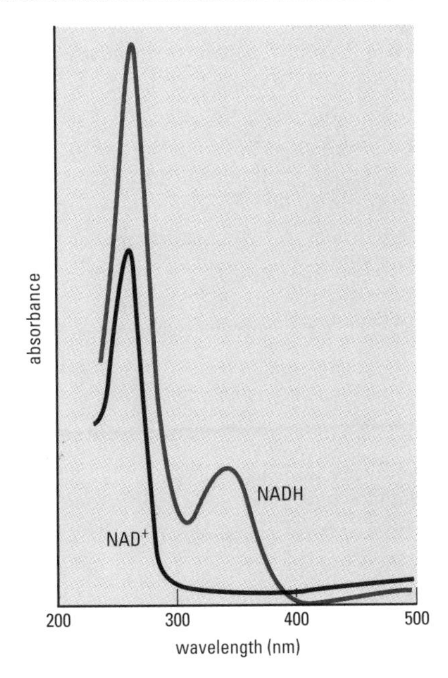

Fig. 21.2 Absorption spectra of nicotinamide adenine dinucleotide in oxidized (NAD^+) and reduced (NADH) form. Note the 340 nm absorption peak, used for quantitative work (p. 114).

differences in the optical properties of cuvettes for reference and test samples will be pronounced. Glass and plastic absorb UV light and quartz cuvettes must be used at wavelengths below 300 nm.

Before taking a measurement, make sure that cuvettes are clean, unscratched, dry on the outside, filled to the correct level and in the correct position in their sample holders. Proteins and nucleic acids in biological samples can accumulate on the inside faces of glass/quartz cuvettes, so remove any deposits using acetone on a cotton bud, or soak overnight in 1 mol l^{-1} nitric acid. Corrosive and hazardous solutions must be used in cuvettes with tightly fitting lids, to prevent damage to the instrument and to reduce the risk of accidental spillage.

Basic instruments use photocells similar to those used in simple colorimeters or photodiode detectors. In many cases, a different photocell must be used at wavelengths above and below 550–600 nm, due to differences in the sensitivity of such detectors over the visible waveband. The detectors used in more sophisticated instruments, give increased sensitivity and stability when compared to photocells.

Digital displays are increasingly used in preference to needle-type meters, as they are not prone to parallax errors and misreading of the absorbance scale. Some digital instruments can be calibrated to give a direct readout of the concentration of the test substance.

Types of UV/visible spectrophotometer

Basic instruments are single beam spectrophotometers in which there is only one light path. The instrument is set to zero absorbance using a blank solution, which is then replaced by the test solution, to obtain an absorbance reading. An alternative approach is used in double beam spectrophotometers, where the light beam from the monochromator is split into two separate beams, one beam passing through the test solution and the other through a reference blank. Absorbance is then measured by an electronic circuit which compares the output from the reference (blank) and sample cuvettes. Double beam spectrophotometry reduces measurement errors caused by fluctuations in output from the light source or changes in the sensitivity of the detection system, since reference and test solutions are measured at the same time (Box 21.2). Recording spectrophotometers are double beam instruments, designed for use with a chart recorder, either by recording the difference in absorbance between reference and test solutions across a predetermined waveband to give an absorption spectrum (Fig. 21.2), or by recording the change in absorbance at a particular wavelength as a function of time (e.g. in an enzyme assay, see Chapter 35).

Quantitative spectrophotometric analysis

A single (purified) substance in solution can be quantified using the Beer–Lambert relationship (eqn [21.3]), provided its absorptivity is known at a particular wavelength (usually at the absorption maximum for the substance, since this will give the greatest sensitivity). The molar absorptivity is the absorbance given by a solution with a concentration of 1 mol l^{-1} (= 1 kmol m^{-3}) of the compound in a light path of 1 cm. The appropriate value may be available from tabulated spectral data (e.g. Anon., 1963), or it can be determined experimentally by measuring the absorbance of known concentrations of the substance (Box 21.1) and plotting a standard curve. This should confirm that the relationship is linear over the desired concentration range and the slope of the line will give the molar absorptivity.

Box 21.1 Using a spectrophotometer

1. **Switch on and select the correct lamp** for your measurements (e.g. deuterium for UV, tungsten for visible light).

2. **Allow up to 15 min for the lamp to warm up** and for the instrument to stabilize before use.

3. **Select the appropriate wavelength:** on older instruments a dial is used to adjust the monochromator, while newer machines have microprocessor-controlled wavelength selection.

4. **Select the appropriate detector:** some instruments choose the correct detector automatically (on the basis of the specified wavelength), while others have manual selection.

5. **Choose the correct slit width** (if available): this may be specified in the protocol you are following, or may be chosen on the manufacturer's recommendations.

6. **Insert appropriate reference blank(s)**: single beam instruments use a single cuvette, while double beam instruments use two cuvettes (a matched pair for accurate work). The reference blanks should match the test solution in all respects apart from the substance under test, i.e. they should contain all reagents apart from this substance. *Make sure that the cuvettes are positioned correctly, with their polished (transparent) faces in the light path, and that they are accurately located in the cuvette holder(s).*

7. **Check/adjust the 0% transmittance**: most instruments have a control which allows you to zero the detector output in the absence of any light (termed dark current correction). Some microprocessor-controlled instruments carry out this step automatically.

8. **Set the absorbance reading to zero**: usually via a dial, or digital readout.

9. **Analyse your samples**: replace the appropriate reference blank with a test sample, allow the absorbance reading to stabilize (5–10 s) and read the absorbance value from the meter/readout device. For absorbance readings greater than 1 (i.e. $<$ 10% transmission), the signal-to-noise ratio is too low for accurate results. Your analysis may require a calibration curve or you may be able to use the Beer–Lambert relationship (eqn [21.3]) to determine the concentration of test substance in your samples.

10. **Check the scale zero at regular intervals** using a reference blank, e.g. after every ten samples.

11. **Check the reproducibility of the instrument**: measure the absorbance of a single solution several times during your analysis. It should give the same value.

Problems (and solutions): inaccurate/unstable readings are most often due to incorrect use of cuvettes, e.g. dirt, fingerprints or test solution on outside of cuvette (wipe the polished faces using a soft tissue before insertion into the cuvette holder), condensation (if cold solutions aren't allowed to reach room temperature before use), air bubbles (which scatter light and increase the absorbance; tap gently to remove), insufficient solution (causing refraction of light at the meniscus), particulate material in the solution (check for 'cloudiness' in the solution and centrifuge before use, where necessary) or incorrect positioning in light path (locate in correct position).

Measuring absorbances in colorimetric analysis – if any final solution has an absorbance that is too high to be read with accuracy on your spectrophotometer (i.e. A > 2), it is bad practice to dilute the solution so that it can be measured. This dilutes both the sample molecules and the colour reagents to an equal extent. Instead, you should dilute the original sample and re-assay.

The specific absorptivity is the absorbance given by a solution containing $10\,g\,l^{-1}$ (i.e. 1% w/v) of the compound in a light path of 1 cm. This is useful for substances of unknown molecular weight, e.g. proteins or nucleic acids, where the amount of substance in solution is expressed in terms of its mass, rather than as a molar concentration. For use in eqn [21.1], the specific absorptivity should be divided by 10 to give the solute concentration in $g\,l^{-1}$.

This simple approach cannot be used for mixed samples where several substances have a significant absorption at a particular wavelength. In such cases, it may be possible to estimate the amount of each substance by measuring the absorbance at several wavelengths, e.g. protein estimation in the presence of nucleic acids (p. 168).

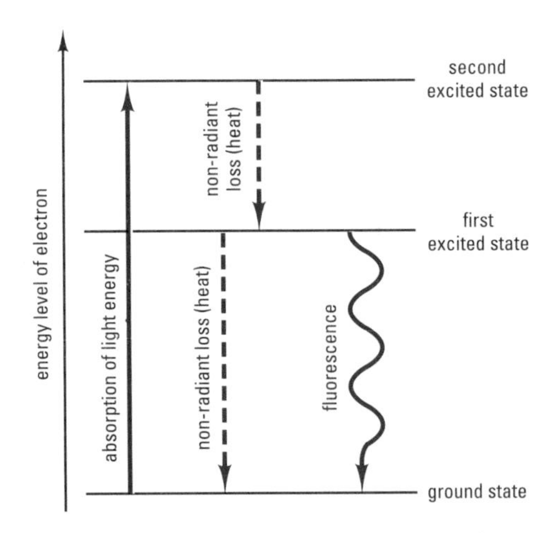

Fig. 21.3 Energy levels and energy transitions in fluorescence.

Table 21.1 Examples of compounds with intrinsic fluorescence

Drugs
Aspirin, morphine, barbiturates, propanalol, ampicillin, tetracyclines

Vitamins
Riboflavin, vitamins A, B6 and E, nicotinamide

Pollutants
Naphthalene, anthracene, benzopyrene

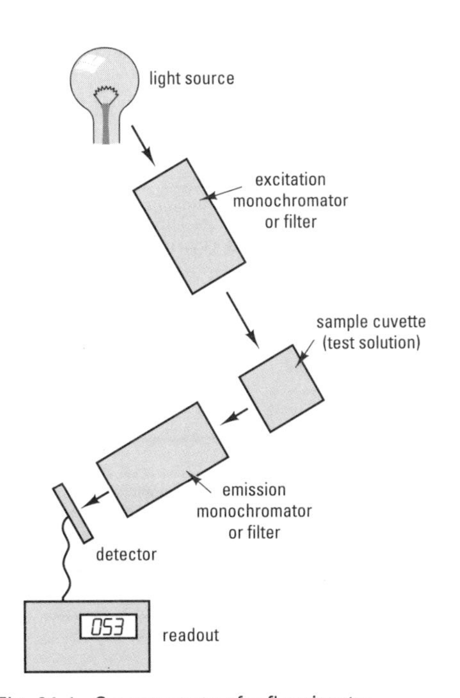

Fig. 21.4 Components of a fluorimeter (fluorescence spectrophotometer). Note that sample cuvettes for fluorimetry must have clear sides all round.

Fluorescence

With most molecules, after electrons are raised to a higher energy level by absorption of electromagnetic radiation, they soon fall back to the ground state by radiationless transfer of energy (heat) to the solvent. However, with some molecules, the events shown in Fig. 21.3 may occur, i.e. electrons may lose only part of their energy by non-radiant routes and the rest may be emitted as electromagnetic radiation, a phenomenon known as fluorescence. Since not all of the energy that was absorbed is emitted (due to non-radiant loss), the wavelength of the fluorescent light is longer than the absorbed light (longer wavelength = lower energy). Thus, a fluorescent molecule has both an absorption spectrum and an emission spectrum.

Fluorescence spectrophotometry

The principal components of a fluorescence spectrophotometer (fluorimeter) are shown in Fig. 21.4. The instrument contains two monochromators, one to select the excitation wavelength and the other to monitor the light emitted, usually at 90° to the incident beam (though light is actually emitted in all directions). As an example, the wavelengths used to measure the highly fluorescent compound amidomethylcoumarin are 388 nm (excitation) and 440 nm (emission). Some examples of biomolecules with intrinsic fluorescence are given in Table 21.1.

Compared with UV/visible spectrophotometry, fluorescence spectroscopy has certain advantages, including:

- Enhanced sensitivity (up to 1000-fold), since the emitted light is detected against a background of zero, in contrast to spectrophotometry where small changes in signal are measured against a large 'background' (see Eqn. 21.3).
- Increased specificity, because not one, but two specific wavelengths are required for a particular compound.

However, there are also certain drawbacks:

- Not all compounds show intrinsic fluorescence, limiting its application. However, some non-fluorescent compounds may be coupled to fluorescent dyes, or fluorophores (e.g. proteins may be coupled to fluorescamine).
- The light emitted can be less than expected due to quenching, i.e. when substances in the sample (e.g. oxygen) either interfere with energy transfer, or absorb the emitted light (in some instances, the sample molecules may self-quench if they are present at high concentration).

The sensitivity of fluorescence has made it invaluable in techniques in which specific antibodies are linked to a fluorescent dye, including:

- fluorescence immunoassay (FIA);
- immunohistochemistry, which requires the use of a fluorescence microscope;
- flow cytometry – a fluorescence-activated cell sorter (FACS) uses cell surface protein-specific monoclonal antibodies labelled with fluorescent dyes to separate and enumerate cells such as lymphocytes.

Phosphorescence and luminescence

A phenomenon related to fluorescence is phosphorescence, which is the emission of light following inter-system crossing between electron orbitals (e.g. between excited singlet and triplet states). Light emission in phosphorescence usually continues after the exciting energy is no longer applied and, since more energy is lost in inter-system crossing, the emission wavelengths are generally longer than with fluorescence. Phosphorescence has limited applications in biomolecular sciences.

Luminescence (or chemiluminescence) is another phenomenon in which light is emitted, but here the energy for the initial excitation of electrons is provided by a chemical reaction rather than by electromagnetic radiation. An example is the action of the enzyme luciferase, extracted from fireflies, which catalyses the following reaction:

$$\text{luciferin} + \text{ATP} + O_2 \Rightarrow \text{oxyluciferin} + \text{AMP} + \text{PP}_i + CO_2 + \text{light} \quad [21.4]$$

The light produced is either yellow–green (560 nm) or red (620 nm). This system can be used in biomolecular analysis of ATP, either to determine ATP concentration in a biological sample, or to follow a coupled reaction (p. 193). Measurement can be performed using the photomultiplier tubes of a scintillation counter (p. 102) to detect the emitted light, with calibration of the output using a series of standards of known ATP content.

Atomic spectroscopy

Atoms of certain metals will absorb and emit radiation of specific wavelengths when heated in a flame, in direct proportion to the number of atoms present. Atomic spectrophotometric techniques measure the absorption or emission of particular wavelengths of UV and visible light, to identify and quantify such metals.

Flame atomic emission spectrophotometry (or flame photometry)

The principal components of a flame photometer are shown in Fig. 21.5. A liquid sample is converted into an aerosol in a nebulizer (atomizer) before being introduced into the flame, where a small proportion (typically less than 1 in 10 000) of the atoms will be raised to a higher energy level, releasing this energy as light of a specific wavelength, which is passed through a filter to a photocell detector. Flame photometry is used to measure the alkali metal ions K^+, Na^+, and Ca^{++} in biological fluids.

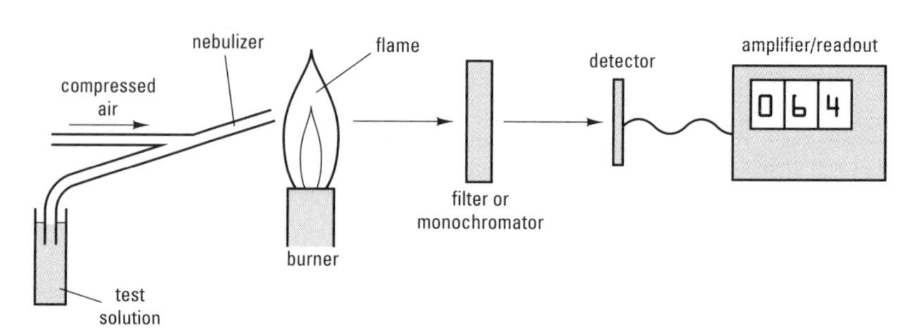

Fig. 21.5 Components of a flame photometer.

When using a flame photometer:

- Allow time for the instrument to stabilize. Switch on the instrument, light the flame and wait at least 5 min before analysing your solutions.
- Check for impurities in your reagents. For example, if you are measuring K^+ in an acid digest of some biological material, check the K^+ content of a reagent blank, containing everything except the biological material, processed in exactly the same way as the samples. Subtract this value from your sample values to obtain the true K^+ content.
- Quantify your samples using a calibration curve (p. 166). Calibration standards should cover the expected concentration range for the test solutions – your calibration curve may be non-linear (especially at concentrations above $1\ \text{mmol}\ l^{-1}$, i.e. $1\ \text{mol}\ m^{-3}$ in SI units).
- Analyse all solutions in duplicate, so that reproducibility can be assessed.
- Check your calibration. Make repeated measurements of a standard solution of known concentration after every six or seven samples, to confirm that the instrument calibration is still valid.
- Consider the possibility of interference. Other metal atoms may emit light which is detected by the photocell, since the filters cover a wider waveband than the emission line of a particular element. This can be a serious problem if you are trying to measure low concentrations of a particular metal in the presence of high concentrations of other metals (e.g. Na^+ in sea water), or other substances which form complexes with the test metal, suppressing the signal (e.g. phosphate).

Atomic absorption spectrophotometry (or flame absorption spectrophotometry)

This technique is applicable to a broad range of metal ions, including those of Pb, Cu, Zn, etc. It relies on the absorption of light of a specific wavelength by atoms dispersed in a flame. The appropriate wavelength is provided by a cathode lamp, coated with the element to be analysed, focused through the flame and onto the detector. When the sample is introduced into the flame, it will decrease the light detected in direct proportion to the amount of metal present. Practical advantages over flame photometry include:

- improved sensitivity,
- increased precision,
- decreased interference.

Newer variants of this method include flameless atomic absorption spectrophotometry and atomic fluorescence spectrophotometry, both of which are more sensitive than the flame absorption technique.

Plotting calibration curves in quantitative analysis – do not force your calibration line to pass through zero if it clearly does not. There is no reason to assume that the zero value is any more accurate than any other reading you have made.

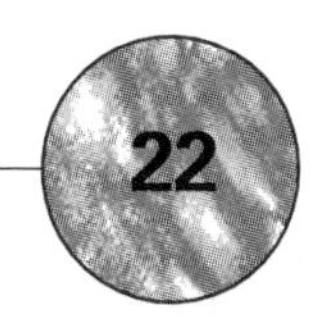

22 Advanced spectroscopy and spectrometry

Identifying compounds – the techniques described in this chapter can often provide sufficient information to identify a compound with a low probability of error.

The following techniques are unlikely to be encountered in the early stages of your course, though they may be presented as a demonstration. Alternatively, you may be shown spectra for interpretation and analysis. The various types of spectroscopy, and their relationship with the electromagnetic spectrum, are shown in Table 22.1.

Infra-red (IR) and Raman spectroscopy

Definitions

Spectroscopy – any technique involving the production and subsequent recording of a spectrum of electromagnetic radiation, usually in terms of wavelength or energy.

Spectrometry – any technique involving the measurement of a spectrum, e.g. of electromagnetic radiation, molecular masses, etc.

Wavenumber – the reciprocal of the wavelength (expressed as cm^{-1}): a term used widely in IR spectroscopy, but rarely in other types of analysis.

Both of these techniques involve the measurement of frequencies produced by the vibration of chemical bonds (bending and stretching). The IR/Raman region is generally considered to be from 800–2500 nm (for near-IR) and up to 16 000 nm (for mid-IR). Near-IR spectroscopy involves recording the spectrum in that region in a manner analogous to UV/visible spectroscopy, and quantitative analysis is possible. However, the most widely used technique is mid-IR spectroscopy, which allows identification of groups or atoms in a sample compound, but is inappropriate for quantitative measurement. A peak at a particular frequency can be identified by reference to libraries or computer databases of IR spectra, e.g. a peak at a wavenumber of 1730–1750 cm^{-1} corresponds to a carbonyl group – ($\sim$C=O), which is present in fatty acids and proteins. The 1400–600 cm^{-1} region is known as the 'fingerprint' region because no two compounds give identical spectra.

Origins of spectra – the IR spectrum is due to changes in charge displacement in bonds. The Raman spectrum is due to changes in polarizability in bonds.

The value of IR spectroscopy is greatly enhanced by Fourier transformation (FT), named after the mathematician J.B. Fourier. FT is a procedure for interconverting frequency functions and time or distance functions. In FT-IR, information is obtained from an interferometer, which splits the incident beam so that it passes through both the sample and a reference. When the beam is recombined, interference patterns arise because the two path lengths are different. The interference pattern has the same relationship to a normal spectrum as a hologram has to a picture, and integral computers use FT to convert the pattern into a spectrum in under a minute. The overall result is a greatly enhanced signal : noise ratio.

Table 22.1 The electromagnetic spectrum and types of spectroscopy

Type of radiation	Origin	Wavelength	Type of spectroscopy
γ-rays	Atomic nuclei	<0.1 nm	γ-ray spectroscopy
X-rays	Inner shell electrons	0.1–1.0 nm	X-ray fluorescence (XRF)
Ultraviolet (UV)	Ionization	10–200 nm	UV spectroscopy
UV/visible	Valency electrons	200–800 nm	UV/visible spectroscopy
Infra-red	Molecular vibrations	0.8–25 μm	Infrared spectroscopy (IR) and Raman
Microwaves	Electron spin alignment	400 μm–30 cm	Electron spin resonance (ESR)
Radiowaves	Nuclear spin resonance	>100 cm	Nuclear magnetic resonance (NMR)

Using IR – the rapid analysis possible with gaseous samples makes IR ideal for studying CO_2 metabolism in photosynthesis and respiration (p. 206).

Applications of IR and Raman spectroscopy

The principal use of IR and Raman spectroscopy is in the identification of drugs (e.g. penicillin), small peptides, pollutants and food contaminants. When an IR spectrometer is coupled to a gas–liquid chromatograph, it can be used for the analysis of drug metabolites.

Nuclear magnetic resonance (NMR)

Electromagnetic radiation (at radiofrequencies of 1–500 MHz) is used to identify and monitor compounds. This is possible because of differences in the magnetic states of atomic nuclei, involving very small transitions in energy levels. The atomic nuclei of isotopes of many elements are magnetic because they are charged and have spin. Typical magnetic nuclei are ^{1}H, ^{13}C, ^{14}N, ^{15}N, ^{19}F, and ^{31}P. When these nuclei interact with a uniform external magnetic field, they behave like tiny compass needles and align themselves in a direction either parallel or antiparallel to the field. The two orientations have different energies, with the parallel direction having a lower energy than the antiparallel (Fig. 22.1). The energy difference between the two levels (ΔE) corresponds to a precise electromagnetic frequency (ν), according to similar quantum principles to those for the excitation of electrons (p. 111). When a sample containing an isotope with a magnetic nucleus is placed in a magnetic field and exposed to an appropriate radiofrequency, transitions between the energy levels of magnetic nuclei will occur when the energy gap and the applied frequency are in *resonance* (i.e. when they are matched exactly). Differences in energy levels, and hence resonance frequencies (ν_o), depend on the magnitude of the applied magnetic field (B_o) and the magnetogyric ratio (λ), according to the equation:

$$\nu_o = \lambda B_o / 2\pi \qquad [22.1]$$

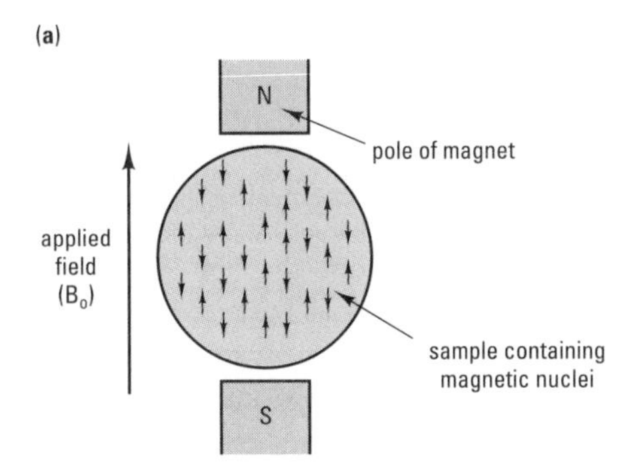

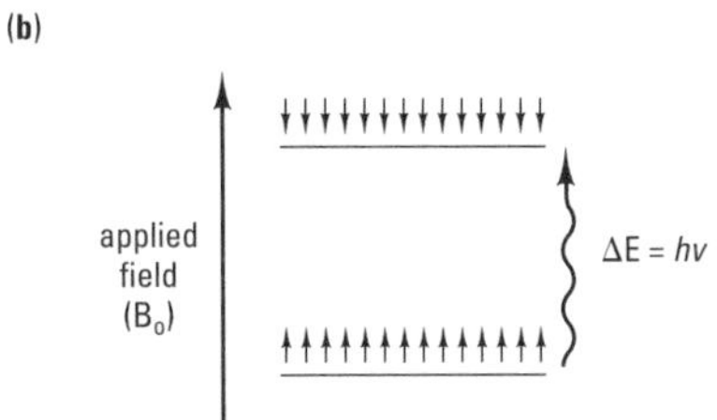

Fig. 22.1 Effect of an applied magnetic field, B_o, on magnetic nuclei. (a) Nuclei in magnetic field have one of two orientations – either with the field or against the field (in the absence of an applied field, the nuclei would have random orientation). (b) Energy diagram for magnetic nuclei in applied magnetic field.

Example For an external magnetic field of 2.5 T (Tesla), ΔE for ^{1}H is 6.6×10^{-26} J, and, since $\Delta E = h\nu$, the corresponding frequency (ν) is 100 MHz; for ^{13}C in the same field, ΔE is 1.7×10^{-26} J, and ν is 25 MHz.

The magnetogyric ratio varies from one isotope to another, so NMR is performed at different frequencies for different nuclei at any given value of B_o. The principal components of an NMR spectrometer are shown in Fig. 22.2.

For magnetic nuclei in a given molecule, an NMR spectrum is generated because, in the presence of the applied field, different nuclei experience different local magnetic fields depending on the arrangement of electrons in their vicinity. The effective field (B) at the nucleus can be expressed as:

$$B = B_o(1 - \sigma) \qquad [22.2]$$

where σ (the shielding constant) expresses the contribution of the small secondary field generated by nearby electrons. The magnitude of σ is dependent

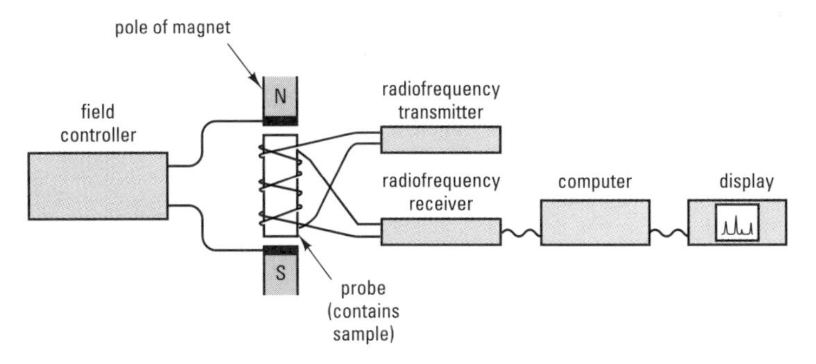

Fig. 22.2 Components of an NMR spectrometer.

Measuring chemical shifts – ppm is *not* a concentration term in NMR, but is used to reflect the very small frequency changes that occur relative to the reference standard, measured in proportional terms.

on the electronic environment of a nucleus, so nuclei in different environments give rise to different resonance frequencies, according to the equation:

$$\nu_o = \lambda B_o(1 - \sigma)/2\pi \qquad [22.3]$$

The separation of resonance frequencies from a reference value is termed the *chemical shift*, and is expressed in dimensionless terms, as parts per million (ppm). By convention, the chemical shift is positive if the sample nucleus is less shielded than the reference and negative if it is more shielded.

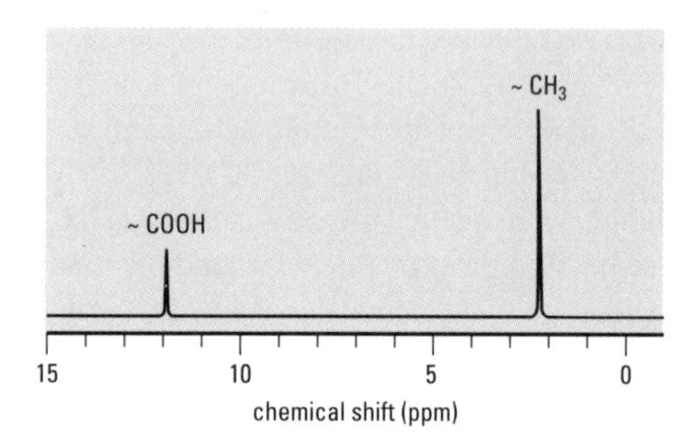

Fig. 22.3 ^{1}H NMR spectrum of acetic acid (CH_3COOH). The relative areas of the two signals are 1:3 and the frequencies (chemical shifts) are expressed in terms of ppm, relative to the reference signal (tetramethyl silane, TMS).

Understanding NMR spectra of biomolecules

Fig. 22.3 illustrates the principle underlying this technique – the different chemical environments of the hydrogen atoms in a simple biomolecule such as acetic acid result in two different ^{1}H resonances or 'signals', one corresponding to those of protons in $\sim CH_3$, and the other from that in $\sim COOH$. Furthermore the relative intensities of the NMR signals, as measured by their areas, are proportional to the number of contributing nuclei, so the relative areas of the peaks due to the protons in $\sim CH_3$ and (undissociated) $\sim COOH$ would be 3 : 1. Similarly, with ^{31}P NMR of a biologically important molecule such as ATP, there are signals corresponding to the α, β, and γ phosphates, the P nuclei of which are in different chemical environments (Fig. 22.4). Thus every molecule that contains one or more magnetic nuclei has its own characteristic NMR fingerprint that may be used for identification and analysis. Spectra such as those shown in Figs 22.3 and 22.4 can be obtained using FT of a large number of individual responses to radiowave pulses. Other factors such as the spin-lattice relaxation time (Gadian, 1995) can affect signal intensities (peak sizes), and resonances may be split into several lines due to spin–spin coupling (interactions between neighbouring nuclei).

In terms of resolution, narrow signals are obtained only from molecules that are fairly mobile, so most high resolution studies are carried out using solutions. However, since many metabolites in biological samples (cells, tissues, etc.) are in aqueous solution and freely mobile, they can give rise to high resolution spectra, as shown in Fig. 22.4 for the ^{31}P-containing compounds, ATP, phosphocreatine, and inorganic phosphate. On the other hand ^{31}P signals from immobile molecules such as DNA or phospholipids are very broad, and could be as wide as the whole scale shown in Fig 22.4.

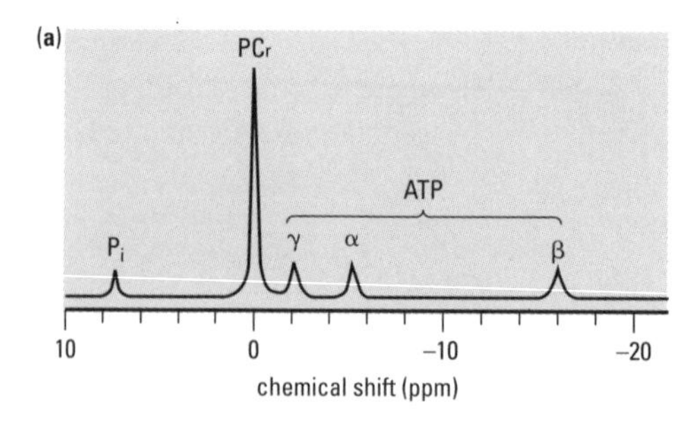

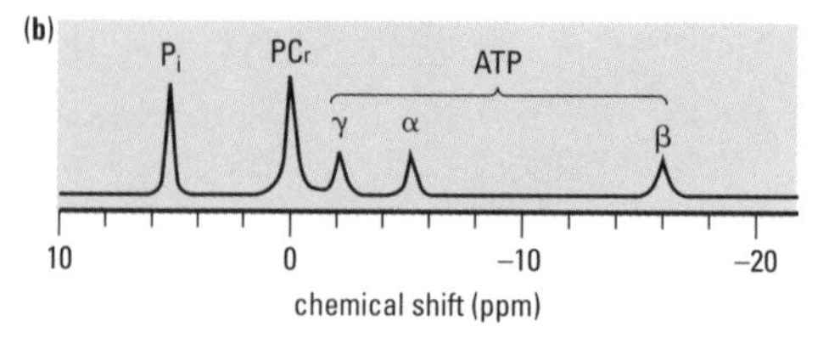

Fig. 22.4 Typical ^{31}P NMR spectrum from intact skeletal muscle (a) at rest and (b) after a fatiguing series of contractions. In (b) ATP levels are preserved at the expense of phosphocreatinine (PCr). Also, hydrolysis of ATP during contraction results in a large phosphate (P_i) peak which is shifted to the right compared to (a), reflecting the decrease in intracellular pH that accompanies glycolysis.

Using NMR – in contrast to most conventional metabolic studies, NMR is non-invasive and the time course of metabolic reactions can be followed using a single experimental subject or preparation, eliminating variation between samples.

Biomolecular applications of NMR

The sensitivity of NMR has improved dramatically with the development of more powerful magnets. Details of the major uses of various magnetic nuclei are given in Table 22.2. The principal applications include:

- studies on the structure and function of macromolecules and biological systems, such as membranes;
- metabolic investigations on living organisms, including humans, since NMR can be used to obtain a 'fingerprint' of a particular molecule and changes in the intensity of spectra can be used for kinetic studies (e.g. Fig. 22.4). This involves use of 'surface coils' as sources of radiation, or placing the organism within the core of an electromagnet;
- measurement of intracellular pH by determination of the chemical shift of the phosphate peak, as this changes with pH in a predictable manner;
- magnetic resonance imaging (MRI), which is a form of proton NMR that uses a field gradient (as opposed to a uniform field) to produce signals that are translated by computers into anatomical images (Gadian, 1995).

Electron spin resonance (ESR)

This technique is based on energy transitions of spinning electrons in a magnetic field. As with NMR, the low energy state occurs when the electromagnetic field generated by the spinning electron is parallel to the externally generated field, whilst the high energy state occurs when the electron-generated field is antiparallel. ESR is very useful for studying metalloproteins and can be used to monitor the activity of such proteins (e.g. cytochrome oxidase) in intact mitochondria or chloroplasts. It can also be used to detect free radicals, for example in irradiated foodstuffs.

Understanding mass spectrometry – since this technique does not involve the production and measurement of electromagnetic spectra and is not based on quantum principles, it should not be referred to as a spectroscopic technique.

Mass spectrometry (MS)

This technique involves the disintegration of organic compounds into fragment ions in a gas phase. These ions are accelerated to specific velocities using an electric field and then separated on the basis of their different masses. Each fragment of a particular mass is detected sequentially with time.

Table 22.2 The relative merits and disadvantages of various magnetic nuclei in biomolecular studies

Nucleus	Relative sensitivity	Natural abundance	Comments
^{1}H	100	99.98%	Multiple, but specific, spectral lines are obtained for individual biomolecules. For mixtures, the ubiquitous occurrence of ^{1}H gives complex, overlapping signals that are often difficult to interpret. Gives a large solvent peak with aqueous samples (can be avoided by using D_2O as solvent). Mainly used for structural studies of pure macromolecules. Essential for MRI.
^{31}P	6.6	100%	Very useful for studies on living systems, with narrow resonance peaks and a wide range of chemical shifts for different molecules. Spectra are simpler and easier to interpret than for ^{1}H, but are not as distinctive: different compounds may give similar ^{31}P spectra. Several important P-containing compounds (including ATP, ADP and inorganic phosphate) can be detected in intact cells – useful in bioenergetic studies.
^{13}C	0.016	1.1%	Gives narrow signals and a wide range of chemical shifts. Resolution is better than for ^{1}H, and a wide range of organic biomolecules can be detected. Low natural abundance gives low sensitivity, extending the time required to accumulate spectra. However, low natural abundance also means that specific metabolites can be selected for ^{13}C isotope enrichment, allowing particular metabolic pathways to be investigated, e.g. carbon assimilation.

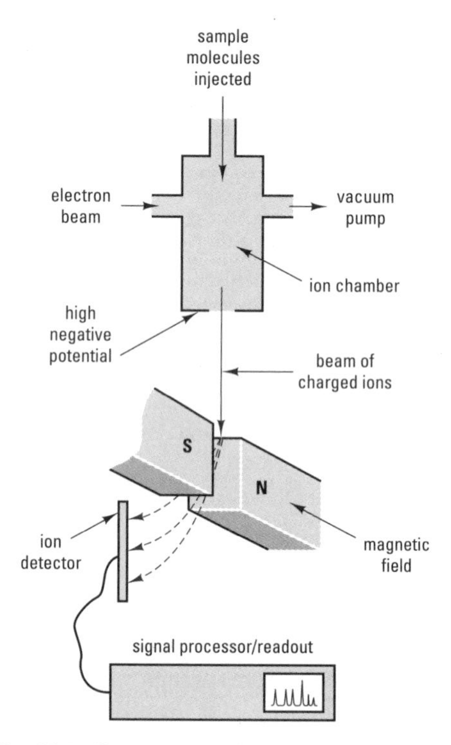

Fig. 22.5 Components of an electron-impact mass spectrometer.

Example IRMS is used in the diagnosis of gastric ulcers caused by *Helicobacter pylori*. This bacterium produces urease, degrading urea to CO_2 and NH_3. A patient drinks a solution of ^{13}C-labelled urea and, if the organism is present in the stomach, $^{13}CO_2$ will appear at up to 5% v/v in the patient's breath, compared with the expected natural abundance value of 1.1% v/v.

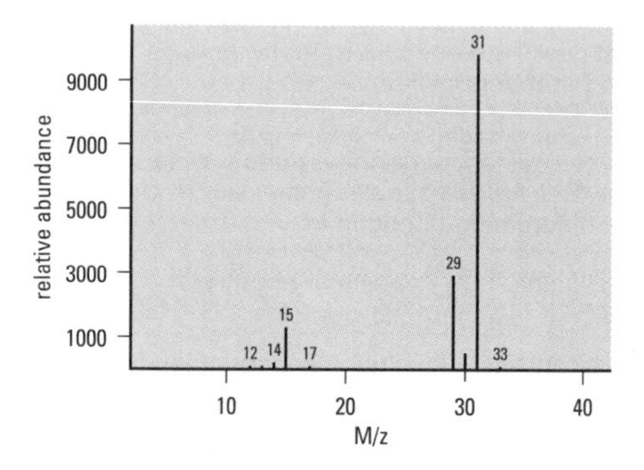

Fig. 22.6 Mass spectrum for methanol. M/z = mass/charge ratio.

The most widely used method in MS is electron impact ionization (EI), where the electron source is a heated metal, such as a tungsten filament, subjected to an appropriate potential gradient. The stream of emitted electrons may then interact with biomolecules (M) in the sample by either:

1. electron removal – where an electron in a bond within the sample molecule is 'knocked out' by bombarding electrons, leaving the bond with only one unpaired electron and resulting in the production of a cationic free radical, i.e. $M + e^- \rightarrow M^{\stackrel{+}{\bullet}} + 2e^-$
2. electron capture – where addition of an extra electron results in the production of an anionic free radical, i.e. $M + e^- \rightarrow M^{\stackrel{-}{\bullet}}$.

Where it is not known whether cations or anions are formed, the radical is given the symbol ⌐, i.e. as $M^{\stackrel{⌐}{\bullet}}$. The first ions formed (parent ions) are unstable, and rapidly undergo further disintegration to give smaller fragments and daughter ions, subsequently separated in the mass spectrometer (Fig. 22.5), to give a mass spectrum such as that shown in Fig. 22.6.

For *in vivo* metabolic studies, the technique of isotope ratio mass spectroscopy (IRMS) is useful since it removes the need to use radioactive isotopes. This technique exploits the ability of MS to distinguish between isotopes such as ^{13}C and ^{12}C. For example, compounds containing ^{13}C have a greater mass than the same compound containing ^{12}C and can be differentiated in the mass spectrum. By selectively labelling a key metabolite with a non-radioactive isotope, the fate of this metabolite can be followed by MS analysis of sequential samples. Normally, the materials used are combusted in oxygen to give gases such as $^{13}CO_2$ and $^{15}NO_2$, followed by exposure of such gases to an EI source; useful metabolic data can be obtained by determining the isotope ratio, e.g. of $^{13}CO_2^{\stackrel{⌐}{\bullet}}$ and $^{12}CO_2^{\stackrel{⌐}{\bullet}}$.

Pyrolysis-mass spectrometry (PY-MS)

This is another variant of mass spectroscopy that is useful for biomolecular applications. The molecules in a sample are volatilized by heating to a precisely controlled high temperature for a specific time period. The volatile material is then removed by a vacuum, ionized by EI, and subjected to MS. Complex mixtures of pyrolysis products are produced, but interpretation is made easier by computer-based multivariate analysis. A useful application of PY-MS is in microbial identification, especially with small samples of organisms that are difficult to culture, e.g. mycobacteria.

Fast atom bombardment–mass spectrometry (FAB-MS)

This is particularly useful for biomolecular applications because it can be used with aqueous solutions. The solution containing the analyte is mixed with glycerol and applied to a probe which is inserted into a vacuum chamber. The mixture is bombarded with a high velocity stream of atoms (usually argon or xenon) rather than electrons, which induces fragmentation of the biomolecules and allows production of the mass spectrum. One problem with FAB-MS is the suppression effect phenomenon – with a mixed sample, not all components may be equally accessible to the atomic bombardment depending on the way they are distributed in the sample–glycerol mixture. This can be overcome by coupling FAB-MS to techniques that can give high resolution separation of components, e.g. gas–liquid chromatography (GC, p. 130), high performance liquid chromatography (HPLC, p. 129), or capillary electrophoresis (p. 152). Johnstone and Rose (1996) give further information on biomolecular aspects of MS.

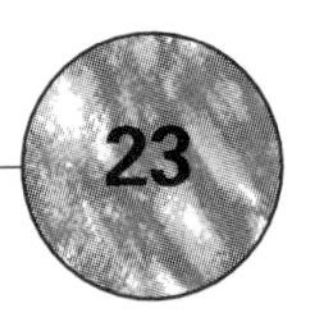

Centrifugation

Particles suspended in a liquid will move at a rate which depends on:

- the applied force – particles in a liquid within a stationary test tube will move in response to gravitational fields, such as the earth's gravity;
- the density difference between the particles and the liquid – particles less dense than the liquid will float upwards while particles denser than the liquid will sink;
- the size and shape of the particles;
- the viscosity of the medium.

For most biological particles (cells, organelles or molecules) the rate of flotation or sedimentation in response to the earth's gravity is too slow to be of practical use in separation.

KEY POINT **A centrifuge is an instrument designed to produce a centrifugal force far greater than the earth's gravity, by spinning the sample about a central axis (Fig. 23.1). Particles of different size, shape or density will thereby sediment at different rates, depending on the speed of rotation and their distance from the central axis.**

Table 23.1 Relationship between speed (r.p.m.) and acceleration (relative centrifugal field, RCF) for a typical bench centrifuge with an average radius of rotation, r_{av} = 115 mm

r.p.m.	RCF*
500	30
1000	130
1500	290
2000	510
2500	800
3000	1160
3500	1570
4000	2060
4500	2600
5000	3210
5500	3890
6000	4630

*RCF values rounded to nearest 10

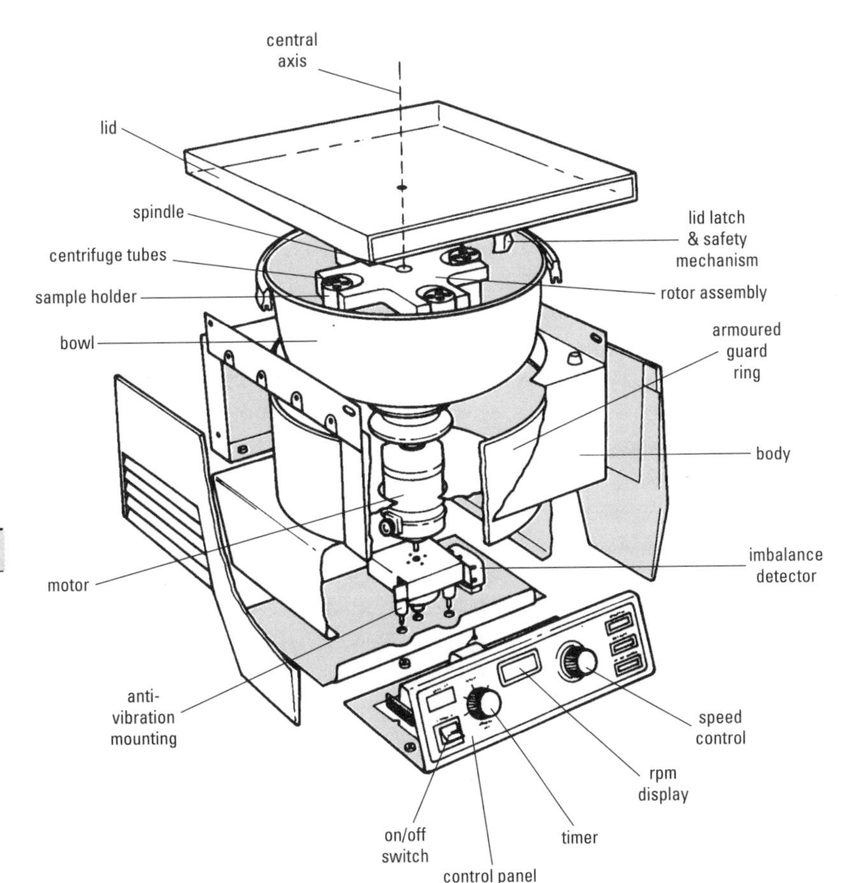

Fig. 23.1 Principal components of a low-speed bench centrifuge.

To convert RCF to acceleration in SI units, multiply by $9.80\,\mathrm{m\,s^{-2}}$.

Examples Suppose you wanted to calculate the RCF of a bench centrifuge with a rotor of $r_{av} = 95$ mm running at a speed of 3000 r.p.m. Using eqn [23.1] the RCF would be: $1.118 \times 95 \times (3)^2 = 956\,g$.

You might wish to calculate the speed (r.p.m.) required to produce a relative centrifugal field of 2000 g using a rotor of $r_{av} = 85$ mm. Using eqn [23.2] the speed would be: $945.7\sqrt{(2000/85)} = 4587$ r.p.m.

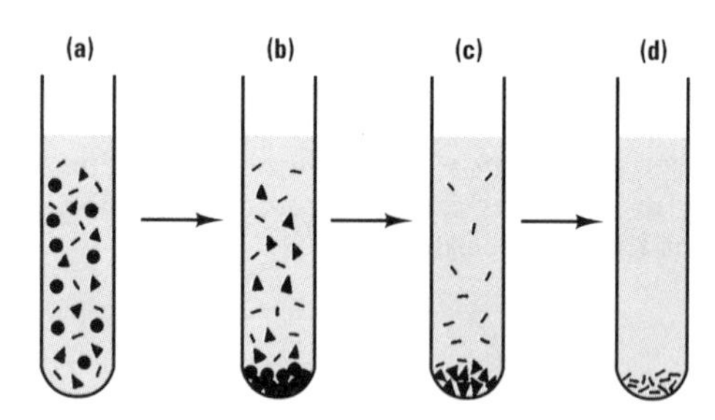

Fig. 23.2 Differential sedimentation. (a) Before centrifugation, the tube contains a mixed suspension of large, medium and small particles. (b) After low-speed centrifugation, the pellet is predominantly composed of the largest particles. (c) Further high-speed centrifugation of the supernatant will give a second pellet, predominantly composed of medium-sized particles. (d) A final ultracentrifugation step pellets the remaining small particles. Note that all of the pellets apart from the final one will have some degree of cross-contamination.

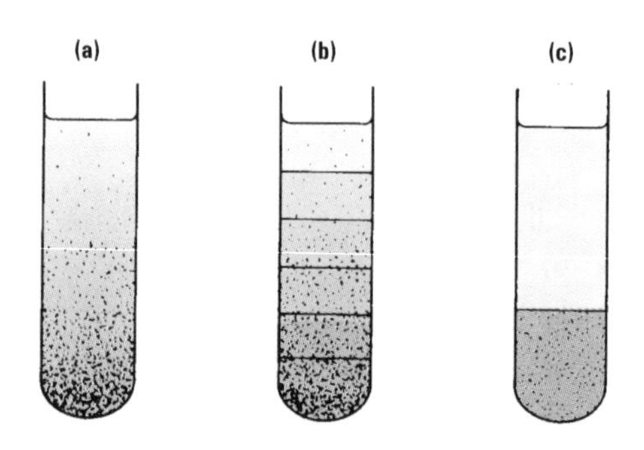

Fig. 23.3 Density gradients. (a) A continuous (linear) density gradient. (b) A discontinuous (stepwise) density gradient, formed by layering solutions of decreasing density on top of each other. (c) A single-step density barrier, designed to allow selective sedimentation of one type of particle.

How to calculate centrifugal acceleration

The acceleration of a centrifuge is usually expressed as a multiple of the acceleration due to gravity ($g = 9.80\,\mathrm{m\,s^{-2}}$), termed the relative centrifugal field (RCF, or 'g value'). The RCF depends on the speed of the rotor (n, in revolutions per minute, r.p.m.) and the radius of rotation (r, in mm) where:

$$\mathrm{RCF} = 1.118\,r\left(\frac{n}{1\,000}\right)^2 \qquad [23.1]$$

This relationship can be rearranged, to calculate the speed (r.p.m.) for specific values of r and RCF:

$$n = 945.7\sqrt{\left(\frac{\mathrm{RCF}}{r}\right)} \qquad [23.2]$$

However, you should note that RCF is not uniform within a centrifuge tube: it is highest near the outside of the rotor (r_{max}) and lowest near the central axis (r_{min}). In practice, it is customary to report the RCF calculated from the average radius of rotation (r_{av}), as shown in Fig. 23.5. It is also worth noting that RCF varies as a *squared* function of the speed: thus the RCF will be doubled by an increase in speed of approximately 41% (Table 23.1).

Centrifugal separation methods

Differential sedimentation (pelleting)

By centrifuging a mixed suspension of particles at a specific RCF for a particular time, the mixture will be separated into a pellet and a supernatant (Fig. 23.2). The successive pelleting of a suspension by spinning for a fixed time at increasing RCF is widely used to separate organelles from cell homogenates. The same principle applies when cells are harvested from a liquid medium.

Density gradient centrifugation

The following techniques use a density gradient, a solution which increases in density from the top to the bottom of a centrifuge tube (Fig. 23.3).

- Rate-zonal centrifugation. By layering a sample onto a shallow pre-formed density gradient, followed by centrifugation, the larger particles will move faster through the gradient than the smaller ones, forming several distinct zones (bands). This method is time dependent, and centrifugation must be stopped before any band reaches the bottom of the tube (Fig. 23.4).
- Isopycnic centrifugation. This technique separates particles on the basis of their buoyant density. Several substances form density gradients during centrifugation (e.g. sucrose, CsCl, Ficoll®, Percoll®, Nycodenz®). The sample is mixed with the appropriate substance and then centrifuged – particles form bands where their density corresponds to that of the medium (Fig. 23.4). This method requires a steep gradient and sufficient time to allow gradient formation and particle redistribution, but is unaffected by further centrifugation.

Bands within a density gradient can be sampled using a fine Pasteur pipette, or a syringe with a long, fine needle. Alternatively, the tube may be punctured and the contents (fractions) collected dropwise in several tubes. For more accurate work, an upward displacement technique can be used (see Ford and Graham, 1991 for specific details).

Working with silicone oil – the density of silicone oil is temperature-sensitive so work in a location with a known, stable temperature or the technique may fail.

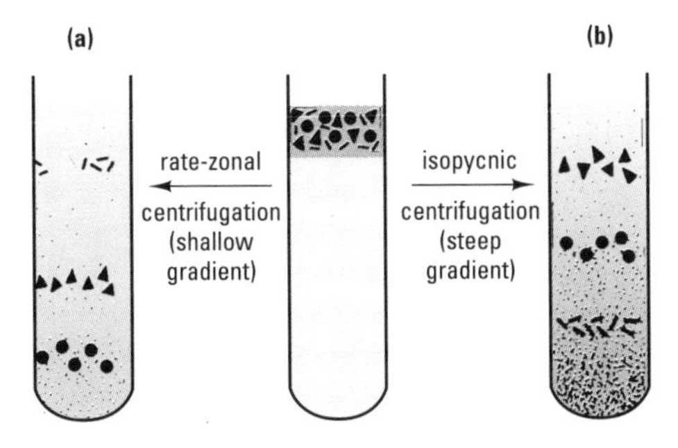

Fig. 23.4 Density gradient centrifugation. The central tube shows the position of the sample prior to centrifugation, as a layer on top of the density gradient medium. Note that particles sediment on the basis of size during rate-zonal centrifugation (a), but form bands in order of their densities during isopycnic centrifugation (b). ●, large particles, intermediate density; ▲, medium-sized particles, low density; ▬ small particles, high density.

Density barrier centrifugation

A single step density barrier (Fig. 23.3c) can be used to separate cells from their surrounding fluid, e.g. using a layer of silicone oil adjusted to the correct density using dinonyl phthalate. Blood cell types can be separated using a density barrier of e.g. Ficoll®.

Types of centrifuge and their uses

Low-speed centrifuges

These are bench-top instruments for routine use, with a maximum speed of 3 000–6 000 r.p.m. and RCF up to 6 000 *g* (Fig. 23.1). They are used to harvest cells, larger organelles (e.g. nuclei, chloroplasts) and coarse precipitates (e.g. antibody–antigen complexes, p. 96). Most modern machines also have a sensor that detects any imbalance when the rotor is spinning and cuts off the power supply (Fig. 23.1). However, some of the older models do not, and must be switched off as soon as any vibration is noticed, to prevent damage to the rotor or harm to the operator.

Microcentrifuges (microfuges)

These are bench-top machines, capable of rapid acceleration up to 12 000 r.p.m. and 10 000 *g*. They are used to sediment small sample volumes (up to 1.5 ml) of larger particles (e.g. cells, precipitates) over short time-scales (typically, 0.5–15 min). They are particularly useful for the rapid separation of cells from a liquid medium, e.g. silicone oil microcentrifugation. Box 23.1 gives details of operation for a low-speed centrifuge.

Continuous flow centrifuges

Useful for harvesting large volumes of cells from their growth medium. During centrifugation, the particles are sedimented as the liquid flows through the rotor.

High-speed centrifuges

These are usually larger, free-standing instruments with a maximum speed of up to 25 000 r.p.m. and RCF up to 60 000 *g*. They are used for microbial cells, many organelles (e.g. mitochondria, lysosomes) and protein precipitates. They often have a refrigeration system to keep the rotor cool at high speed. You would normally use such instruments only under direct supervision.

Recording usage of high-speed centrifuges and ultracentrifuges – most departments have a log book (for samples/speeds/times): make sure you record these details, as the information is important for servicing and replacement of rotors.

Ultracentrifuges

These are the most powerful machines, having maximum speeds in excess of 30 000 r.p.m. and RCF up to 600 000 *g*, with sophisticated refrigeration and vacuum systems. They are used for smaller organelles (e.g. ribosomes, membrane vesicles) and biological macromolecules. You would not normally use one, though your samples may be run by a member of staff.

Changing a rotor – if you ever have to change a rotor, make sure that you carry it properly (don't knock/drop it), that you fit it correctly (don't cross-thread it, and tighten to the correct setting using a torque wrench) and that you store it correctly (clean it after use with a neutral detergent and don't leave it lying around). Disinfect contaminated rotors with a non-corrosive disinfectant, e.g. Virkon®.

Rotors

Many centrifuges can be used with tubes of different size and capacity, either by changing the rotor, or by using a single rotor with different buckets/adaptors.

- Swing-out rotors: sample tubes are placed in buckets which pivot as the rotor accelerates (Fig. 23.5a). Swing-out rotors are used on many low-speed centrifuges: their major drawback is their extended path length and the resuspension of pellets due to currents created during deceleration.

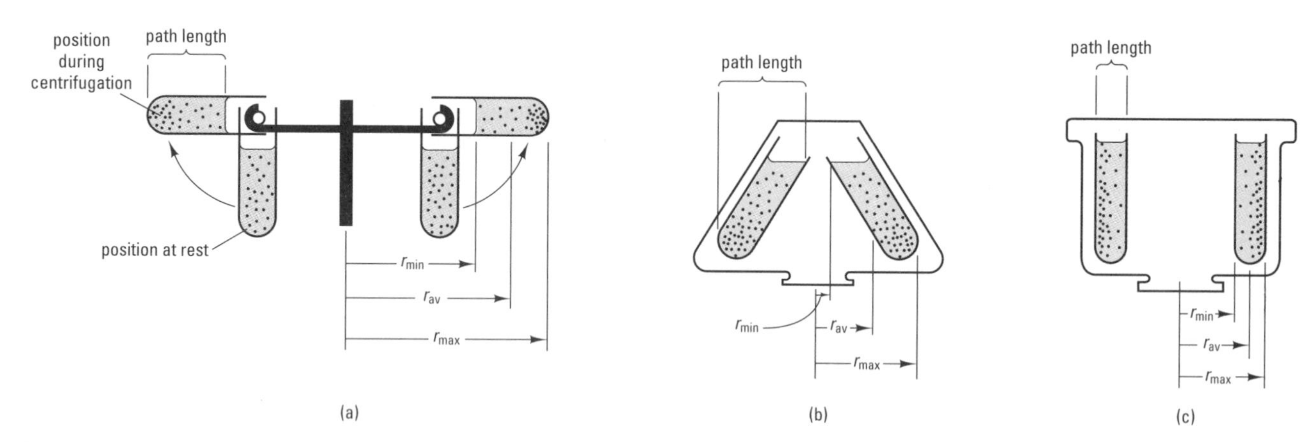

Fig. 23.5 Rotors: (a) swing-out rotor; (b) fixed angle rotor; (c) vertical tube rotor.

- Fixed-angle rotors: used in many high-speed centrifuges and microcentrifuges (Fig. 23.5b). With their shorter path length, fixed rotors are more effective at pelleting particles than swing-out rotors.
- Vertical tube rotors: used for isopycnic density gradient centrifugation in high-speed centrifuges and ultracentrifuges (Fig. 23.5c). They cannot be used to harvest particles in suspension as a pellet is not formed.

Box 23.1 How to use a low-speed bench centrifuge

1. **Choose the appropriate tube size and material for your application,** with caps where necessary. Most low-speed machines have four-place or six-place rotors – use the correct number of samples to *fill* the rotor assembly whenever possible.
2. **Fill the containers to the appropriate level:** do not overfill, or the sample may spill during centrifugation.
3. **It is vital that the rotor is balanced during use.** Therefore, *identical* tubes must be prepared, to be placed opposite each other in the rotor assembly. This is particularly important for density gradient samples, or for samples containing materials of widely differing densities, e.g. soil samples, since the density profile of the tube will change during a run. However, for low-speed work using small amounts of particulate matter in aqueous solution, it may be sufficient to counterbalance a sample with a second tube filled with water, or a saline solution of similar density to the sample.
4. **Balance each pair of sample tubes** (plus the corresponding caps, where necessary) to within 0.1 g using a top-pan balance; add liquid dropwise to the lighter tube, until the desired weight is reached. Alternatively, use a set of scales. For small sample volumes (up to 10 ml) added to disposable, lightweight plastic tubes, accurate pipetting of your solution may be sufficient for low-speed use.
5. **For centrifuges with swing-out rotors,** check that each holder/bucket is correctly positioned in its locating slots on the rotor and that it is able to swing freely. All buckets must be in position on a swing-out rotor, even if they do not contain sample tubes – buckets are an integral part of the rotor assembly.
6. **Load the sample tubes into the centrifuge.** Make sure that the outside of the centrifuge tubes, the sample holders and sample chambers are dry: any liquid present will cause an imbalance during centrifugation, in addition to the corrosive damage it may cause to the rotor. For sample holders where rubber cushions are provided, make sure that these are correctly located. Balanced tubes must be placed opposite each other – use a simple code if necessary, to prevent mix-ups.
7. **Bring the centrifuge up to operating speed** by gentle acceleration. Do not exceed the maximum speed for the rotor and tubes used.
8. **If the centrifuge vibrates at any time during use, switch off** and find the source of the problem.
9. **Once the rotor has stopped spinning, release the lid and remove all tubes.** If any sample has spilled, make sure you clean it up thoroughly using a non-corrosive disinfectant, e.g. Virkon®, so that it is ready for the next user.
10. **Close the lid (to prevent the entry of dust) and return all controls to zero.**

Centrifuge tubes

Safe working practice with centrifuge tubes – *never* be tempted to use a tube or bottle which was not designed to fit the machine you are using (e.g. a general-purpose glass test-tube, or a screw-capped bottle), or you may damage the centrifuge and cause an accident.

These are manufactured in a range of sizes (from 1.5 ml up to 1 000 ml) and materials. The following aspects may influence your choice:

- Capacity. This is obviously governed by the volume of your sample. Note that centrifuge tubes must be completely full for certain applications, e.g. for high-speed work.
- Shape. Conical-bottomed centrifuge tubes retain pellets more effectively than round-bottomed tubes, while the latter may be more useful for density gradient work.
- Maximum centrifugal force. Detailed information is supplied by the manufacturers. Standard Pyrex® glass tubes can only be used at low centrifugal force (up to 2 000 *g*).
- Solvent resistance. Glass tubes are inert, polycarbonate tubes are particularly sensitive to organic solvents (e.g. ethanol, acetone), while polypropylene tubes are more resistant. See manufacturer's guidelines for detailed information.
- Sterilization. Disposable plastic centrifuge tubes are often supplied in sterile form. Glass and polypropylene tubes can be repeatedly sterilized. Cellulose ester tubes should *not* be autoclaved. Repeated autoclaving of polycarbonate tubes may lead to cracking/stress damage.
- Opacity. Glass and polycarbonate tubes are clear, while polypropylene tubes are more opaque.
- Ability to be pierced. If you intend to harvest your sample by puncturing the tube wall, cellulose acetate and polypropylene tubes are readily punctured using a syringe needle.
- Caps. Most fixed-angle and vertical tube rotors require tubes to be capped, to prevent leakage during use and to provide support to the tube during centrifugation. For low-speed centrifugation, caps must be used for any hazardous samples. Make sure you use the correct caps for your tubes.

Using microcentrifuge tubes – the integral push-on caps of microcentrifuge tubes must be correctly pushed home before use or they may come off during centrifugation.

Balancing the rotor

For the safe use of centrifuges, the rotor must be balanced during use, or the spindle and rotor assembly may be permanently damaged; in severe cases, the rotor may fail and cause a serious accident.

KEY POINT ***It is vital that you balance your loaded centrifuge tubes before use.*** **As a general rule, *balance all sample tubes to within 1% or better,* using a top-pan balance or scales. Place balanced tubes opposite each other.**

Safe practice

Given their speed of rotation and the extremely high forces generated, centrifuges have the potential to be extremely dangerous, if used incorrectly. For safety reasons, all centrifuges are manufactured with an armoured casing that should contain any fragments in cases of rotor failure. Machines usually have a safety lock to prevent the motor from being switched on unless the lid is closed and to stop the lid from being opened while the rotor is moving. Don't be tempted to use older machines without a safety lock, or centrifuges where the locking mechanism is damaged/inoperative. Be particularly careful to make sure that hair and clothing are kept well away from moving parts.

Balancing tubes – *never* balance centrifuge tubes 'by eye' – use a balance. Note that a 35 ml tube full of liquid at an RCF of 3 000 *g* has an effective weight greater than a large adult man.

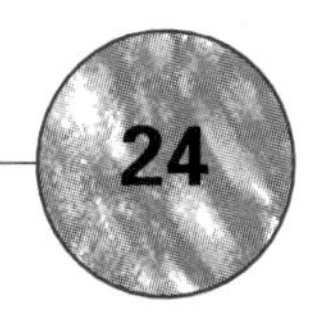

24 Chromatography – separation methods

Chromatography is often a three-way compromise between:

- separation of analytes
- time of analysis
- volume of eluent

Chromatography is used to separate the individual constituents within a sample on the basis of differences in their physical characteristics, e.g. molecular size, shape, charge, volatility, solubility and/or adsorptivity. The essential components of a chromatographic system are:

- A stationary phase, either a solid, a gel or an immobilized liquid, held by a support matrix.
- A chromatographic bed: the stationary phase may be packed into a glass or metal column, spread as a thin layer on a sheet of glass or plastic, or adsorbed on cellulose fibres (paper).
- A mobile phase, either a liquid or a gas which acts as a solvent, carrying the sample through the stationary phase and eluting from the chromatographic bed.
- A delivery system to pass the mobile phase through the chromatographic bed.
- A detection system to monitor the test substances (Chapter 25).

Individual substances interact with the stationary phase to different extents as they are carried through the system, enabling separation to be achieved.

KEY POINT In a chromatographic system, those substances which interact strongly with the stationary phase will be retarded to the greatest extent while those which show little interaction will pass through with minimal delay, leading to differences in distances travelled or elution times.

Types of chromatographic system

Chromatographic systems can be categorized according to the form of the chromatographic bed, the nature of the mobile and stationary phases and the method of separation.

Thin-layer chromatography (TLC) and paper chromatography

Here, you apply the sample as a single spot near one end of the sheet, by microsyringe or microcapillary. This sheet is allowed to dry fully, then it is transferred to a glass tank containing a shallow layer of solvent (Fig. 24.1). Remove the sheet when the solvent front has travelled across 80–90% of its length.

Fig. 24.1 Components of a TLC system.

You can express movement of an individual substance in terms of its relative frontal mobility, or R_F value, where:

$$R_F = \frac{\text{distance moved by substance}}{\text{distance moved by solvent}} \qquad [24.1]$$

Alternatively, you may express movement with respect to a standard of known mobility, as R_X, where:

$$R_X = \frac{\text{distance moved by test substance}}{\text{distance moved by standard}} \qquad [24.2]$$

The R_F (or R_X) value is a constant for a particular substance and solvent system (under standard conditions) and closely reflects the partitioning of the substance between the stationary and mobile phases. Tabulated values are available for a range of biological molecules and solvents (e.g. Stahl, 1965). However, you should analyse one or more reference compounds on the same sheet as your unknown sample, to check their R_F values.

Using a TLC system – it is essential that you allow the solvent to pre-equilibrate in the chromatography tank for at least 2 h before use, to saturate the atmosphere with vapour. Deliver drops of sample with a blunt-ended microsyringe. Make sure you know exactly where each sample is applied, so that R_F values can be calculated.

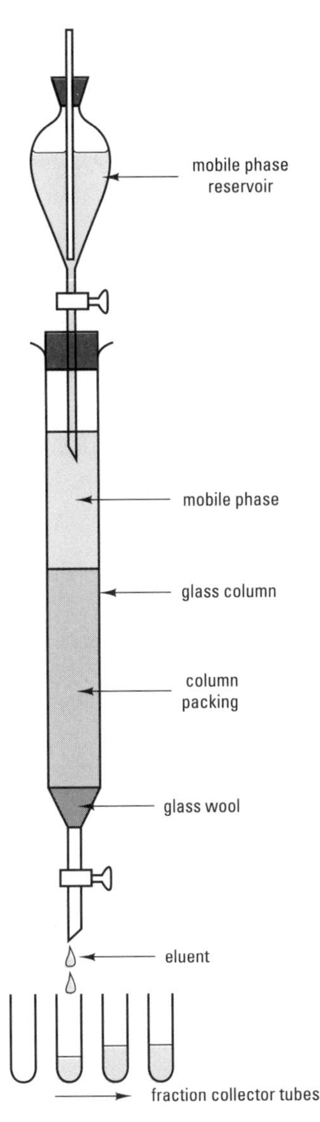

Fig. 24.2 Equipment for column chromatography (gravity feed system).

HPLC is a versatile form of chromatography, used with a wide variety of stationary and mobile phases, to separate individual compounds of a particular class of molecules on the basis of size, polarity, solubility or adsorption characteristics.

Problems with peaks – non-symmetrical peaks may result from column overloading, co-elution of solutes, poor packing of the stationary phase, or interactions between the substance and the support material.

Column chromatography

Here, you pack a glass column with the appropriate stationary phase and equilibrate the mobile phase by passage through the column, either by gravity (Fig. 24.2), or using a low pressure peristaltic pump. You can then introduce the sample to the top of the column, to form a discrete band of material. This is then flushed through the column by the mobile phase. If the individual substances have different rates of migration, they will separate within the column, eluting at different times as the mobile phase travels through the column.

You can detect eluted substances by collecting the mobile phase as it elutes from the column in a series of tubes (discontinuous monitoring), either manually or with an automatic fraction collector. Fractions of 2–5% of the bed volume are usually collected and analysed, e.g. by chemical assay. You can now construct an elution profile (or chromatogram) by plotting the amount of substance against either time, elution volume or fraction number, which should give a symmetrical peak for each substance (Fig. 24.3).

You can express the migration of a particular substance at a given flow rate in terms of its retention time (t), or elution volume (V_e). The separation efficiency of a column is measured by its ability to distinguish between two similar substances, assessed in terms of:

- selectivity, as measured by the difference in retention times of the two peaks (i.e. as $t_a - t_b$);
- resolution (R_s), quantified in terms of the retention time and the base width (W) of each peak:

$$R_s = \frac{2(t_a - t_b)}{W_a + W_b} \qquad [24.3]$$

where the subscripts a and b refer to substances a and b respectively (Fig. 24.3). For most practical purposes, R_s values of 1 or more are satisfactory, corresponding to 98% peak separation for symmetrical peaks.

High performance liquid chromatography (HPLC)

Column chromatography originally used large 'soft' stationary phases that required low pressure flow of the mobile phase to avoid compression; separations were usually time-consuming and of low resolution ('low performance'). Subsequently, the production of small, incompressible, homogeneous particulate support materials and high pressure pumps with reliable, steady flow rates have enabled high performance systems to be developed. These systems operate at pressures up to 10 MPa, forcing the mobile phase through the column at a high flow rate to give rapid separation with reduced band broadening, due to smaller particle size.

HPLC columns are usually made of stainless steel, and all components, valves, etc., are manufactured from materials which can withstand the high pressures involved. The choice of solvent delivery system depends on the type of separation to be performed:

- Isocratic separation: a single solvent (or solvent mixture) is used throughout the analysis.
- Gradient elution separation: the composition of the mobile phase is altered using a microprocessor-controlled gradient programmer, which mixes appropriate amounts of two different substances to produce the required gradient.

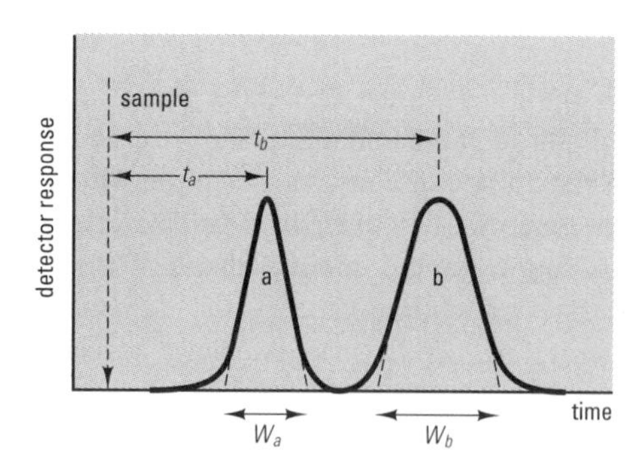

Fig. 24.3 Peak characteristics in a chromatographic separation, i.e. a chromatogram. For symbols, see Eqn 24.3, p. 129.

Most HPLC systems are linked to a continuous monitoring detector of high sensitivity, e.g. proteins may be detected spectrophotometrically by monitoring the absorbance of the eluent at 280 nm as it passes through a flow cell (cuvette). Other detectors can be used to measure changes in fluorescence, current or potential (Chapter 25). Most detection systems are non-destructive, which means that you can collect eluent with an automatic fraction collector for further study (Fig. 24.4).

The speed, sensitivity and versatility of HPLC makes this the method of choice for the separation of many small molecules of biological interest, normally using reverse phase partition chromatography (p. 131). Separation of macromolecules (especially proteins and nucleic acids) usually requires 'biocompatible' systems in which stainless steel components are replaced by titanium, glass or fluoroplastics, using lower pressures to avoid denaturation, e.g. the Pharmacia FPLC® system. Such separations are carried out using ion-exchange, gel permeation and/or hydrophobic interaction chromatography (pp. 131–4).

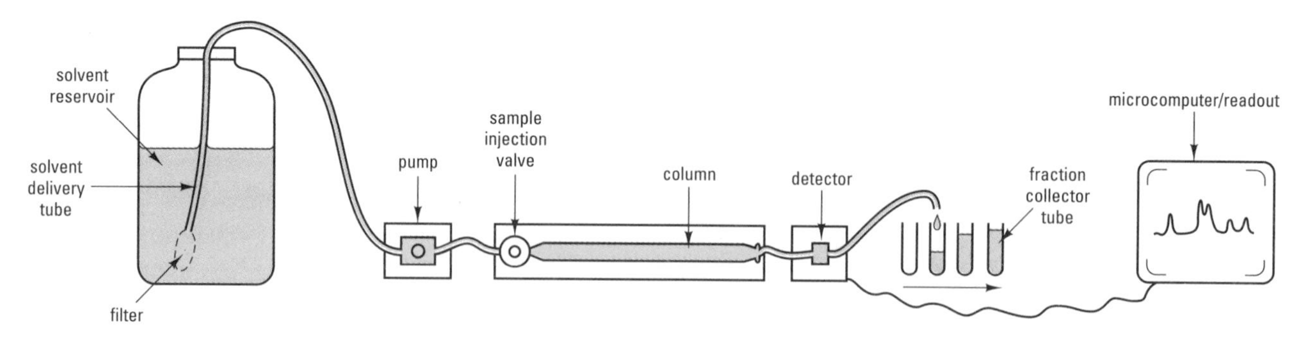

Fig. 24.4 Components of an HPLC system.

Gas chromatography (GC)

Modern GC uses capillary chromatography columns (internal diameter 0.1–0.5 mm) up to 50 m in length (Fig. 24.5). The stationary phase is generally a cross-linked silicone polymer, coated as a thin film on the inner wall of the capillary: at normal operating temperatures, this behaves in a similar manner to a liquid film, but is far more robust. The mobile phase ('carrier gas') is usually nitrogen or helium. Selective separation is achieved as a result of the differential partitioning of individual compounds between the carrier gas and silicone polymer phases. The separation of most biomolecules is influenced

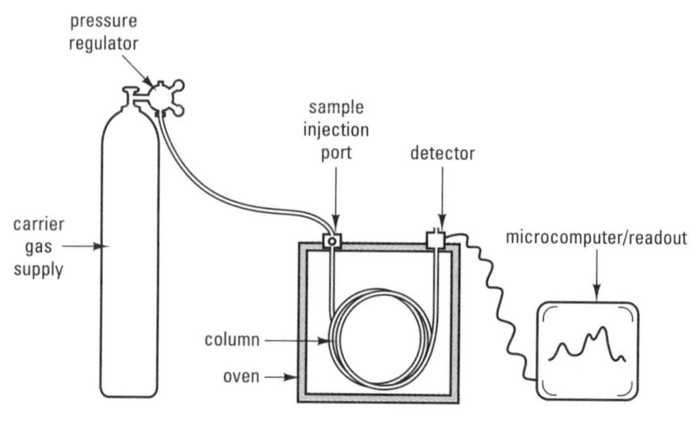

Fig. 24.5 Components of a GC system.

by the temperature of the column, which may be constant during the analysis ('isothermal' – usually 50–250 °C) or, more commonly, may increase in a pre-programmed manner (e.g. from 50 °C to 250 °C at 10 °C per min). Samples are injected onto the 'top' of the column, through a sample injection port containing a gas-tight septum. The output from the column can be monitored by flame ionization, electron capture or thermal conductivity (Chapter 25).

Spectrometric detection systems include mass spectrometry (GC-MS) and infra-red spectroscopy (GC-IR). GC can only be used with samples capable of volatilization at the operating temperature of the column, e.g. short chain fatty acids. Other substances may need to be chemically modified to produce more volatile compounds, e.g. long chain saturated fatty acids (Chapter 31) are usually analysed as methyl esters while monosaccharides (Chapter 32) are converted to their trimethylsilyl derivatives.

Applications of gas–liquid chromatography – GLC is used to separate volatile, non-polar compounds: substances with polar groups must be converted to less polar derivatives prior to analysis, in order to prevent adsorption on the column, resulting in poor resolution and peak tailing.

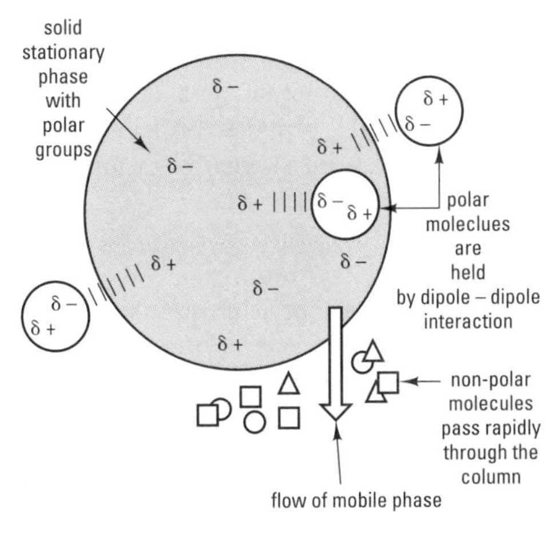

Fig. 24.6 Adsorption chromatography (polar stationary phase).

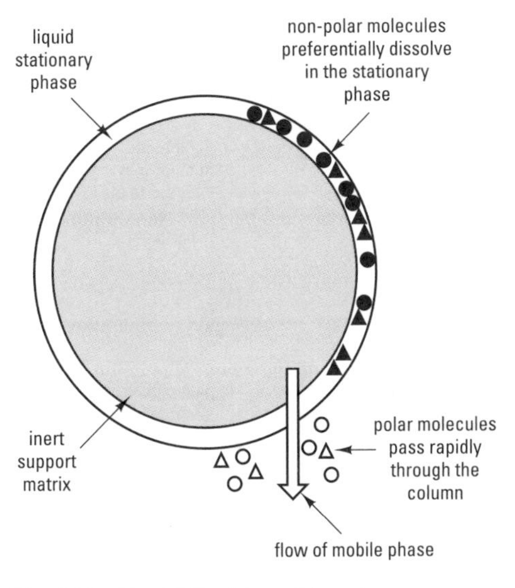

Fig. 24.7 Liquid–liquid partition chromatography, e.g. reverse phase HPLC.

Separation methods

Adsorption chromatography

This is a form of solid–liquid chromatography. The stationary phase is a porous, finely divided solid which adsorbs molecules of the test substance on its surface due to dipole–dipole interactions, hydrogen bonding and/or van der Waals interactions (Fig. 24.6). The range of adsorbents is limited, e.g. polystyrene-based resins (for non-polar molecules), silica, aluminium oxide and calcium phosphate (for polar molecules). Most adsorbents must be activated by heating to 110–120 °C before use, since their adsorptive capacity is significantly decreased in the presence of bound water. Adsorption chromatography can be carried out in column or thin-layer form, using a wide range of organic solvents.

Partition chromatography

This is based on the partitioning of a substance between two liquid phases, in this instance the stationary and mobile phases. Substances which are more soluble in the mobile phase will pass rapidly through the system while those which favour the stationary phase will be retarded (Fig. 24.7). In normal phase partition chromatography the stationary phase is a polar solvent, usually water, supported by a solid matrix (e.g. cellulose fibres in paper chromatography) and the mobile phase is an immiscible, non-polar organic solvent. For reverse-phase partition chromatography the stationary phase is a non-polar solvent (e.g. a C_{18} hydrocarbon, such as octadecylsilane) which is chemically bonded to a porous support matrix (e.g. silica), while the mobile phase can be chosen from a wide range of polar solvents, usually water or an aqueous buffered solution containing one or more organic solvents, e.g. acetonitrile. Solutes interact with the stationary phase through non-polar interactions and so the *least* polar solutes elute last from the column. Solute retention and separation are controlled by changing the composition of the mobile phase (e.g. % v/v acetonitrile). Reverse-phase high performance liquid chromatography (RP–HPLC) is used to separate a broad range of non-polar, polar and ionic biomolecules, including peptides, proteins, oligosaccharides and vitamins for ionic and ionizable solutes.

Ion-exchange chromatography

Here, separations are carried out using a column packed with a porous matrix which has a large number of ionized groups on its surfaces, i.e. the stationary phase is an ion-exchange resin. The groups may be cation or anion exchangers, depending upon their affinity for positive or negative ions. The

Selecting a separation method – it is often best to select a technique that involves direct interaction between the substance(s) and the stationary phase, (e.g. ion-exchange, affinity or hydrophobic interaction chromatography), due to their increased capacity and resolution compared to other methods, (e.g. partition or gel permeation chromatography).

Maximizing resolution in IEC (and HIC) – keep your columns as short as possible. Once the sample components have been separated, they should be eluted as quickly as possible from the column in order to avoid band broadening resulting from diffusion of sample ions in the mobile phase.

Using a gel-permeation system – keep your sample volume as small as possible, in order to minimize band broadening due to dilution of the sample during passage through the column.

net charge on a particular resin depends on the pK_a of the ionizable groups and the pH of the solution, in accordance with the Henderson–Hasselbalch equation (p. 30).

For most practical applications, you should select the ion-exchange resin and buffer pH so that the test substances are strongly bound by electrostatic attraction to the ion-exchange resin on passage through the system, while the other components of the sample are rapidly eluted (Fig. 24.8). You can then elute the bound components by raising the salt concentration of the mobile phase, either stepwise or as a continuous gradient, so that exchange of ions of the same charge occurs at oppositely charged sites on the stationary phase. Weakly bound sample molecules will elute first, while more strongly bound molecules will elute at a higher concentration. Computer-controlled gradient formers are available: if two or more components cannot be resolved using a linear salt gradient, an adapted gradient can be used in which the rate of change in salt concentration is decreased over the range where these components are expected to elute.

Ion-exchange chromatography can be used to separate mixtures of a wide range of ionic biomolecules, including amino acids, peptides, proteins and nucleotides. Electrophoresis (Chapters 26 and 27) is an alternative means of separating charged biomolecules.

Gel permeation chromatography (GPC) or gel filtration

Here, the stationary phase is in the form of beads of a cross-linked gel containing pores of a discrete size (Fig. 24.9). The size of the pores is controlled so that at the molecular level, the pores act as 'gates' that will exclude large molecules and admit smaller ones. However, this gating effect is not an all or nothing phenomenon: molecules of intermediate size partly enter the pores. A column packed with such beads will have within it two effective volumes that are potentially available to sample molecules in the mobile phase, i.e. V_i, the volume surrounding the beads and V_{ii}, the volume within the pores. If a sample is placed at the top of such a column, the

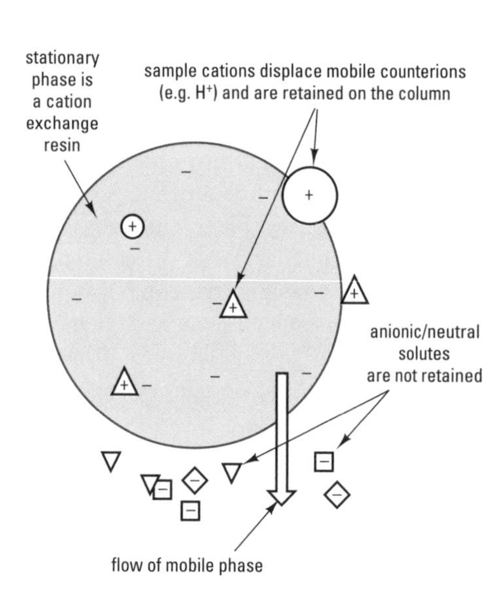

Fig. 24.8 Ion exchange chromatography (cation exchanger).

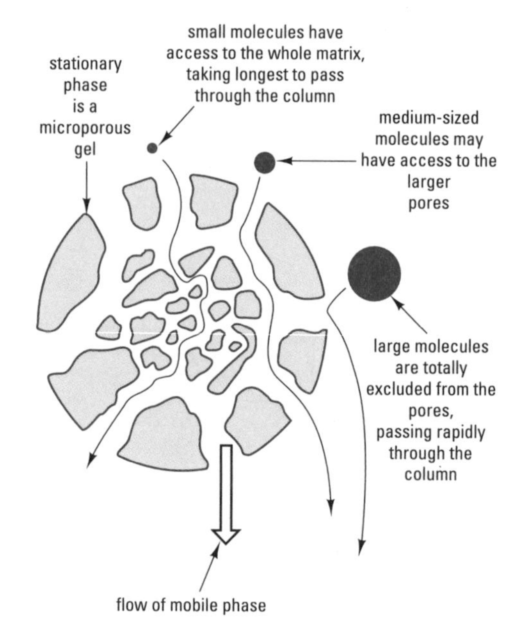

Fig. 24.9 Gel permeation chromatography.

Table 24.1 Fractionation ranges of selected gel permeation chromatography media

M_r	Medium
50–1000	Sephadex G15 Biogel P-2
1000–5000	Sephadex G-25
1500–30 000	Sephadex G-50 Biogel P-10
4000–150 000	Sephadex G-100
5000–250 000	Sephadex G-200
20 000–1 500 000	Sephacryl S 300
60 000–20 000 000	Sepharose 4B

mobile phase will carry the sample components down the column, but at different rates according to their molecular size. A very large molecule will have access to all of V_i but to none of V_{ii}, and will therefore elute in the minimum possible volume (the 'void volume', or V_o, equivalent to V_i). A very small molecule will have access to all of V_i and all of V_{ii}, and therefore it has to pass through the total liquid volume of the column (V_t, equivalent to $V_i + V_{ii}$) before it emerges. Molecules of intermediate size have access to all of V_i but only part of V_{ii}, and will elute at a volume between V_o and V_t, in order of decreasing size depending on their access to V_{ii}.

Cross-linked dextrans (e.g. Sephadex®), agarose (e.g. Sepharose®) and polyacrylamide (e.g. Bio-gel®) can be used to separate mixtures of macromolecules, particularly enzymes, antibodies and other globular proteins. Selectivity in GPC is solely dependent on the stationary phase, with the mobile phase being used solely to transport the sample components through the column. Thus, it is possible to estimate the molecular mass of a sample component by calibrating a given column using molecules of known molecular mass and similar shape. A plot of elution volume (V_e) against $\log_{10}$ molecular mass is approximately linear. A further application of GPC is the general separation of low molecular mass and high molecular mass components, e.g. 'desalting' a protein extract by passage through a Sephadex® G-25 column is faster and more efficient than dialysis.

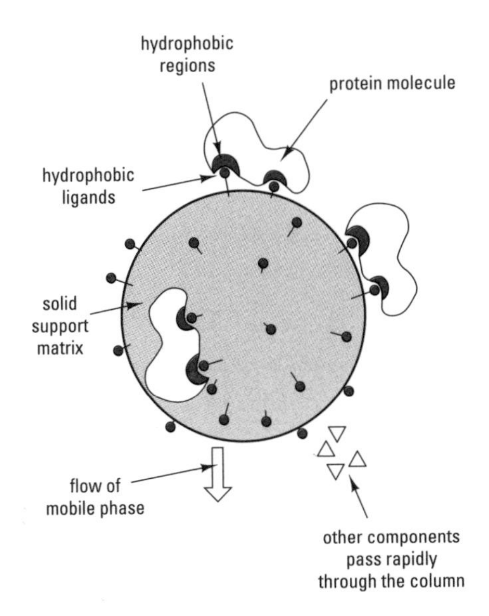

Fig. 24.10 Hydrophobic interaction chromatography. Hydrophobic interactions between ligands and hydrophobic amino acid residues release 'structured' water, making the interactions thermodynamically favourable.

Hydrophobic interaction chromatography (HIC)

This technique is used to separate proteins, and exploits the fact that many proteins have hydrophobic sites, with hydrophobic amino acid residues (e.g. leucine, isoleucine, valine, phenylalanine) on their surfaces. Proteins differ in the nature and extent of these hydrophobic regions. The underlying principle is similar to that of RP-HPLC in that it involves hydrophobic interactions between the sample components and a non-polar stationary phase. However, RP-HPLC is only useful for analytical separations of proteins where retention of biological activity is not required, and is unsuitable for separation of native proteins for several reasons:

- the stationary phase in RP-HPLC columns tends to be densely packed with hydrophobic groups, leading to tight protein binding, possibly through multisite attachment;
- the use of polar organic solvents may lead to protein denaturation;
- the high pressures used to obtain rapid flow rates in HPLC may also denature proteins.

The groups used on HIC stationary phases are both less densely packed and less hydrophobic than those used in RP-HPLC, and this results in milder adsorption of proteins (octyl or phenyl groups are commonly used in HIC, rather than octadecyl groups). Furthermore, retention and elution can be achieved using aqueous solutions so that an individual protein can be isolated with its 3-D conformation intact.

Using HIC for purification – HIC can be used immediately after salt precipitation, when the salt concentration is high. If the desired biomolecule is then eluted using a gradient of reducing salt concentration, it may be possible to follow HIC directly by IEC (which requires an initial buffer of low ionic strength) without changing the buffer.

Separations are based on interactions between the three components of the system, i.e. the hydrophobic stationary phase, the hydrophobic sample molecules, and the aqueous stationary phase. In an aqueous environment, hydrophobic groups tend to associate, and this results in certain proteins binding to the stationary phase, where the strength of binding is related to the degree of hydrophobicity of the protein (Fig. 24.10). This tendency is promoted by the presence of certain salts, most commonly ammonium sulphate, that produce 'salting out' effects. Elution of components can be achieved by a variety of means:

- reducing the ammonium sulphate concentration of the mobile phase, to decrease the 'salting out' effect;
- changing the salt in the mobile phase to one that does not promote salting out;
- including non-ionic detergents (e.g. Triton X-100) to reduce hydrophobic interactions;
- including aliphatic alcohols, reducing the polarity of the mobile phase;
- changing the pH;
- reducing the temperature.

Avoiding protein precipitation in HIC – certain protein components may precipitate in high ionic strength buffers. If this occurs, dilute your sample and inject larger volumes.

Affinity chromatography (AC)

Affinity chromatography enables biomolecules to be purified on the basis of their biological specificity rather than by differences in physico–chemical properties, and a high degree of purification (>1000-fold) can be expected. It is especially useful for isolating small quantities of material from large amounts of contaminating substances. The technique involves the immobilization of a complementary binding substance (the ligand) onto a solid matrix in such a way that the specific binding affinity of the ligand is preserved. When a biological sample is applied to a column packed with this affinity support matrix, the molecule of interest will bind specifically to the ligand, while contaminating substances will be washed through with buffer (Fig. 24.11). Elution of the desired molecule can be achieved by changing the pH or ionic strength of the buffer, to weaken the non-covalent interactions between the molecule and the ligand, or by the addition of other substances that have greater affinity for the ligand.

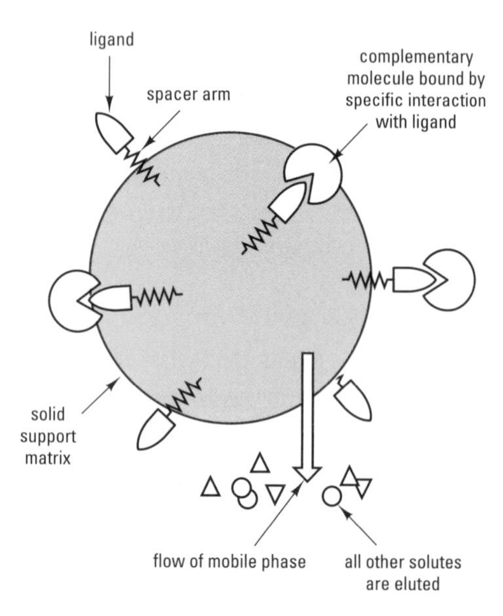

Fig. 24.11 Affinity chromatography.

In AC, the ligand must show specific, but reversible, interaction with the molecule to be purified. It must also contain a reactive functional group, independent of the biospecific site, that will allow covalent attachment to the matrix. The support matrix must be free of non-specific adsorption effects and have sufficient reactive functional groups for the attachment of ligands. Agarose is an ideal support matrix for use in AC. If the ligand is small and the molecule to be purified is large, binding may be restricted due to the proximity of the matrix surface. This problem may be overcome by the introduction of a 'spacer arm' (e.g. a hexane group) between the ligand and the matrix (Fig. 24.11).

Examples Biospecific molecules used in affinity chromatography include:

- enzymes and inhibitors/cofactors/substrates;
- hormones and receptor proteins;
- antibodies and antigens;
- complementary base sequences in DNA and RNA (p. 180).

A potential disadvantage of the specificity of AC is that a new ligand may have to be sought for each individual separation – a potentially time-consuming and expensive process. It may be more practical to use 'group-specific adsorbents' which contain ligands that have an affinity for a class of biochemically related substances. Examples of group-specific adsorbents include:

- Lectins, which are a group of proteins produced from moulds, plants and animals, that bind reversibly with specific sugar residues. They are very useful for the purification of sugar-containing macromolecules such as glycoproteins, serum lipoproteins, and membrane proteins such as receptors. Different lectins show different specificities, e.g. concanavalin A and lentil lectin bind to sugars having –OH groups at C-3, C-4 and C-5 (i.e. mannose and glucose), while wheat germ lectin binds to N-acetyl glucosamine residues. Once bound, substances can be resolved by eluting the column with a salt gradient, or by including free sugars to act as

Elution of substances from an affinity system – make sure that your elution conditions do not affect the interaction between the ligand and the stationary phase, or you may elute the ligand from the column.

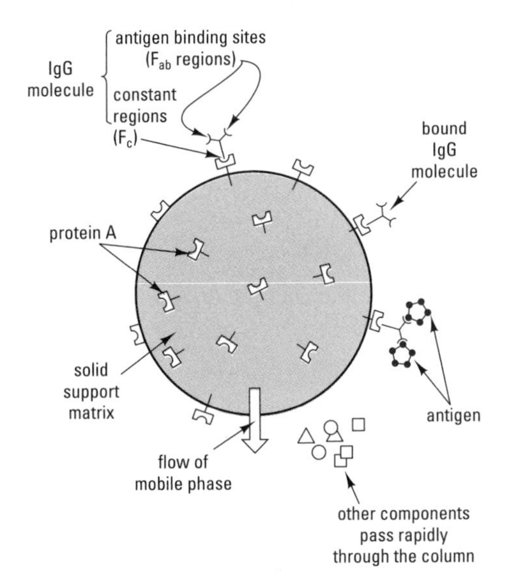

Fig. 24.12 Affinity chromatography using protein A as a ligand to bind IgG antibodies via the F_c region.

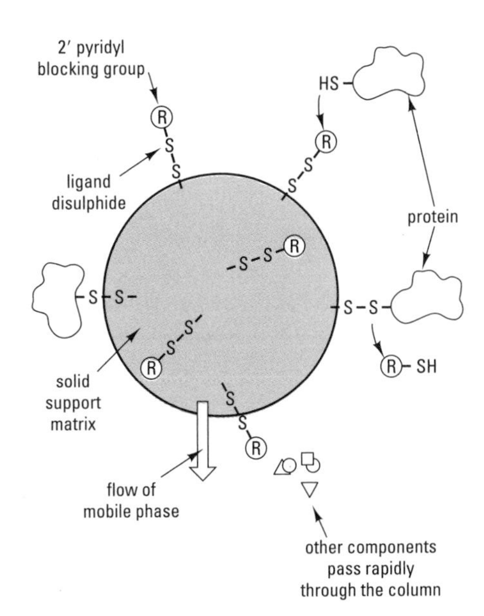

Fig. 24.13 Covalent chromatography. A protein with a free surface thiol group reduces the ligand disulphide, releasing the blocking group and forming a new disulphide.

competitive binding agents. Lectins are sometimes called agglutinins because of their ability to agglutinate different types of erythrocyte (e.g. A, B and O) by binding to the specific receptors on their surfaces.

- Protein A, which is a commercially available surface protein from the bacterium *Staphylococcus aureus* that has a specific binding capacity for the constant (F_c) region of IgG-type antibodies from most mammalian species (Fig. 24.12). As well as being useful in the purification of IgG, protein A can also be used for the isolation of other molecules, as long as an antibody can be raised against them. This is possible because interaction of Protein A with IgG occurs via the F_c segment of immunoglobulins of this class, irrespective of their F_{ab} components (see Fig. 18.1). Immobilization of a specific IgG antibody onto a Protein A-agarose column produces an affinity support that can be used to purify the desired antigen.
- Immobilized dyes, which can be used to purify a wide range of enzymes and proteins. One of the most widely used dyes is Cibacron Blue F3G-A, available as an affinity support as Blue Sepharose®. This dye has some structural similarities to nucleotide cofactors such as NAD^+ and $NADP^+$, and is useful for the purification of enzymes that require such cofactors. Chemical modification of such dyes increases their specificity, and it is likely that computer modelling will play a role in the future development of new dye-based affinity supports. Dyes are also used in the related non-chromatographic technique of affinity precipitation. This makes use of two dye molecules linked via a spacer molecule; when this 'bis-ligand' is added to a solution containing the molecule to be purified, the ligand specifically binds two desired molecules, forming an insoluble complex. Alternatively, a heterobifunctional ligand can be used – a combination of a specific ligand and a group that can be used to initiate precipitation once the desired molecule has bound to the ligand. Such interactions are equivalent to the precipitin reaction in immunology (p. 96).
- Poly(U)-agarose is an affinity support that can be used for the isolation of mRNA because of the biospecific hybridization of poly(U) with the poly(A) 'tail' sequence characteristic of mRNAs. It can also be used to isolate proteins and enzymes that bind to RNA, such as reverse transcriptases.

Covalent chromatography

This is a variant of affinity chromatography which involves the formation of fairly strong, but reversible, covalent bonds between the affinity support and the molecule to be purified. One type of covalent chromatography is used for the purification of proteins containing thiol (–SH) groups (Fig. 24.13). If the sample is applied to a column containing matrix-attached disulphide 2′-pyridyl groups, thiol-containing proteins displace the 2′-pyridyl group on the support and become immobilized via disulphide bridges. Elution of bound components can be achieved by including thiol reagents such as cysteine, glutathione, 2-mercaptoethanol or dithiothreitol (DTT) in the mobile phase. Another form of covalent chromatography uses immobilized boronic acid, which binds certain carbohydrate groups (e.g. in glycoproteins).

Metal chelate chromatography

This exploits the ability of certain metal ions, especially Cu^{2+}, Zn^{2+}, Hg^{2+} and Cd^{2+}, to bind to proteins by forming co-ordination complexes with imidazole groups of histidine, indole groups of tryptophan, or thiol groups of cysteine residues. The metal ion may be immobilized by chelation with iminodiacetic acid covalently bound to agarose. Proteins that bind to the metal ions can be eluted using free metal-binding ligands (e.g. amino acids) in the mobile phase.

Optimizing chromatographic separations

Remember that you cannot *quantify* a particular analyte without first *identifying* it: the presence of a single peak on a chromatogram does not prove that a single type of analyte is present.

In any chromatographic technique, sample components leave the column at different elution volumes/different times, and are then monitored by a suitable detector (Chapter 25). The responses of the detector are recorded on a chart or a screen in the form of a *chromatogram*. Ideally the sample biomolecules will be completely separated and detection of components will result in a series of discrete individual peaks corresponding to each type of biomolecule (e.g. Fig. 24.3). However, to minimize the possibility of overlapping peaks, or of peaks composed of more than one component, it is important to maximize the separation efficiency of the technique, which depends on:

- the selectivity, as measured by the relative retention times of the two components, or by the volume of mobile phase between the peak maxima of the two components after they have passed through the column; this depends on the ability of the chromatographic method to separate two components with similar properties;
- the band-broadening properties of the chromatographic system, which influence the width of the peaks; these are mainly due to the effects of diffusion.

Separating small biomolecules – these will diffuse faster than large molecules, so they should be separated using faster flow rates.

The separation efficiency, or resolution of two adjacent components, can be defined in terms of selectivity and peak broadening, using Eqn. 24.3. In practical terms, good resolution is achieved when there is a large 'distance' (either time or volume) between peak maxima, and the peaks are as narrow as possible. The resolution of components is also affected by the relative amount of each substance: for systems showing low resolution, it can be difficult to resolve small amounts of a particular component in the presence of larger amounts of a second component. If you cannot obtain the desired results from a poorly resolved chromatogram, other chromatographic conditions, or even different methods, should be tried in an attempt to improve resolution. For liquid chromatography, changes in the following factors may improve resolution:

- Stationary phase particle size – the smaller the particle, the greater the area available for partitioning between the mobile phase and the stationary phase. This partly accounts for the high resolution observed with HPLC and FPLC compared with low pressure methods.
- The slope of the salt gradient in eluting IEC or HIC columns, e.g. using computer-controlled adapted gradients (p. 187).
- In low pressure liquid chromatography, the flow rate of the mobile phase must be optimized because this influences two band broadening effects which are dependent on diffusion of sample molecules, i.e. (i) the flow rate must be slow enough to allow effective partitioning between the mobile phase and the stationary phase, and (ii) it must be fast enough to ensure that there is minimal diffusion along the column once the molecules have been separated. To allow for these opposing influences, a compromise flow rate must be used.

Learning from experience – if you are unable to separate your biomolecule using a particular method, do not regard this as a failure, but instead, think about what this tells you about either the substance(s) or the sample.

- If you prepare your own columns, they must be packed correctly, with no channels present that might result in uneven flow and eddy diffusion.

Quantifying biomolecules – note that quantitative analysis often requires assumptions about the *identity* of separated components (p. 165) and that further techniques may be required to provide *information* about the nature of the biomolecules present, e.g. mass spectrometry (see Chapter 22).

Quantitative analysis

Most detectors and chemical assay systems give a linear response with increasing amounts of the test substance over a given 'working range' of concentration. Alternative ways of converting the measured response to an amount of substance are:

- External standardization: this is applicable where the sample volume is sufficiently precise to give reproducible results (e.g. HPLC, column chromatography). You measure the peak areas (or heights) of known amounts of the substance to give a calibration factor or calibration curve which can be used to calculate the amount of test substance in the sample.
- Internal standardization: where you add a known amount of a reference substance (not originally present in the sample) to the sample, to give an additional peak in the elution profile. You determine the response of the detector to the test and reference substances by analysing a standard containing known amounts of both substances, to provide a response factor (r), where

$$r = \frac{\text{peak area (or height) of test substance}}{\text{peak area (or height) of reference substance}} \quad [24.4]$$

Use this response factor to quantify the amount of test substance (Q_t) in a sample containing a known amount of the reference substance (Q_r), from the relationship:

$$Q_t = \frac{\text{peak area (or height) of test substance}}{\text{peak area (or height) of reference substance}} \times (Q_r/r) \quad [24.5]$$

When using external standardization, samples and standards should be analysed more than once, to confirm the reproducibility of the technique.

When using an internal standard, you should add an internal standard to the sample at the first stage in the extraction procedure, so that any loss or degradation of test substance during purification is accompanied by an equivalent change in the internal standard, as long as the extraction characteristics of the internal standard and the test substance are very similar.

Internal standardization should be the method of choice wherever possible, since it is unaffected by small variations in sample volume (e.g. for GC microsyringe injection). The internal standard should be chemically similar to the test substance(s) and must give a peak that is distinct from all other substances in the sample. An additional advantage of an internal standard which is chemically related to the test substance is that it may show up problems due to changes in detector response, incomplete derivatization, etc. A disadvantage is that it may be difficult to fit an internal standard peak into a complex chromatogram.

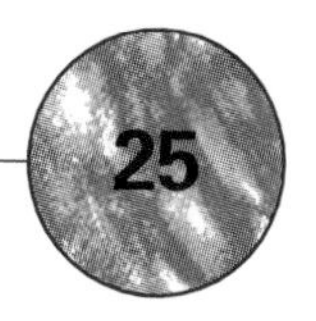

25 Chromatography – detection and analysis

Once the individual components in a sample mixture have been separated by a given chromatographic technique (Chapter 24), a suitable detection system is required to monitor and record the elution of the components of the mixture as they pass through a 'flow cell'.

KEY POINT The most appropriate detector depends on the type of chromatography and the application: ideally, the detector should show high sensitivity, a low detection limit and minimal noise or drift. These terms are defined in Chapter 29.

Liquid chromatography detectors

UV/visible detectors

UV detection of proteins – note that the absorbance at 280 nm is mostly due to tryptophan and tyrosine residues; if a protein contains low amounts of these amino acids, its absorbance at this wavelength will be low.

These are widely used and have the advantages of versatility, sensitivity and stability. Such detectors are of two types: fixed wavelength and variable wavelength. Fixed-wavelength detectors are simple to use, with low operating costs. They usually contain a mercury lamp as a light source, emitting at several wavelengths between 254 and 578 nm; a particular wavelength is selected using suitable cut-off filters. The most frequently used wavelengths for analysis of biomolecules are 254 nm (for nucleic acids) and 280 nm (for proteins). Variable-wavelength detectors use a deuterium lamp and a continuously adjustable monochromator for wavelengths of 190–600 nm. For both types of detector, sensitivity is in the absorbance range 0.001–1.0 (down to $\cong$ 1 ng), with noise levels as low as 4×10^{-5}. Note that sensitivity is partly influenced by the path length of the flow cell (typically 10 mm). Monitoring at short wavelength UV (e.g. below 240 nm) may give increased sensitivity but decreased specificity, since many biological molecules absorb in this range. Additional problems with short wavelength UV detection include instrument instability, giving a variable baseline, and absorption by components of the eluting buffer (e.g. TRIS, which absorbs at 206 nm).

An important development in chromatographic monitoring is diode array detection (DAD). The incident light comprises the whole spectrum of light from the source, which is passed through a diffraction grating and the diffracted light detected by an array of photodiodes. A typical DAD can measure the absorbance of each sample component at 1–10 nm intervals over the range 190–600 nm. This gives an absorbance spectrum for each eluting substance which may be used to identify the compound and give some indication as to its purity. An example of a three-dimensional diode array spectrum is shown in Fig. 25.1.

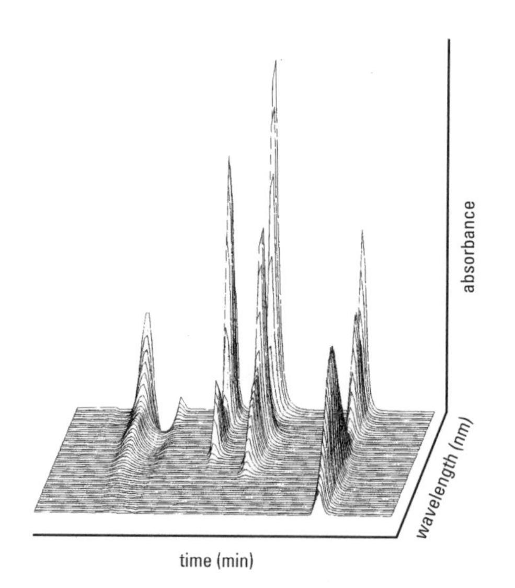

Fig. 25.1 Diode array detector absorption spectra of the eluent from an HPLC separation of a mixture of four steroids, taken every 15 seconds.

Fluorescence detectors

Overcoming interference with fluorescence detectors – use a dual flow cell to offset background fluorescence due to components of the mobile phase.

Many biomolecules, including some vitamins, nucleotides and porphyrins show natural fluorescence (Table 21.1), or can be made to fluoresce by pre-column or post-column derivatization with a fluorophore. Fluorescence detection is more sensitive than UV/visible detection (p. 115), and may allow analysis in the pg range. A fluorescence detector consists of a light source

Maximizing sensitivity with fluorescence detectors – the concentration of other sample components, e.g. pigments, must not be so high that they cause quenching of fluorescence.

(e.g. a xenon lamp), a diffraction grating to supply light at the excitation wavelength, and a photomultiplier to monitor the emitted light (usually arranged to be at right angles to the excitation beam). The use of instruments with a laser light source can give an extremely narrow excitation waveband, and increased sensitivity and specificity.

Electrochemical detectors

These offer very high sensitivity and specificity, with the possibility of detection of fg amounts of electroactive compounds such as catecholamines, vitamins, thiols, purines, ascorbate and uric acid. The two main types of detector, amperometric and coulometric, operate on similar principles, i.e. by measuring the change in current or potential as sample components pass between two electrodes within the flow cell. One of these electrodes acts as a reference (or counter) electrode (e.g. calomel electrode), while the other – the working electrode – is held at a voltage that is high enough to cause either oxidation or reduction of sample molecules. In the oxidative mode, the working electrode is usually glassy carbon, while in reductive mode a mercury electrode is used. In either case, a current flow between the electrodes is induced and detected.

Optimizing electrochemical detection – the mobile phase must be free of any compounds that might give a response; all constituents must be of the highest purity.

Gas chromatography detectors

The most commonly used detectors for GC analysis of biomolecules are:

Flame ionization detector (FID)

This is a widely used detector, being particularly useful for the analysis of a broad range of organic biomolecules. It involves passing the exit gas stream from the column through a hydrogen flame that has a potential of >100 V applied across it (Fig. 25.2). Most organic compounds, on passage through this flame, produce ions and electrons that create a small current across the electrodes, and this is amplified for measurement purposes. The FID is very sensitive (down to ≈0.1 pg), with a linear response over a wide concentration range. One drawback is that the sample is destroyed during analysis.

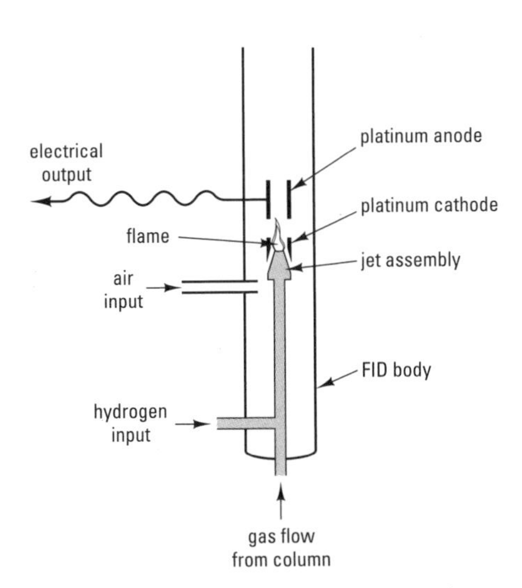

Fig. 25.2 Components of a flame ionization detector (FID).

Thermal conductivity detector (TCD)

This simple detector is based on changes in the thermal conductivity of the gas stream brought about by the presence of separated sample molecules. The detector elements are two electrically heated platinum wires, one in a chamber through which only the carrier gas flows (the reference detector cell), and the other in a chamber that takes the gas flow from the column (the sample detector cell). In the presence of a constant gas flow, the temperature of the wires (and therefore their electrical resistance) is dependent on the thermal conductivity of the gas. Analytes in the gas stream are detected by temperature-dependent changes in resistance dependent on the thermal conductivity of each separated molecule; the size of the signal is directly related to concentration of the analyte.

The advantages of TCD include its applicability to a wide range of organic and inorganic biomolecules and its non-destructive nature, since the sample can be collected for further study. Its major limitation is its low sensitivity (down to ≈10 ng), compared with other systems.

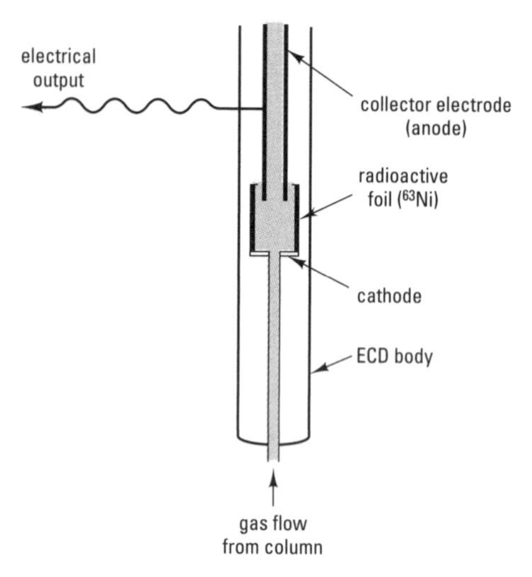

Fig. 25.3 Components of an electron capture detector (ECD).

Electron capture detector (ECD)

This highly sensitive detector (Fig. 25.3) is useful for the detection of certain compounds with electronegative functional groups, e.g. halogens, peroxides and quinones. The gas stream from the column passes over a β-emitter (p. 100) such as ^{63}Ni, which provides electrons that cause ionization of the carrier gas (e.g. nitrogen). When carrier gas alone is passing the β-emitter, its ionization results in a constant current flowing between two electrodes placed in the gas flow. However, when electron-capturing sample molecules are present in the gas flow, a decrease in current is detected. An example of the application of the ECD is in detecting and quantifying chlorinated pesticides.

Interfacing GC or HPLC with mass spectrometry

Mass spectrometry (p. 122) used in conjunction with chromatographic methods can provide a powerful tool for identifying the components of complex mixtures, e.g. aqueous pollutants. The procedure requires computer control of instruments and data storage/analysis. One drawback is the limited capacity of the mass spectrometer – due to its vacuum requirements – compared with the volume of material leaving the chromatography column. For capillary GC, the relatively small output can be fed directly into the ionization chamber of the mass spectrometer. For packed column GC, a 'jet separator' is used to remove most of the carrier gas from the molecules to be analysed before they enter the spectrometer. Similarly, in HPLC, devices have been developed for solving the problem of large solvent volumes, e.g. by splitting the eluent from the column so only a small fraction reaches the mass spectrometer.

The computer-generated outputs from the mass spectrometer are similar to chromatograms obtained from other methods, and show peaks corresponding to the elution of particular components. However, it is then possible to select an individual peak and obtain a mass spectrum for the component in that peak to aid in its identification (p. 122). This has helped to identify hundreds of components present in biological systems, including flavour molecules in food, drug metabolites and water pollutants.

Coupling capillary GC columns with FT-IR spectrometers (p. 118) provides another powerful means of separating and identifying compounds in complex biological mixtures.

Recording and interpreting chromatograms

Recording detector output

For analytical purposes, the detector output is usually connected to a computer-based data acquisition and analysis system. This consists of a personal computer (PC) with data acquisition hardware to convert the analogue detector signal to digital format, plus software to control the data acquisition process, store the signal information and display the resulting chromatogram. The software will also detect peaks and calculate their retention times and sizes (areas) for quantitative analysis. The software often incorporates functions to control the chromatographic equipment, enabling automatic operation. In sophisticated systems, the detector output may be compared with that from a 'library' of chromatograms for known compounds, to suggest possible identities of unknown sample peaks.

Interpreting chromatograms – *never* assume that a single peak is a guarantee of purity: there may be more than one compound with the same chromatographic characteristics.

In simpler chromatographic systems, you may need to use a chart recorder for detector output. Two important settings must be considered before using a chart recorder:

1. The baseline reading – this should be set only after a suitable quantity of mobile phase has passed through the column (prior to injection of the sample) and stability is established. The chart recorder is usually set a little above the edge of the chart paper grid, to allow for baseline drift.
2. The detector range – this must be set to ensure that the largest peaks do not go off the top of the chart. Adjustment may be based on the expected quantity of analyte, or by trial-and-error process. Use the maximum sensitivity that gives intact peaks. If peaks are still too large, you may need to reduce the amount of sample used, or prepare and analyse a diluted sample.

Using a fraction collector – make sure you can relate individual fractions to the position of peaks on the chromatogram. Most fraction collectors send a signal to the recorder each time a fraction is changed.

Interpretation of chromatograms

Make sure you know the direction of the horizontal axis (usually, either volume or time) – it may run from right to left or *vice versa*, and make a note of the detector sensitivity on the vertical axis.

Ideally, the baseline should be 'flat' between peaks, but it may drift up or down due to a number of factors including:

- changes in the composition of the mobile phase (e.g. in gradient elution, p. 132);
- tailing of material from previous peaks;
- carry-over of material from previous samples; this can be avoided by efficient cleaning of columns between runs – allow sufficient time for the previous sample to pass through the column before you introduce the next sample;
- loss of stationary phase from the column (column 'bleed'), caused by extreme elution conditions;
- air bubbles (in liquid chromatography); if the buffers used in the mobile phase are not effectively degassed, air bubbles may build up in the flow cell of the detector, leading to a gradual upward drift of the baseline, followed by a sharp fall when the accumulated air is released. Small air bubbles that do not become trapped may give spurious small peaks.

Avoiding problems with air bubbles in liquid chromatography – always ensure that buffers are effectively degassed by vacuum treatment before use, and regularly clean the flow cell of the detector.

A peak close to the origin may be due to non-retained sample molecules, flowing at the same rate as the mobile phase, or to artefacts, e.g. air (GC) or solvent (HPLC) in the sample. Whatever its origin, this peak can be used to measure the void volume of the column (p. 129). No peaks from genuine sample components should appear before this type of peak.

Peaks can be denoted on the basis of their elution volume (used mainly in liquid chromatography) or their retention times (mainly in GC). If the peaks are not narrow and symmetrical, they may contain more than one component. Where peaks are more curved on the trailing side compared with the leading side, this may indicate too great an association between the component and the stationary phase, or overloading of the column.

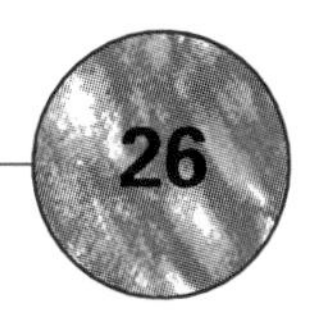

26 Principles and practice of electrophoresis

KEY POINT **Electrophoresis is a separation technique based on the movement of charged molecules in an electric field. Dissimilar molecules move at different rates and the components of a mixture will be separated when an electric field is applied. It is a widely used technique, particularly for the analysis of complex mixtures or for the verification of purity (homogeneity) of isolated biomolecules.**

While electrophoresis is mostly used for the separation of charged macromolecules, techniques are available for high resolution separations of small molecules such as amino acids (for example, by capillary electrophoresis, p. 152). This chapter deals mainly with the electrophoretic separation of proteins, which has many applications including clinical diagnosis. However, the principles apply equally to other molecules (separation of nucleic acids is considered in Chapter 40).

Definition

Electrophoretic mobility – the rate of migration of a particular type of molecule in response to an applied electrical field.

The electrophoretic mobility of a charged molecule depends on:

- Net charge – negatively charged molecules (anions) migrate towards the anode (+), while positively charged molecules (cations) migrate towards the cathode (−); highly charged molecules move faster towards the electrode of opposite charge than those with lesser charge.
- Size – frictional resistance exerted on molecules moving in a solution means that smaller molecules migrate faster than large molecules.
- Shape – the effect of friction also means that the shape of the molecule will affect mobility, e.g. globular proteins compared with fibrous proteins, linear DNA compared with circular DNA.
- Electrical field strength – mobility increases with increasing field strength (voltage), but there are practical limitations to using high voltages, especially due to heating effects.

The combined influence of net charge and size means that mobility (μ) is determined by the charge : density or the charge : mass ratio, according to the formula:

$$\mu = \frac{q\,\mathrm{E}}{r} \qquad [26.1]$$

where q is the net charge on the molecule, r is the molecular radius and E is the field strength.

Table 26.1 pK_a values of ionizable groups in selected amino acid residues of proteins

Group/residue	pK_a^1
Terminal carboxyl	3.1
Aspartic acid	4.4
Glutamic acid	4.4
Histidine	6.5
Terminal amino	8.0
Cysteine	8.5
Tyrosine	10.0
Lysine	10.0
Arginine	12.0

[1]Note that these are typical values: the pK_a will change with temperature and ionic strength. Acidic residues will tend to be negatively charged at pH values above their pK_a, while basic residues will tend to be positively charged below their pK_a.

Electrophoresis and the separation of proteins

The net charge of a sample molecule determines its direction of movement and significantly affects its mobility. The net charge of a protein molecule is pH dependent, and is determined by the relative numbers of positively and negatively charged amino acid side chains at a given pH (Table 26.1). The degree of ionization of each amino acid side chain is pH dependent, resulting in a variation of net charge on the protein at different pH values (Fig. 26.1). Since an individual protein will have a unique content of ionizable amino acids, each protein will have a characteristic 'titration curve' when net charge is plotted against pH. Thus, electrophoresis is always carried out at *constant* pH and a suitable buffer must be present along with the sample in order to maintain that pH (Chapter 6). If the proteins shown in Fig. 26.1 were subjected to electrophoresis at pH 9.0, and if the proteins were of similar size

Understanding electrophoresis – this is, in essence, an incomplete form of electrolysis, since the applied electrical field is switched off well before sample molecules reach the electrodes.

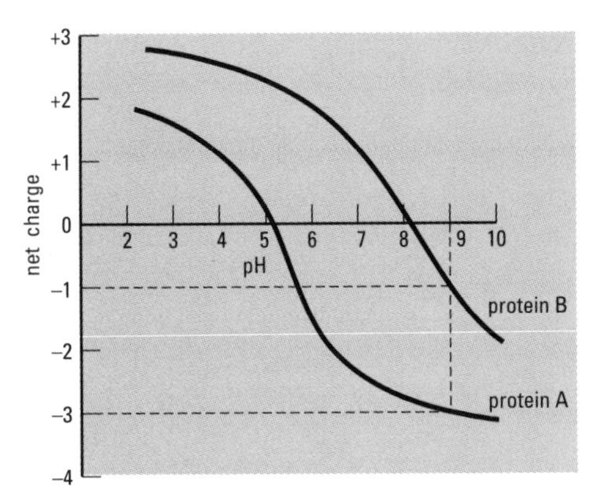

Fig. 26.1 Titration curves for two proteins, A and B containing different proportions of acidic and basic amino acid residues.

and shape, then the rate at which protein A (net charge, −3) migrates towards the anode would be faster than that for protein B (net charge, −1). Separation of proteins is usually carried out at alkaline pH, where most proteins carry a net negative charge.

Basic apparatus

Most types of electrophoresis using supporting media (described below) are simple to carry out and the apparatus can be easily constructed, although inexpensive equipment is commercially available. High-resolution techniques such as 2D-electrophoresis and capillary electrophoresis require more sophisticated equipment, both for separation and analysis (p. 152).

Simple electrophoretic separations can be performed either vertically (Fig. 26.2) or horizontally (Fig. 40.2). The electrodes are normally made of platinum wire, each in its own buffer compartment. In vertical electrophoresis, the buffer solution forms the electrical contact between the electrodes and the supporting medium in which the sample separation takes place. In horizontal electrophoresis electrical contact can be made by buffer-soaked paper 'wicks' dipping in the buffer reservoir and laid upon the supporting medium. The buffer reservoir normally contains a divider acting as a barrier to diffusion (but not to electrical current), so that localized pH changes which occur in the region of the electrodes (as a result of electrolysis, p. 158) are not transmitted to the supporting medium or the sample. Individual samples are spotted onto a solid supporting medium containing buffer or are applied to 'wells' formed in the supporting medium. The power pack used for most types of electrophoresis should be capable of delivering ≈500 V and ≈100 mA.

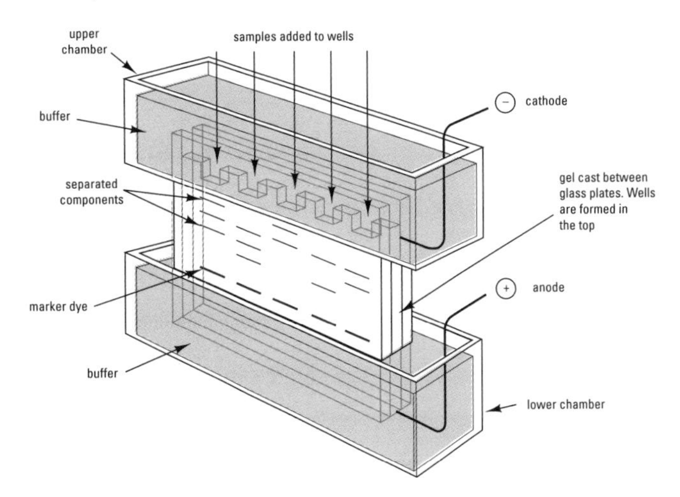

Fig. 26.2 Apparatus for vertical slab electrophoresis (components move downwards from wells, through the gel matrix).

Using a supporting medium

The effects of convection currents (resulting from the heating effect of the applied field) and the diffusion of molecules within the buffer solution can be minimized by carrying out the electrophoresis in a porous supporting medium. This contains buffer electrolytes and the sample is added in a discrete location or zone. When the electrical field is applied, individual

Minimizing diffusion – make the sample zone as narrow as possible, and fix and stain the protein bands as soon as possible after the run.

Definition

Ohm's law V = IR, where V = voltage, I = current and R = resistance.

Optimizing electrophoresis – attempting to minimize heat production using very low currents is not practical, since it leads to long separation times and therefore to increased diffusion.

sample molecules remain in sharp zones as they migrate at different rates. After separation, post-electrophoretic diffusion of selected biomolecules (e.g. proteins) can be avoided by 'fixing' them in position on the supporting medium, e.g. using trichloracetic acid (TCA).

The heat generated during electrophoresis is proportional to the square of the applied current and to the electrical resistance of the medium: even when a supporting medium is used, heat production will lead to zone broadening by increasing the rate of diffusion of sample components and buffer ions. Heat denaturation of sample proteins may also occur, resulting in loss of biological activity e.g. with enzymes (p. 195). Another problem is that heat will reduce buffer viscosity, leading to a decrease in resistance. If the electrophoresis is run at constant voltage, Ohm's law dictates that as resistance falls, the current will increase, leading to further heat production. This can be avoided by using a power pack that provides constant power. In practice, most electrophoresis equipment incorporates a cooling device; even so, distortions of an electrophoretic zone from the ideal 'sharp, linear band' can often be explained by inefficient heat dissipation.

Types of supporting media

These can be sub-divided into:

- Inert media – these provide physical support and minimize convection: separation is based on charge density only (e.g. cellulose acetate).
- Porous media – these introduce molecular sieving as an additional effect: their pore size is of the same order as the size of molecules being separated, restricting the movement of larger molecules relative to smaller ones. Thus, separation depends on both the charge density and the size of the molecule.

Definition

Electro-osmotic flow – the osmotically driven mass flow of water resulting from the movement of ions in an electrophoretic system.

With some supporting media, e.g. cellulose acetate, a phenomenon called electro-endosmosis or electro-osmotic flow (EOF) occurs. This is due to the presence of negatively charged groups on the surface of the supporting medium, attracting cations in the electrophoresis buffer solution and creating an electrical double layer. The cations are hydrated (surrounded by water molecules) and when the electric field is applied, they are attracted towards the cathode, creating a flow of solvent that opposes the direction of migration of anionic biomolecules towards the anode. The EOF can be so great that weakly anionic biomolecules (e.g. antibodies) may be carried towards the cathode.

Where necessary, EOF can be avoided by using supporting media such as agarose or polyacrylamide (p. 145), but it is not always a hindrance to electrophoretic separation. Indeed, the phenomenon of EOF is used in the high resolution technique of capillary electrophoresis (p. 152).

Cellulose acetate

Handling cellulose acetate – the fragile strips must be carefully handled, avoiding touching the flat surfaces with your fingers.

Acetylation of the hydroxyl groups of cellulose produces a less hydrophilic structure than cellulose in the form of paper: as a result it holds less water and diffusion is reduced, with a corresponding increase in resolution. Cellulose acetate is often used in the electrophoretic separation of plasma proteins in clinical diagnosis – it can be carried out quickly (≈45 min) and its resolution is adequate to detect gross differences in various types of protein (e.g. paraproteins in myeloma). Cellulose acetate has a fairly uniform pore structure and the pores are large enough to allow unrestricted passage of all but the largest of molecules as they migrate through the medium.

Agarose

Agarose is a neutral, linear polysaccharide in agar (from seaweed), consisting of repeating galactose and 3,6-anhydrogalactose. Powdered agarose is mixed with electrophoresis buffer at concentrations of 0.5–3.0% w/v, boiled until the mixture becomes clear, poured onto a glass plate, then allowed to cool until it forms a gel. Gelation is due to the formation of hydrogen bonds both between and within the agarose polymers, resulting in the formation of pores. The pore size depends on the agarose concentration. Low concentrations produce gels with large pores relative to the size of proteins, allowing them to migrate relatively unhindered through the gel, as determined by their individual charge densities. Low concentrations of agarose gel are suitable for techniques such as immunoelectrophoresis (p. 97) and isoelectric focusing (p. 151), where charge is the main basis of separation. The smaller pores produced by higher concentrations of agarose may result in molecular sieving.

When agarose gels are used for the separation of DNA, the large fragment size means that molecular sieving is observed, even with low concentration gels. This is the basis of the electrophoretic separation of nucleic acids (see Chapter 40).

Advantages of polyacrylamide gels – in addition to their versatility in terms of pore size, these gels are chemically inert, stable over a wide range of pH, ionic strength and temperature, and transparent.

Safe working and PAGE – *both acrylamide and bisacrylamide are extremely potent neurotoxins, so you must wear plastic gloves when handling solutions containing these reagents.* Although the polymerized gel is non-toxic, it is still advisable to wear gloves when handling the gel, because some monomer may still be present.

Preparing polyacrylamide gels – most solutions used for gel preparation can be made in advance, but the ammonium persulphate solution must be prepared immediately before use.

Polyacrylamide

Polyacrylamide gel electrophoresis (PAGE) has a major role in protein analysis, both for one-dimensional and two-dimensional separations. The gel is formed by polymerizing acrylamide monomer into long chains and crosslinking these chains using *N,N′*-methylene bisacrylamide (often abbreviated to 'bis'). The process is shown in Fig. 26.3. In most protocols, polymerization is initiated by free radicals produced by ammonium persulphate in the presence of *N,N,N′,N′*-tetramethylethylenediamine (TEMED). The photodecomposition of riboflavin can also be used as a source of free radicals.

The formation of polyacrylamide from its acrylamide monomers is extremely reproducible under standard conditions, and electrophoretic separations are correspondingly precise. The pore size, and hence the extent of molecular sieving, depends on the total concentration of monomer (% T), i.e. acrylamide plus bisacrylamide in a fixed ratio. This means that pores in the gel can be 'tailored' to suit the size of biomolecule to be separated: gels containing 3% acrylamide have large pores and are used in methods where molecular sieving should be avoided (e.g. in isoelectric focusing, p. 151), while higher concentrations of acrylamide (5–30% T) introduces molecular sieving to various degrees depending on the size of the sample components, i.e. with 30% acrylamide gels, molecules as small as M_r 2000 may be subject to molecular sieving. Gels of <2.5% are necessary for molecular sieving of molecules of $M_r > 10^6$, but such gels are almost fluid and require 0.5% agarose to make them solid. Note that a gel of 3% will separate DNA by molecular sieving, due to the large size of the nucleic acid molecules (p. 230).

Before you embark on a particular PAGE protein separation, you will need to think about the general strategy for that separation and make certain choices, including whether to use:

- Rod or slab gels – flat slab gels are formed between glass plates, using plastic spacers 0.75–1.5 mm thick: rod gels are made in narrow bore tubes. For most separations using several samples, a slab gel saves time because up to 25 samples can be separated under identical conditions in a single gel, while rod gels can only be used for individual samples.

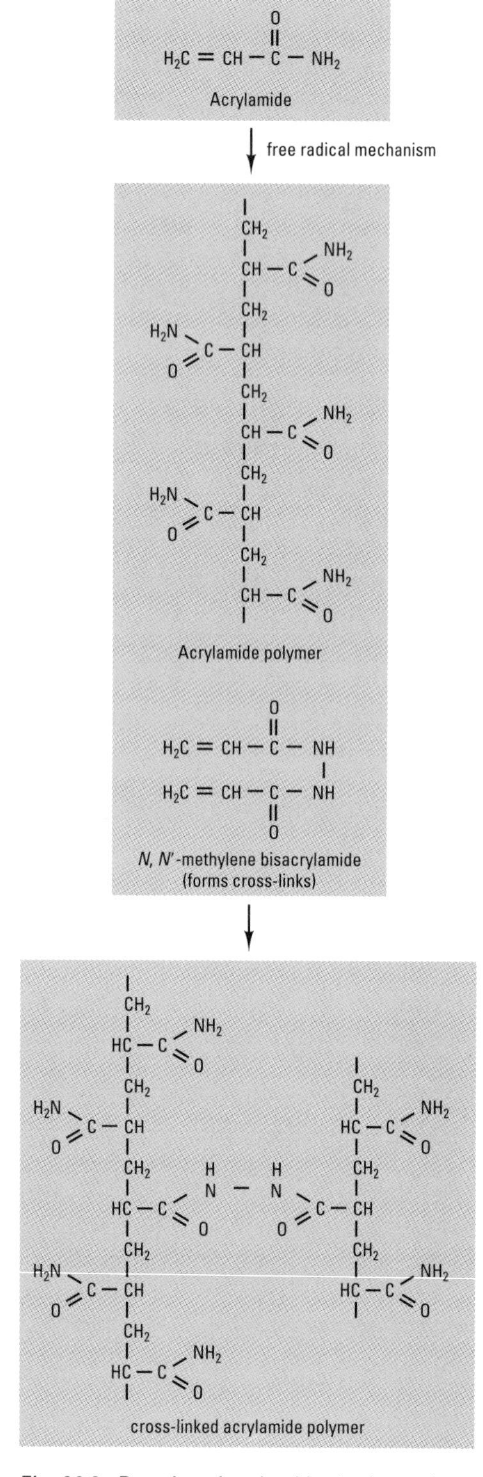

Fig. 26.3 Reactions involved in the formation of polyacrylamide gels.

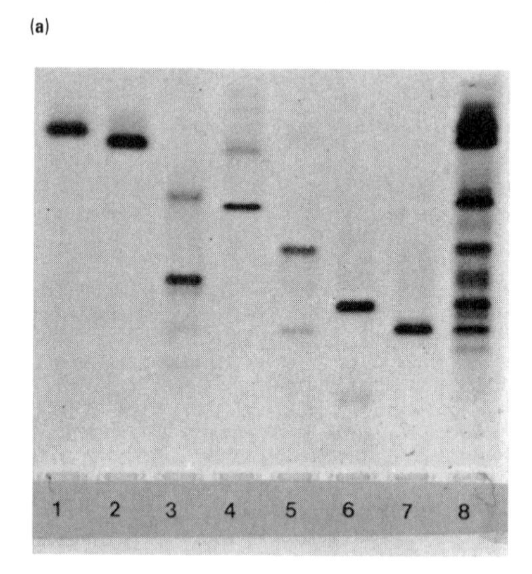

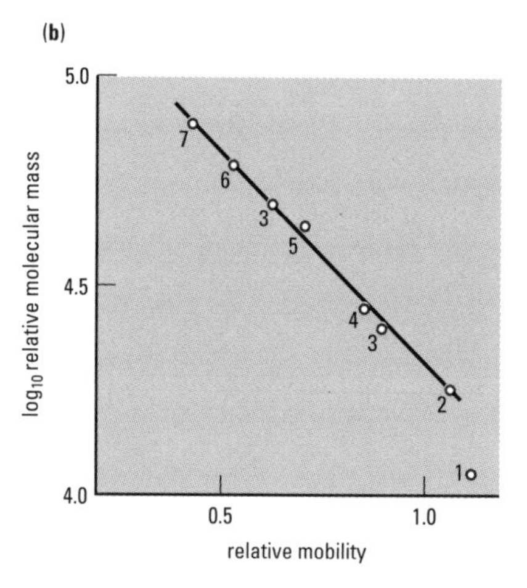

Fig. 26.4 Determination of relative molecular mass (M_r) of proteins by SDS-PAGE: (a) gel samples: 1, cytochrome c; 2, myoglobin; 3, γ-globulin; 4, carbonic anhydrase; 5, ovalbumin; 6, albumin; 7, transferrin; 8, mixture of samples 1–7 (photo courtesy of Pharmacia Biotech). (b) plot of log M_r against distance travelled through gel.

Table 26.2 Molecular masses of standard proteins used in electrophoresis

Protein	M_r	$\log_{10}M_r$
Cytochrome c	11 700	4.068
Myoglobin	17 200	4.236
γ-globulin (light chain)	23 500	4.371
Carbonic anhydrase	29 000	4.462
Ovalbumin	43 000	4.634
γ-globulin (heavy chain)	50 000	4.699
Human albumin	68 000	4.832
Transferrin	77 000	4.886
Myosin (heavy chain)	212 000	5.326

Following the progress of PAGE – add bromophenol blue solution (0.002% w/v) to the sample in the ratio 1:25 (dye:sample). This highly ionic, small M_r dye migrates with the electrophoretic front.

Choosing a pH for protein electrophoresis – many proteins have isoelectric points in the range pH 4–7 and in response to electrophoresis with buffers in the region pH 8.0–9.5, most proteins will migrate towards the anode.

Rectangular slab gels are also easier to read, by densitometry, and photograph. However, rod gels are useful in preliminary separations, for determining a suitable pH and gel concentration, and for applications where the gel is sliced in order to extract and assay proteins of interest. (In 2D-electrophoresis, a rod may be used for one dimension, and a slab for the second, p. 152).

- Dissociating or non-dissociating conditions – the most widely used PAGE protein separation technique uses an ionic detergent, usually sodium dodecyl sulphate (SDS), which dissociates proteins into their individual polypeptide sub-units and gives a uniform net charge along each denatured polypeptide. This technique, known as SDS-PAGE, requires only μg amounts of sample and is quick and easy to carry out. On the other hand, if it is necessary to preserve the native protein conformation and biological activity, non-dissociating conditions are used, i.e. no SDS is added. In SDS-PAGE the sample protein is normally heated to 100 °C for 2 min, in buffer containing 1% (w/v) SDS and 1% (w/v) 2-mercaptoethanol, the latter to cleave any disulphide bonds. The resultant polypeptides bind to SDS in a constant weight ratio, with 1.4 g of SDS per g of protein. As a result, the intrinsic net charge of each polypeptide is 'swamped' by the negative charge imposed by SDS, and there is a uniform negative charge per unit length of polypeptide. Since the polypeptides now have identical charge densities, when they are subject to PAGE (with SDS present) using a gel of appropriate pore size, molecular sieving will occur and they will migrate strictly according to polypeptide size. This not only gives effective separation, but the molecular mass of a given polypeptide can be determined by comparing its mobility to polypeptides of known molecular mass run under the same conditions (Fig. 26.4). Several manufacturers (e.g. Pharmacia®, Sigma®) supply molecular mass standard kits which may include polypeptides of M_r 11 700 to 212 000 (Table 26.2), together with details of their preparation and use. Where necessary, the treated sample can be concentrated by ultrafiltration and the buffer composition can be altered by diafiltration (p. 191).
- Continuous or discontinuous buffer systems – a continuous system is where the same buffer ions are present in the sample, gel and buffer reservoirs, all at the same pH. The sample is loaded directly onto a gel (the 'separating gel' or 'resolving gel') that has pores small enough to introduce molecular sieving. In contrast, discontinuous systems have different buffers in the gel compared to the reservoirs, both in terms of buffer ions and pH. The sample is loaded onto a large-pore 'stacking gel', previously polymerized on top of a small-pore separating gel (Fig. 26.5). Discontinuous systems are more time-consuming to prepare, but have the advantage over continuous systems in that relatively large volumes of dilute sample can be used and good resolution is still obtained. This is because the individual proteins in the sample concentrate into very narrow zones during their migration through the large-pore gel and stack up according to their charge densities, prior to separation in the small-pore gel.
- pH and buffer for the separation – PAGE can be carried out between pH 3 and 10. The pH is not critical for continuous SDS-PAGE, since SDS-treated polypeptides are negatively charged over a wide pH range. However, when using non-dissociating systems, the pH can be critical, particularly if the biological activity of the molecules is to be retained.

Table 26.3 Preparation of gels for PAGE and SDS-PAGE. The gel solutions are made by mixing the components in the proportions and in the order shown. Figures are ml of each solution required to give the stated 90 gel strength.

Solution (added in order shown)	PAGE 3.5% gel (T = 3.6%)	PAGE 7.5% gel (T = 7.7%)	SDS-PAGE 5% gel (T = 5.1%)	SDS-PAGE 10% gel (T = 10.2%)
1. Distilled water	19.3	7.5	14.9	—
2. TRIS-glycine buffer, pH 8.9, 0.1 mol $^{-1}$	33.0	33.0	—	—
3. Imidazole buffer, pH 7.0, 0.1 mol l^{-1} plus 0.2% w/v SDS	—	—	33.0	33.0
4. Acrylamide solution 22.2% w/v and 0.6% w/v bis	10.4	22.2	14.8	29.7
5. Ammonium persulphate solution, 0.15% w/v	3.2	3.2	3.2	3.2
6. TEMED	0.1	0.1	0.1	0.1
Final volume (ml)	66.0	66.0	66.0	66.0

Separating protein mixtures – for high resolution, a combination of dissociating and discontinuous PAGE in slab gels is the system of choice.

If your polyacrylamide gels fail to set – polymerization is inhibited by oxygen, so all solutions should be degassed, and the surfaces of the polymerization mixture exposed to air should be overlayed with water; if your gels still do not polymerize, the most common cause is the use of 'old' ammonium persulphate stock solution. If acid buffers are used, polymerization may be delayed because TEMED is required in the free base form.

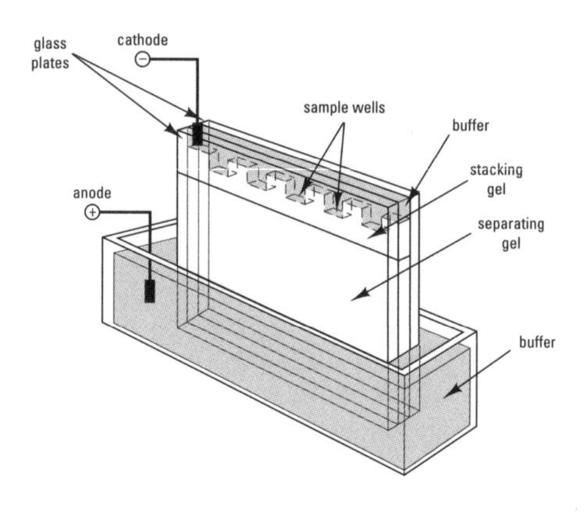

Fig. 26.5 Apparatus for discontinuous electrophoresis.

- Gel concentration – at one extreme, a gel with a very high percentage T might totally exclude the sample components, while a gel with a very low percentage T might lead to all SDS-treated proteins migrating at the same rate, i.e. with the electrophoretic front. A sensible approach is to set up a series of rod gels in the range 5–15% T and observe the separation and resolution obtained. Alternatively, a *gradient gel* can be used, in which the percentage T increases, and hence pore size decreases, in the direction of protein migration. A useful gradient for a preliminary experiment would be 5–20% T. Such gels are able to resolve protein mixtures with a wide range of molecular masses. Furthermore, as proteins migrate into regions of ever-decreasing pore size, the movement of the leading edge of a zone will become increasingly restricted. This allows the trailing edge to catch up, resulting in considerable zone sharpening. Gradient formers are fairly simple to make, and are commercially available.

Practical details of the preparation of PAGE and SDS-PAGE gels are given in Table 26.3 (See Westermeier, 1993 and Gersten, 1996 for further details).

Post-electrophoretic procedures – handling of the supporting medium, staining and analysis

For protein electrophoresis, the following stages are appropriate: details for nucleic acids are given on p. 230.

Handling

All types of supporting medium should be handled carefully: wearing gloves is advisable, for safety and to avoid transfer of proteins from the skin. Cellulose acetate strips should be immediately transferred to a fixing and staining solution. Agarose gels should be dried quickly before staining. Polyacrylamide gels in vertical slabs must be freed carefully from one of the glass plates in which they are formed, taking care to lever the glass at a point well away from the part of the gel containing the wells: once free, the gels should be immediately transferred to fixing or staining solution.

Rod gels are recovered by a process called 'rimming', a technique that takes a little practice to master. To remove the gel from the tube, hold the tube in one hand and, using a syringe with a long blunt needle, squirt water between the gel and the inner wall of the tube: while squirting, move the needle gently up and down and rotate the tube. If the gel does not become free, perform the same procedure at the other end of the tube, until the gel is released. If necessary, a Pasteur pipette bulb applied to one end can be used to push the gel out of the tube.

Handling gels – avoid touching gels with paper as it sticks readily and is difficult to remove without tearing.

Avoiding overloaded gels and band distortion – determine the protein concentration of the sample beforehand. Around 100 μg of a complex mixture, or 1–10 μg of an individual component will be sufficient, but bear in mind that underloading may result in bands being too faint to be detected.

Optimizing resolution – keep the sample volume as small as possible; methods for concentrating protein solutions are considered on p. 190. For vertical slab gels and for rod gels, include 10% w/v sucrose or glycerol to increase density and allow buffer solution to be overlayed on the sample without dilution.

Fixing, staining and destaining

To prevent the separated proteins from diffusing, they are usually fixed in position. Proteins on cellulose acetate strips can be fixed and stained using Ponceau S (0.2% w/v in 3% v/v aqueous trichloracetic acid, TCA).

Fixing is also required for most types of gel electrophoresis: 3% v/v TCA is often used. The most widely used stain for protein separations in gels is Coomassie Blue R-250 (where R = reddish hue): the detection limit is ≈0.2 μg and staining is quantitative up to 20 μg for some proteins. It is normal for background staining of the gel to occur, and removal of background colour ('destaining') can be achieved either by diffusion or electrophoresis. To destain by diffusion, transfer the gel to isopropanol : acetic acid : water (12.5 : 10 : 77.5 v/v/v) and allow to stand for 48 h, or change the solution several times to speed up the staining process. Electrophoretic destaining can remove Coomassie Blue, which is anionic: stained gels are placed between porous plastic sheets with electrodes on each side, and the tank is filled with 7% acetic acid. Passing a current of up to 1.0 A destains the gel in ≈30 min.

If you need greater sensitivity (e.g. for ng to fg amounts), or when using high resolution techniques such as 2D-electrophoresis (p. 152), silver staining can be used. Depending on the protocol chosen and the proteins being stained, the silver technique can be 5 to 200-fold more sensitive than Coomassie Blue. The method involves a fixation step (e.g. with TCA), followed by exposure to silver nitrate solution and development of the stain. The silver ions are thought to react with basic and thiol groups in proteins, and subsequent reduction (e.g. by formaldehyde at alkaline pH, or by photodevelopment) leads to deposits of silver in the protein bands. Most proteins stain brown or black, but lipoproteins may stain blue, and some glycoproteins stain yellow or red. Some proteins lacking in amino acids with reducing groups (e.g. those lacking cysteine residues) may stain negatively, i.e. the bands are more transparent than the background staining of the gel. Although many protocols have been published, silver stain kits are commercially available, e.g. from Bio-Rad.

Although silver staining has clear advantages in terms of sensitivity, for routine work it is more laborious and expensive to carry out than the Coomassie Blue method. It also requires high purity water, otherwise significant background staining occurs. Another feature is that the staining can be non-specific, since DNA and polysaccharides may stain on the same gel as proteins.

Other methods for the detection of separated components include:

- autoradiography for proteins labelled with ^{32}P or ^{125}I (p. 104);
- fluorescence for proteins pre-labelled with fluorescent dyes (p. 169);
- periodic acid-Schiff (PAS) stain using dansyl hydrazine for glycoproteins.

Origins of 'Western' blotting – following the description of 'Southern' blotting of DNA by Ed Southern (p. 233), other points of the compass have been used to describe other forms of blotting, with 'Western' blotting for proteins.

Blotting

The term 'blotting' refers to the transfer of separated proteins from the gel matrix to a thin sheet such as nitrocellulose membrane (commercially available from e.g. Millipore, Amersham, etc.). The proteins bind to this membrane, and are immobilized. Blotting of proteins is usually achieved by electrophoretic transfer, and this process is normally referred to as Western blotting (see also Southern blotting and Northern blotting for DNA and RNA respectively, p. 233). Its major advantage is that the immobilized proteins on the surface of the membrane are readily accessible to detection reagents, and staining and destaining can be achieved in less than 5 min. Use of labelled antibodies to detect specific proteins (immunoblotting) can take less than 6 h. In addition, it is easy to dry and store Western blots for long periods, for further analysis.

Example For detection of lactate dehydrogenase (LDH): when a solution containing lactate, NAD^+, phenazine methasulphate (PMS) and methyl thiazolyl tetrazolium (MTT) is added to a gel containing LDH, a series of redox reactions occurs in the enzyme-containing regions, starting with the oxidation of lactate to pyruvate, and proceeding via NAD^+, PMS and MTT to the eventual reduction of MTT to formazan dye. After incubation at 37 °C in the dark (since MTT is light-sensitive), LDH is detected by the appearance of blue-black bands on the gel.

Detecting enzymes in gels – minimize zone spreading by incorporating substrates, etc. in a thin agarose indicator gel poured over the separating gel.

Measuring peak areas – a valid 'low-tech' alternative to computer-based systems is to cut out and weigh peaks from a recorder chart.

Detection of enzymes

If you need to detect enzyme activity you should use a non-denaturing gel. The gel matrix will hinder diffusion of the enzyme, but will allow access to the small molecular weight substrates, co-factors and dyes necessary to localize enzyme activity *in situ*. Most methods for detecting enzymes on gels are modifications of protocols originally developed by histochemists, e.g.:

- NAD^+-requiring oxidoreductases can be detected by incubating the gels with substrate, NAD^+ and a solution of a tetrazolium salt which, when reduced, forms an insoluble coloured formazan dye.
- Transferases and isomerases can be detected by coupling their reactions to an oxidoreductase-requiring reaction, visualized as described above.
- Hydrolases can be detected using appropriate chromogenic or fluorogenic substrates.
- Phosphatases can be detected by precipitating any phosphate released from the substrate with Ca^{2+}.

Recording and quantification of results

A number of expensive, dedicated instruments are available for the analysis of gels, e.g. laser densitometers. Alternatively, gel scanning attachments can be purchased for standard spectrophotometers, allowing measurement and recording of the absorbance of the Coomassie Blue stained bands at 560–575 nm: for instruments connected to a computer, quantification of individual components can be achieved by integrating the areas under the peaks.

You can photograph gels using a conventional camera (fine grain film) or Polaroid® camera, or using a photocopier or digital camera. The gel should be placed on a white glass transilluminator. A red filter will increase contrast with bands stained with Coomassie Blue. If the gels themselves need to be retained, they can be preserved in 7% acetic acid. Alternatively, they can be dried using a commercially available gel drier.

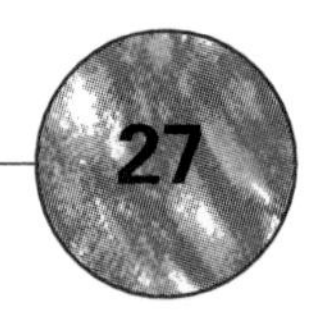

27 Advanced electrophoretic techniques

Although the resolution obtained using the basic electrophoretic techniques described in Chapter 26 is adequate for many biomolecular applications, a number of advanced techniques are available that give very high resolution and which can be used with very small amounts of sample material.

Isoelectric focusing (IEF)

In contrast to electrophoresis, which is carried out at constant pH, IEF is carried out using a pH gradient. The gradient is formed using small molecular mass ampholytes, which are analogues and homologues of polyamino-, polycarboxylic acids that collectively have a range of isoelectric points (pI values) between pH 3 and 10. The mixture of ampholytes (p. 26), either in a gel or in free solution, is placed between the anode in acid solution (e.g. H_3PO_4), and the cathode in alkaline solution (e.g. NaOH). When an electric field is applied, each ampholyte migrates to its own pI and forms a stable pH gradient which will persist for as long as the field is applied. When a protein sample is applied to this gradient separation is achieved, since individual proteins will migrate to their isoelectric points. The net charge on the protein when first applied will depend on the specific 'titration curve' for that protein (Fig. 27.1). As an example, consider two proteins, X and Y, having pI values of pH 5 and pH 8 respectively, which are placed together on the gradient at pH 6 (Fig. 27.2). At that pH, protein X will have a net negative charge, and will migrate towards the anode, progressively losing charge until it reaches its pI (pH 5) and stops migrating. Protein Y will have a net positive charge at pH 6, and so will migrate towards the cathode until it reaches its pI (pH 8).

Fig. 27.1 Titration curves for two proteins, X and Y.

Using a polyacrylamide gel as a supporting medium and a narrow pH gradient, proteins differing in pI by 0.01 units can be separated. Even greater resolution is possible in free solution (e.g. in capillary electrophoresis, p. 152). Such resolution is possible because protein molecules that diffuse away from the pI will acquire a net charge (negative at increased pH, positive at decreased pH) and immediately be focused back to their pI. This focusing effect will continue for as long as the electric field is applied.

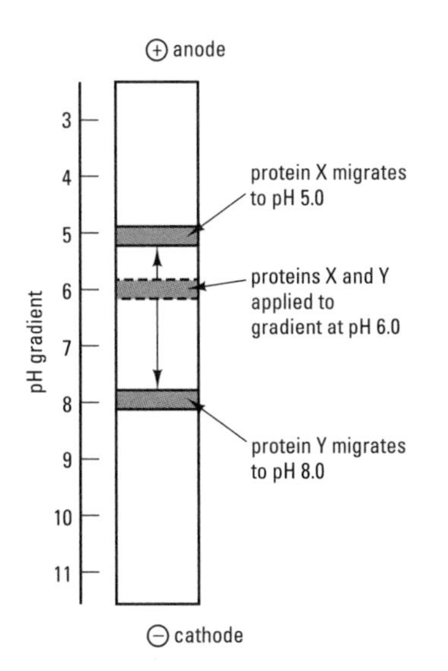

Fig. 27.2 The migration of two proteins, X and Y, in response to a pH gradient.

A useful variant of IEF is in obtaining *titration curves* for proteins. A pH gradient is set up, and the sample applied in a line at a right angle to the gradient. The net charge on a given protein will vary according to its position on the gradient – when electrophoresis is carried out at right angles to the pH gradient, the protein will migrate at a velocity and direction governed by that charge. When stained, each protein will appear as a continuous curved line, corresponding to its titration curve (Fig. 27.3). This technique can be usefully performed during protein purification, prior to ion exchange chromatography (p. 131): by obtaining the titration curve for a protein of interest and those of major contaminants, the mobile phase pH that gives optimal separation can be selected.

In IEF, it is important that electro-osmotic flow (EOF) is avoided, as this would affect the ability of the proteins to remain stationary at their pIs. For gel IEF, polyacrylamide minimizes EOF, while capillary IEF uses narrow bore tubing with an internal polymer coating (p. 154).

Two-dimensional electrophoresis

Avoiding streaking in 2D electrophoresis – ensure that the sample contains no particulate material (e.g. from protein aggregation); filter or centrifuge before use.

Maximizing resolution in 2D electrophoresis of proteins – try to minimize nucleic acid contamination of your sample (see Chapter 34), as they may interact with polypeptides/proteins, affecting their movement in the gel.

Freezing rod gels – be sure to mark the identity and orientation of the gel before freezing, e.g. by inserting a fine wire in one end of the gel. Most recipes for the one-dimensional gel include urea, which forms crystals on freezing.

The most commonly used version of this high resolution technique involves separating proteins by charge in one dimension using IEF in polyacrylamide gel, followed by separation by molecular mass in the second dimension using denaturing SDS-PAGE (p. 145). The technique allows up to one thousand proteins to be separated from a single sample. Typically, the first dimension IEF run (pH 3–10) is carried out in rod gels, varying in diameter from 1–5 mm: with narrow rods a thin thread is sometimes polymerized within the gel to aid handling during transfer to the second dimension gel. At 1 kV with 2 mA current, the pH gradient takes ≈30 min to reach equilibrium, and separation of the applied sample takes a further 60 min. After the IEF stage, the rods must be removed from the glass tubes in which they were formed. This can be done by rimming (p. 148), or by freezing the gels at −20 °C and cracking the glass tubes in a vice. The gel must be allowed to warm up before it is run in the second dimension.

It is common for the second-dimension SDS-PAGE separation to be carried out on a discontinuous slab gel 0.5–1.5 mm thick, which includes a low percentage T stacking gel and a separating gel with an exponential gradient of 10–16% T. The separating gel can be prepared in advance, but the stacking gel must be formed shortly before addition of the rod gel from the one-dimensional run.

After equilibration with the buffer used in SDS-PAGE, the one-dimensional rod gel is loaded onto the second-dimensional gel (still between the glass plates in which it was formed) and sealed in position using acrylamide or agarose. Before the sealing gel sets, a well should be formed in it at one end to allow addition of molecular mass markers. The second-dimension run at 100–200 V until the dye front is ≈1 cm from the bottom edge of the slab. After running, the gel is processed for the detection of polypeptides, e.g. using Coomassie Blue or silver stain. Analysis of the complex patterns that result from 2D-electrophoresis requires computer-aided gel scanners to acquire, store and process data from a gel, such as that shown in Fig. 27.4. These systems can compare, adjust and match up patterns from several gels, allowing both accurate identification of spots and quantification of individual proteins. Allowance is made for the slight variations in patterns found in different runs, using internal references ('landmarks'), which are either added standard proteins or particular spots known to be present in all samples.

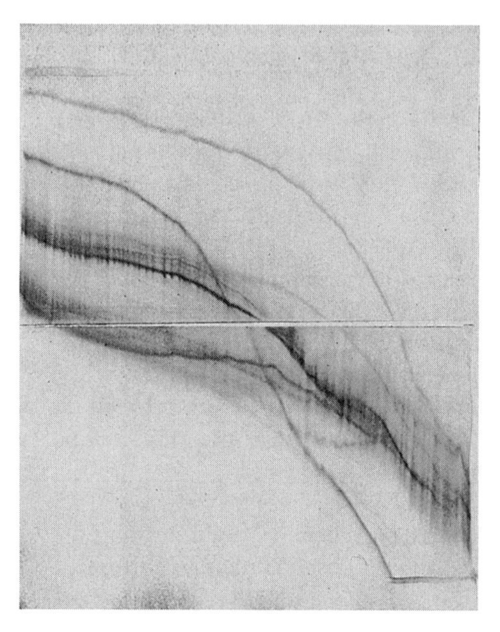

Fig. 27.3 Titration curves of bovine muscle proteins produced by electrofocusing-electrophoresis (photo courtesy of Pharmacia Biotech).

Capillary electrophoresis

The technique of capillary electrophoresis (CE) combines the high resolving power of electrophoresis with the speed and versatility of HPLC (p. 130). The technique largely overcomes the major problem of carrying out electrophoresis without a supporting medium, i.e. poor resolution due to convection currents and diffusion. A capillary tube has a high surface area : volume ratio, and consequently the heat generated as a result of the applied electric current is rapidly dissipated. A further advantage is that very small samples (5–10 nl) can be analysed. The versatility of CE is demonstrated by its use in the separation of a range of biomolecules, e.g. amino acids, proteins, nucleic acids, drugs, vitamins, organic acids and inorganic ions; CE can even separate neutral species, e.g. steroids, aromatic hydrocarbons.

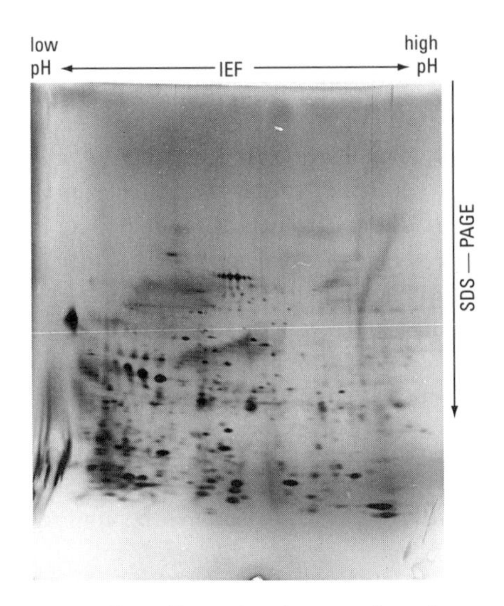

Fig. 27.4 Two-dimensional separation of proteins from 100× concentrated urine (2.5 μg total protein; silver stain. T. Marshall and K.M. Williams).

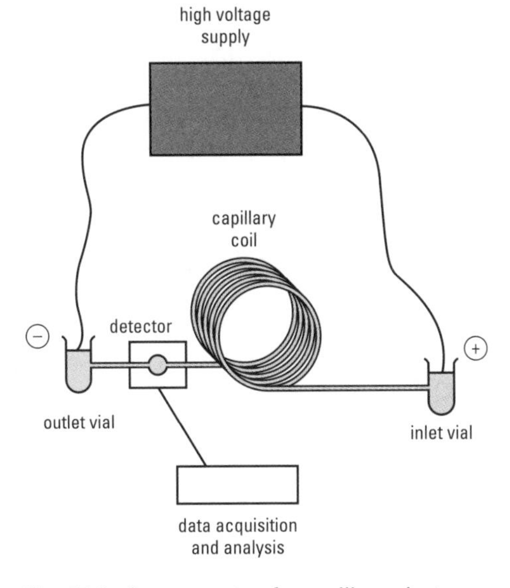

Fig. 27.5 Components of a capillary electrophoresis system.

The components of a typical CE apparatus are shown in Fig. 27.5. The capillary is made of fused silica and externally coated with a polymer for mechanical strength. The internal diameter is usually 25–50 μm, a compromise between efficient heat dissipation and the need for a light path that is not too short for detection using UV/visible spectrophotometry. A gap in the polymer coating provides a window for detection purposes. Samples are injected into the capillary by a variety of means, e.g. electrophoretic loading or displacement. In the former, the inlet end of the capillary is immersed in the sample and a pulse of high voltage is applied. The displacement method involves forcing the sample into the capillary, either by applying pressure in the sample vial using an inert gas, or by introducing a vacuum at the outlet. The detectors used in CE are similar to those used in chromatography (p. 138), e.g. UV/visible spectrophotometric systems. Fluorescence detection is more sensitive, but this may require sample derivitization. Electrochemical and conductivity detection is also used in some applications, e.g. conductivity detection of inorganic cations such as Na^+, K^+.

Electro-osmotic flow (EOF), described on p. 144, is essential to the most commonly used types of CE. The existence of EOF in the capillary is the result of the net negative charge on the fused silica surface at pH values over 3.0. The resulting solvent flow towards the cathode is greater than the attraction of anions towards the anode, so they will flow towards the cathode (note that the detector is situated at the cathodic end of the capillary). The greater the net negative charge on an anion, the greater is its resistance to the EOF and the lower its mobility. Separated components migrate towards the cathode in the order: (1) cations, (2) neutral species, (3) anions.

Capillary zone electrophoresis (CZE)

This is the most widely used form of CE, and is based on electrophoresis in free solution and EOF, as discussed above. Separations are due to the charge:mass ratio of the sample components, and the technique can be used for almost any type of charged molecule, and is especially useful for peptide separation and confirmation of purity.

Micellar electrokinetic chromatography (MEKC)

This technique involves the principles of both electrophoresis and chromatography. Its main strength is that it can be used for the separation of neutral molecules as well as charged ones. This is achieved by including surfactants (e.g. SDS, Triton X-100) in the electrophoresis buffer at concentrations that promote the formation of spherical micelles, with a hydrophobic interior and a charged, hydrophilic surface. When an electric field is applied, these micelles will tend to migrate with or against the EOF depending on their surface charge. Anionic surfactants like SDS are attracted by the anode, but if the pH of the buffer is high enough to ensure that the EOF is faster than the migration velocity of the micelles, the net migration is in the direction of the EOF, i.e. towards the cathode. During this migration, sample components partition between the buffer and the micelles (acting as a pseudo-stationary phase); this may involve both hydrophobic and electrostatic interactions. For neutral species it is only the partitioning effect that is involved in separation; the more hydrophobic a sample molecule, the more it will interact with the micelle, and the longer will be its migration time, since the micelle resists the EOF. The versatility of MEKC enables it to be used for separations of biomolecules as diverse as amino acids and polycyclic hydrocarbons.

Capillary gel electrophoresis (CGE)

The underlying principle of this technique is directly comparable with that of conventional PAGE, i.e. the capillary contains a polymer that acts as a molecular sieve. As charged sample molecules migrate through the polymer network, larger molecules are hindered to a greater extent than smaller ones and will tend to move more slowly. CGE differs from CZE and MEKC in that the inner surface of the capillary is polymer-coated in order to prevent EOF; this means that for most applications (e.g. polypeptide or oligonucleotide separations) sample components will migrate towards the anode at a rate determined by their size. The technique also differs from conventional PAGE in that a 'polymer network' is used rather than a gel: the polymer network may be polyacrylamide or agarose.

CGE offers the following advantages over conventional electrophoresis:

- efficient heat-dissipation means that a high electrical field can be applied, giving shorter separation times;
- detection of the separated components as they move towards the anodic end of the capillary (e.g. using a UV/visible detector) means that staining is unnecessary;
- automation is feasible.

Capillary isoelectric focusing

This is used mainly for protein separation. Here, the principles of IEF are valid as long as EOF is prevented by using capillaries that are polymer-coated on their inner surface. Sample components migrate to their isoelectric points and become stationary. Once separated (<10 min), the components must be mobilized so that they flow past the detector. This is achieved by changing the NaOH solution in the cathodic reservoir with an NaOH/NaCl solution. When the electric field is re-applied, Cl^- enters the capillary, causing a decrease in pH at the cathodic end and the subsequent migration of sample components.

Choosing a detector for capillary electrophoresis – most types of HPLC detector are suitable for CE and related applications (see Chapter 24).

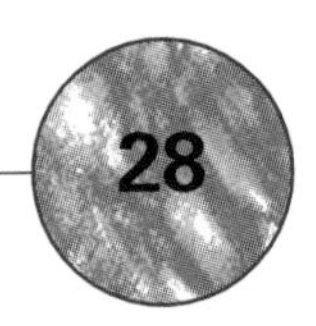

28 Electroanalytical techniques

Electrochemical methods are used to quantify a broad range of different biomolecules, including ions, gases, metabolites, drugs and hormones.

Definitions

Oxidation – loss of electrons by an atom or molecule (or gain of O atoms, loss of H atoms, increase in positive charge, or decrease in negative charge).
Reduction – gain of electrons by an atom or molecule (or loss of O atoms, gain of H atoms, decrease in positive charge or increase in negative charge).

KEY POINT **The basis of all electrochemical analysis is the transfer of electrons from one atom or molecule to another atom or molecule in an obligately coupled oxidation–reduction reaction (a redox reaction).**

It is convenient to separate such redox reactions into two half-reactions and, by convention, each is written as:

$$\text{oxidized form} + \text{electron(s)}\ (ne^-) \underset{oxidation}{\overset{reduction}{\rightleftharpoons}} \text{reduced form} \qquad [28.1]$$

You should note that the half-reaction is reversible: by applying suitable conditions, reduction *or* oxidation can take place. As an example, a simple redox reaction occurs when metallic zinc (Zn) is placed in a solution containing copper ions (Cu^{2+}), as follows:

$$Cu^{2+} + Zn \rightarrow Cu + Zn^{2+} \qquad [28.2]$$

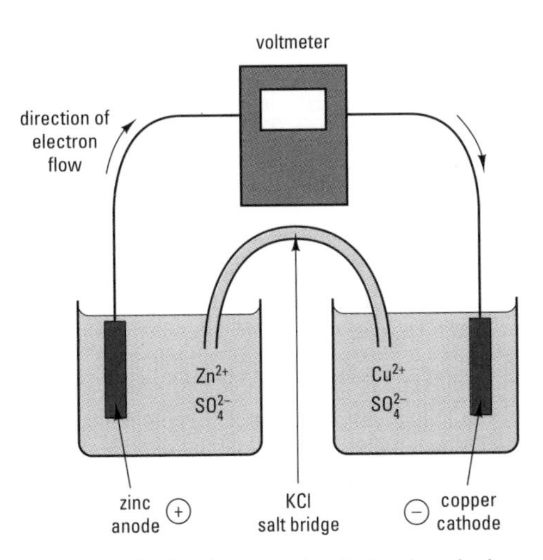

Fig. 28.1 A simple galvanic electrochemical cell. The KCl salt bridge allows migration of ions between the two compartments but prevents mixing of the two solutions.

The half-reactions are (i) $Cu^{2+} + 2e^- \rightarrow Cu$ and (ii) $Zn^{2+} + 2e^- \rightarrow Zn$. The oxidizing power of (i) is greater than that of (ii), so in a coupled system, the latter half-reaction proceeds in the opposite direction to that shown above, i.e. as $Zn - 2e^- \rightarrow Zn^{2+}$. When Zn and Cu electrodes are placed in separate solutions containing their ions, and connected electrically (Fig. 28.1), electrons will flow from the Zn electrode to Cu^{2+} via the Cu electrode due to the difference in oxidizing power of the two half-reactions.

By convention, the electrode potential of any half-reaction is expressed relative to that of a standard hydrogen electrode (half-reaction $2H^+ + 2e^- \rightarrow H_2$) and is called the standard electrode potential, E^o. Table 28.1 shows the values of E^o for selected half-reactions. With any pair of half-reactions from this series, electrons will flow from that having the lowest electrode potential to that of the highest. E^o is determined at pH = 0. It is often more appropriate to express standard electrode potentials at pH 7 for biological systems, and the symbol $E^{o\prime}$ is used: in all circumstances, it is important that the pH is clearly stated.

The arrangement shown in Fig. 28.1 represents a simple galvanic cell where two electrodes serve as the interfaces between a chemical system and an electrical system. For analytical purposes, the magnitude of the potential (voltage) or the current produced by an electrochemical cell is related to the concentration (strictly the activity, a, p. 22) of a particular chemical species. Electrochemical methods offer the following advantages:

Definition

Galvanic cell – an electrochemical cell in which reactions occur spontaneously at the electrodes when they are connected externally by a conductor, producing electrical energy.

- excellent detection limits, and wide operating range (10^{-1}–$10^{-8}\ mol\,l^{-1}$);
- measurements may be made on very small volumes (μl) allowing small amounts (pmol) of sample to be measured in some cases;
- miniature electrochemical sensors can be used for certain *in vivo* measurements, e.g. pH, glucose, oxygen content.

Table 28.1 Standard electrode potentials* (E^o) for selected half-reactions.

Half-reaction	E^o at 25 °C (V)
$Cl_2 + 2e^- \rightleftharpoons 2Cl^-$	+1.36
$O_2 + 4H^+ + 4e^- \rightleftharpoons 2H_2O$	+1.23
$Br_2 + 2e^- \rightleftharpoons 2Br^-$	+1.09
$Ag^+ + e^- \rightleftharpoons Ag$	+0.80
$Fe^{3+} + e^- \rightleftharpoons Fe^{2+}$	+0.77
$I_3^- + 2e^- \rightleftharpoons 3I^-$	+0.54
$Cu^{2+} + 2e^- \rightleftharpoons Cu$	+0.34
$Hg_2Cl_2 + 2e^- \rightleftharpoons 2Hg + 2Cl^-$	+0.27
$AgCl + e^- \rightleftharpoons Ag + Cl^-$	+0.22
$Ag(S_2O_3)_2^{3-} + e^- \rightleftharpoons Ag + 2S_2O_3^{2-}$	+0.01
$2H^+ + 2e^- \rightleftharpoons H_2$	+0.00
$AgI + e^- \rightleftharpoons Ag + I$	−0.15
$PbSO_4 + 2e^- \rightleftharpoons Pb + SO_4^{2-}$	−0.35
$Cd^{2+} + 2e^- \rightleftharpoons Cd$	−0.40
$Zn^{2+} + 2e^- \rightleftharpoons Zn$	−0.76

*From Milazzo *et al.*, 1978

Using a calomel electrode – always ensure that the KCl solution is saturated by checking that KCl crystals are present.

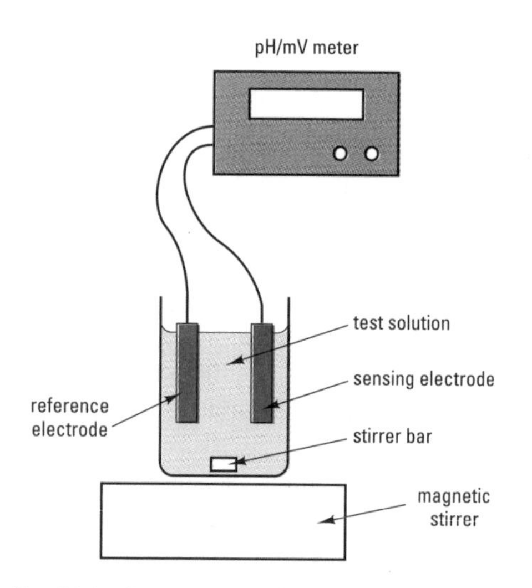

Fig. 28.2 Components of a potentiometric cell.

Using ISEs, including pH electrodes – standards and samples must be measured at the same temperature, as the Nernst equation shows that the measured potential is temperature-dependent.

Potentiometry and ion-selective electrodes

Operating principles

These systems involve galvanic cells (p. 155) and are based on measurement of the potential (voltage) difference between two electrodes in solution when no net current flows between them: no net electrochemical reaction occurs and measurements are made under equilibrium conditions. These systems include methods for measuring pH, ions, and gases such as CO_2 and NH_3. A typical potentiometric cell is shown in Fig. 28.2. It contains two electrodes:

1. a 'sensing' electrode, the half-cell potential of which responds to changes in the activity (concentration) of the substance to be measured; the most common type of indicator electrodes are ion-selective electrodes (ISE);
2. a 'reference' electrode, the potential of which does not change, forming the second half of the cell.

To assay a particular analyte, the potential difference between these electrodes is measured by a mV meter (e.g. a standard pH meter).

Reference electrodes for potentiometry are of three main types:

1. The standard hydrogen electrode, which is the reference half-cell electrode, defined as 0.0 V at all temperatures, against which values of E^o are expressed. H_2 gas at 1 atmosphere pressure is bubbled over a platinum electrode immersed in an acid solution with an activity of unity. This electrode is rarely used for analytical work, since it is unstable and other reference electrodes are easier to construct and use.
2. The calomel electrode (Fig. 28.3), which consists of a paste of mercury covered by a coat of calomel (Hg_2Cl_2), immersed in a saturated solution of KCl. The half-reaction: $Hg_2Cl_2 + 2e^- \rightarrow 2Hg + 2Cl^-$ gives a stable standard electrode potential of +0.24 V.
3. The silver/silver chloride electrode. This is a silver wire coated with AgCl and immersed in a solution of constant chloride concentration. The half-reaction: $AgCl + e^- \rightarrow Ag + Cl^-$ gives a stable, standard electrode potential of +0.20 V.

KEY POINT **Ion-selective electrodes (ISEs) are based on measurement of a potential across a membrane which is selective for a particular analyte.**

An ISE consists of a membrane, an internal reference electrode, and an internal reference electrolyte of fixed activity. The ISE is immersed in a sample solution that contains the analyte of interest, along with a reference electrode. The membrane is chosen to have a specific affinity for a particular ion, and if activity of this ion in the sample differs from that in the reference electrolyte, a potential develops across the membrane that is dependent on the ratio of these activities. Since the potentials of the two reference electrodes (internal and external) are fixed, and the internal electrolyte is of constant activity, the measured potential, E, is dependent on the membrane potential and is given by the Nernst equation:

$$E = K + 2.303\frac{RT}{zF}\log[a] \qquad [28.3]$$

where K represents a constant potential which is dependent on the reference electrode, z represents the net charge on the analyte, $[a]$ the activity of analyte in the sample and all other symbols and constants have their usual meaning (p. xi). For a series of standards of known activity, a plot of E against $\log[a]$ should be linear over the working range of the electrode, with a slope of $2.303\,RT/zF$ (0.059 V at 25 °C). Although ISEs strictly measure *activity*, the

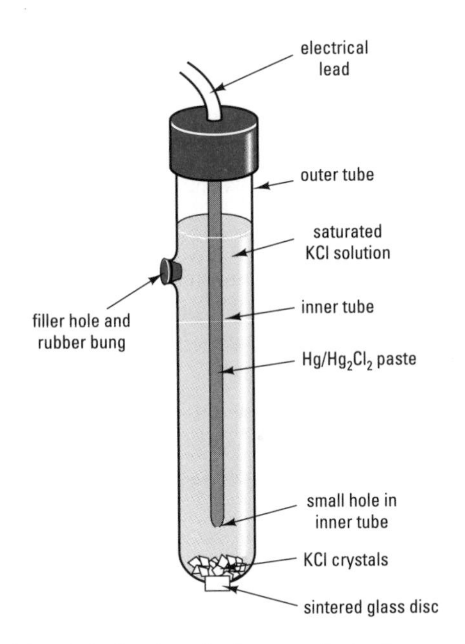

Fig. 28.3 A calomel reference electrode.

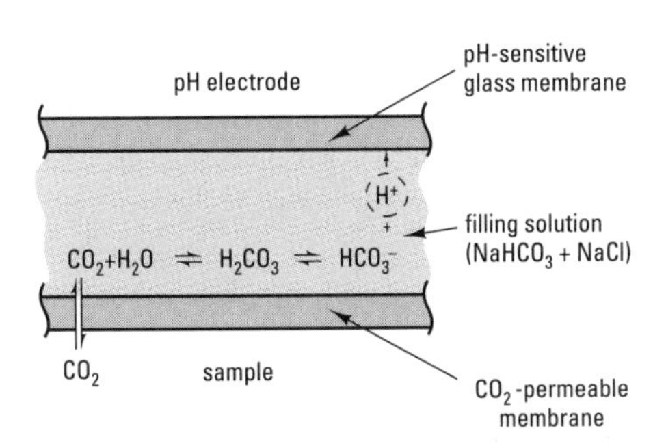

Fig. 28.4 Underlying principles of a gas-sensing electrode.

Using CO_2 electrodes – applications include measurement of blood PCO_2 and in enzyme studies where CO_2 is utilized or released: calibration of the electrode is accomplished using 5% v/v and 10% v/v mixtures of CO_2 in an inert gas equilibrated against the measuring solution.

potential differences can be approximated to concentration as long as (i) the analyte is in dilute solution (p. 22), (ii) the ionic strength of the calibration standards matches that of the sample, e.g. by adding appropriate amounts of a high ionic strength solution to the standards, and (iii) the effect of binding to sample macromolecules (e.g. proteins, nucleic acids) is minimal. Potentiometric measurements on undiluted biological fluids, e.g. K^+ and Na^+ levels in plasma, tissue fluids or urine, are likely to give lower values than flame emission spectrophotometry, since the latter procedure measures total ion levels, rather than just those in aqueous solution.

All of the various types of membrane used in ISEs operate by incorporating the ion to be analysed into the membrane, with the accompanying establishment of a membrane potential. The scope of electrochemical analysis has been extended to measuring gases and non-ionic compounds by combining ISEs with gas-permeable membranes, enzymes, and even immobilized bacteria or tissues.

Glass membrane electrodes

The most widely used ISE is the glass membrane electrode for pH measurement (p. 29). The membrane is thin glass (50 μm wall thickness) made of silica which contains some Na^+. When the membrane is soaked in water, a thin hydrated layer is formed on the surface in which negative oxide groups (Si-O^-) in the glass act as ion exchange sites. If the electrode is placed in an acid solution, H^+ exchanges with Na^+ in the hydrated layer, producing an external surface potential: in alkaline solution, H^+ moves out of the membrane in exchange for Na^+. Since the inner surface potential is kept constant by exposure to a fixed activity of H^+, a consistent, accurate potentiometric response is observed over a wide pH range. Glass electrodes for other cations (e.g. Na^+, NH_4^+) have been developed by changing the composition of the glass, so that it is predominantly sensitive to the particular analyte, though the specificity of such electrodes is not absolute. The operating principles and maintenance of such electrodes are broadly similar to those for pH electrodes (p. 29).

Gas-sensing glass electrodes

Here, an ISE in contact with a thin external layer of aqueous electrolyte (the 'filling solution') is kept close to the glass membrane by an additional, outer membrane that is selectively permeable to the gas of interest. The arrangement for a CO_2 electrode is shown in Fig. 28.4: in this case the outer membrane is made of CO_2-permeable silicone rubber. When CO_2 gas in the sample selectively diffuses across the membrane and dissolves in the filling solution (in this case an aqueous $NaHCO_3$/NaCl mixture), a change in pH occurs due to the shift in the equilibrium:

$$CO_2 + H_2O \rightleftharpoons H_2CO_3 \rightleftharpoons H^+ + HCO_3^- \quad [28.4]$$

The pH change is 'sensed' by the internal ion-selective pH electrode, and its response is proportional to the partial pressure of CO_2 of the solution (PCO_2). A similar principle operates in the NH_3 electrode, where a Teflon® membrane is used, and the filling solution is NH_4Cl.

Liquid and polymer membrane electrodes

In this type of ISE, the liquid is a water-insoluble viscous solvent containing a soluble ionophore, i.e. an organic ion-exchanger, or a neutral carrier molecule, that is specific for the analyte of interest. When this liquid is

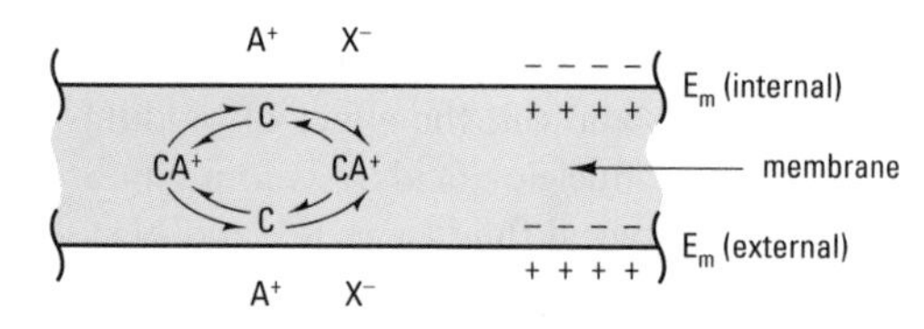

Fig. 28.5 Underlying principles of a liquid membrane ion-selective electrode. A^+ = analyte; C = neutral carrier ionophore; E_m = surface potential; membrane potential = E_m(internal) – E_m(external).

soaked into a thin membrane such as cellulose acetate, it becomes effectively immobilized. The arrangement of analyte (A^+) and ionophore in relation to this membrane is shown in Fig. 28.5. The potential on the inner surface of the membrane is kept constant by maintaining a constant activity of A^+ in the internal solution, so the potential change measured is that which results from A^+ in the sample interacting with the ionophore in the outer surface of the membrane.

A relevant example of a suitable ionophore is the antibiotic valinomycin (p. 203), which specifically binds K^+. Other ionophores have been developed for measurement of e.g. NH_4^+, Ca^{2+}, Cl^-. In addition, electrodes have been developed for organic species by using specific ion-pairing reagents in the membrane that interact with ionic forms of the organic compound, e.g. with drugs such as 5,5-diphenylhydantoin.

Definition

Ionophore – a compound that enhances membrane permeability to a specific ion: an ionophore may be incorporated into an ISE to detect that ion.

Solid-state membrane electrodes

These contain membranes made from single crystals or pressed pellets of salts of the analyte. The membrane material must show some permeability to ions and must be virtually insoluble in water. Examples include:

- the fluoride electrode, which uses LaF_3 impregnated with Eu^{2+} (the latter to increase permeability to F^-). A membrane potential is set up when F^- in the sample solution enters spaces in the crystal lattice;
- the chloride electrode, which uses a pressed pellet membrane of Ag_2S and AgCl.

Voltammetric methods

Definitions

Electrolytic cell – an electrochemical cell in which reactions are driven by the application of an external voltage greater than the spontaneous potential of the cell.

Electrolysis – a non-spontaneous chemical reaction resulting from the application of a potential to an electrode.

Voltammetric methods are based on measurements made using an electrochemical cell in which electrolysis is occurring. Voltammetry, sometimes also called amperometry, involves the use of a potential applied between two electrodes (the working electrode and the reference electrode) to cause oxidation or reduction of an electroactive analyte. The loss or gain of electrons at an electrode surface causes current to flow, and the size of the current (usually measured in mA or μA) is directly proportional to the concentration of the electroactive analyte. The materials used for the working electrode must be good conductors and electrochemically inert, so that they simply transfer electrons to and from species in solution. Suitable materials include Pt, Au, Hg and glassy carbon.

Two widely used devices that operate on the voltammetric principle are the oxygen electrode and the glucose electrode. These are sometimes referred to as amperometric sensors.

Oxygen electrodes

The Clark (Rank) oxygen electrode

These instruments measure oxygen in solution using the polarographic principle, i.e. by monitoring the current flowing between two electrodes when a voltage is applied. The most widespread electrode is the Clark type (Fig. 28.6), manufactured by Rank Bros, Cambridge, UK, which is suitable for measuring O_2 concentrations in cell, organelle and enzyme suspensions. Pt and Au electrodes are in contact with a solution of electrolyte (normally saturated KCl). The electrodes are separated from the medium by a Teflon® membrane, permeable to O_2. When a potential is applied across the electrodes, this generates a current proportional to the O_2 concentration. The reactions can be summed up as:

$4Ag \rightarrow 4Ag^+ + 4e^-$	(at silver anode)
$O_2 + 2e^- + 2H^+ \rightleftharpoons H_2O_2$	(in electrolyte solution; O_2 replenished by diffusion from test solution)
$H_2O_2 + 2e^- + 2H^+ \rightleftharpoons 2H_2O$	(at platinum cathode)

Setting up and using a Clark (Rank) oxygen electrode

Box 28.1 describes the steps involved: if you are setting up from scratch, perform steps 1–13; if a satisfactory membrane is already in place, start at step 7.

The temperature of the incubation vessel can be controlled by passing water (e.g. from a water bath) through the outer chamber. Cells or organelles may be present in the solution added to the incubation chamber or can be added via the hole in the stopper using a syringe, as can chemicals such as metabolic substrates or inhibitors. Take care not to introduce air bubbles and remove any that appear by gently raising and lowering the stopper. The electrode can be used repeatedly, providing the membrane is satisfactory: remove solutions carefully (e.g. using a pipette, or vacuum line and trap). Keep water in the chamber when not in use. Replace the membrane if:

- the reading becomes noisy;
- the electrode will not zero after adding sodium dithionite;
- the response becomes too slow (check by switching off stirrer – oxygen concentration should drop rapidly as the available oxygen is consumed).

If the Teflon® membrane becomes contaminated or dirty, it should be changed. While replacing the Teflon® membrane, make sure that the electrodes are clean: use a mild abrasive cleaning paste.

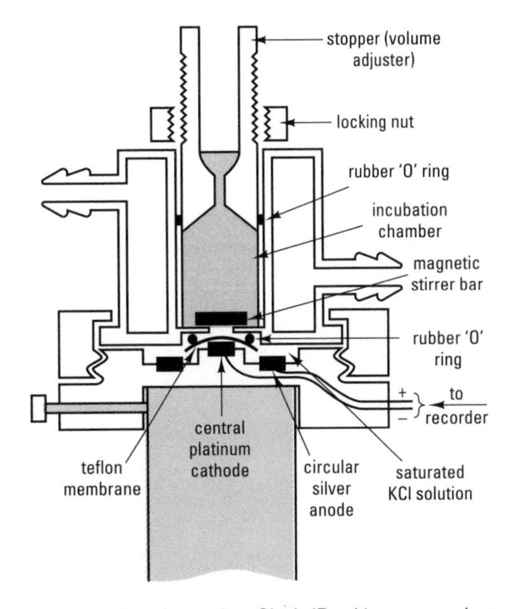

Fig. 28.6 Transverse section through a Clark (Rank) oxygen electrode.

Oxygen probes

Clark-type oxygen electrodes are also available in probe form for immersion in the test solution (Fig. 28.7) e.g. for field studies, allowing direct measurement of oxygen status *in situ*, in contrast to chemical assays. The main point to note is that the solution must be stirred during measurement, to replenish the oxygen consumed by the electrode ('boundary layer' effect).

Examples At 37°C, air-saturated distilled water contains O_2 at 0.212 mol m^{-3} (Table 28.2), so in 5×10^{-6} m^3 (5 ml) of medium, there will be 1.06 μmol O_2. If the deflection between zero and fully saturated solution is 67 divisions, each division will correspond to 1.06/67 = 0.0158 μmol O_2 (15.8 nmol per division).

For the above calibration, for a trace with a gradient of −14 units over 2 cm of chart paper at a speed of 0.2 cm min^{-1} (10 min), the O_2 consumption rate would be $14/10 \times 15.8 = 22.1$ nmol min^{-1}.

If there were 1.2×10^9 cells in the electrode, the O_2 consumption rate on a cellular basis would be $22.1 \times 10^{-9}/1.2 \times 10^9 = 18.4$ amol O_2 min^{-1} cell^{-1}.

Converting chart traces to rates of O_2 consumption or evolution

1. Calibrate the chart recorder using the zero and fully saturated values as reference points. The concentration of O_2 in saturated aqueous solution changes as a function of temperature and salt concentration (see Table 28.2) so multiply the appropriate value by the volume of the solution in the incubation chamber (in m^3) to obtain the number of moles of O_2 present. Divide by the number of chart divisions to find the number of moles per chart division.
2. Convert the gradient of the trace into a rate of oxygen consumption or evolution. Only use portions of the trace showing a stable rate for at least 5 min. Work out the gradient as divisions per unit time, taking account of the speed of the chart recorder. Use the calibration to convert this to a rate of O_2 consumption or evolution.
3. Express the rate on a cellular, chlorophyll or protein basis. Find the number of cells in the suspension (e.g. by using a haemocytometer, p. 83) and divide the figure obtained in (2) by this. Alternatively, find the chlorophyll or protein concentration (pp. 113 and 168 respectively) and divide by this value.

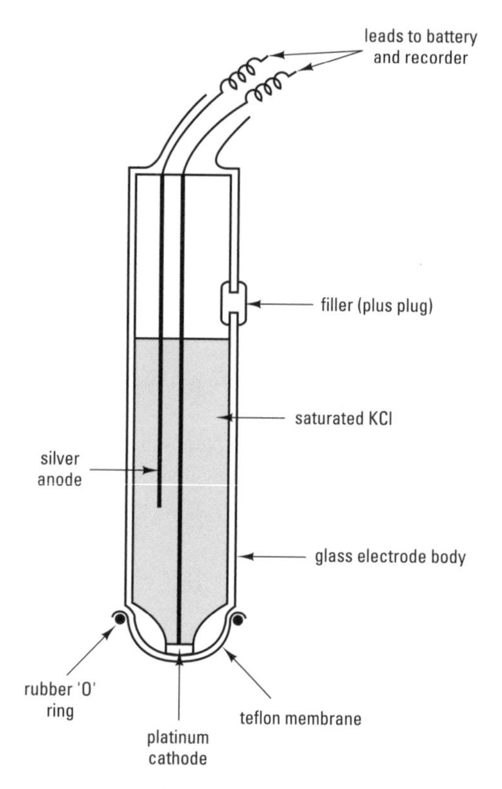

Fig. 28.7 A Clark-type oxygen probe.

Table 28.2 O_2 saturation values for distilled and sea water at standard atmospheric pressure and a range of temperatures. (Green and Carritt, 1967)

Temperature (°C)	O_2 saturation concentration ($mmol\,l^{-1}$)	
	Distilled water	Sea water (salinity 35‰)
0	0.460	0.359
2	0.435	0.342
4	0.413	0.326
6	0.392	0.311
8	0.373	0.298
10	0.355	0.285
12	0.339	0.273
14	0.324	0.261
16	0.310	0.251
18	0.297	0.241
20	0.285	0.232
22	0.274	0.224
24	0.263	0.215
26	0.253	0.208
28	0.244	0.200
30	0.235	0.193
37	0.212	0.174

Note: Tabulated values assume atmospheric pressure = 101.3 kPa; for more accurate work, a correction for any deviation from this value can be made by multiplying the appropriate figures by the ratio of the real pressure to the assumed value.

Box 28.1 How to set up a Clark (Rank) oxygen electrode

1. **Detach the base of the incubation vessel** (see Fig. 28.1) by unscrewing the locking ring.
2. **Add enough saturated KCl to cover the electrodes.**
3. **Cut a 1 mm square hole in the centre of a 10 × 10 mm square of lens tissue** and place this so that the hole is over the central platinum cathode.
4. **Cut a 10 × 10 mm square of Teflon® membrane and place over the lens tissue;** seal by gently lowering the incubation vessel and tightening the locking ring, making sure that the rubber O-ring is correctly positioned over the membrane.
   (a) Do not overtighten the locking ring.
   (b) Take care not to trap air bubbles beneath the membrane.
   (c) Make sure that the membrane does not become twisted.
5. **Clamp the electrode over the magnetic stirrer base using the clamping screw.**
6. **Connect the electrode leads to the polarizing unit/recording device** (silver anode to positive, platinum cathode to negative). In most cases, the output will be passed to a chart recorder, to give a readout of oxygen status as a function of time.
7. **Add air-saturated experimental solution and a small Teflon®-coated magnetic stirrer bar to the chamber.** The volume of the incubation chamber can be adjusted by moving the locking nut on the stopper. To adjust, add the appropriate amount of liquid to the chamber using a pipette, insert the stopper and screw the locking nut until the solution just fills the incubation chamber.
8. **Gently push the stopper (volume adjuster) into position,** making sure that no air bubbles are trapped in the chamber, and switch on the stirrer.
9. **Adjust the sensitivity control of the recorder** to give a suitable reading and allow to stabilize – this may take 5–10 min. This is the air-saturated reading.
10. **Add a few crystals of sodium dithionite to absorb all of the O_2 in solution;** this gives a zero reading (an alternative is to bubble N_2 through the solution for several minutes).
11. **Rinse the incubation chamber thoroughly and add fresh experimental solution.** Make sure that all traces of sodium dithionite are removed.
12. **Carry out your experiment.**
13. **Remove the solution and check the calibration** (steps 7 to 10). If the recorder deflections are different, the electode's sensitivity or the temperature may have changed and you may need to repeat the measurement.

> **Definition**
>
> **Biosensor** – a device for measuring a substance, combining the selectivity of a biological reaction with that of a sensing electrode.

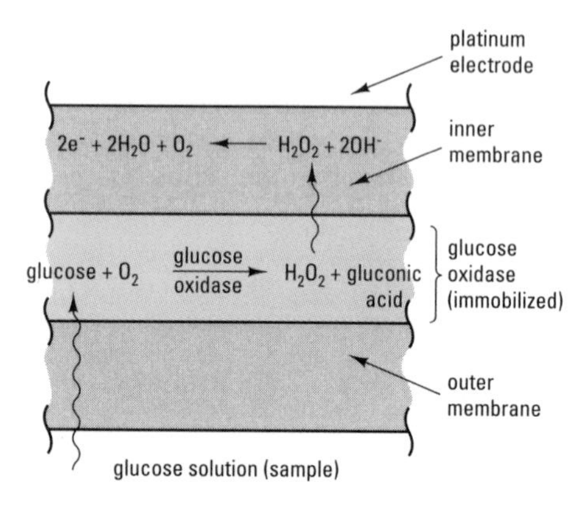

Fig. 28.8 Underlying principles of a glucose electrode.

The glucose electrode

The glucose electrode is a simple type of biosensor, whose basic design is shown in Fig. 28.8. It consists of a Pt electrode, overlaid by two membranes. Sandwiched between these membranes is a layer of the immobilized enzyme glucose oxidase. The outer membrane is glucose-permeable and allows glucose in the sample to diffuse through to the glucose oxidase layer, where it is converted to gluconic acid and H_2O_2. The inner membrane is selectively permeable to H_2O_2, which is oxidized to O_2 at the surface of the Pt electrode. The current arising from this release of electrons is proportional to the glucose concentration in the sample within the range 10^{-7}–10^{-3} mol l^{-1}.

Electrochemical detectors used in chromatography operate by voltammetric principles and currents are produced as the mobile phase flows over electrodes set at a fixed potential: to achieve maximum sensitivity, this potential must be set at a level that allows electrochemical reactions to occur in all analytes of interest.

Coulometric methods

Here, the charge required to completely electrolyse a sample is measured: the time required to titrate an analyte is measured at constant current and related to the amount of analyte using Faraday's law. There are few biological applications of this technique, though it is sometimes used for determination of Cl^- in serum and body fluids.

Assaying biomolecules and studying metabolism

29 Analysis of biomolecules: fundamental principles

Definitions

Accuracy – the closeness of an individual measurement, or a mean value based on a number of measurements, to the true value.

Concentration range – the range of values from the detection limit to the upper concentration at which the technique becomes inaccurate or imprecise.

Detection limit – the minimum concentration of an analyte that can be detected at a particular confidence level.

Drift – 'baseline' movement in a particular direction: drift can be a problem between analyses (e.g. using a spectrophotometer for colorimetric analysis), or for a single analysis (e.g. when separating biomolecules by chromatography).

Noise – random fluctuations in a continuously monitored signal.

Precision – the extent of mutual agreement between data values for an individual sample.

Replicate – repeated measurement.

Selectivity – the extent to which a method is free from interference due to other substances in the sample.

Sensitivity – the ability to discriminate between small differences in analyte concentration.

Validation – the process whereby the accuracy and precision of a particular analytical method is checked in relation to specific standards, using an appropriate reference material containing a known amount of analyte.

Table 29.1 Some examples of colorimetric assays

Analyte	Reagent/wavelength
Amino acids	ninhydrin/540 nm (p. 167)
Proteins	biuret/520 nm (p. 168)
	Folin–Ciocalteau/600 nm (p. 169)
Carbohydrates	anthrone/625 nm (p. 177)
Reducing sugars	dinitrosalicylate/540 nm (p. 178)
DNA	diphenylamine/600 nm (p. 183)
RNA	orcinol/660 nm (p. 184)

Biochemical analysis involves the characterization of biological components within a sample using appropriate laboratory techniques. Most analytical methods rely on one or more chemical or physical properties of the test substance (the analyte) for detection and/or measurement. There are two principal approaches:

1. Qualitative analysis – where a sample is assayed to determine whether a biomolecule is present or absent. As an example, a blood sample might be analysed for a particular drug or a specific antibody (p. 95), or a bacterial cell might be probed for a nucleic acid sequence (p. 233).
2. Quantitative analysis – where the quantity of a particular biomolecule in a sample is determined, either as an amount (e.g. as g, or mol) or in terms of its concentration in the sample (e.g. as $g\,l^{-1}$, or $mol\,m^{-3}$). For example, a blood sample might be analysed to determine its pH ($-\log_{10}[H^+]$), alcohol concentration in mg ml^{-1}, or glucose concentration in $mmol\,l^{-1}$. Skoog and Leary (1992) give details of methods.

Your choice of approach will be determined by the purpose of the investigation and by the level of accuracy and precision required (p. 41). Many of the basic quantitative methods described in Chapters 30–33 rely on chemical reactions of the analyte and involve assumptions about the nature of the test substance and the lack of interfering compounds in the sample: such assumptions are unlikely to be wholly valid at all times. If you need to make more exacting measurements of a particular analyte, it may be necessary to separate it from the other components in the sample, e.g. using chromatography (Chapter 24), ultracentrifugation (Chapter 23), or electrophoresis (Chapter 26), and then identify the separated components, e.g. using spectroscopic methods (Chapter 22). However, each stage in the separation and purification procedure may introduce further errors and/or loss of sample, as described in detail for protein purification in Chapter 34.

KEY POINT **In general, you should aim to use the simplest procedure that satisfies the purpose of your investigation – there is little value in using a complex, time-consuming or costly analytical procedure to answer a simple problem where a high degree of accuracy is not required.**

Most of the routine methods based on chemical analysis are destructive, since the analyte is usually converted to another substance which is then assayed, e.g. in colorimetric assays of the major types of biomolecules (Table 29.1). In contrast, many of the analytical methods based on physical properties are non-destructive (e.g. the intrinsic absorption and emission of electromagnetic radiation in spectrophotometry, p. 111, or nuclear magnetic resonance techniques, p. 119). Non-destructive methods are often preferred, as they allow the further characterization of a particular sample. Most biological methods are destructive: bioassays are often sensitive to interference and require validation (p. 84).

Validity and quantitative analysis

Before using a particular procedure, you should consider its possible limitations in terms of:

- measurement errors, and their likely magnitude: these might include processing errors (e.g. in preparing solutions and making dilutions), instrumental errors (e.g. a pH electrode that has not been set up correctly), calibration errors (e.g. converting a digital readout analyte concentration) and errors due to the presence of interfering substances;
- sampling errors: these may occur if the material used for analysis is not representative, e.g. due to biological differences between the individual organisms used in the sampling procedure (p. 51).

Evaluating a new method – results from a novel technique can be compared with an established 'standardized' technique by measuring the same set of samples by each method and analysing the results by correlation (p. 275).

Replication will allow you to make quantitative estimates of several potential sources of error: for example, repeated measurements of the same sample can provide information on the precision of the analytical method, e.g. by calculating the coefficient of variation (p. 266), while measurements of several different samples can provide information on biological and sample variability, e.g. by calculating the standard deviation (Chapter 46). Analyses of different sets of samples at different times (e.g. on different days) can provide information on 'between batch' variability, as opposed to 'within batch' variability (based on a single set of analytical data).

The reliability of a particular method can be assessed by measuring 'standards' (sometimes termed 'controls'). These are often prepared in the laboratory by adding a known amount of analyte to a real sample (this is often termed 'spiking' a sample), or by preparing an artificial sample containing a known amount of analyte along with other relevant components (e.g. the major sample constituents and possible interfering substances). In many instances, several standards (including a 'blank' or 'zero') are assayed to construct a 'standard curve', which is then used to convert sample measurements to amounts of analyte (e.g. Fig. 18.4). Such standard curves form the basis of many routine laboratory assays: while hand-drawn linear calibration curves are sufficient for basic assays, more complex curves are often fitted to a particular mathematical function using a computer program (e.g. bioassays, p. 84). Standards can also be used to check the calibration of a particular method: a mean value based on repeated measurements of an individual 'standard' can be compared with the true value using a modified *t* test (p. 273, Equation 47.1), in which there is only one standard error term, i.e. that associated with the measured values.

Interpreting results from 'spiked' samples – remember that such procedures tell you nothing about the extraction efficiency of biomolecules from a particular sample, e.g. during homogenization (Chapter 17).

Validation of a particular method can be important in certain circumstances, e.g. in a forensic science laboratory, or in a clinical biochemistry laboratory, where particular results can have important implications. Such laboratories operate strict validation procedures, which include: (i) adherence to standard operating procedures for each analytical method; (ii) calibration of assays using certified reference materials containing known amount of analyte and traceable to a national reference laboratory; (iii) effective quality assurance and quality control systems; (iv) detailed record-keeping, covering all aspects of the analysis and recording of results. Although such rigour is not required for routine analysis, the general principles of standardization, calibration, assessment of performance and record-keeping are equally valid for all analytical work.

Criteria for the selection of a particular analytical method:

- the required level of accuracy and precision;
- the number of samples to be analysed;
- the amount of each sample available for analysis;
- the physical form of the samples;
- the expected concentration range of the analyte in the samples;
- the sensitivity and detection limit of the technique;
- the likelihood of interfering substances;
- the speed of analysis;
- the ease and convenience of the procedure;
- the skill required by the operator;
- the cost and availability of the equipment.

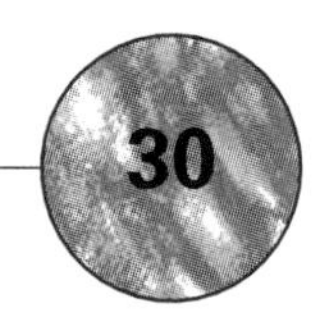

30 Assaying amino acids and proteins

COOH	COO⁻
H_2N-C-H	H_3N^+-C-H
R	R
unionized form	zwitterionic form

Fig. 30.1 Structure of α-amino acids. R = side chain (see Table 30.1 for examples).

Table 30.1 The 20 amino acids incorporated into protein, grouped according to their side chains

Name	Three letter code	Capital letter code
Aliphatic side chains		
Glycine	gly	G
Alanine	ala	A
Valine	val	V
Leucine	leu	L
Isoleucine	ile	I
Aromatic side chains		
Phenylalanine	phe	F
Tyrosine	tyr	Y
Tryptophan	trp	W
S-containing side chains		
Cysteine	cys	C
Methionine	met	M
Side chains with –OH groups[1]		
Serine	ser	S
Threonine	thr	T
Basic side chains		
Histidine	his	H
Lysine	lys	K
Arginine	arg	R
Acidic side chains		
Aspartate	asp	D
Glutamate	glu	E
Amide side chains		
Asparagine	asn	N
Glutamine	gln	Q
Cyclic structure (imino acid)		
Proline	pro	P

[1]tyr also has an OH group.

Definitions

Simple protein – a protein composed entirely of amino acid residues.

Conjugated protein – a protein containing a non-amino acid component (the prosthetic group), in addition to a polypeptide component (the apoprotein).

Peptide – a short polymer, containing up to 50 amino acid residues.

Proteins are linear polymers, formed by the linkage of α-amino acids (Fig. 30.1) via peptide bonds to create polypeptide chains. The 20 amino acids most commonly incorporated into proteins are listed in Table 30.1. Others may be found when a protein is hydrolysed, due to post-translational modification (e.g. hydroxyproline).

KEY POINT **In all amino acids apart from glycine, the α-carbon atom is asymmetrical, resulting in stereoisomers. Most life scientists use the D- and L- nomenclature (Fischer convention) when referring to stereoisomers of amino acids, although the R- and S- convention can be a useful alternative when considering enzymatic reaction mechanisms.**

While most of the chemical properties of stereoisomers are identical, enzymes (including those involved in protein synthesis) can distinguish between them and, with the exception of a few bacterial components, only the L-isomers of amino acids are normally used in protein synthesis.

Amino acids and their properties

There is no agreed formal classification of amino acids but, in terms of protein structure and function, the side chains of amino acid residues can be grouped into:

- those that carry a charge (e.g. asp, glu, lys, arg, his) – they may participate in electrostatic interactions and charge–dipole (hydrophilic) interactions;
- those that contain O, N, or S (e.g. ser, met, thr, asn, gln, cys, tyr) – they can participate in hydrogen bonding and hydrophilic interactions;
- those that are hydrocarbon in nature (e.g. ala, leu, val, ile, phe, trp) – they may participate in hydrophobic interactions;
- cysteine – this amino acid can form intra- and inter-polypeptide disulphide bridges.

Detection and quantification of amino acids

The primary amino group of amino acids will react with ninhydrin to give a purple coloured product – this reaction can be used for qualitative assay, e.g. to detect individual amino acids in chromatography (p. 138), or for quantitative colorimetric assay by measuring the absorbance at 570 nm using a spectrophotometer (p. 114). Note that different amino acids give different amounts of coloured product on reaction with ninhydrin, so careful standardization is needed. The secondary amino groups of the cyclic imino acids hydroxyproline and proline give yellow products with ninhydrin and are assayed at 440 nm.

Proteins and their properties

The diversity of protein structure must be taken into account when considering a suitable analytical method. Globular proteins (e.g. albumin, antibodies) are relatively soluble in dilute salt solutions while fibrous proteins (e.g. collagens, keratin) are typically insoluble, unless hydrolysed (e.g. using 6 mol l^{-1} HCl). Conjugated proteins (e.g. metalloproteins, glycoproteins, nucleoproteins) also include non-amino acid components, which may influence the choice of analytical method.

Chemical properties of proteins

The features most commonly exploited for quantitative analysis are:

- the peptide bond, which forms a violet-coloured complex with Cu^{2+} in the biuret reaction (Box 30.1);
- the phenolic group of tyr and the indole group of trp, which react with the oxidizing agents phosphotungstic and phosphomolybdic acids in the Folin–Ciocalteau reagent to produce a blue colour. This is combined with the biuret reaction in the Lowry method (Box 30.1);

Box 30.1 Methods of determining the amount of protein in an aqueous solution

Most assays for protein content do not give an absolute value, but require standard solutions, containing appropriate concentrations of a particular protein, to be analysed at the same time, so that a standard curve can be constructed. Bovine serum albumin (BSA) is commonly used as a protein standard. However, you may need an alternative standard if the protein you are assaying has an amino acid composition which is markedly different from that of BSA, depending on your chosen method. For all of the following methods, the amounts are appropriate for semi-micro cuvettes (1.5 ml volume, path length 1 cm). Appropriate controls (blanks) must be analysed, to assess for possible interference, (e.g. due to buffers, etc.) in your protein extract.

Biuret method

This is based on the specific reaction between cupric ions (Cu^{++}) in alkaline solution and two adjacent peptide bonds, as found in proteins. As such, it is not significantly affected by differences in amino acid composition between proteins.

1. **Prepare protein standards over an appropriate range** (typically, between 1–10 mg ml^{-1}).
2. **Add 1 ml of each standard solution to separate test tubes. Prepare a reagent blank, using 1 ml of distilled water, or an appropriate solution.**
3. **Add 1 ml of each unknown solution to separate test tubes.**
4. **Add 1 ml of biuret reagent (1.5 g $CuSO_4.5H_2O$,** 6.0 g sodium potassium tartrate in 300 ml of 10 w/v NaOH) to all standard and unknown tubes and to the reagent blank.
5. **Incubate at 37 °C for 15 min.**
6. **Read the absorbance of each solution at 520 nm against the reagent blank.** The colour is stable for several hours.

The main limitation of the biuret method is its lack of sensitivity – it is unsuitable for solutions with a protein content of less than 1 mg ml^{-1}.

Direct measurement of UV absorbance (Warburg–Christian method)

Proteins absorb electromagnetic radiation maximally at 280 nm (due to the presence of aromatic amino acids, especially tyrosine and tryptophan) and this forms the basis of the method. The principal advantages of this approach are its simplicity and the fact that the assay is non-destructive. The most common interfering substances are nucleic acids, which can be corrected for by measuring the absorbance at 260 nm (where nucleic acids have a stronger absorption and proteins have a weaker absorption): a pure solution of protein will have a ratio of absorption (A_{280}/A_{260}) of approximately 1.8, decreasing with increasing nucleic acid contamination. Note also that any free aromatic amino acids in your solution will absorb at 280 nm, leading to an overestimation of protein content. The simplest procedure, which includes a correction for small amounts of nucleic acid, is as follows (use quartz cuvettes throughout):

1. **Measure the absorbance of your solution at 280 nm (A_{280}):** if A_{280} is greater than 1, dilute by an appropriate amount and re-measure (see p. 114).
2. **Repeat at 260 nm (A_{260}).**
3. **Estimate the approximate protein content using the following relationship:**

$$[\text{protein}]\ \text{mg ml}^{-1} = 1.45\,A_{280} - 0.74\,A_{260} \qquad [30.1]$$

This equation is based on the work of Warburg and Christian (1942), for the enzyme enolase. For other proteins, it should not be used for quantitative work, since it gives only a rough approximation of the amount of protein present, due to variations in aromatic amino acid composition. A better approach is to establish the specific absorptivity of your particular protein at 280 nm (e.g. using the pure substance and a spectrophotometer) and then use the Beer–Lambert relationship (p. 111) to determine the amount of protein in your samples – this is only feasible if you have a purified sample, free from significant nucleic acid and amino acid contamination.

- dye binding to hydrophobic regions, e.g. in the Bradford assay (Box 30.1);
- the primary amino groups of lys residues, the guanidino group of arg residues and the N-terminal amino acid residue will react with ninhydrin, allowing colorimetric assay similar to that described above for isolated amino acids. However, a more sensitive method uses the reaction of these groups with fluorescamine to give an intensely fluorescent product, which you can measure by spectrofluorimetry (p. 115).

Physical properties of proteins

The characteristics that can be used for separation and analysis include:

- absorbance of aromatic amino acid residues (phe, tyr and trp) at 280 nm (Warburg–Christian method, Box 30.1);
- prosthetic groups of conjugated proteins – these often have characteristic absorption maxima, e.g. the haem group of haemoglobin absorbs strongly at 415 nm, allowing quantitative assay;

Exploiting the biological properties of proteins – relevant analytical methods include immunoassay (p. 95), enzymatic analysis (p. 192) and affinity chromatography (p. 134), based on specific biological interactions.

Box 30.1 – continued

Lowry (Folin-Ciocalteau) method

This is a colorimetric assay, based on a combination of the biuret method, described above, and the oxidation of tyrosine and tryptophan residues with Folin and Ciocalteau's reagent to give a blue–purple colour. The method is extremely sensitive (down to a protein content of 20 $\mu g\ ml^{-1}$), but is subject to interference from a wide range of non-protein substances, including many organic buffers (e.g. TRIS, HEPES, etc.), EDTA, urea and certain sugars. Choice of an appropriate standard is important, as the intensity of colour produced for a particular protein is dependent on the amount of aromatic amino acids present.

1. **Prepare protein standards within an appropriate range for your samples** (the method can be used from 0.02–1.00 $mg\ ml^{-1}$).
2. **Add 1 ml of each standard solution to separate test tubes. Prepare a reagent blank, using 1 ml of distilled water, or an appropriate solution.**
3. **Add 1 ml of each of your unknown solutions to separate test tubes.**
4. **Then, add 5 ml of 'alkaline solution' (prepared by mixing 2% w/v Na_2CO_3 in 0.1 $mol\ l^{-1}$ NaOH, 1% w/v aqueous $CuSO_4$ and 2% w/v aqueous Na K tartrate in the ratio 100 : 1 : 1. Mix thoroughly and allow to stand for at least 10 min.**
5. **Add 0.5 ml of Folin–Ciocalteau reagent (commercial reagent, diluted 1 : 1 with distilled water on the day of use). Mix rapidly and thoroughly and then allow to stand for 30 min.**
6. **Read the absorbance of each sample at 600 nm.**

Dye-binding (Bradford) method

Coomassie brilliant blue combines with proteins to give a dye–protein complex with an absorption maximum of 595 nm. This provides a simple and sensitive means of measuring protein content, with few interferences. However, the formation of dye–protein complex is affected by the number of basic amino acids within a protein, so the choice of an appropriate standard is important. The method is sensitive down to a protein content of approximately 5 $\mu g\ ml^{-1}$ but the relationship between absorbance and concentration is often non-linear, particularly at high protein content.

1. **Prepare protein standards over an appropriate range (between 5 and 100 $\mu g\ ml^{-1}$).**
2. **Add 100 μl of each standard solution to separate test tubes. Prepare a reagent blank, using 100 μl of distilled water, or an appropriate solution** (note that these small volumes must be accurately dispensed, e.g. using a calibrated pipettor, p. 9).
3. **Add 100 μl of your unknown solutions to separate test tubes.**
4. **Add 5.0 ml of Coomassie brilliant blue G250 solution (0.1 $g\ l^{-1}$).**
5. **Mix and incubate for at least 5 min: read the absorbance of each solution at 595 nm.**

Other methods are less widely used. They include determination of the total amount of nitrogen in solution (e.g. using the Kjeldahl technique) and calculating the protein content, assuming a nitrogen content of 16%. An alternative approach is to precipitate the protein (e.g. using trichloracetic acid, tannic acid, or salicylic acid) and then measure the turbidity of the resulting precipitate (using a nephelometer, or a spectrophotometer, p. 114).

Definition

Lipoproteins – globular, micelle-like particles consisting of a non-polar core of triacylglyceryl and cholesteryl esters surrounded by a coating of proteins, phospholipids and cholesterol: involved in the transport of lipids in blood.

- density – most proteins have a density of $1.33\,kg\,l^{-1}$, but lipoproteins have a lower density. This can be used to separate lipoproteins from other classes of protein, or to sub-divide the various classes of lipoproteins;
- net charge – proteins differ in the types and number of amino acids with ionizable groups in their side chains. The ionization of these side chains is pH dependent, resulting in variation of net charge with pH (p. 143). This property is exploited in the techniques of electrophoresis (Chapter 26), isoelectric focusing (Chapter 27), and ion exchange chromatography (Chapter 24);
- water solubility and surface hydrophobicity – these are exploited in salt fractionation (p. 188) and hydrophobic interaction chromatography (p. 133).

Determining the primary structure of a protein

The term 'primary structure' refers to the linear sequence of amino acid residues along the polypeptide: ultimately this determines the three-dimensional shape of the molecule and hence its biological properties.

Amino acid analysis

Purifying a protein – the practical procedures involved in the purification of a particular protein from a biological sample are considered in detail in Chapter 34.

Once a particular protein has been purified (Chapter 34), the first step in determining its primary structure is to hydrolyse the polypeptide (e.g. by treating with $6\,mol\,l^{-1}$ HCl at 110°C for 24–72 h) and then determine the constituent amino acids. This can be achieved by several techniques, including thin layer chromatography and high-performance liquid chromatography (Chapter 24). Often a dedicated amino acid analyser is used for quantitative assays: a polystyrene resin-based cation exchange column is used to separate the amino acids on the basis of ion-exchange and hydrophobic interactions.

Sequence analysis

Having obtained the amino acid composition of the protein, the next step is to determine the order of the amino acid residues along the polypeptide chain. This may be achieved by:

1. Cleavage of the polypeptide at specific peptide bonds resulting in smaller fragments (peptides); e.g. using cyanogen bromide, which cleaves peptide bonds formed by the carboxyl group of met, or by proteolytic enzymes, e.g. trypsin (cleaving after lys and arg residues) and chymotrypsin (cleaving after phe, tyr, and trp).
2. Separation of the peptides, e.g. by column chromatography.
3. Determination of the sequence of each peptide by a process called Edman degradation. This involves the selective removal of the N-terminal amino acid by treating the fragments with phenyl isothiocyanate ($C_6H_5N{=}C{=}S$) followed by acid hydrolysis; the substituted phenylthiohydantoin formed can then be identified by chromatography. The process can be repeated as many as 50 times, releasing the N-terminal residue on each occasion, and allowing the sequence of each fragment to be established.
4. Matching of the sequenced, overlapping peptide fragments, to determine the overall sequence of amino acid residues in the original protein.

Investigating the secondary and tertiary structure of a protein – computer-based molecular modelling programs can be used to predict the three-dimensional conformation of a protein from information based on the primary structure (Chapter 49).

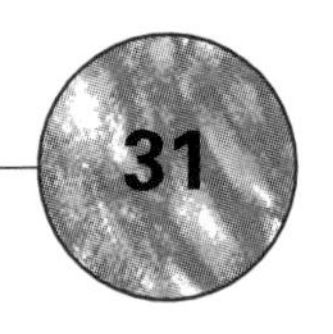

31 Assaying lipids

The term lipid is used to describe a broad group of compounds with a wide variety of chemical structures, physical properties and biological functions. Lipids in biological systems are often combined with either proteins (e.g. in lipoproteins) or polysaccharides (e.g. in lipopolysaccharides). This chapter can provide only a brief outline of the techniques used in lipid analysis (for further details, see Kates, 1986, or Gunstone *et al.*, 1986).

Table 31.1 Some examples of fatty acids

No. of carbon atoms	Systematic name	Trivial name
Saturated fatty acids (no C=C bonds)		
12	*n*-Dodecanoic	Lauric
14	*n*-Tetradecanoic	Mystiric
16	*n*-Hexadecanoic	Palmitic
18	*n*-Octadecanoic	Stearic
20	*n*-Eicosanoic	Arachidic
22	*n*-Docosanoic	Behenic
Mono-unsaturated fatty acids (one C=C bond)		
12	*cis*-9-Dodecenoic	Lauroleic
14	*cis*-9-Tetradecenoic	Myristoleic
16	*cis*-9-Hexadecenoic	Palmitoleic
18	*cis*-9-Octadecenoic	Oleic
20	*cis*-9-Eicosenoic	Gadoleic
22	*cis*-9-Docosenoic	Erucic

Note: palmitic, stearic, palmitoleic and oleic acids are quantitatively the most common fatty acids in the majority of organisms. Major polyunsaturated fatty acids include linoleic acid (C_{18}, 2 double bonds), linolenic acid (C_{18}, three double bonds) and arachidonic acid (C_{20}, four double bonds).

KEY POINT **The defining feature of lipids is their relative insolubility in water: consequently, they are extracted from biological material using organic solvents, e.g. acetone, ether and chloroform. Because of their diversity and complexity, they are often referred to by more than one name, and the various types of non-systematic names can be confusing.**

Biological lipids are often sub-divided into two main types, each of which contains fatty acids as a major structural component:

Simple, or neutral lipids

These are esters of fatty acids and an alcohol (e.g. Fig. 31.1). Fatty acids are straight-chain carboxylic acids, typically with an even number of carbon atoms and chain lengths of C_{12}–C_{22}, which may be saturated or unsaturated (Table 31.1). The greater the chain length and the fewer the number of double bonds, the higher the melting point, making most long chain saturated fatty acids solids at room temperature. Glycerol is the most common alcohol found in simple lipids, though higher M_r alcohols occur in waxes and cyclic alcohols (sterols, e.g. cholesterol) occur in bile acids, steroid hormones and vitamins (e.g. vitamin K). While glycerol is a liquid at room temperature, cholesterol remains solid up to 150 °C.

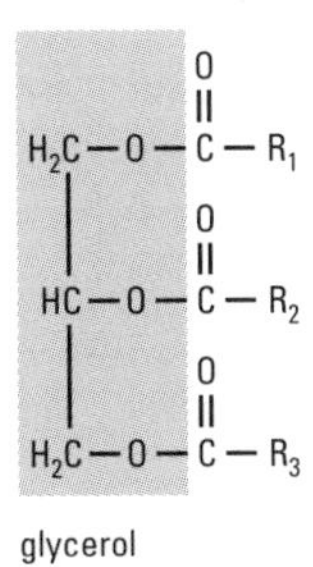

Fig. 31.1 Example of a neutral lipid: the structure of a triacylglycerol. R = remainder of fatty acid.

Neutral fats (triglycerides or triacylglycerol) are esters of fatty acids and glycerol, as shown in Fig. 31.1. Many animal triglycerides ('fats') contain mainly saturated fatty acids and are solids at room temperature, while plant triglycerides ('oils') often have shorter chain lengths and a greater degree of unsaturation and are liquids at room temperature. Waxes are esters of fatty acids with alcohols of higher M_r than glycerol. The major biological functions of simple lipids include (i) energy storage, e.g. oils in plant seeds, (ii) insulation, e.g. subcutaneous fat deposits in whales, and (iii) waterproofing, e.g. waxes in the cuticles of plant leaves.

Complex, compound or conjugated lipids

These are acyl esters of glycerol, or the amino alcohol sphingosine, that also include a hydrophilic group (e.g. a phosphoryl or carbohydrate group). They are often described in terms of this hydrophilic group, e.g. phospholipids contain a phosphoryl group while glycolipids contain a carbohydrate group.

KEY POINT **An important feature of complex lipids is their amphipathic nature, i.e. each molecule has a polar (hydrophilic) and a non-polar (hydrophobic) region.**

The two major types of complex lipid are:

1. Phospholipids – the most common types (phosphoglycerides) are based on phosphatidic acid, with two fatty acids esterified to glycerol. Most phospholipids also contain an amino alcohol or a similar group, attached to the phosphoryl group. The principal phospholipid classes are: (i) phosphatidyl cholines (or lecithins), which form stable emulsions with water and dissolve completely on addition of bile salts. These are insoluble in acetone, a feature that enables lecithins to be separated from most other lipids; (ii) phosphatidyl ethanolamines (or cephalins) – unlike lecithins, they are insoluble in ethanol and methanol; (iii) phosphatidyl serine; (iv) phosphatidyl inositol; and (v) plasmalogens.
2. Sphingolipids – these incorporate the amino dialcohol, sphingosine, rather than glycerol. Fatty acids are linked to sphingosine via an amide bond to form ceramides, which include: (i) cerebrosides (glycosphingolipids); (ii) sulphatides (sulphated cerebrosides); (iii) gangliosides (glycosphingolipids containing sialic acid residues); and (iv) phosphosphingolipids, including sphingomyelins, which are esters of a ceramide and phosphoryl choline.

Complex lipids have important roles in biological membranes: phospholipids are major structural components while sphingolipids are involved in cell–cell recognition and similar membrane features, e.g. glycosphingolipids are determinants of human ABO blood groups.

Extraction and analysis of lipids

Solvent extraction

Lipid extraction is carried out in a suitable organic solvent, using a homogenizer to disrupt cell and tissue structure (Chapter 17). There are two possible approaches:

Safe working with solvents – chloroform and benzene must be used with care, due to their high toxicity: they are often replaced by dichloromethane and toluene respectively. All mixing and pouring steps should be carried out in a spark-free fume cupboard. Note that ethers may form explosive peroxides on prolonged storage. Lipid extracts extracted in flammable solvents must be stored in a spark-proof refrigerator, not in routine lab fridges.

1. Total extraction of all lipids, followed by separation of the different lipid classes. A commonly used solvent mixture is methanol: chloroform : water (2 : 1 : 0.5, v/v/v). Adding an equal volume of aqueous 1% w/v NaCl solution to the extract results in the formation of two layers: the lipids are present in the lower layer, while the upper aqueous layer contains other biomolecules (e.g. proteins). Hexane : isopropanol (3 : 2, v/v) can be used instead of chloroform : methanol, since it is less toxic and it dissolves very little non-lipid material.
2. Selective lipid extraction. Neutral lipids, e.g. within storage tissue, can be extracted by relatively non-polar solvents including hexane, diethyl ether and chloroform. Extraction of membrane lipids (e.g. phospholipids) requires disruption of the membrane using more polar solvents (e.g. methanol or ethanol), with selective precipitation by adding cold, dry acetone. Glycolipids can be extracted using acetone.

Definition

Emulsion – a colloidal mixture where one liquid is dispersed (but not dissolved) in another liquid, e.g. lipids often aggregate to form micelles in aqueous solutions.

The principal disadvantages of solvent extraction methods include; (i) the requirement for large volumes of potentially hazardous solvents; and (ii) the possible formation of an emulsion, with incomplete extraction of component lipids.

Adsorption chromatography

Silica gel, octadecylsilane-bonded silica or ion-exchange resins can be used to bind solvent-extracted lipids by a combination of polar, ionic and van der Waals forces. In practice, a glass column is packed with a slurry of adsorbent

Preventing oxidative rancidity – all lipids will oxidize (become rancid) when exposed to air in daylight, due to hydrolysis and/or photo-oxidation. Keep extracted lipids in the dark, and add an antioxidant such as butylated hydroxytoluene for longer term storage. Minimize oxidation by flushing vessels with nitrogen gas and bubbling solvents with (oxygen-free) nitrogen during analysis.

in an appropriate organic solvent, and the lipid extract (dissolved in the same solvent) is applied to the top of the column. Lipids can then be selectively eluted; a mixture can be broadly separated into neutral lipids, glycolipids and phospholipids using solvents of increasing polarity, e.g. chloroform → acetone → methanol. It is possible to further separate the lipid sub-fractions on silica gel columns, as follows:

- Neutral lipids can be separated and eluted using hexane containing increasing proportions of diethyl ether (0 → 100% v/v) in the order: hydrocarbons, cholesterol esters, triacylglycerols, free fatty acids, cholesterol, diacylglycerols, then monoacylglycerols.
- Glycolipids and sulpholipids can be separated by first eluting the glycolipids with chloroform : acetone (1 : 1 v/v), then using acetone to elute the sulpholipids.
- Phospholipids can be eluted using chloroform containing increasing proportions of methanol (5% → 50% v/v) in the order: phosphatidic acid, phosphatidyl ethanolamine, phosphatidyl serine, phosphatidyl choline, phosphatidyl inositol, sphingomyelin.

Thin layer chromatography (TLC) of lipids

This can be used to separate lipid mixtures into their constituents, or to quantify particular lipids, as part of an analytical procedure. The principle of the technique is described in Chapter 24.

Understanding silica gel codes – silica gel is manufactured as particles of 10–15 μm diameter, containing pores of diameter 40, 60, 80, 100, or 150 Å (Ångstrom p. 47), where Å = 10^{-10} m (thus, silica gel 60 has a pore size of 60 Å). Silica gel G contains calcium sulphate to assist binding to the glass support, while silica gel H has no binder.

Silica gel G60 is the most frequently used stationary phase, acting as a polar absorbent. When the separation is carried out with a non-polar mobile phase, non-polar lipids will migrate more rapidly (i.e. they will have high R_f values, p. 128), while polar lipids will migrate more slowly. By increasing the polarity of the solvent, the R_f values of the polar lipids can be increased. Your choice of solvent will depend on the lipids in the extract:

- For a broad range of neutral lipids, a typical solvent system is hexane : diethyl ether : glacial acetic acid at 80 : 20 : 2 (v/v/v), separating in the following order of decreasing R_f: steryl esters, wax esters, fatty acids, methyl esters, triacylglycerols, fatty acids, fatty alcohols, sterols, 1,2-diacylglycerols, monoacylglycerols.
- For polar lipids, silica gel H is preferred, as silica gel G prevents the separation of acidic phospholipids. Most solvent systems for polar lipids are based on chloroform : methanol : water e.g. at 65 : 25 : 4 (v/v/v). In this system, the relative order of migration is: monogalactosyldiacylglycerol, cerebrosides, phosphatidic acid, cardiolipin, lysophosphatidyl ethanolamine, phosphatidyl ethanolamine and digalactosyldiacylglycerol, sulphatides, phosphatidyl choline, phosphatidyl inositol, sphingomyelin and phosphatidyl serine.

KEY POINT **No single solvent system will completely separate all lipid components in a single TLC procedure. However, two-dimensional TLC may separate up to 200 individual constituents from a sample.**

Safe working with lipid stains – these often contain corrosive or toxic reagents, so all manipulations should be performed in a fume cupboard.

Lipids can also be separated by reversed-phase TLC (RP-TLC), where the silica gel is made non-polar (e.g. by silanization), and highly polar solvents are used as the mobile phase: here, the polar lipids will have the highest R_f values and the non-polar lipids the lowest R_f values. Lipids separated by TLC can be located by staining. Several stains are non-specific, enabling almost all types of lipid to be visualized, while others will locate particular lipid classes. Staining can be carried out by immersion of the TLC plate in the stain, or by spraying. Non-specific staining methods include:

- Spraying with iodine solution (e.g. 1–3% w/v in chloroform) – most lipids appear as brown spots on a yellow background, though glycolipids stain weakly by this method.
- Treatment with a strong oxidizing agent followed by charring (e.g. spray with 5% v/v sulphuric acid in ethanol, followed by heating in an oven at 180 °C for 30–60 min) – lipids appear as black deposits; the detection limit is 1–2 μg. Scanning densitometry (Chapter 25) can provide quantitative information.
- Fluorescent stains (e.g. the widely used 2′,7′-dichlorofluorescein, DCF, at 0.1–0.2% w/v in ethanol). Under UV, lipids appear as bright yellow spots on a yellow–green background; the limit of detection is 5 μg. Other fluorescent stains such as 1-anilo-8-naphthalene sulphonate (ANS, 0.1% w/v in water) can detect ng quantities of lipid.

Quantitative assay of lipids and their components

While TLC can be used to quantify particular lipids, it is more common to assay the compounds released on hydrolysis of simple or complex lipids, namely the alcohols, fatty acids or other components, rather than the native lipid. Alkali hydrolysis of lipids containing fatty acids results in the formation of a soap (i.e. saponification): the incubation of tripalmitin with KOH yields potassium palmitate and glycerol. Acid hydrolysis of triacylglycerols releases 'free' fatty acids and glycerol.

Basic information about the relative size of the fatty acid component of oils and fats is given by the saponification value, determined by titration against 0.8 mol l^{-1} KOH; the lower the saponification value, the higher the M_r of the fatty acids. The degree of unsaturation of the fatty acids is given by the iodine number; the higher the iodine number, the greater the content of unsaturated fatty acids.

Definitions

Saponification value – the amount of KOH (in mg) required to completely saponify 1 g of fat. Typically within the range 150–300.

Iodine number – the amount of iodine (g) absorbed by 100 g of fat, due to the reaction of iodine with C=C bonds within the fat. Typically within the range 30–200.

To obtain the iodine number for a particular fat, the free iodine remaining after reaction is titrated against 0.1 mol l^{-1} sodium thiosulphate using a trace amount of starch as an indicator, giving a titration volume for the test solution, V_t. A blank containing no fat is also titrated, to establish the volume of sodium thiosulphate required to titrate the initial free iodine, V_o. This allows the amount of iodine that reacts with the fat to be calculated according to the formula:

$$\text{iodine number (in g)} = \frac{127 \times 0.1(V_o - V_t)}{m \times 100} \qquad [31.1]$$

where m is the mass of test fat (in g), and V_o and V_t are expressed in ml.

Measurement of glycerol content

Glycerolipids can be quantified by measuring the glycerol released on hydrolysis. A widely used method involves a coupled enzyme assay (p. 193), with the following reactions:

$$\text{triacylglycerol} \xrightarrow{\textit{lipase}} \text{glycerol} + \text{fatty acids} \qquad [31.2]$$

$$\text{glycerol} + \text{ATP} \xrightleftharpoons{\textit{glycerol kinase, } Mg^{2+}} \text{glycerol-3-phosphate} + \text{ADP} \qquad [31.3]$$

$$\text{ADP} + \text{phosphoenolpyruvate} \xrightleftharpoons{\textit{pyruvate kinase}} \text{ATP} + \text{pyruvate} \quad [31.4]$$

$$\text{pyruvate} + \text{NADH} + \text{H}^+ \xrightleftharpoons{\textit{lactate dehydrogenase}} \text{lactate} + \text{NAD}^+ \quad [31.5]$$

The glycerol concentration is determined by measuring the decrease in absorbance of NADH at 340 nm, compared to that of a blank with no added lipase. Details of the conversion of A_{340} to [NADH] are given on p. 113.

Measurement of cholesterol content

Cholesterol in cholesterol esters can be estimated after hydrolysis to free cholesterol using cholesterol esterase. The subsequent assay is as follows:

$$\text{cholesterol} + O_2 \xrightarrow{\textit{cholesterol oxidase}} \text{cholest-4-en-3-one} + H_2O_2 \quad [31.6]$$

The hydrogen peroxide produced as a result of the action of cholesterol oxidase can be measured amperometrically (p. 158) or colorimetrically, via the peroxidase-catalysed reaction:

$$2H_2O_2 + \text{phenol} + \text{4-aminoantipyrene} \xrightarrow{\textit{peroxidase}} \text{quinoneimine} + 4H_2O \quad [31.7]$$

This is the basis of many commercially available cholesterol testing kits. An alternative approach is to measure the cholest-4-en-3-one directly, via its absorbance maximum at 240 nm.

Example Fatty acids are derivatized to their methyl esters by boiling for 2–3 min with boron tetrafluoride (BF_3) solution (14% w/v in methanol), prior to GC analysis, to increase their volatility and stability in the gas phase. The separated components are normally detected by flame ionization (p. 139).

Gas chromatography (GC) of lipids and lipid components

This technique is used for the quantitative and qualitative analysis of a broad range of lipids. Volatile lipids may be analysed without modification, while non-volatile lipids must first be converted to a more volatile form, either by degradation (e.g. phospholipids), or derivatization (p. 131). The most effective GC columns are support-coated open tubular (SCOT) capillary columns, with a thin film (0.1–10 μm) of the stationary phase coated onto the internal wall (p. 130). The choice of stationary phase depends on the components to be separated, e.g. some non-polar stationary phases cannot resolve methyl esters of saturated and monounsaturated fatty acids. Many stationary phases are based on silicone greases or polysiloxane, ranging from the non-polar dimethyl polysiloxanes (e.g. OV-101) to the polar trifluoropropyl methyl polysiloxanes (e.g. OV-210). Other polar stationary phases are based on polyethylene glycol (e.g. Carbowax 20M).

32 Assaying carbohydrates

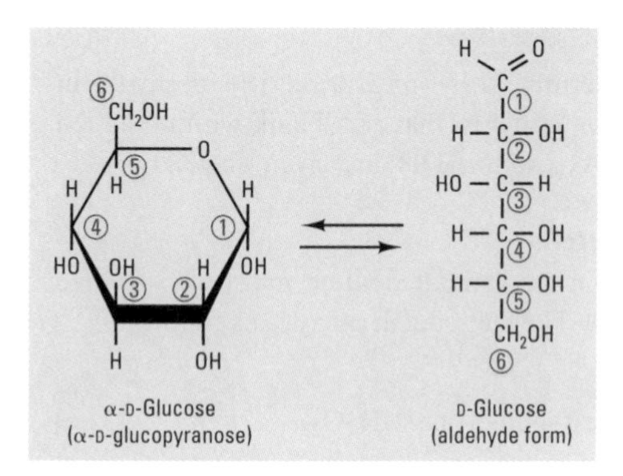

Fig. 32.1 Ring and straight chain forms of glucose. Note that β-D-glucose has H and OH groups reversed at C-1 in the pyranose form.

These are compounds with a formula based on $C_x(H_2O)_x$. They play a key role in energy metabolism, and are essential constituents of cell walls and membranes. They may exist individually, or as heteropolymers, e.g. linked to protein in glycoproteins (where carbohydrates form the minor component), or proteoglycans (where they form the major component).

KEY POINT The identification and quantitative analysis of carbohydrates in biological samples can be difficult, due to structural and physico–chemical similarities between related compounds. Several routine analytical methods cannot distinguish between isomeric forms of a particular carbohydrate.

Monosaccharides

The simplest carbohydrates are the monosaccharide sugars, which are polyhydroxy aldehydes and polyhydroxy ketones (so-called aldoses and ketoses), typically with three to seven carbon atoms per molecule. The common names end with the suffix '-ose', e.g. glucose (Fig. 32.1), which contains six carbon atoms and is a hexose. The simplest aldose is the three-carbon compound (triose) glyceraldehyde (Fig. 32.2a), and the other aldoses can be considered to be derived from glyceraldehyde by the addition of successive secondary alcohol groups (H–C–OH). In a similar manner, all ketoses can be considered to be structurally related to the triose dihydroxyacetone (Fig. 32.2b).

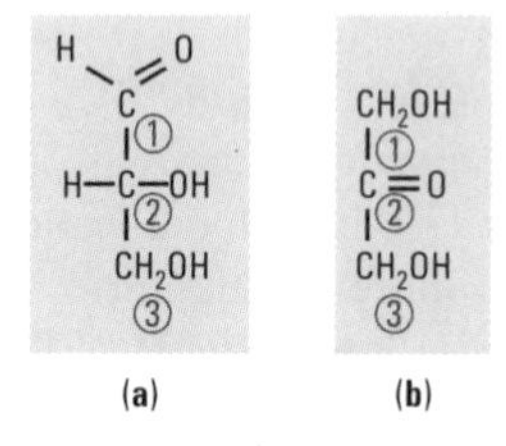

Fig. 32.2 Structure of (a) glyceraldehyde and (b) dihydroxyacetone.

Monosaccharides are assigned as D or L isomers according to a convention based on D-glyceraldehyde as the reference compound, with the carbon atoms numbered from the end of the chain containing the reactive group. Most of the carbohydrates found in biological systems are D isomers. While trioses and tetroses exist in linear form, pentoses and larger monosaccharides can be represented either as a linear structure (Fischer form) or as a ring structure (Haworth form), as in Fig. 32.1. The cyclic structure results from the reaction of the carbonyl group (aldehyde or ketone) at one end of the molecule with a hydroxyl group at the other end of the chain, forming a hemiacetal or hemiketal, as shown for glucose (Fig. 32.1) and fructose (Fig. 32.3).

The formation of a hemiacetal or hemiketal creates another asymmetric carbon atom, so that two ring forms exist – one with the –OH group on this asymmetric carbon positioned below the plane of the ring (α) and another with the –OH group above the plane of the ring (β). These different isomeric forms are called anomers and are particularly difficult to separate by chromatographic methods. The six-membered ring structure shown for glucose has a structure similar to that of pyran and is termed glucopyranose. A few monosaccharides have a five-membered ring structure similar to that of furan, e.g. fructofuranose.

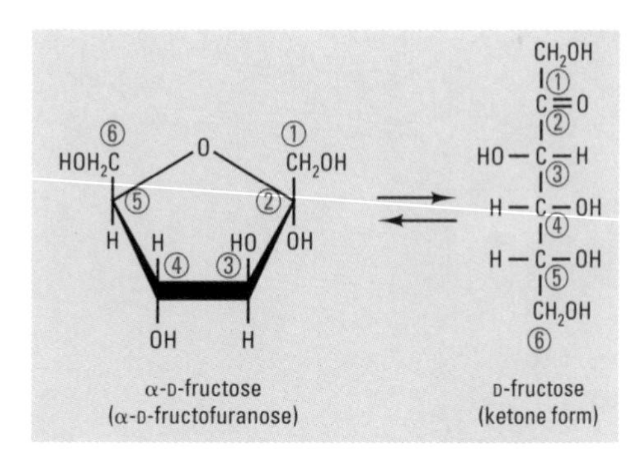

Fig. 32.3 Ring and straight chain forms of fructose. Note that β-D-fructose has H and CH_2OH groups reversed at C-2 in the furanose form.

While the open chain form is present in very small amounts in aqueous solution it has a very reactive carbonyl group which is responsible for the reducing properties of sugars: several analytical methods are based on this feature (see p. 177).

Studying the biological roles of glycosides – these compounds are often formed in plants during the detoxification of certain compounds, or to control plant hormone activity.

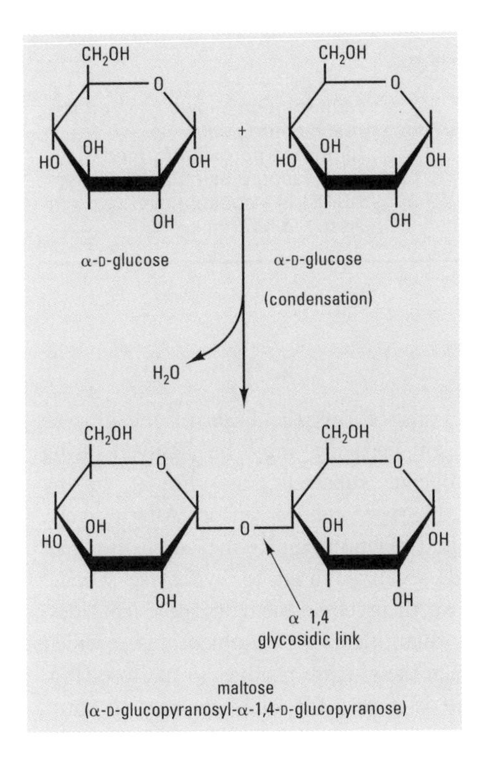

Fig. 32.4 Formation of a glycosidic link in the disaccharide maltose.

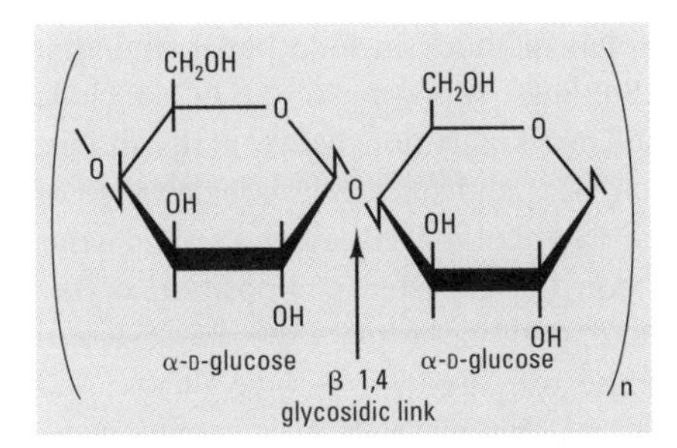

Fig. 32.5 The repeating structure of cellulose ($\beta 1 \rightarrow 4$ glycosidic links) $n \approx 5000$.

Glycosides

These are formed when a covalent bond, or glycosidic link, is created between the hemiacetal or hemiketal group of a carbohydrate (e.g. a monosaccharide) and the hydroxyl group of a second compound (e.g. a polyhydroxy alcohol, or another monosaccharide). If the sugars are not joined via their reactive groups, they will still show reducing properties, as in the disaccharide maltose (Fig. 32.4). In contrast, sucrose is a non-reducing disaccharide, since the hemiacetal and hemiketal groups of glucose and fructose are involved in the formation of the glycosidic link.

Polymeric carbohydrates have important biological roles – oligosaccharides contain up to 10 sugars, while polysaccharides are larger polymers with M_r of up to several million, e.g. glycogen, amylopectin, cellulose. The structure of polysaccharides is not fully defined, in contrast to proteins (Chapter 30) and nucleic acids (Chapter 33). Polysaccharide structure is the result of the separate actions of a number of biosynthetic enzymes, producing a range of molecules that may vary in the number of monosaccharide residues and the types of glycosidic bond. The terminology used to describe the glycosidic links in such compounds denotes the anomer involved (α or β) and the C atoms involved in the link e.g. $\alpha\ 1 \rightarrow 4$ in maltose (Fig. 32.4), $\beta\ 1 \rightarrow 4$ in cellulose (Fig. 32.5).

Extraction and analysis of carbohydrates

While most low M_r carbohydrates are soluble in water, ethanol:water (80% v/v) is more often used, since polysaccharides and other biological macromolecules are insoluble in aqueous ethanol. In contrast, polysaccharides are more diverse, and the isolation procedures vary greatly, e.g. boiling water, mild acid or mild alkali can be used to solubilize storage polysaccharides, e.g. starch, while more vigorous treatment is required for structural polysaccharides, e.g. 24% w/v KOH for cellulose. Among the techniques used to purify extracted polysaccharides are gel permeation chromatography (p. 132) and ultracentrifugation (p. 125).

Identification and quantification of carbohydrates

This can be achieved by a variety of procedures, including:

Chemical methods Several monosaccharide assay methods are based on the reductive capacity of the aldehyde or ketone groups (Table 32.1). A widely used method for quantitative analysis is that based on reduction of 3,5-dinitrosalicylate. This method is also suitable for glycosides, provided the reducing carbonyl groups are not involved in the glycosidic links. Certain polysaccharides react with iodine in acid solution to form coloured complexes: starch gives a blue colour, while glycogen gives a red–brown colour. The anthrone method (typically using $0.1\ g\,l^{-1}$ anthrone in H_2SO_4, at 100 °C for 10 min, then assayed at 630 nm) is an alternative assay for estimating total carbohydrate content. Careful choice of calibration standards is required for quantitative work – try to use a standard that matches the likely composition of the samples. While such chemical methods can give a general indication of the relative amount of carbohydrate in a sample, they can provide little useful information on the types of carbohydrate present.

Enzymatic methods These offer a higher degree of specificity in monosaccharide assay, and may allow differentiation between the various stereoisomeric and anomeric forms. The glucose-specific method based on

Table 32.1 Methods for carbohydrate analysis

Method	Principle	Comments
Chemical assay		
Benedict's test (and Fehling's test)	Reduction of Cu^{2+} to Cu^{+} in presence of reducing sugar; alkaline solution plus heat results in formation of Cu_2O; solution turns from blue, through yellow, to red.	Usually presence/absence test; quantitative assay involves measurement of Cu_2O formed.
Dinitrosalicylate (DNS)	Reduction of DNS (yellow) to orange-red derivative; alkaline solution plus heat (100 °C, 10 min).	Quantitative: read at 540 nm.
Enzymatic assay		
Glucose oxidase (coupled reaction)	β-D-glucose + O_2 $\xrightarrow{\text{glucose oxidase}}$ gluconic acid + H_2O_2 H_2O_2 + reduced dye $\xrightarrow{\text{peroxidase}}$ H_2O + oxidized dye	Mutarotation allows reaction to reach completion; hydrogen peroxide formed may be measured using peroxidase; ABTS[1] is a suitable dye. Assay at 437 nm.

[1]ABTS = 2,2′-azino-di-[3-ethylbenzthiazoline]-6-sulphonate

glucose oxidase shown in Table 32.1 also forms the basis of the glucose electrode (see Chapter 28). Hydrogen peroxide may be assayed using peroxidase and a suitable chromogenic substrate, as shown, or by electrochemical methods, e.g. using an amperometric sensor. Alternatively, the consumption of oxygen in the initial reaction can be measured using an oxygen electrode (p. 158). Glycosidases can be used to hydrolyse specific disaccharides or polysaccharides into their constituent monosaccharides, which can then be identified and quantified, e.g. α-glucosidase, which hydrolyses the $\alpha(1 \rightarrow 4)$ linkage between the glucose residues in maltose (Fig. 32.4). However, you should note that it is rare for such enzymes to show absolute specificity for a particular substrate, so you should be alert to the possibility of interference due to related compounds in the sample, or to impurities in the enzyme preparation.

Measuring carbohydrate migration in TLC systems – the distance migrated by glucose is taken as a reference ($R_f = 100$), and the migration of other carbohydrates is given as the R_g value, where R_g is:

$$\frac{\text{distance moved by carbohydrate}}{\text{distance moved by glucose}} \times 100$$

Chromatographic methods Traditional methods include paper and thin layer chromatographic procedures. Suitable supports for TLC include microcrystalline cellulose (in which the sugars partition between the mobile phase and the cellulose-bound water complex) and silica gel. A wide range of traditional solvent systems can be used (Stahl, 1965 gives details) and your choice of mobile phase will depend upon the expected composition of the mixture, e.g. cellulose with ethyl acetate : pyridine : water (100 : 35 : 25 v/v/v) gives a good separation of pentoses and hexoses, and will resolve glucose and galactose, as well as some disaccharides. Staining methods usually exploit the reducing properties of carbohydrates. Of the high resolution techniques, HPLC (p. 129) is the preferred method for the analysis of simple monosaccharide mixtures, and for oligosaccharide analysis and purification. Ion-exchange columns are often used for these purposes, with refractive index, or electrochemical detection of separated components (Chapter 24). GC is more suitable for complex monosaccharide mixtures, and can analyse sub-nanomolar amounts of carbohydrates and their derivatives, e.g. polyols, including glycerol (p. 171). However, a preliminary step is necessary to produce volatile derivatives of the carbohydrates in the mixture, e.g.

Safe working with trimethylsilylating reagents – these compounds are extremely reactive and must be handled with care, using gloves and a fume hood. Since these reagents react violently with water, samples are dried and then redissolved in an organic solvent before trimethylsilylation.

methylation or, more often, trimethylsilation (adding hexamethyldisilazane and trimethylchlorosilane at 2:1 v/v at room temperature rapidly produces trimethylsilyl ethers – TMS derivatives). Efficient separations can be obtained with a non-polar stationary phase, e.g. methylpolysiloxy gum (OV-1): the use of SCOT columns (p. 175) can enable over 30 components to be resolved. A disadvantage of GC methods is that the carbohydrates are first converted to a derivative form, then quantified by destructive means, preventing further analysis.

Capillary electrophoresis (Chapter 27) is a powerful technique for the separation of carbohydrates, though it is usual to modify uncharged carbohydrates by reductive amination to form primary amines to allow their separation within the capillary.

Characterization of polysaccharides

Polysaccharides are characterized according to the relative proportions of the constituent sugar residues, the various types of glycosidic links and the M_r. To investigate a given type of polysaccharide, its glycosidic links must be hydrolysed (e.g. by heating with concentrated HCl at 60 °C for 30 min), followed by separation, identification and quantification of the individual components. The position of glycosidic linkages can be determined by methylation of all free hydroxyl groups of the polysaccharide followed by complete hydrolysis to give a mixture of partially methylated monosaccharides. These are then reduced to alditols, and acetylated. The partially methylated alditol acetates can be identified by GC-mass spectrometry (p. 140), allowing the types of glycosidic links to be deduced from the positions of the acetylated groups. The α and β configuration of the glycosidic linkages can be determined by enzymic assay or NMR spectroscopy (p. 119).

Lignins are a specialized class of structural polymer, found in some plant tissues and composed mainly of residues of the amino acids phenylalanine and tyrosine (p. 167) together with some carbohydrate constituents. In contrast to polysaccharides, lignins are difficult to characterize as they are extremely resistant to hydrolysis.

33 Assaying nucleic acids and nucleotides

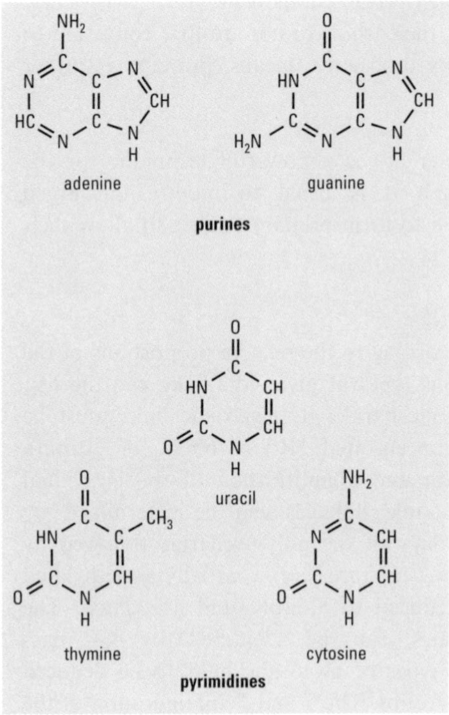

Nucleic acids are nitrogen-containing compounds of high M_r, often found within nucleic acid–protein (nucleoprotein) complexes in cells. The two main groups of nucleic acids are:

1. Deoxyribonucleic acid (DNA) – found in chromosomes and the principal molecule responsible for the storage and transfer of genetic information (Chapter 38).
2. Ribonucleic acid (RNA) – involved with the DNA-directed synthesis of proteins in cells. Three principal types of RNA exist: messenger RNA (mRNA), ribosomal RNA (rRNA) and transfer RNA (tRNA). In some viruses, RNA acts as the genetic material. In eukaryotes, mRNA molecules initially synthesized in the nucleus from genomic DNA (nascent mRNA) will contain several sequences – called introns – that are not transcribed into protein. These are successively excised in the nucleus, leaving only coding sequences (exons) in the mRNA that migrates to the cytoplasm, to be translated at the ribosome.

KEY POINT **Nucleic acids are important in the transmission of information within cells, and one of the most important aspects of nucleic acid analysis is to 'decipher' the coded information within these molecules (see also Chapter 40).**

adenine guanine
purines
uracil
thymine cytosine
pyrimidines

Fig. 33.1 Nitrogenous bases in nucleic acids.

The structure of nucleic acids

Nucleic acids are polymers of nucleotides (polynucleotides), where each nucleotide consists of:

- a nitrogenous base, of which there are five main types. Two have a purine ring structure, i.e. adenine (A) and guanine (G), and three have a pyrimidine ring i.e. thymine (T), uracil (U) and cytosine (C), as shown in Fig. 33.1. Their carbon atoms are numbered C-1, C-2, etc;
- a pentose sugar, which is ribose in RNA and deoxyribose in DNA. The carbon atoms are denoted as C-1′, C-2′, etc. and deoxyribose has no hydroxyl group on C-2′ (Fig. 33.2). The C-1′ of the sugar is linked either to the N-9 of a purine or the N-1 of a pyrimidine;
- a phosphate group, which links with the sugars to form the sugar–phosphate backbone of the polynucleotide chain.

A compound with sugar and base only is called a nucleoside (Fig. 33.2) and the specific names given to the various nucleosides and nucleotides are listed in Table 33.1. The individual nucleotides within nucleic acids are linked by phosphodiester bonds between the 3′ and 5′ positions of the sugars (Fig. 33.3).

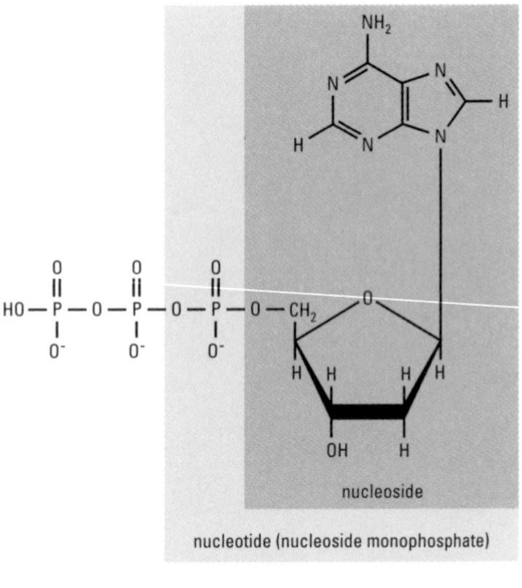

nucleoside
nucleotide (nucleoside monophosphate)

Fig. 33.2 A nucleoside triphosphate – deoxyadenosine 5′ triphosphate (dATP).

KEY POINT **RNA and DNA differ both in the nature of the pentose sugar residue, and in their base composition: both types contain adenine, guanine and cytosine, but RNA contains uracil while DNA contains thymine.**

Table 33.1 Nomenclature of nucleosides and nucleotides

Base	Nucleoside	Nucleotide
Adenine	Adenosine	Adenylic acid
Guanine	Guanosine	Guanylic acid
Uracil[1]	Uridine	Uridylic acid
Cytosine	Cytidine	Cytidylic acid
Thymine[2]	Thymidine	Thymidylic acid

[1]in DNA [2]in RNA

Differences also exist in the conformation of the two types of nucleic acid. DNA typically exists as two interwoven helical polynucleotide chains, with their structure stabilized by hydrogen bonds between matching base pairs on the adjacent strands: A always pairs with T (two hydrogen bonds), and G with C (three hydrogen bonds). This complementarity is important since it stabilizes the DNA duplex (double helix) and provides the basis for replication and transcription (Chapter 40). While most double stranded (ds) DNA molecules are in this form, i.e. as a double helix, some viral DNA is single stranded (ss). Intact DNA molecules are very large indeed, with high M_r values (e.g. 10^9). In the main, RNA is single stranded and in the form of a gentle right handed helix stabilized by base-stacking interactions, although some sections of RNA (i.e. tRNA) have regions of self-complementarity, leading to base pairing. Typical values for M_r of RNA range from 10^4 for tRNA to 10^6 for other types.

Minimizing damage to chromosomes – these vary in size from 0.3–200 megabase pairs (Mb), so some breaks in DNA inevitably occur during manipulation. Shear effects can be minimized by using wide-mouthed pipettes, gentle mixing, by avoiding rotamixing, and by precipitating DNA with ethanol at −20 °C.

Extraction and purification of nucleic acids

The types of nucleic acid most commonly isolated are chromosomal DNA, plasmid DNA and mRNA. Irrespective of the source, extraction and purification involves the following stages, in sequence:

- disruption of cells to release their contents;
- removal of non-nucleic acid components (e.g. protein), leaving DNA and/or RNA;
- concentration of the remaining nucleic acids.

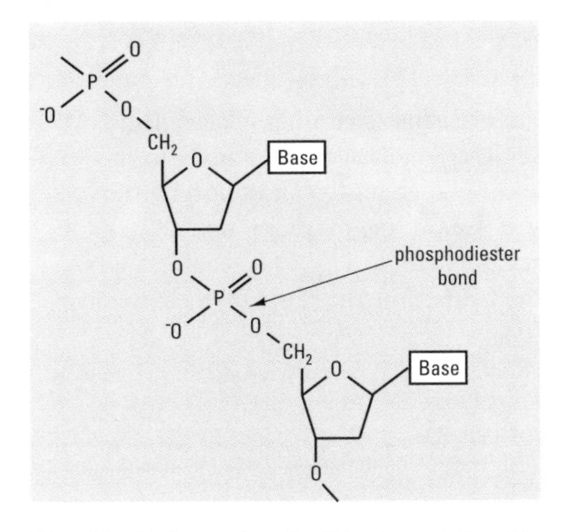

Fig. 33.3 Linkage of nucleotides in nucleic acids.

DNA isolation procedures

Specific details for the preparation of plasmid DNA from bacterial cells are given in Chapter 42. For other sources of DNA, such as mammalian tissue or plant material, the following steps are required:

1. Homogenization – tissues can be disrupted by the methods described in detail in Chapter 17, e.g. by lysis in a buffered solution containing the detergent sodium dodecyl sulphate (SDS) or Triton X-100.
2. Enzymic removal of protein and RNA – using proteinase (e.g. proteinase K at 0.1 mg/ml) and ribonuclease (typically at 0.1 μg ml^{-1}) for 1–2 h.
3. Phenol-chloroform extraction – to remove any remaining traces of contaminating protein.
4. Precipitation of nucleic acids – usually by adding twice the volume of ethanol.
5. Solubilization in an appropriate volume of buffer (pH 7.5) – ribonuclease is often added to remove any traces of contaminating RNA.

Working with precipitated DNA – this can be spooled from solution by winding it around a glass rod.

Density gradient centrifugation (p. 124) is an alternative approach to the separation of DNA from contaminating RNA, as described below.

RNA isolation procedures

Each mammalian cell contains about 10 pg of RNA, made up of rRNA (80–85%), tRNA (10–15%) and mRNA (1–5%). While rRNA and tRNA components are of discrete sizes, mRNA is heterogeneous and varies in length from several hundred to several thousand nucleotides.

Avoiding RNase degradation of RNA – endogenous RNases can be inhibited by including diethyl pyrocarbonate (DEP) (at 0.1% v/v g ml^{-1}) in solutions used for RNA extraction.

KEY POINT RNA is more difficult to purify than DNA, partly because of degradation during the extraction process due to the action of contaminating ribonucleases, and partly because the rigorous treatment required to dissociate the RNA from protein in ribosomes may fragment the polyribonucleotide strands.

RNA can be prepared either from the cytoplasm of cells (to give mainly rRNA, tRNA, and fully processed mRNA), or from whole cells, in which case nascent mRNA from the nucleus will also be present (p. 180):

- Preparation of cytoplasmic RNA involves lysis of cells or protoplasts with a hypotonic buffer, leaving the nuclei intact. Cell debris and nuclei are then removed by centrifugation (Chapter 23), and sodium dodecyl sulphate (SDS) is added to the supernatant (the cytoplasmic fraction) to inhibit ribonuclease. Proteinase K can be added to release rRNA from ribosomes. Phenol–chloroform extraction removes contaminating proteins, as for DNA preparation (p. 242), and the RNA present in the aqueous phase can then be precipitated by addition of twice the volume of ethanol.
- Preparation of whole-cell RNA requires more vigorous cell lysis, e.g. with a solution containing 6 $mol\,l^{-1}$ guanidinium chloride and 2-mercaptoethanol: this effectively denatures any ribonuclease present. Caesium chloride is added to the extract to give a final concentration of 2.4 $mol\,l^{-1}$, and the solution is processed by density gradient centrifugation (Chapter 23) at 100 000 g for 18 h, using a cushion of 5.7 $mol\,l^{-1}$ CsCl. DNA and protein remain in the upper layers of CsCl, while RNA forms a pellet at the base of the tube. The RNA pellet is redissolved in buffer, and precipitated in cold ethanol. After resuspension of the cellular RNA, mRNA can be purified by affinity chromatography using poly (U)-Sepharose® (p. 134).

Avoiding contamination of glassware by RNases – autoclave all glassware before use, to denature RNases. These enzymes are present in skin secretions: use gloves at all times and use plasticware wherever possible.

Separating nucleic acids

Electrophoresis is the principal method used for separating nucleic acids. At alkaline pH values, linear DNA and RNA molecules have a uniform net negative charge per unit length due to the charge on the phosphoryl group of the backbone. Electrophoresis using a supporting medium that acts as a molecular sieve (e.g. agarose or polyacrylamide, p. 144) enables DNA fragments or RNA molecules to be separated on the basis of their relative sizes (see Chapter 26 for further details).

Quantitative analysis of nucleic acids

Measuring nucleic acid content

The concentration of reasonably pure samples of DNA or RNA can be measured by spectrophotometry (p. 114). In contrast, measurement of the nucleic acid content of whole cell or tissue homogenates requires chemical methods, since the homogenates will contain many substances that would interfere with the spectroscopic methods. The principles involved in each technique are as follows:

Measuring nucleic acids by spectrophotometry – ideally, the nucleic acid extracts should be prepared to give A_{260} values of between 0.10 and 0.50, for maximum accuracy and precision.

- Spectrophotometry – DNA and RNA both show absorption maxima at ≈260 nm, due to the conjugated double bonds present in their constituent bases. At 260 nm, an A_{260} value of 1.0 is given by a 50 $\mu g\,ml^{-1}$ solution of dsDNA, or a 40 $\mu g\,ml^{-1}$ solution of ssRNA. If the absorbance at 260 nm is also measured, protein contamination can be quantified. Pure nucleic acids give A_{260}/A_{280} ratios of 1.8–2.0, and a value below 1.6 indicates significant protein contamination. Further purification steps are required

Safe working with ethidium bromide – this compound is highly toxic and mutagenic. Avoid skin contact (wear gloves) and avoid ingestion. Use a safe method of disposal (e.g. adsorb from solution using an appropriate adsorbant).

for contaminated samples, e.g. by repeating the phenol–chloroform extraction step. RNA contamination of a DNA preparation is indicated if A_{260} decreases when the sample is treated with 2.5 μl of RNase at 20 μg μl^{-1}. DNA contamination of an RNA preparation might be suspected if the sample is very viscous, and this can be confirmed by electrophoresis.

- Spectrofluorimetry – this is the best approach for samples where the DNA concentration is too low to allow direct assay by the spectrophotometric method described above. The method uses the fluorescent dye ethidium bromide, which binds to dsDNA by insertion between stacked base pairs, a phenomenon termed intercalation. The fluorescence of ethidium bromide is enhanced 25-fold when it interacts with dsDNA. Since ssDNA gives no significant enhancement of fluorescence, dsDNA can be quantified in the presence of denatured DNA. The concentration of dsDNA in solution $[dsDNA]_x$ can be calculated by comparing its fluorescence (excitation wavelength, 525 nm; emission wavelength, 590 nm) with that of a standard dsDNA solution of known concentration $[dsDNA]_{std}$ using the following relationship:

$$[\text{dsDNA}]_x = \frac{[\text{dsDNA}]_{\text{std}} \times \text{fluorescence of unknown}}{\text{fluorescence of standard}} \qquad [33.1]$$

- Chemical methods – these are mostly based on colorimetric reactions with the pentose groups of nucleic acids. The total DNA concentration can be measured by the diphenylamine reaction (Fig. 33.4), which is specific for 2-deoxypentoses. The diphenylamine reaction involves heating 2 ml of DNA solution with 4 ml of freshly prepared diphenylamine reagent (diphenylamine, at 10 g l^{-1} in glacial acetic acid, plus 25 ml concentrated sulphuric acid) for 10 min in a boiling water bath. The acids cleave some of the phosphodiester bonds, and hydrolyse the glycosidic links between the deoxyribose and purines. Deoxyribose residues are converted to ω-hydroxylevulinyl aldehyde (Fig. 33.4), which reacts with the diphenylamine to produce a blue pigment, assayed at 600 nm. By constructing a DNA standard curve ranging from 0–400 μg ml^{-1}, the DNA concentration of the unknown can be determined.

  RNA concentration can be measured by the orcinol reaction, which is a general assay for pentoses. The orcinol reaction involves heating 2 ml of RNA solution with 3 ml of orcinol reagent (prepared by dissolving 1 g of $FeCl_3.6H_2O$ in 1 litre of concentrated HCl, and adding 35 ml of 6% w/v orcinol in ethanol) in a boiling water bath for 20 min. The acid cleaves some phosphodiester bonds and hydrolyses the glycosidic links between the ribose and purines. The hot acid also converts the ribose to furfural (Fig. 33.5), which reacts with orcinol in the presence of ferric ions to produce green-coloured compounds, assayed at 660 nm. A standard curve for RNA ranging from 0–400 μg ml^{-1} is used to determine the RNA concentration of the unknown. The orcinol reaction is less specific than the diphenylamine reaction, as deoxyribose reacts to some extent and DNA gives about 10% of the colour given by the same concentration of RNA. If the DNA concentration of the extract is known (e.g. from a diphenylamine assay), the contribution of DNA to A_{660} can be measured for a standard prepared to have the same DNA concentration as the sample: the A_{660} due to DNA can then be subtracted from the result of the orcinol reaction, and the remaining A_{660} value is then used to determine RNA concentration from the RNA standard curve.

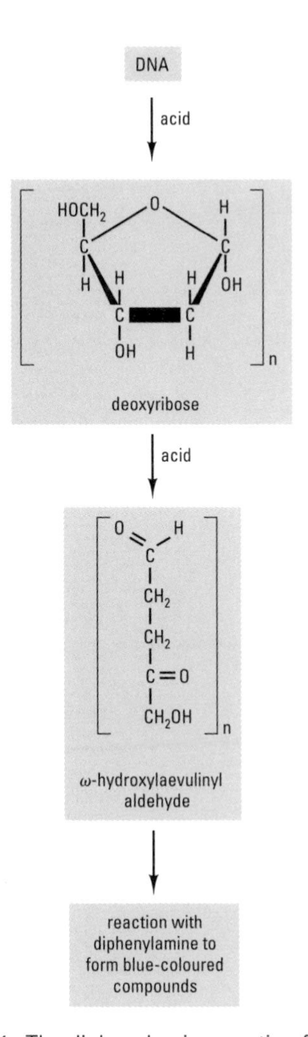

Fig. 33.4 The diphenylamine reaction for assay of DNA.

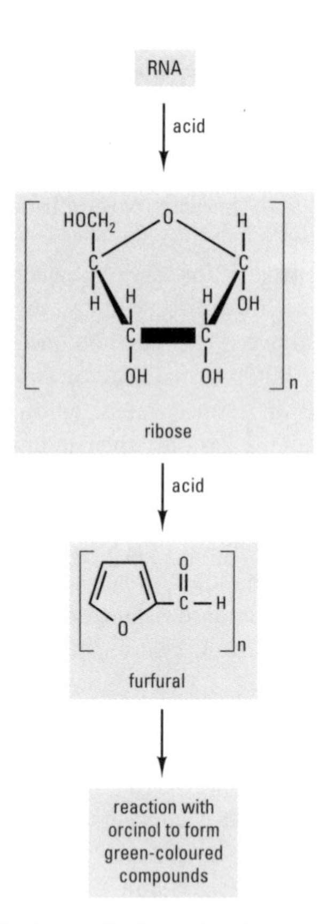

Fig. 33.5 The orcinol reaction for assay of RNA.

Assaying the relative proportions of base pairs in DNA

Double stranded DNA has a lower molar absorptivity at 260 nm than single stranded DNA, due to electron interactions between the stacked base pairs and hydrogen bonding between the complementary bases of dsDNA. When a dsDNA solution is heated slowly, there is little change in absorbance until a temperature is reached where the hydrogen bonds are broken and the DNA strands separate. At this so-called 'melting temperature' (T_m), the A_{260} value increases sharply (Fig. 33.6). This temperature-dependent increase in absorbance is often referred to as the 'hyperchromic effect'. The actual value of the T_m (and hence the stability of dsDNA) is dependent on the base-pair composition of the DNA being studied: GC base pairs have three double bonds while AT base pairs have two, and the higher the % GC content, the higher the value of the T_m. The length of the molecule also affects the T_m, since short dsDNA molecules do not show a hyperchromic effect. Studies on a number of DNA samples have shown the following relationship, for GC contents between 30% and 70%:

$$\%\,GC = 2.44(T_m - 69.3) \qquad [33.2]$$

The relative proportion of AT base pairs can then be calculated by subtracting the % GC value from 100, giving % AT.

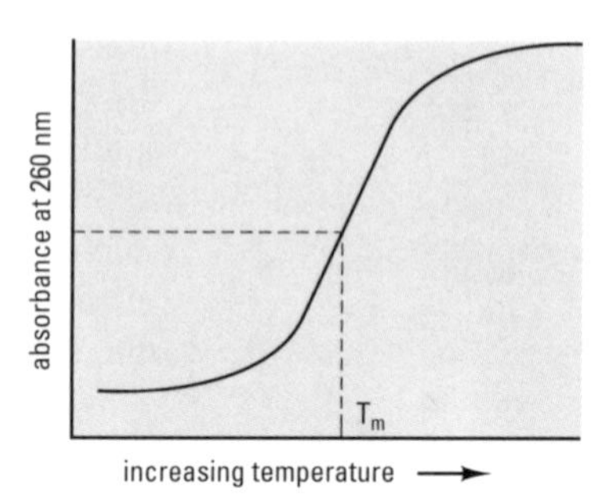

Fig. 33.6 Effect of temperature on the absorbance of DNA in solution. T_m is the temperature for the mid-point of the absorbance change.

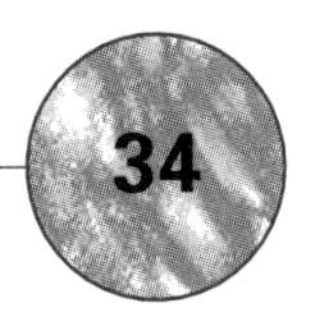

34 Protein purification

Much of the current knowledge about metabolic and physiological events has been gained from *in vitro* studies of purified proteins. Such studies range from investigations into the kinetics and regulation of enzymes (Chapter 35) to the determination of the structure of a protein and its relationship to function. In addition, certain purified proteins have a role as therapeutic agents, e.g. Factor VIII in the treatment of haemophilia.

KEY POINT **The purification of most proteins involves a series of procedures, based on differential solubility and/or chromatography, that selectively separate the protein of interest from contaminating proteins and other material. It is rare for purification to be a one-step process.**

Deciding on your objectives

Purity, yield and cost are the major considerations, and the relative importance of these factors will depend on the purpose of the purification. For studies on the biological activity of a protein (e.g. an enzyme), only μg amounts may be required and, as long as there are no interfering substances present, 100% purity may not be necessary. For structural studies, mg amounts may be needed and the protein must be pure. A protein produced for industrial applications (e.g. an enzyme for use in starch degradation) needs to be produced in large amounts (g or kg), but purity usually is not essential. In contrast, commercially produced proteins for therapeutic use need to be free of any significant contamination.

Preventing loss of protein during purification – silicone-coated glass containers prevent the adsorption of proteins to the surface of a glass vessel.

The cost of the purification will depend on the nature of the source material, the number of purification steps required, and the price of materials for the separation techniques employed. Inevitably, some protein will be lost at each stage of the purification procedure.

KEY POINT **You should aim to use the smallest number of purification steps that will give you the yield and purity required for your application.**

Preliminary considerations

Finding out about the protein

Rather than approaching every purification step on a 'trial and error' basis, try to find out as much as you can about the physical and biological properties of the desired protein before you begin your practical work. Even if the protein to be purified is novel, it is likely that similar proteins will be described in the literature. Information on the isoelectric point (pI) and M_r, of both the protein and of the likely major contaminants, is particularly useful for ion-exchange and gel permeation chromatography (p. 131). Knowledge of other factors such as metal ion and co-enzyme requirements, presence of thiol groups, and known inhibitors can indicate useful chromatographic steps, and may allow you to take steps to preserve the tertiary structure and biological activity of the desired protein during the purification process.

Definition

Isoelectric point (pI) – the pH value at which a protein has a net charge of zero, i.e. where positive and negative charges on amino acid side chains are balanced.

Source material for protein purification – an important advance has been the use of genetically engineered organisms, designed to produce large amounts of a particular protein (p. 242).

Choosing the source material

If you have a choice of starting material, use a convenient source in which the protein of interest is abundant. This can vary from animal tissues (e.g. heart, kidney or liver) obtained from an abattoir or from laboratory animals, to plant material, or to microbial cells from a laboratory culture.

Homogenization and solubilization

Unless the protein is extracellular, the cells in the source material need to be disrupted and homogenized by one of the methods described in Chapter 17. Following homogenization, proteins from the cytosol or extracellular fluid will normally be present in a soluble form, but membrane-bound proteins and those within organelles will require further treatment. Isolation of organelles by differential centrifugation (p. 124) will provide a degree of purification, while membrane-bound proteins will need organic solvents or detergents to render them soluble. Once the protein of interest is in a soluble form, any particulate material in the extract should be removed, e.g. by centrifugation or filtration.

Storing protein solutions during a purification procedure – overnight storage at 4 °C is acceptable, but it is advisable to include bacteriostatic agents (e.g. azide at 0.5% w/v) and protease inhibitors: freezing may be required for longer term storage – use liquid nitrogen or dry ice/methanol for rapid freezing. If freezing and thawing might denature the sample, include 20% v/v glycerol in buffers, to allow storage at −20 °C in liquid form.

Preserving enzyme activity during purification

Once an enzyme has been released from its intracellular environment, it will encounter potentially adverse conditions that may result in denaturation and permanent inactivation, e.g. a lower pH, an oxidizing environment, or exposure to lysosomal proteases. During homogenization, and in some or all of the purification steps, buffers should contain reagents that can counteract these potentially damaging effects (see Chapter 17). Fewer precautions may be needed with extracellular enzymes. For most proteins, the initial procedures should be carried out at 4 °C, to minimize the risk of proteolysis.

Devising a strategy for protein purification

Any successful protein purification scheme will exploit the unique properties of the desired protein in terms of its size, net charge, hydrophobic nature, biological activity, etc. The chromatographic techniques that separate on the basis of these properties are detailed in Chapter 24. However, the *order* in which the various purification steps are carried out needs some thought, and each stage needs to be considered in relation to the following factors:

- Capacity – the amount of material (volume or concentration) that the technique can handle. High capacity techniques such as ammonium sulphate precipitation (see later) and ion exchange chromatography (p. 131) should be used at an early stage, to reduce the sample volume.
- Resolution – the efficiency of separating one component from another. At later stages of the purification, the sample volume will be considerably reduced, but any contaminating proteins may well have very similar properties to those of the protein of interest. Therefore, a high resolution (but low capacity) technique will be required, such as covalent chromatography or immobilized metal affinity chromatography (IMAC).
- Yield – the amount of protein recovered at each step, expressed as a percentage of the initial amount. For example, yields of >80% can be obtained with ammonium sulphate precipitation. With certain types of affinity chromatography, where harsh elution conditions need to be employed, yields may be quite low (e.g. <20%). Some of the other possible reasons for a decreased yield are considered in Table 34.1.

Table 34.1 Principal causes of decreased yield in protein purification

Cause	Possible solution
Denaturation	Include EDTA or reducing agents in buffers; avoid extreme temperatures
Inhibition	Check buffer composition for possible inhibitors
Proteolysis	Include protease inhibitors in buffers
Non-elution	Alter salt concentration or pH of the eluting buffer
Co-factor loss	Recombine fractions on a trial-and-error basis

Measuring yield – note that yields of >100% may be obtained if an inhibitor is lost during purification.

An ideal purification procedure will have the following features:

1. an initial step that has a high capacity, but not necessarily a high resolution;
2. a series of chromatographic steps, each of which exploits a different property of the protein to increase purity;
3. a minimum number of manipulations and changes in conditions (e.g. in buffer composition) between steps.

The following series of steps would meet the above conditions:

Step 1 Ammonium sulphate precipitation – a high capacity, low resolution technique that yields a protein solution with a greatly reduced volume and a high concentration of ammonium sulphate.

Step 2 Hydrophobic interaction chromatography (HIC) – an absorption technique (p. 133) using a starting buffer with a high ammonium sulphate concentration. During development of the column, the ionic strength of the mobile phase is reduced in a gradient elution.

Step 3 Ion exchange chromatography (IEC) – another absorption technique (p. 131). Here the starting buffer has a low ionic strength, and a gradient of increasing ionic strength is used to separate components.

Step 4 Gel permeation chromatography (GPC) – a low capacity method (p. 132) where separation is independent of the composition of the mobile phase.

The sample from step 1 may be able to be applied directly to the HIC column, and that from step 2 may be used for IEC without changing the buffer. A concentration step (p. 188) would be required before step 4. The above scheme is very much an idealized approach: in practice, buffers may need to be changed by dialysis or ultrafiltration, as described later. However, the principle of using the minimum number of manipulations to obtain the desired purification remains valid.

Monitoring purification

Calculating total protein content – multiply the protein concentration by the total volume, making sure that the units are consistent, e.g. mg ml^{-1} × ml. A similar procedure is required to determine the total amount of enzyme.

At each step, the separated material is usually collected as a series of 'fractions'. Each fraction must be assayed for the protein of interest, and fractions containing that protein are pooled prior to the next step. The assay performed will depend on the properties of the protein of interest but specific enzyme assays or immunoassays are most commonly used. The protein concentration (p. 168) and the volume of the pooled fraction must be determined, and these values, together with those obtained for the amount of the protein of interest, are used to determine the purity and yield after each step. For enzymes, the biological activity is measured and used to calculate the specific activity (the enzyme activity per unit mass of protein) as described on p. 192. By determining the specific activity of the enzyme at each step, the degree of purification, or purification factor (n-fold purification) can be obtained from the following relationship:

$$\text{Purification factor} = \frac{\text{specific activity after a particular step}}{\text{specific activity of initial sample}} \qquad [34.1]$$

Using immunoassays – note that the inactive or denatured forms of the protein may be detected along with the biologically active form.

Increased purification usually represents a decrease in total protein relative to the biological activity of the protein of interest, though in some instances it may reflect the loss of an inhibitor during a purification step.

Table 34.2 Example of a record of purification for an enzyme

Step	Procedure	Total protein (mg)	Total enzyme (IU)	Specific activity	Purification factor	Enzyme yield (%)
Initial sample	—	210	1984	9.43	—	100.0
Step 1	'Salting out'	112	1740	15.53	1.6	87.7
Step 2	HIC	30	1701	56.70	6.0	85.7
Step 3	IEC	6	1604	267.33	28.4	80.8
Step 4	GPC	3	1550	516.66	54.8	78.1

Assaying enzymes – if possible, use a preliminary screening technique to detect active fractions prior to quantitative assay, e.g. by using tetrazolium dyes in microtitre plates to detect oxidoreductases (p. 150).

Measuring the mass of proteins – this is sometimes expressed in daltons (d), or kilodaltons (kd), where 1 dalton = 1 atomic mass unit. M_r is an alternative expression (p. 21).

Calculation of the yield of enzyme at each step is straightforward, since:

$$\text{Yield} = \frac{\text{total enzyme activity after a particular step}}{\text{total enzyme activity in initial sample}} \times 100\ (\%) \qquad [34.2]$$

Note that the yield equation uses the total amount of enzyme and is therefore unaffected by the volume of the solutions involved. You should make a record of the progress of your purification procedure at each step, as in Table 34.2.

Monitoring progress using electrophoresis

Polyacrylamide gel electrophoresis (PAGE) of sub-samples from each step will give an indication of the number of proteins still present: ideally, a pure protein will give only one band after silver staining (p. 149); however, this stain is so sensitive that even trace impurities can be detected in what you expect to be a pure sample, so don't be dismayed by the appearance of the gel after staining.

Carrying out SDS-PAGE and isoelectric focusing (or, ideally, 2D-PAGE, p. 152) on sub-samples also gives information on the M_r and pI of any remaining contaminants, and this can be taken into account when planning the next chromatographic step.

Differential solubility separation techniques

Often the volume and protein concentration of the initial soluble extract will be quite high. Application of a differential solubility technique at this stage results in the precipitation of selected proteins. These can be recovered by filtration or centrifugation, washed, and then resuspended in an appropriate buffer. This will reduce the sample volume and may give a small degree of purification, making the sample more suitable for subsequent chromatographic steps.

Ammonium sulphate precipitation ('salting out')

This is the most widely used differential solubility technique, having the advantage that most precipitated enzymes are not permanently denatured, and can be redissolved with restoration of activity. Precipitation depends on the existence of hydrophobic 'patches' on the surface of proteins, inducing a reorganization of water molecules in their vicinity. When ammonium sulphate is added to the extract, it dissolves to give ions that become hydrated, leaving fewer water molecules in association with the protein. As a result, the hydrophobic patches become 'exposed', and hydrophobic interactions between different protein molecules leads to their aggregation and precipitation. The basis of fractionation in this method is that, as the salt concentration of the extract is increased, proteins with larger or more

Table 34.3 Amount of $(NH_4)_2SO_4$ ($g\,l^{-1}$) required for a particular percentage saturation

Final concentration (%) →	20	30	40	50	60	70	80	90	100
Initial concentration (%) ↓	Ammonium sulphate added ($g\,l^{-1}$)								
0	107	166	229	295	366	442	523	611	707
10	54	111	171	236	305	379	458	545	636
20	—	56	115	177	244	316	392	476	565
30		—	57	119	184	253	328	408	495
40			—	59	122	190	262	340	424
50				—	61	127	197	272	353
60					—	63	131	204	283
70						—	66	136	212
80							—	68	141
90								—	71

abundant hydrophobic patches will precipitate before those with smaller or fewer patches. Although the phenomenon of salting out is seen with several salts, $(NH_4)_2SO_4$ is the most widely used because it is highly soluble (saturating at $\approx 4\,mol\,l^{-1}$), inexpensive, and can be obtained in very pure form.

Unless you can obtain information from the literature about the $(NH_4)_2SO_4$ concentration that will precipitate your target protein, fractionation is done on a trial-and-error basis. The $(NH_4)_2SO_4$ concentration is expressed in terms of the percentage saturation value: Table 34.3 shows the amount of $(NH_4)_2SO_4$ required to give 20–100% saturation. For each percentage saturation chosen, the $(NH_4)_2SO_4$ salt should be added slowly while stirring, and the mixture left at 4 °C for 1 h before centrifuging at 3000 *g* for 40 min. For an effective separation, start with the maximum percentage saturation that does not precipitate the protein of interest, then increase the percentage saturation by the minimum amount that will then precipitate it. The proteins precipitated between any two values of percentage saturation (say between 30% and 50%) are referred to as a 'cut'. An alternative approach is to add saturated $(NH_4)_2SO_4$ solution – the volume (V_a) to be added to an initial volume of solution (V_i) with an initial saturation S_i, to give a final saturation S_f, is given by the equation:

$$V_a = \frac{V_i(S_f - S_i)}{1 - S_f} \quad [34.3]$$

where S_i and S_f are expressed as fractional saturation, e.g. $S_f = 0.5 \equiv 50\%$ saturation, and both volumes are expressed in the same terms, e.g. ml.

Example To prepare a solution of 40% $(NH_4)_2SO_4$ saturation ($S_f = 0.4$) from 100 ml (V_i) containing no added $(NH_4)_2SO_4$ using Eqn. 34.3:
$V_a = 100(0.4 - 0) \div (1 - 0.4) = 67$ ml of saturated $(NH_4)_2SO_4$ solution, giving a volume of 167 ml.
For a 40% to 60% 'cut' of this sample ($V_i = 167$ ml), using Eqn 34.3:
$V_a = 167(0.6 - 0.4) \div (1 - 0.6) = 83.5$ ml of saturated $(NH_4)_2SO_4$ solution, to give a final volume of 250.5 ml.

Precipitation by changing pH

Proteins are least soluble at their isoelectric points because, at that pH, there is no longer the repulsion that occurs between positively or negatively charged protein molecules at physiological pH values. If the precipitated proteins are required for further purification, it is essential that the protein of interest is not irreversibly denatured. The method is probably best employed by precipitating contaminating proteins, leaving the desired protein in solution. Use citric acid for <pH 3, acetic acid for <pH 4, and sodium carbonate or ethanolamine for >pH 8.

Purifying bacterial proteins – adjusting the extract to pH 5 can be useful, since many bacterial proteins have pI values in this region.

Heat denaturation

Exposure of most proteins to high temperatures disrupts their conformation through effects on non-covalent interactions such as hydrogen bonds and van der Waals forces. However, different proteins are denatured, and hence precipitated, at different temperatures, and this can provide a basis for the separation of some heat stable proteins. By incubating small aliquots (≈1 ml) of extract for 1 min at a range of temperatures between 45–65 °C, it is possible to determine the temperature that gives maximum precipitation of contaminating protein with minimal inactivation of the desired protein.

Definition

Dielectric constant ($\mathcal{E}$) – a dimensionless measure of the screening effect on the force (F) between two charges (q_1 and q_2) due to the presence of solvent, from the equation $F = (k\ q_1\ q_2) \div (\mathcal{E}\ r^2)$, where r is the distance between the two particles and k is a constant. The dielectric constant for water is high (≈80), while most organic solvents have lower values, in the range 1–10.

Solvent and polymer precipitation methods

Organic solvents (e.g. acetone, ethanol) cause precipitation of proteins by lowering the dielectric constant of the solution. Performing the precipitation at 0 °C minimizes permanent denaturation. Stepwise concentration (% v/v) increments are used, giving 'cuts' of precipitated proteins, as with ammonium sulphate precipitation.

Organic polymers, particularly polyethylene glycol (PEG), also lower the dielectric constant, but at lower concentrations than with acetone or ethanol. The most commonly used PEG preparations have M_r values of 6000 or 20 000, and these can be removed from the sample by ultrafiltration. PEG precipitation does not involve salts, so it may be a useful preliminary step prior to ion exchange chromatography which starts with low salt buffer. Also, since the techniques of PEG and ammonium sulphate precipitation involve different principles, they can be used sequentially.

Concentration by ultrafiltration

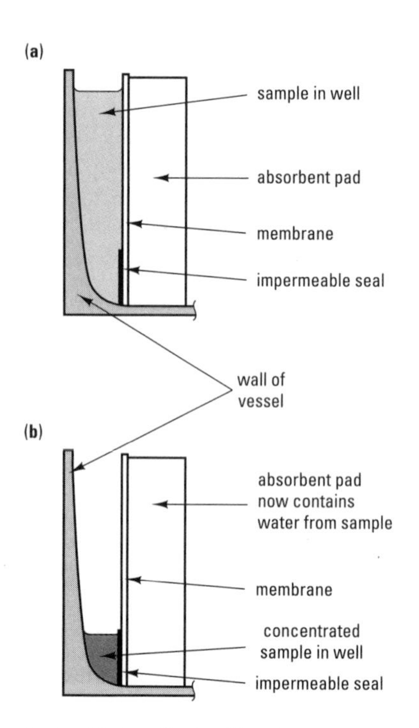

Fig. 34.1 Minicon® ultrafiltration system (a) with sample added, (b) after ultrafiltration.

This involves forcing water and small molecules through a semi-permeable membrane using high pressure or centrifugation. A range of membranes with 'nominal' molecular weight cut-offs between 500 and 300 000 are commercially available (e.g. Amicon®, Millipore®), with pore sizes of 0.1–10 μm. Concentration of small samples (<5 ml) can be achieved using either a membrane backed by an absorbent pad (e.g. Minicon®, Fig. 34.1, available with M_r cut-off from 5000–30 000), or by using a centrifugal concentrator (e.g. Centricon®). Larger volumes (up to 400 ml) can be concentrated using a stirred ultrafiltration chamber (an ultrafiltration 'cell') where the liquid is forced through the membrane using nitrogen or an inert gas.

Ultrafiltration not only concentrates the sample, but also may give a degree of purification. It can also be used to change the buffer composition by diafiltration (see below). Note that molecular weight cut-off values are quoted for globular proteins – fibrous proteins of higher M_r may pass through the ultrafiltration membrane.

Removing salts and changing the buffer

Although you should aim to use the minimum number of manipulations to obtain the desired purification, it may be necessary to remove salts or to change the buffer before the next step will work effectively (e.g. when carrying out IEC, the ionic strength or the pH of the sample may need to be changed before the target protein will bind to the column). Several methods are available, including:

Using Visking® tubing – this must be boiled before use to ensure a uniform pore size and to remove heavy metal contaminants.

- Dialysis: the sample is placed in a bag consisting of semi-permeable membrane (e.g. Visking® tubing) and placed in at least 20 volumes of the required buffer. The membrane allows molecules of $M_r < 20\,000$ to pass freely, while retaining larger molecules (note that dialysis is not suitable for small proteins!). The small molecules diffuse through the membrane until the osmotic pressure between the sample and the dialysis buffer is equalized. Several changes of dialysis buffer may be required to obtain the desired buffer conditions within the sample. Normally dialysis is carried out overnight at 4 °C. Efficient stirring is required – use a magnetic stirrer and stirrer bar.
- Diafiltration: quicker than dialysis and more suitable for larger volumes. It involves addition of the desired buffer to the sample solution, followed by ultrafiltration. Several buffer addition and ultrafiltration steps may be necessary to obtain the desired sample conditions.
- Gel permeation chromatography (gel filtration): a gel filtration medium of small pore size (e.g. Sephadex® G-25) is used to prepare a column of $\approx$5 times the volume of the sample. When the sample passes through the column, the large protein molecules will elute with the void volume, while salt ions are retained. This method is only suitable for small volumes, and it results in dilution of the sample.

Avoiding protein precipitation during dialysis or gel filtration – make sure your buffer pH is either above or below the pI of the proteins.

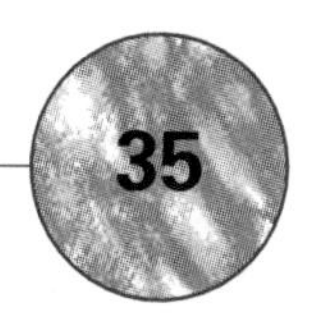

35 Enzyme studies

Enzymes are globular proteins that increase the rate of specific biochemical reactions. Each enzyme operates on a limited number of substrates of similar structure to generate products under well-defined conditions of concentration, pH, temperature, etc. In metabolism, groups of enzymes work together in sequential pathways to carry out complex molecular transformations, e.g. the multi-reaction conversion of glucose to lactate (glycolysis).

Example Enzyme EC 1.1.1.1 is usually known by its trivial name, alcohol dehydrogenase.

KEY POINT Enzymes are categorized according to the chemical reactions they catalyse, leading to a four-figure Enzyme Commission code number and a systematic name for each enzyme. Most enzymes also have a recommended trivial name, often denoted by the suffix 'ase'.

Measuring enzyme reactions

Activity

This is measured in terms of the rate of enzyme reaction. Activity may be expressed directly as amount of substrate utilized per unit time (e.g. nmol min^{-1}, etc.), or in terms of the non-SI international unit (U, or sometimes IU), defined as the amount of enzyme which will convert 1 μmol of substrate to product(s) in 1 min under specified conditions. However, the recommended (SI) unit of enzyme activity is the katal (kat), which is the amount of enzyme which will convert 1 mol of substrate to product(s) in 1 s under optimal conditions, determined from the following equation:

$$\text{enzyme activity (kat)} = \frac{\text{substrate converted (mol)}}{\text{time (s)}} \quad [35.1]$$

Example The hydrolysis of 1 molecule of maltose to give 2 glucose molecules by α-glucosidase means that enzyme activity specified in terms of substrate consumption (nmol maltose) would be half the value expressed with respect to product formation (nmol glucose).

This unit is relatively large (1 kat = 6×10^7 U) so SI prefixes are often used, e.g. nkat or pkat (p. 45). Note that the units are amount of substrate (mol), not concentration ($mol\,l^{-1}$ or $mol\,m^{-3}$).

For enzymes with macromolecular substrates of unknown molecular weight (e.g. deoxyribonuclease, amylase), activity can be expressed as the mass of substrate consumed (e.g. ng DNA min^{-1}), or amount of product formed (e.g. nmol glucose min^{-1}). You must ensure that your units clearly specify the substrate or product used, especially when the enzyme transformations involve different numbers of substrate or product molecules. Specific activity, expressed in terms of the amount or mass of substrate or product, is useful for comparing the purity of different enzyme preparations (e.g. p. 188).

Definition

Specific activity – enzyme activity (e.g. kat, U, ng min^{-1}) expressed per unit mass of protein present (e.g. mg, μg).

The turnover number of an enzyme is the amount of substrate (mol) converted to product in 1 s by 1 mol of enzyme operating under optimum conditions. In practice, this requires information on the molecular weight of the enzyme, the amount of enzyme present and its maximum activity.

Types of assay

The rate of substrate utilization or product formation must be measured under controlled conditions, using some characteristic which changes in direct proportion to the concentration of the test substance.

Example α-Glucosidase activity can be measured using the artificial substrate p-nitrophenol-α-D-glucose, which liberates p-nitrophenol (yellow) with an absorption maximum at 404 nm.

Spectrophotometric assays

Many substrates and products absorb visible or UV light and the change in absorbance at a particular wavelength provides a convenient assay method (p. 114). Artificial substrates are used where no suitable spectrophotometric assay is available for the natural substrates. In most cases, these are chromogenic analogues of the natural substrate, producing coloured products. In other cases, a product may be measured by a colorimetric chemical reaction.

Several assays are based on interconversion of the nicotinamide adenine dinucleotide coenzymes NAD^+ or $NADP^+$ which are reversibly reduced in many enzymic reactions. The reduced form (either NADH or NADPH) can be detected at 340 nm, where the oxidized form has negligible absorbance (p. 114). An alternative approach is to use a coupled enzyme assay, where a product of the test enzyme is used as a substrate for a second enzyme reaction which involves oxidation/reduction of nucleotide coenzymes. Such assays are particularly useful for continuous monitoring of enzyme activity and for reactions where the product from the test substance is too low to detect by other methods, since coupled assays are more sensitive. Note that the reaction of interest (test enzyme) must be the rate-limiting process, not the indicator reaction (second enzyme).

Fluorimetric assays

Certain substrates liberate fluorescent products as a result of enzyme activity, providing a highly sensitive assay method (p. 115). Fluorogenic substrates include fluorescein and methylumbelliferone derivatives. Care is required, since impurities in the enzyme preparation may produce background fluorescence, or quenching (reduction) of the signal.

Radioisotopic assays

These are useful where the substrate and product can be easily separated, e.g. in decarboxylase assays using a ^{14}C-labelled substrate, where gaseous $^{14}CO_2$ is produced.

Electrochemical assays

Enzyme reactions involving acids and bases can be monitored using a pH electrode (p. 29), though the change in pH will affect the activity of the test enzyme. An alternative approach is to measure the amount of acid or alkali required to maintain a constant pre-selected pH in a pH-stat.

An oxygen electrode can be used if O_2 is a substrate or a product (p. 159). Other ion-specific electrodes can monitor ammonia, nitrate, etc.

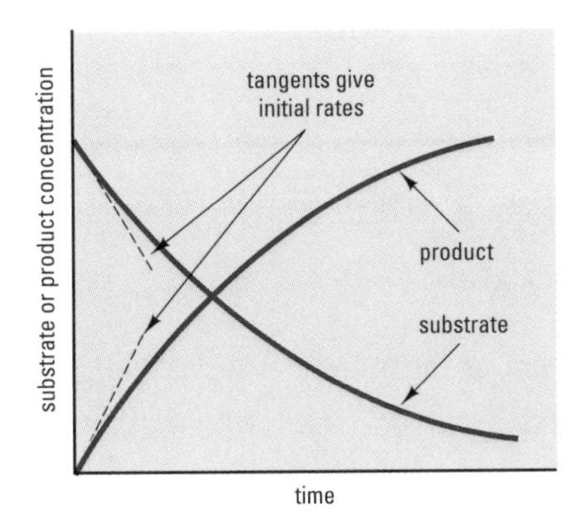

Fig. 35.1 Enzyme reaction progress curve: substrate utilization/product formation as a function of time.

Methods of monitoring substrate utilization/product formation

Continuous assays (kinetic assays)

The change in substrate or product is monitored as a function of time, to provide a progress curve for the reaction (Fig. 35.1). These curves start off in a near-linear manner, decreasing in slope as the reaction proceeds and substrate is used up. The initial velocity of the reaction (v_0) is obtained by drawing a tangent to the curve at zero time and measuring its slope. Continuous monitoring can be used when the test substance can be assayed rapidly (and non-destructively), e.g. using a chromogenic substrate. Reaction rate analysers allow simultaneous addition of reactant(s) or enzyme, mixing and measurement of absorbance: this enables the initial rate to be determined accurately.

Discontinuous assays (fixed time assays)

It is sometimes necessary to measure the amount of substrate consumed or product formed after a fixed time period, e.g. where the test substance is assayed by a (destructive) colorimetric chemical method. It is vital that the time period is kept as short as possible, with the change in substrate concentration limited to around 10%, so that the assay is within the linear part of the progress curve (Fig. 35.1). A continuous assay may be carried out as a preliminary step, in order to determine whether the reaction is approximately linear over the time period to be used in the fixed time assay.

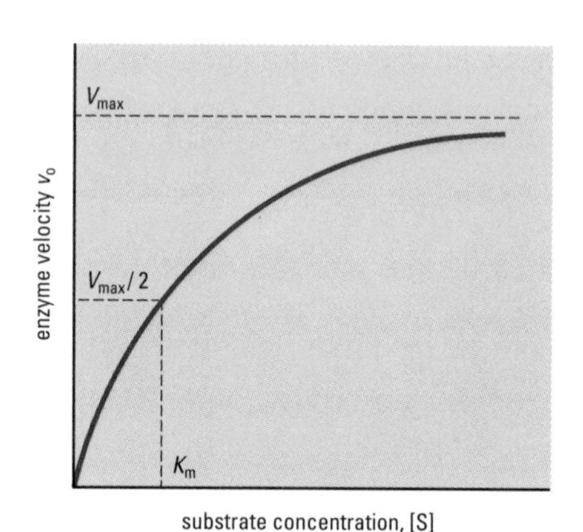

Fig. 35.2 Effect of substrate concentration on enzyme activity.

Enzyme kinetics

For most enzymes, when the initial reaction rate (v_0) using a fixed amount of an enzyme is plotted as a function of the concentration of a single substrate [S] with all other substrates present in excess, a rectangular hyperbola is obtained (Fig. 35.2). At low substrate concentrations v_0 is directly proportional to [S], with a decreasing response as substrate concentration is increased until saturation is achieved. The shape of this plot can be described by a mathematical relationship, known as the Michaelis-Menten equation:

$$v_0 = \frac{V_{max}\,[S]}{K_m + [S]} \qquad [35.2]$$

This equation makes use of two kinetic constants:

- V_{max}, the maximum velocity of the reaction (at infinite substrate concentration).
- K_m, the Michaelis constant, is the substrate concentration where $v_0 = \frac{1}{2}V_{max}$.

V_{max} is a function of the amount of enzyme and is the appropriate rate to use when determining the specific activity of a purified enzyme. The Michaelis constant is expressed in terms of substrate concentration ($mol\,l^{-1}$) and is independent of enzyme concentration. K_m is derived from the individual rate constants of the reaction – for example, with a single substrate (S) and single product (P) enzymic reaction, the process can be described as follows:

$$E + S \underset{k_2}{\overset{k_1}{\rightleftharpoons}} ES \overset{k_3}{\longrightarrow} E + P \qquad [35.3]$$

where E = enzyme, ES = enzyme–substrate complex, and k_1, k_2 and k_3 are rate constants. The Michaelis constant can be expressed as:

$$K_m = \frac{k_2 + k_3}{k_1} \qquad [35.4]$$

For many enzymes, $k_3 \ll k_2$, and Equation 35.4 simplifies to:

$$K_m = \frac{k_2}{k_1} = \frac{[E]\,[S]}{[ES]} \qquad [35.5]$$

When this applies, K_m provides a measure of the affinity of an enzyme for the substrate, and this is an important characteristic of each particular enzyme. Values for mammalian enzymes usually fall within the range $10^{-2}\,mol\,l^{-1}$ to $10^{-5}\,mol\,l^{-1}$. Thus an enzyme with a large K_m usually has a low affinity for its substrate, while an enzyme with a small K_m usually has a high affinity. K_m values can be used to select a substrate concentration that will give maximum reaction velocity (p. 196) or to compare the affinities of

(a)

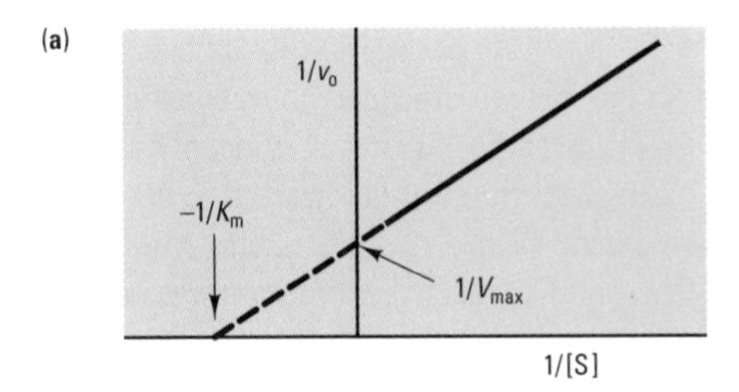

(b)

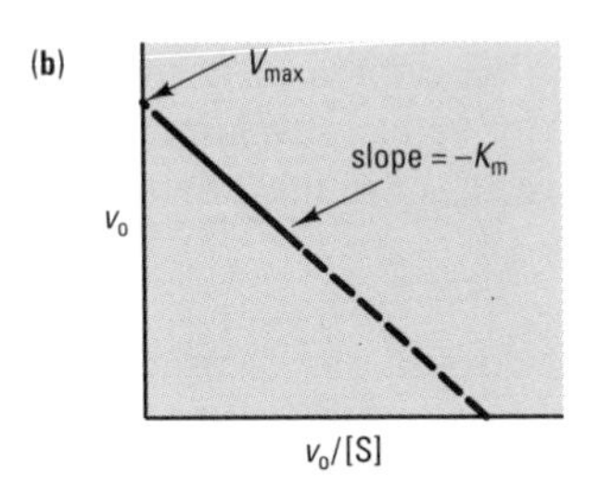

(c)

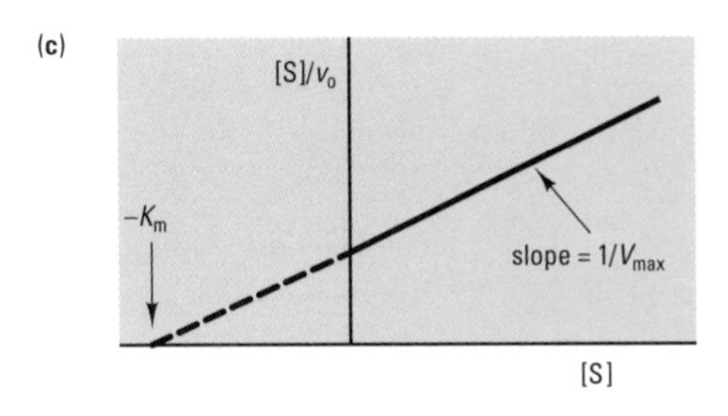

Fig. 35.3 Graphical transformations for determining the kinetic constants of an enzyme. (a) Lineweaver–Burk plot. (b) Eadie–Hofstee plot. (c) Hanes–Woolf plot.

When removing a bottle of freeze dried enzyme from a freezer or fridge, do not open it until it has been warmed to room temperature or water may condense on the contents – this will make weighing inaccurate and may lead to loss of enzyme activity.

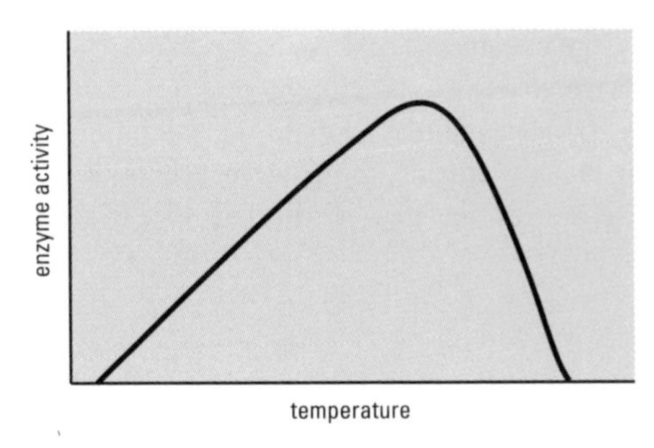

Fig. 35.4 Effect of temperature on enzyme activity.

different substrates for a given enzyme, or the same substrate with different enzymes.

Your first step in determining the kinetic constants for a particular enzyme is to measure the rate of reaction at several substrate concentrations, as in Fig. 35.2. There are various ways to obtain K_m and V_{max} from such data, mostly involving drawing a graph representing a linear transformation of eqn [35.2]:

- The Lineweaver–Burk plot: a graph of the reciprocal of the reaction rate ($1/v_0$) against the reciprocal of the substrate concentration (1/[S]) gives $-1/K_m$ as the intercept of the x axis and $1/V_{max}$ as the intercept of the y axis (Fig. 35.3a). The slope of the plot is most affected by the least accurate values, i.e. those measured at low substrate concentration.
- The Eadie–Hofstee plot: v_0 against v_0/[S], where the intercept on the y axis gives V_{max} and the slope equals $-K_m$ (Fig. 35.3b).
- The Hanes–Woolf plot: [S]/v_0 against [S], giving $-K_m$ as the intercept of the x axis and $1/V_{max}$ from the slope (Fig. 35.3c).
- The direct linear plot (Isenbhal and Cornish) is another option that may give more accurate estimates of kinetic constants than the other plots.

There are several computer packages that will plot the above relationships and calculate the kinetic constants from a given set of data using linear regression analysis (p. 275). While the Eadie–Hofstee and Hanes–Woolf plots distribute the data points more evenly than the Lineweaver–Burk plot, the best approach to such data is to use non-linear regression on untransformed data. This is usually outside the scope of the simpler computer programs, though tailor-made commercial packages can carry out such analyses. Note also that some enzymes, particularly those involved in the control of metabolism, do not show Michaelis–Menten kinetics.

Factors affecting enzyme activity

If you want to measure the maximum rate of a particular enzyme reaction, you will need to optimize the following:

Temperature

Enzyme activity increases with temperature, until an optimum is reached. Above this point, activity decreases as a result of protein denaturation (Fig. 35.4). Note that the optimum temperature for enzyme *activity* may not be the same as that for maximum *stability* (enzymes are usually stored at temperatures near to or below 0 °C, to maximize stability).

pH

Enzymes work best at a particular pH, due to changes in ionization of the substrates or of the amino acid residues within the enzyme (Fig. 35.5). Most enzyme assays are performed in buffer solutions (p. 27), to prevent changes in pH during the assay. Note that some enzymes have different pH optima for different substrates.

Cofactors

Many enzymes require appropriate concentrations of specific cofactors for maximum activity. These are sub-divided into coenzymes (soluble, low molecular weight organic compounds which are actively involved in catalysis by accepting or donating specific chemical groups, i.e. they are co-substrates of the enzyme); examples include NAD^+ and ADP); and activators (inorganic metal ions, required for maximal activity, e.g. Mg^{2+}, K^+).

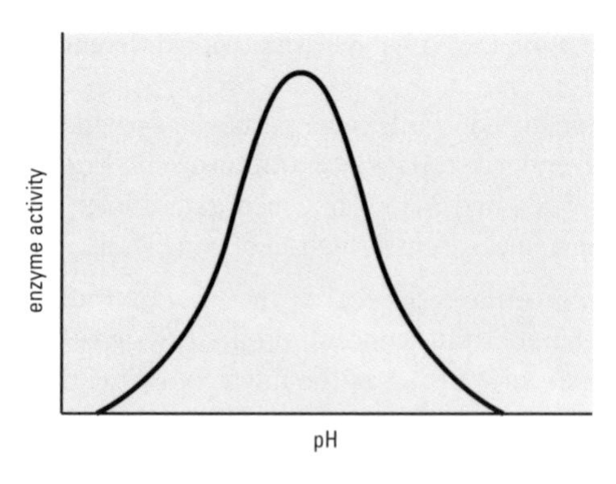

Fig. 35.5 Effect of pH on enzyme activity.

Substrate concentration

A substrate must be present in excess to ensure maximum reaction velocity. The K_m of an enzyme can be used to predict the substrate concentration required for the enzyme to operate at or near its maximum rate – this occurs when all active sites of the enzyme are filled. Using the Michaelis–Menten equation, the fraction of active sites filled (f_{ES}) for a given reaction velocity, v_o, is given by:

$$f_{ES} = \frac{v_o}{V_{max}} = \frac{[S_o]}{[S_o] + K_m} \quad [35.6]$$

When $[S_o] = K_m$, 50% of active sites are filled, and the reaction proceeds at $1/2\ V_{max}$. When $[S_o]$ is 10-fold greater than K_m, 91% of active sites are filled. When $[S_o]$ is 100-fold greater than K_m, 99% of active sites are filled and the reaction proceeds at 99% of V_{max}.

KEY POINT Note that temperature and pH optima are dependent upon reaction conditions, including cofactor and substrate concentrations – you should therefore specify the experimental conditions under which such optima are determined.

Enzyme inhibition

Definition

Inhibition – the reduction in enzyme activity due to the presence of another compound (an inhibitor).

It is important to investigate the inhibition of enzyme activity by specific molecules and ions because:

- enzyme inhibition is an important control mechanism in biological systems (e.g. negative feedback by a product);
- many drugs act by inhibiting enzymes;
- the action of many toxins can be explained by enzyme inhibition.

In terms of their interaction with enzymes, inhibitors can be involved in either reversible or irreversible reactions, and the inhibition can be competitive, non-competitive or uncompetitive, depending on whether inhibition can be reduced by an increase in the concentration of the natural substrate.

Irreversible inhibitors

These are substances, usually not of biological origin, which react covalently with an enzyme (E), preventing substrate binding or catalysis, e.g. iodoacetamide, which binds to thiol groups in enzymes as follows:

$$E{-}CH_2{-}SH + ICH_2CONH_2 \longrightarrow E{-}CH_2{-}S{-}CH_2CONH_2 + HI \quad [35.7]$$

Determining the effects of an inhibitor on enzyme kinetics – measure enzyme activity at $\geqslant$6 substrate concentrations, both in the presence and absence of inhibitor.

The thiol group may be within the active site, forming part of the binding and/or the catalytic sites, or it may be further from the active site and affect the three-dimensional conformation of the enzyme.

The toxicity of heavy metal ions (e.g. Hg^{2+}) is largely due to their irreversible effects on enzyme activity. Other examples include the inhibition of acetylcholinesterase by organophosphate pesticides and nerve agents. In such cases, it is often an active site serine residue that is affected, e.g. with diisopropylfluorophosphate (DFP):

$$E{-}CH_2OH + F{-}\underset{R}{\overset{R}{|}}\!P{=}O \rightarrow E{-}CH_2O{-}\underset{R}{\overset{R}{|}}\!P{=}O + HF \quad [35.8]$$

where R = isopropyl residue.

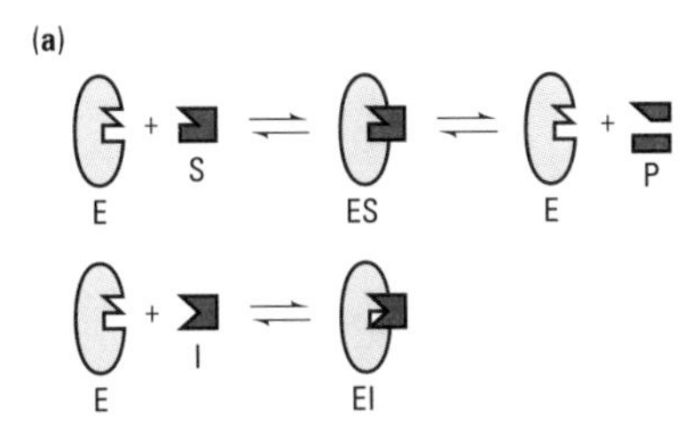

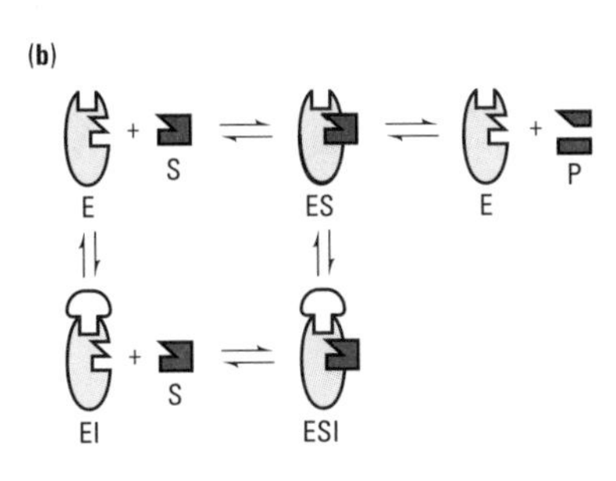

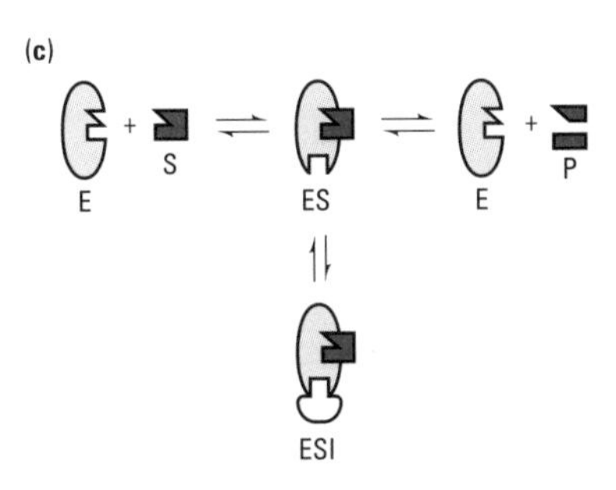

Fig. 35.6 Representation of (a) competitive inhibition, where the inhibitor (I) binds to the same site on the enzyme (E) as the substrate (S); (b) simple linear non-competitive inhibition; (c) uncompetitive inhibition (P = products).

Reversible inhibitors

These inhibitors do not react covalently with an enzyme, but show rapid reversible binding and dissociation. The velocity of the enzyme-catalysed reaction is reduced by the formation of enzyme–inhibitor (EI) or enzyme–substrate–inhibitor (ESI) complexes. Reversible inhibitors are sub-divided according to their effects on K_m and V_{max}:

- Competitive inhibitors bind reversibly to groups in the active site. Often the inhibitor resembles the substrate, and the occupation of the active site by an inhibitor molecule prevents a substrate molecule from binding to the same active site (Fig. 35.6a). The enzyme (E) can bind to the substrate (S), to form an ES complex, or to the inhibitor (I), to form EI, but cannot bind to both (ESI). A competitive inhibitor lowers the rate of catalysis by reducing the proportion of enzyme molecules bound to the substrate. It is possible to reverse competitive inhibition by increasing the substrate concentration; V_{max} is unaffected, but K_m increases (see p. 198). An example of competitive inhibition is the inhibition of succinate dehydrogenase by malonate, an analogue of the natural substrate, succinate.
- Non-competitive reversible inhibition occurs when an inhibitor binds to an enzyme whether or not the active site is occupied by the substrate. This type of inhibition often involves natural inhibitors of enzymes, and is important in the control of metabolism. The inhibitor binds to a site other than the active site (Fig. 35.6b). K_m is unchanged because the affinity of substrate molecules that bind to any uninhibited enzyme is unaffected, but V_{max} decreases due to the concentration of active enzyme molecules being effectively reduced by the presence of inhibitor.
- Uncompetitive inhibition, where the inhibitor binds only after the substrate has bound to the enzyme (Fig. 35.6c). It is most common in reactions involving more than one substrate.

Using enzyme inhibition kinetics to identify the type of inhibitor

Enzyme kinetics can be used to distinguish between the different forms of inhibition and to provide quantitative information on the effectiveness of various inhibitors, by allowing the dissociation constant of the enzyme–inhibitor complex, K_i, to be determined. K_i is an expression that relates to the strength of binding of the inhibitor to the enzyme.

Competitive inhibition

With inhibitor present, the enzyme can react with the substrate (Equation 35.3), but can also react reversibly with the inhibitor (I), to give an inactive enzyme–inhibitor complex (EI), as follows:

$$\mathrm{E} + \mathrm{I} \underset{k_2'}{\overset{k_1'}{\rightleftharpoons}} \mathrm{EI} \qquad [35.9]$$

As a result, the presence of I decreases the amount of free enzyme [E] available for interaction with S, i.e. $[E] = [E_T] - [ES] - [EI]$, where E_T is the total amount of enzyme present. The concentration of the EI complex depends on the concentration of free inhibitor and on the dissociation constant, K_i, where:

$$K_i = \frac{[E]\,[I]}{[EI]} = \frac{k_2'}{k_1'} \qquad [35.10]$$

(a)

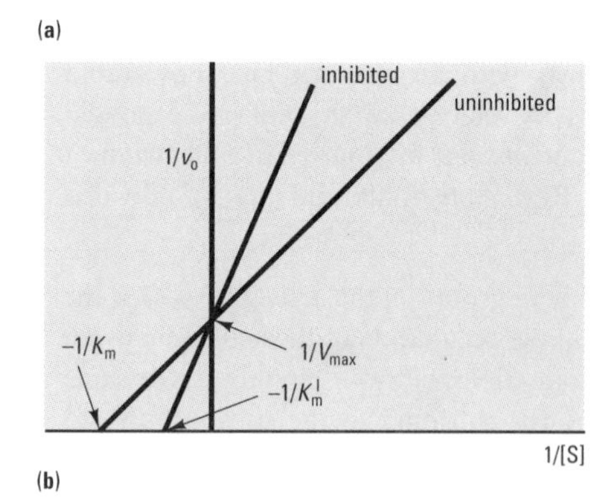

(b)

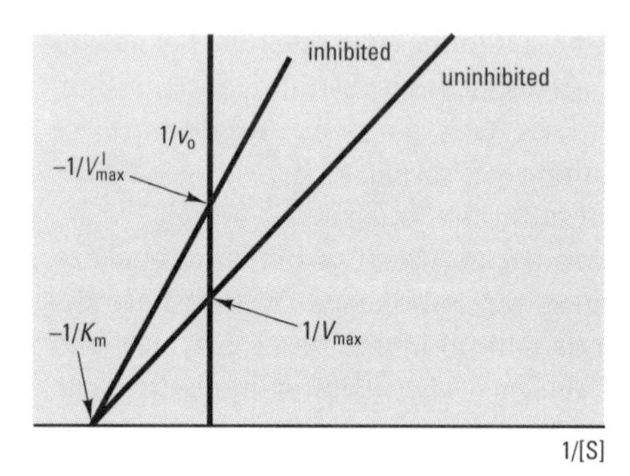

(c)

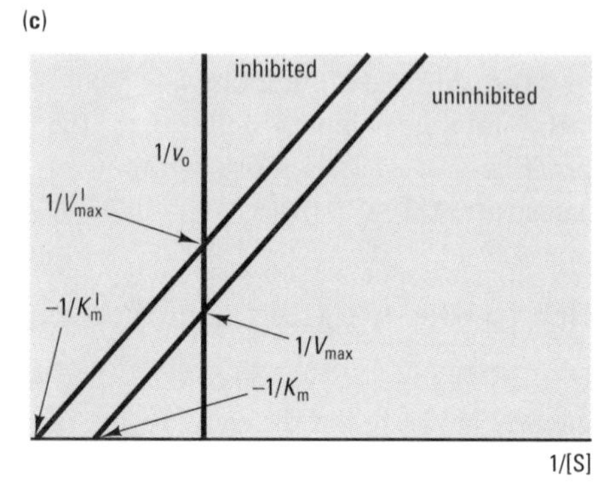

Fig. 35.7 Lineweaver–Burk plots showing the effect of (a) competitive inhibition, (b) simple linear non-competitive inhibition, (c) uncompetitive inhibition.

The Michaelis–Menten equation (Equation 35.2) must be modified to account for the presence of inhibitor and for a competitive inhibitor, and results in the Lineweaver–Burk plot shown in Fig. 35.7. The lines in the presence and absence of inhibitor intersect on the ordinate (y axis), indicating that with competitive inhibition, the inhibitory effect disappears at high substrate concentration (when $1/[S] = 0$, $[S] = \infty$). The value of K_i can be calculated either from knowledge of the K_m values obtained in the presence and absence of inhibitor (intercepts on the abscissa or x axis), since the following relationship applies:

$$\text{slope}_{\text{inhibited}} = \text{slope}_{\text{uninhibited}} \frac{(1+[\text{I}])}{(K_i)} \quad [35.11]$$

Non-competitive inhibition

In this case, not only can I bind E, but I can also bind to the ES complex to give an inactive ESI complex, i.e. $E + I \rightleftharpoons EI$ and $ES + I \rightleftharpoons ESI$. In the simplest case, the binding of S has no effect on the binding of I, and *vice versa*, so the dissociation constant of ESI is the same as that of ES, i.e. the K_m. A typical Lineweaver–Burk plot for a non-competitive inhibitor is shown in Fig. 35.7b. This indicates that this type of inhibitor decreases V_{max} but does not affect K_m. Effectively, this means that the inhibitor removes a certain fraction of active enzyme from operation, no matter what the concentration of substrate. The V_{max} changes by a factor of $(1 + [I])/K_i$ and K_i can be obtained by comparison of the slopes obtained with the uninhibited enzyme and the inhibited enzyme, as described above for a competitive inhibitor.

Uncompetitive inhibition

An uncompetitive inhibitor leads to a Lineweaver–Burk plot as shown in Fig. 35.7c. Parallel lines are obtained, and both K_m and V_{max} are affected. Here K_i pertains to the dissociation of I from ESI. Uncompetitive inhibitors are fairly rare.

Regulatory enzymes

In cell metabolism, groups of enzymes work together in sequential pathways to carry out a given metabolic process. In such enzyme systems, the reaction product of the first enzyme becomes the substrate for the next. Most of the enzymes in each system obey Michaelis–Menten kinetics. However in each system, there is at least one enzyme – often the first in the sequence – that sets the rate of the overall sequence (the flux through the pathway) because it catalyses the slowest or rate-limiting step. These regulatory enzymes exhibit increased or decreased catalytic activity in response to certain signals. By the action of such regulatory ,enzymes, the rate of each metabolic sequence is constantly adjusted to meet changes in the cell's demands for energy and biosynthesis.

The activities of regulatory enzymes are altered by non-covalent binding of various types of signal molecules, which are generally small metabolites or cofactors termed 'modulators'. Such enzymes are generally described as 'allosteric'.

Definition

Allosteric enzyme – an enzyme having more than one shape or conformation induced by the binding of modulators (Greek allos, 'other'; stereos, 'solid' or 'shape').

Properties of allosteric enzymes

The following general characteristics enable allosteric enzymes to be identified:

Example In the multi-stage conversion of L-threonine into L-isoleucine in bacteria, the first enzyme in the sequence (threonine dehydratase) is inhibited by isoleucine, the final product.

- Feedback inhibition: in some multienzyme systems the regulatory enzyme is specifically inhibited by the end product of the pathway whenever the end product increases in excess of the cell's needs. The end product does not bind to the active site but binds to another, specific site, termed the regulatory site. This binding is non-covalent and readily reversible. Thus if the concentration of the end product decreases, the rate of enzyme activity increases.
- Modulators for allosteric enzymes may be either inhibitory or stimulatory. An activator is often the substrate itself, and regulatory enzymes for which substrate and modulator are identical are called homotropic. When a modulator is a molecule other than the substrate the enzyme is heterotropic. Some enzymes have two or more modulators.
- Each enzyme molecule has one or more regulatory or allosteric sites for binding the modulator. Just as the enzyme's active site is specific for its substrate, the allosteric site is specific for its modulator. Enzymes with several modulators generally have different specific binding sites for each. In homotropic enzymes the active site and regulatory site are the same.
- These enzymes are generally larger and more complex than simple enzymes and normally consist of two or more subunits, e.g. aspartate transcarbamoylase, the principal regulatory enzyme in the synthetic pathway for CTP, has six catalytic subunits and six regulatory subunits. A useful technique for preliminary determination of subunit structure is SDS-PAGE (Chapter 26).
- Michaelis–Menten kinetics are not followed. When v_0 is plotted against [S] a sigmoid saturation curve normally results (Fig. 35.8), rather than the hyperbolic curve shown by non-regulatory enzymes (Fig. 35.2). With sigmoidal kinetics, although a value of [S] can be determined at which v_0 is half maximal, this value is not equivalent to K_m. Instead the symbol $[S]_{0.5}$, or $K_{0.5}$, is often used to represent the substrate concentration giving half maximum velocity of the reaction catalysed by an allosteric enzyme. Sigmoid kinetic behaviour generally reflects co-operative interactions between multiple protein subunits, i.e. changes in the structure of one subunit result in structural changes in adjacent subunits that affect substrate binding. In general, allosteric activators cause the curve to become more nearly hyperbolic (Fig. 35.8), with a decrease in $K_{0.5}$, but no change in V_{max}, and therefore v_0 is higher for any value of substrate concentration. However, some allosteric enzymes respond to an activator by an increase in V_{max} with little change in $K_{0.5}$. In contrast, an allosteric inhibitor may produce a more sigmoidal curve with an increase in $K_{0.5}$ (Fig. 35.8).

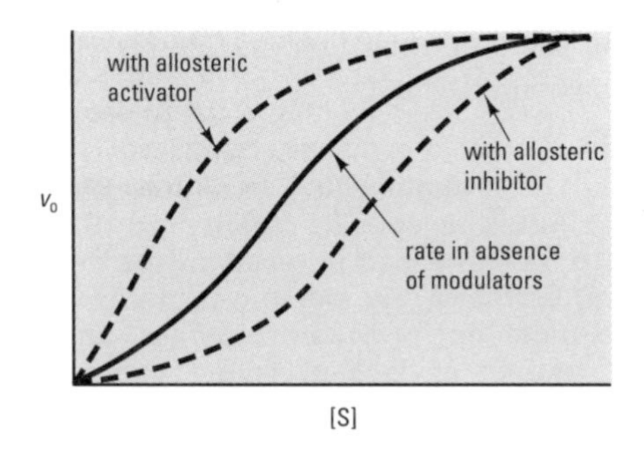

Fig. 35.8 Allosteric effects in enzyme kinetics. The graph shows the rate of enzyme reaction (v_0) against substrate concentration [S].

Covalently modified regulatory enzymes

Some enzymes are regulated by covalent modification, e.g. by phosphorylation of the hydroxyl groups of specific serine, threonine or tyrosine residues. Depending on the enzyme, this may result in activation or inactivation: in most cases, the non-phosphorylated form of the enzyme is relatively inactive. Often these modifications are under hormonal control. Some regulatory enzymes are influenced both by covalent modification and by allosteric effects.

Example The binding of adrenaline to receptors in skeletal muscle triggers a series of reactions that culminates in the phosphorylation of a particular serine residue in the enzyme glycogen phosphorylase: this activates the enzyme, resulting in increased glycogen breakdown.

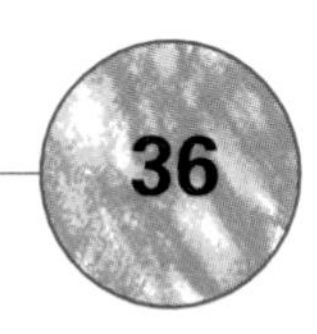

Membrane transport processes

Choosing an experimental system – transport studies can be carried out using whole multicellular organisms (e.g. fish), tissue slices/discs from animals or plants, bacterial cell suspensions or sub-cellular membrane vesicles, prepared by homogenization and fractionation (Chapter 17).

All cells operate a variety of processes whereby solutes move across membranes, including those processes involved in the uptake of nutrients and the loss of metabolic waste products at the plasma membrane, or the intracellular transport of substances across internal membranes, e.g. mitochondria. Membrane transport processes serve to regulate the intracellular environment, creating suitable conditions for internal metabolism. In microbes, the plasma membrane plays the principal role in the selective movement of biomolecules into, or out of, the cell. In multicellular animals and plants, membrane transport processes in the outermost epithelial layers are especially important in determining the solute composition of tissues and organs. Individual organs often have specialized transport functions, e.g. in animals, nutrient absorption by intestinal epithelial cells and urea excretion via the proximal tubular epithelial cells of the kidney or, in plants, the uptake of nutrients by root hair cells.

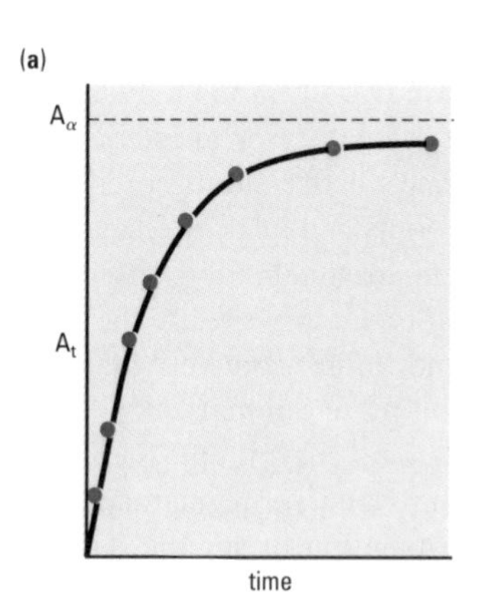

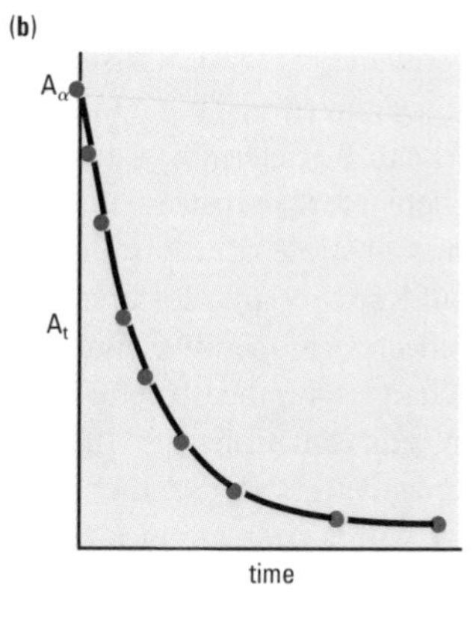

Fig. 36.1 Time course of (a) tracer influx and (b) tracer efflux (A_t). A_α = tracer activity at equilibrium.

Measuring solute transport

Usually, the rate of movement of a solute in a particular biological system is determined using a radioisotope-labelled (p. 105) or a stable isotope-labelled solute (p. 122) as a 'tracer'. Typically, in solute uptake (influx), the labelled solute is added to the external medium and its accumulation is then monitored by assaying samples of cells or tissue after known time intervals, enabling you to plot the accumulation of tracer as a function of time. Loss of solute (efflux) is studied using cells or tissues pre-loaded with labelled solute and then transferred to a tracer-free incubation medium: samples of the medium can be assayed after known time intervals, allowing the time course to be followed without destruction of the biological material. For microbial cell suspensions, filtration or silicone oil microcentrifugation (p. 125) may be used to separate the cells from their incubation medium.

KEY POINT **Determination of the relationship between the amount of tracer and the amount of solute (the 'specific activity', p. 103) enables the uptake or loss of solute to be expressed in terms of the net amount of solute transported per unit biomass (e.g. per mg protein, or per cell) per unit time. If the relationship between biomass and membrane area is known, the net uptake or loss of solute can be expressed per m^2 surface area per unit time.**

Example Suppose 5 ml of a red cell suspension containing 2 mg protein ml^{-1} accumulated 1050 Bq of radiolabelled solute of specific activity 210 Bq $nmol^{-1}$ in 8 min. The net rate of solute accumulation over this time period, expressed in nmol mg $protein^{-1}$ min^{-1} would be $1050 \div 210 \div 10 \div 8 = 0.0625$ ($\equiv$ 62.5 pmol mg $protein^{-1}$ min^{-1}).

Short-term tracer studies will give approximate values for the unidirectional flux rate from the labelled side of the membrane: longer-term studies may show more complex kinetics, due to two-way tracer movement or to secondary transport within the cells.

For non-metabolized solutes that are accumulated to particular levels within individual cells, e.g. inorganic ions such as K^+, Na^+, Cl^-, the time course for tracer accumulation or loss will often show exponential kinetics, as the transport process allows the tracer to equilibrate across the cell membrane (Fig. 36.1). Such exponential plots can be transformed to give linear plots using the relationship ln $(1 - A_t/A_\alpha)$ for influx, where A_α is the activity at equilibrium and A_t is the activity after time = t, while a plot of ln A_t/A_α should be linear when the cell behaves as a single compartment with respect to tracer exchange. The slope of such a plot gives a rate constant for exchange (k) that can be used to quantify the unidirectional solute movement, either influx or efflux (ϕ) from the following equation:

$$\phi = \frac{k\,[C_i]}{M_a} \qquad [36.1]$$

where $[C_i]$ is the internal solute content, determined either from tracer equilibration or by chemical assay (p. 165), and M_a is the membrane surface area. For some cells and tissues, an initial, short-term component is observed, due to tracer exchange between the extracellular space and the bulk medium – this can be subtracted from the longer-term component, allowing correction for extracellular tracer content.

Metabolized solutes pose additional problems, since they do not simply equilibrate across membranes – for example, glucose may be used as a source of carbon or energy after being transported into a cell. Transport studies of metabolized solutes may need to account for such processes, e.g. by separating and assaying the various labelled metabolites. A simpler alternative is to use a non-metabolized analogue of the solute, enabling transmembrane movement to be studied in the absence of metabolism and simplifying the interpretation of tracer studies.

Example 3-0-methyl glucose and 2-deoxyglucose are non-metabolized analogues of glucose, while methyl-β-D-thiogalactoside is a non-metabolized analogue of lactose.

Identifying the transport mechanism for a particular solute

Many studies are carried out to provide information on the nature of the transport process. Solute movement is often sub-divided into a number of categories:

Simple diffusion – passive transport

In accordance with Fick's first law of diffusion, an uncharged molecule can move through the lipid bilayer of a membrane with a net flux, J, determined by its lipid solubility and by the difference in concentration (strictly, the difference in activity) across the membrane, according to the relationship:

$$J = P([C_o] - [C_i]) \qquad [36.2]$$

where P is the permeability coefficient for the substance while $[C_o]$ and $[C_i]$ are external and internal solute concentrations respectively. The permeability coefficients of various solutes correlate with their partition coefficients between non-polar organic solvents and water, reflecting their relative solubility in the lipid phase. At a practical level, the flux rate calculated from Equation 36.2 will apply only to the initial state, since diffusion will reduce the concentration gradient and hence the flux rate.

Definition

Fick's first law of diffusion – each solute diffuses in a direction that eliminates its concentration gradient and at a rate proportional to the size of the gradient.

At a practical level, a number of characteristics can be used to recognize simple diffusion, including:

- The measured rates of transport are consistent with the lipid permeabilities of artificial lipid bilayers (e.g. values in Stein, 1990).
- The rate of transport is directly proportional to the concentration gradient (Eqn. 36.2), giving a straight line relationship with no evidence of saturation, in contrast to *all* other transport systems.
- Movement is insensitive to inhibitors, including structural analogues of the solute, metabolic inhibitors and chemical inactivators.
- Passive diffusion lacks specificity – chemically similar solutes with comparable lipid solubilities will show similar rates of transmembrane movement.

Passive diffusion of water (osmosis) – this is a special case, due to the high permeability of biological membranes to water and the high concentration of water on each side of the membrane.

Simple diffusion can account for the movement of only a limited number of low M_r, uncharged substances, including O_2, CO_2, NH_3 and non-polar hydrocarbons. There is negligible transbilayer diffusion of all other polar and ionic solutes, due to their hydrophilic, lipophobic nature and to the presence of a surrounding shell of water molecules, preventing movement through the hydrophobic interior of the membrane.

KEY POINT **All solutes not transported across membranes by simple diffusion are moved by protein-mediated transporters, often termed 'permeases', 'porters', or 'translocators'.**

Facilitated diffusion – passive transport

Here, the transmembrane movement is the result of specific membrane proteins (often termed 'uniporters'). While net movement is energetically 'downhill', as in simple diffusion, the rate of movement is more rapid than would be predicted from Eqn. 36.2, since the individual uniporter molecules act as solute-specific channels across the membrane. Examples include glucose transport across the erythrocyte membrane, maltodextrin transport across the outer membrane of Gram-negative bacteria and the stretch activated ion channels in stomatal membranes. The diagnostic features of facilitated diffusion include:

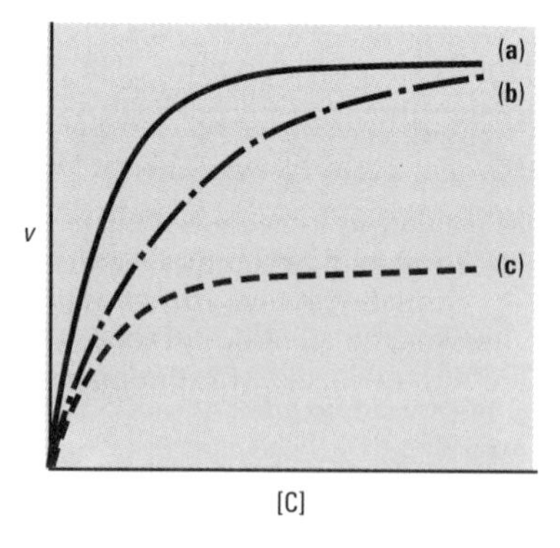

Fig. 36.2 Relationship between the rate of transport (V) and solute concentration [C] for (a) uninhibited 'control', (b) with a competitive inhibitor, (c) with a non-competitive inhibitor.

- The measured rates of transport are far greater than can be accounted for by simple diffusion and Fick's first law: for example, the erythrocyte glucose system operates at more than 10 000 times the rate predicted from permeability coefficients of synthetic lipid bilayers.
- The system exhibits specificity, transporting the solute but not chemically related compounds with a different three-dimensional shape.
- The rate of transport should show saturation kinetics similar to those observed for enzymes (Fig. 36.2), with a hyperbolic relationship between the rate of transport (V) and solute concentration [C] according to the relationship:

$$v = \frac{V_{max}\,[C]}{K_s + [C]} \qquad [36.3]$$

where V_{max} is the maximum rate of transport and K_s, the half saturation constant, is a measure of the affinity of the transport system for the solute – a low K_s generally indicates a high-affinity transport system while a high K_s shows that the transport system has a low affinity for the solute.

Determining kinetic parameters in transport studies – since Eqn. 36.3 is equivalent to the Michaelis–Menten equation, the same approaches can be used to determine V_{max} and K_s (see p. 194).

- Movement is sensitive to competitive inhibitors – the presence of structurally similar compounds will reduce the rate of solute transport. Competitive inhibitors decrease the affinity of the permease (increased K_s), but V_{max} is not affected (Fig. 36.2).
- Facilitated diffusion systems are susceptible to inactivation by non-competitive inhibitors that act as protein denaturants, e.g. heavy metal salts. Such irreversible inactivation confirms that the transport process is mediated by a protein. At low concentrations of inhibitor, only some of the permease molecules will be inactivated, giving a decreased V_{max} but unchanged K_s (Fig. 36.2).

If facilitated diffusion of an uncharged solute occurs, then net solute movement will occur down the transmembrane concentration gradient.

However, for charged solutes (ions), net movement will depend on the transmembrane electrical potential (the membrane potential), as well as the concentration gradient. The transmembrane equilibrium potential for a particular ion, E_n, can be calculated from the Nernst equation, as:

$$E_n = 2.303 \frac{RT}{zF} \log \frac{[C_o]}{[C_i]} \qquad [36.4]$$

Example A membrane potential of −100 mV (interior negative) can account for a 50-fold passive accumulation of a monovalent cation and a 50-fold passive exclusion of a monovalent anion.

where all symbols have their usual meanings (p. *xi*, see also Chapter 28). A comparison of the equilibrium potential and the membrane potential will show whether the ion is in electrochemical equilibrium across the membrane, or whether it has been transported against the electrochemical potential gradient.

KEY POINT **The transmembrane electrical potential, responsible for the facilitated diffusion of ions, is usually established as a direct result of the action of one or more ion-pumping, active transport systems, e.g. H^+-ATPases.**

Measuring membrane potentials – depending on the cell type, this can be carried out using microelectrodes, lipophilic cations, or fluorescent dyes (see p. 212).

Ionophores (Table 36.1) can be viewed as simple models of facilitated diffusion systems: they can be used to study the effects of an increase in the permeability of membranes to a single solute or group of solutes, to create transmembrane movements of ions, or to monitor specific ions in solution.

Table 36.1 Ionophores and inhibitors of transmembrane solute movement[1]

Ionophores	**Effect**
valinomycin	electrogenic carrier: increased permeability to K^+ and Rb^+
gramicidin A	electrogenic channel: uniport of monovalent ions, particularly H^+
monensin	electroneutral carrier: exchange of Na^+ or Li^+ for H^+
nigericin	electroneutral carrier: exchange of K^+ or Rb^+ for H^+
A-23187	electroneutral carrier: exchange of divalent cations (e.g. Ca^{++}, Mg^{++}) for $2H^+$
2,4-dinitrophenol (DNP)	electrogenic carrier: increased H^+ permeability (protonophore)
carbonylcyanide m-chlorophenyl-hydrazone (CCCP)	electrogenic carrier: increased H^+ permeability (protonophore)
carbonylcyanide p-trifluoromethoxy-phenylhydrazone (FCCP)	electrogenic carrier: increased H^+ permeability (protonophore)
nonactin	increased permeability of monovalent cations
amphotericin B	anion channel (weak selectivity)
nystatin	anion channel (weak selectivity)
Transport inhibitors[2]	**Effect**
oligomycin	inhibition of F_0F_1-type H^+-ATPases (mitochondria, bacteria, chloroplasts)
ouabain	inhibition of Na^+-K^+-ATPases (mammalian cell plasma membrane)
digitoxigenin	inhibition of Na^+-K^+-ATPases (mammalian cell plasma membrane)
vanadate	inhibition of E_1E_2-type ATPases (plant plasma membrane)
N,N′-dicyclohexylcarbodiimide (DCCD)	inhibition of F_0F_1 and V-type ATPases (mitochondria and plant tonoplast)
N-ethylmaleimide (NEM)	inhibition of plant tonoplast V-type ATPases
phlorizin	inhibition of Na^+-glucose symporters (intestinal epithelial cells)
furosemide	inhibition of cotransporters (mammalian cell membrane)
cytochalasin B	inhibition of glucose transporters (mammalian cell membrane)
amiloride	inhibition of Na^+/H^+ antiporters (mammalian and plant cells)
4,4′-diisothiocyano-2,2′-stilbene-disulphonate (DIDS)	inhibition of mammalian anion antiporters.

[1] *Safety note:* all of these inhibitors are highly toxic and must be handled carefully, observing the appropriate safety precautions.
[2] Details of electron transport inhibitors are given in Chapter 37.

Active transport

The principal criterion to establish active transport is that movement of the solute occurs against its electrochemical potential gradient. The energy for such 'uphill' transport can be provided by one of two mechanisms:

1. Primary active transport. The movement of solute molecules is coupled to the hydrolysis of ATP, via a membrane-bound solute-specific ATPase, or to a redox chain. For example, the Na^+-K^+ ATPase of mammalian cells is a primary active transporter, moving $3\,Na^+$ ions outwards and $2\,K^+$ ions inwards for every molecule of ATP hydrolysed, i.e. it is *electrogenic*, with a net charge separation across the membrane. Some of the most important primary active transport processes are those which create a transmembrane H^+ gradient, or proton-motive force (Δp), either via a H^+-translocating ATPase or an electron transport chain (p. 212). In many membranes, Δp provides the major driving force for the movement of other solutes (see Nicholls & Ferguson, 1992 for details). Some Gram-negative bacteria use a variation of phosphoryl-driven active transport to accumulate certain sugars – this is known as 'group translocation' since it involves the simultaneous phosphorylation of each sugar molecule as it is transported. Because the cell membrane is impermeable to sugar phosphates, the bacterial 'phosphotransferase' system can concentrate these solutes within cells.
2. Secondary active transport, or secondary transport. Here, the movement of solute is coupled to the 'downhill' movement of another solute, either via a 'symporter' (co-transporter) that simultaneously transports the two solutes in the same direction, or an 'antiporter' (counter-transporter) that simultaneously moves the two solutes in opposite directions across the membrane (Fig. 36.3). These systems are 'active' only in the sense that the 'downhill' movement of the other solute can only occur after primary active transport has established a transmembrane solute gradient (e.g. for H^+, or Na^+). Thus, lactose uptake in bacterial cells is coupled to the transmembrane proton gradient, via the lactose–H^+ symporter, while Na^+ efflux from plant cells is mediated by an Na^+–H^+ antiporter. In both instances, the H^+ gradient is created by a primary active transport system that pumps protons out of the cell, generating a proton-motive force that can be utilized by secondary transport systems. Similarly, the Na^+–K^+ primary active transport system creates an Na^+ gradient that can be utilized for nutrient uptake in mammalian cells, e.g. via the Na^+–glucose symporter (a secondary transporter). Primary and secondary transport systems can be envisaged as operating as a 'circuit' for the movement of ions across membranes (Fig. 36.3).

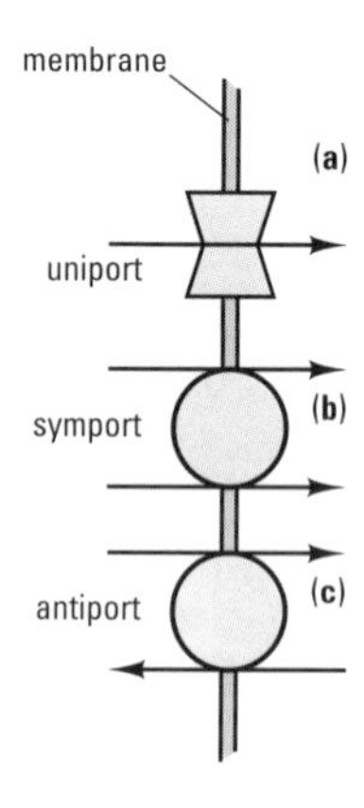

Fig. 36.3 Diagrammatic representation of various membrane transport systems, indicating the direction of movement of transported solute molecules.

Interpreting inhibitor experiments – while short-term effects may be unique to a particular inhibitor, in the longer term more general, non-specific effects are likely. For example, inhibitors of primary active transport will eventually affect secondary transport systems, due to their effects on ion gradients and membrane potentials.

In addition to the characteristics of specificity, saturation and inhibition shared by facilitated diffusion, active transport systems can be identified by the following features:

- The transmembrane distribution of the solute is not consistent with Eqn. 36.2 (where $J = 0$ at equilibrium, i.e. $[C_o] = [C_i]$) for an uncharged solute, or Eqn. 36.4 for an ion.
- The active component will be sensitive to metabolic inhibitors, e.g. ATPase inhibitors, for primary active transport, or specific inhibitors of secondary transporters (see Table 36.1).

- For secondary transport systems, dissipation of the primary ion gradient using a suitable ionophore will inhibit transport of the solute. Conversely, artificially generated ion gradients can be used to drive solute movement, either in whole cells or in membrane vesicle preparations.

Studying membrane transport using patch-clamp techniques

In this technique, part of a membrane (a 'patch') is sealed to the end of a heat-polished micropipette of diameter $\approx 1\ \mu m$ and the flow of current across the membrane patch is then measured for different bathing solutions and/or transmembrane voltages, set at particular values ('clamped') by a feedback amplifier. The size and orientation of the membrane patch can be controlled, as shown in Fig. 36.4. Since the membrane patch can be relatively small, the operation of ion-transporting proteins can be seen as changes in current flow, due to the opening or closing of individual membrane 'channels' (voltage-sensitive transporters). While the physiological factors controlling the *in vivo* operation of such channels remain unclear, patch clamping provides a powerful means of studying channel-mediated ion movements *in vitro*, enabling investigation of:

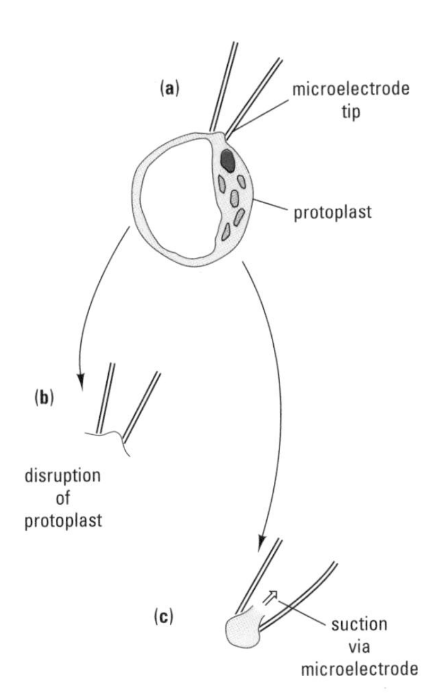

Fig. 36.4 Patch-clamp techniques (a) whole plant cell, (b) inside-out membrane patch, (c) outside-out membrane patch.

- the voltage required to produce 'channel' opening;
- the membrane conductance under conditions where the ion concentrations on both sides of the membrane are known;
- temporal effects, including opening and closing times, frequency of opening, etc.;
- the ion selectivity of particular 'channels';
- the effects of inhibitors, ionophores, etc.

When whole cells are used, the flow of ions through primary active ion transport systems (ion 'pumps') can be studied, since changes in current flow represent the sum of many transporters, acting together. Patch-clamp systems have provided a considerable amount of information on the transport properties of individual membranes, e.g. the plasma membrane and tonoplast of plant cells (see Flowers and Yeo 1992 for details).

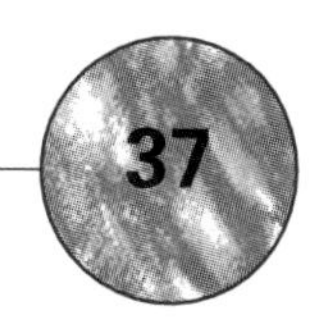

Photosynthesis and respiration

Techniques for investigating photosynthesis and respiration are considered together, since:

- Gas exchange processes in these two processes may be studied using similar techniques, and interpreted in similar stoichiometric terms, e.g. by relating changes in O_2 or CO_2 to those of other reactants or products.
- Membrane-bound electron transport systems are involved in the production of a transmembrane H^+ gradient that drives ATP synthesis via an ATPase embedded in the same membrane. Measurement of electron transport in photosynthesis and respiration often makes use of redox dyes as artificial electron donors or acceptors, with spectrophotometric analysis of the reactions.
- Some electron transport inhibitors can affect both photosynthesis and respiration, while others are specific to one or the other process.

KEY POINT **Photosynthesis and respiration are complex, multi-stage metabolic pathways involving a large number of cellular components: they may be studied as a whole, or as isolated components and individual metabolic reactions, to gain a deeper insight into the underlying processes.**

acceptor
FD
$NADP^+ + H^+$
$2e^-$
NADPH
acceptor
PQ
cyt b/f
PC
PSI
H_2O
$2e^-$
PSII
$2H^+ + \frac{1}{2}O_2$

Fig. 37.1 Non-cyclic electron transport from H_2O to $NADP^+$ in plant photosynthesis (Z scheme) showing the principal components: PS II, photosystem II; PQ, plastoquinone; cyt, cytochrome; PC, plastocyanin; PS I, photosystem I; FD, ferredoxin; $NADP^+$, nicotinamide adenine dinucleotide phosphate.

Measuring photosynthetic pigments – equations for spectrophotometric analysis of chlorophylls are given on p. 113.

Definitions

C_3 plants – those in which phosphoglyceric acid is the first stable product.

C_4 plants – those in which malate or aspartate are the principal products, via oxaloacetic acid, in a reaction catalysed by phosphoenolpyruvate (PEP) carboxylase.

CAM plants – those in which malic acid is produced, via oxaloacetic acid, by PEP carboxylase activity at night, with little or no net assimilation of CO_2 during the day (crassulacean acid metabolism).

Photosynthesis

This is the conversion of light energy to chemical energy by photoautotrophic plants and bacteria. The overall process is often summarized in terms of the synthesis of carbohydrate, $(CH_2O)_n$, driven by the energy within photons of light (hv), as:

$$nCO_2 + 2nH_2O^* + 4nhv \rightarrow (CH_2O)_n + nO_2^* + nH_2O \qquad [37.1]$$

where the asterisks show that all of the O_2 is derived from the photolysis of water. Carbohydrate synthesis occurs as a result of two distinct processes, namely (i) light reactions and (ii) dark reactions. In the former, light energy is absorbed by pigments within membrane-bound reaction centres (photosystems). The water-splitting light reactions generate electrons, protons and molecular oxygen. Electron and proton movement via a series of membrane-bound redox carriers leads to the reduction of $NADP^+$ and the generation of ATP (photophosphorylation), driven by a transmembrane H^+ gradient, or proton-motive force (p. 204). The light reactions of plant-type photosynthesis are often represented by the 'Z scheme' (Fig. 37.1).

The dark reactions use the products of the light reactions to 'fix' CO_2, via a series of soluble enzymes known as the Calvin cycle, or reductive pentose phosphate cycle. The primary carboxylating enzyme in temperate green plants (C_3 plants) and cyanobacteria is ribulose bisphosphate carboxylase ('Rubisco'). This enzyme catalyses the addition of CO_2 to ribulose bisphosphate (C_5), producing two molecules of phosphoglyceric acid (C_3) which are then reduced to triose phosphate (C_3) by enzymes that utilize the NADPH and ATP generated by the light reactions. Thus both light and dark reactions operate in a coupled manner when photosynthetically active cells are illuminated with visible light.

Definitions

Net photosynthesis – the net rate of CO_2 uptake or O_2 production, including respiratory and photorespiratory gas exchange.

Gross photosynthesis – the rate of CO_2 uptake or O_2 production, allowing for gas exchange due to respiratory and photorespiratory activity.

Correcting for photorespiration and respiration – measuring the rate of O_2 consumption or CO_2 production in the first few minutes after the plant material has been transferred from the light to darkness will give you the best estimate of photorespiratory and respiratory gas exchange. Use this rate to convert net photosynthesis to gross photosynthesis.

Preventing oxygen supersaturation in oxygen electrode studies – lower the O_2 content of your experimental solutions by bubbling them with N_2 before use, so that the O_2 evolved during the experiment remains in solution.

Measuring photosynthetic activity

Eqn. 37.1 shows that the rate of photosynthesis can be measured in terms of the amount of oxygen evolved or carbon dioxide fixed. However, several other metabolic processes may lead to concurrent changes in oxygen and/or carbon dioxide status, particularly:

- Photorespiration – light-dependent O_2 uptake, due to the oxygenase activity of Rubisco, leading to the production of glycollic acid (C_2) and the subsequent release of CO_2 due to the operation of a 'scavenging' pathway where two molecules of glycollate are converted to one of phosphoglyceric acid (C_3). The relative rates of photosynthesis and photorespiration depend upon the concentration of O_2 and CO_2 at the active site of Rubisco – in C_3 plants, photorespiration is highest under conditions of high temperature, high light and low water availability while C_4 plants show low levels of photorespiration.
- Respiration – generation of ATP via carbohydrate oxidation, as in non-photosynthetic cells (p. 210). In most photoautotrophs, respiratory activity is likely to be lower during the day than at night. However, plant tissues often respond to wounding or environmental stress by *increasing* their rate of respiration – this may be significant if you are measuring photosynthetic activity in tissue fragments, isolated cells or protoplasts. If you wish to calculate gross rates of photosynthesis, you must make appropriate allowances for both respiration and photorespiration.

Measurement of oxygen production This is most often used with aquatic systems, including algae and photosynthetic bacteria, higher plant cells, protoplasts, chloroplast suspensions (p. 87) or isolated thylakoids. Oxygen in solution can be determined by end-point chemical analysis (e.g. the Winkler method, often used in ecological and field studies) or, more conveniently, by continuous monitoring using an oxygen electrode and chart recorder, to give the rate of net oxygen production (Chapter 28) under various conditions, e.g. light, temperature, etc. For photosynthesis–irradiance (P–I) curves using an oxygen electrode, you should note that the electrode assembly can act as a lens, so the light within the chamber can be higher than that measured at the outside surface. The rate of gross photosynthesis is obtained by correcting for oxygen uptake when the electrode assembly is transferred to darkness (e.g. using a thick black cloth).

Measurement of carbon dioxide uptake For higher plant studies, this is most easily achieved using an infra-red gas analyser, or IRGA. Most modern instruments are portable and generally incorporate: (i) a cuvette, with a transparent window that attaches to a whole leaf or a known area of leaf providing a gas-tight seal, and within which the air is stirred by a fan; (ii) a gas supply system that allows control over input gases; (iii) miniaturized infra-red analysers (p. 118) to detect differences in CO_2 and H_2O content of input and output gas streams; (iv) systems for measuring (and possibly controlling) other environmental variables, e.g. light and temperature; (v) an on-board microprocessor for calculating results as rates of photosynthesis and transpiration and estimating the leaf's internal pCO_2 (C_i) and for exporting these and other data (e.g. time, leaf temperature, photosynthetic photon flux density, etc.) to PCs. Portable IRGAs are particularly useful for the rapid construction of photosynthesis–irradiance (P–I) and photosynthesis–pCO_2 (P–C_i) curves used to estimate photosynthetic efficiency, photosynthetic capacity and compensation points.

Definitions

Photosynthetic efficiency – the rate of photosynthesis when limited either (i) by supply of photons or (ii) by supply of CO_2, determined from the initial gradient of a P–I or P–C_i curve respectively.

Photosynthetic capacity – the rate of photosynthesis under saturating photon flux density (PFD) or CO_2 supply, determined from the upper asymptote of a P–I or P–C_i curve respectively.

Compensation point – PFD or pCO_2 at which net photosynthesis = 0, i.e. where gross photosynthesis, respiration and photorespiration are balanced: from the intercept of a P–I or P–C_i curve (*x* axis).

Interpreting ^{14}C fixation data – short-term studies of a few minutes may estimate gross photosynthesis, while longer term studies will give a rate closer to net photosynthesis, since some of the fixed ^{14}C will be respired.

Studying the release of photoassimilated ^{14}C in aquatic photoautotrophs – the loss of glycollate and other metabolites can be quantified by acidifying a known amount of medium, to drive off unfixed ^{14}C, then counting directly, or after an appropriate separation procedure.

Measurement of radiocarbon (^{14}C) fixation The tracer may be supplied as $^{14}CO_2$ for gas exchange studies or, more readily, as $H^{14}CO_3^-$ for studies in aqueous systems, since there will be an interconversion of soluble CO_2, HCO_3^- and $H_2CO_3^{2-}$ in accordance with Eqn. 28.4. In aqueous systems, the plant material is incubated in medium containing ^{14}C-labelled bicarbonate for a known time period, then removed and prepared for liquid scintillation counting (p. 102). Microalgal cells and photosynthetic bacteria can be separated from the experimental medium by filtration, or silicone oil microcentrifugation (p. 125) – mild acid treatment (e.g. 50 mmol l^{-1} HCl) will ensure that any unfixed ^{14}C is released as $^{14}CO_2$. To express the results in terms of the amount of C assimilated, you will need to calculate (i) the total inorganic C content of the medium, i.e. $CO_2 + HCO_3^- + H_2CO_3^{2-}$, obtained from pH and alkalinity measurements or by IR spectroscopy of the CO_2 produced when a known amount of the medium is acidified and (ii) the relationship between the total inorganic C content and the amount of ^{14}C tracer added (i.e. the specific activity of the experimental solution, p. 103).

One of the advantages of studying the photoassimilation of ^{14}C is that the radiotracer can be used to follow the fate of fixed carbon, by separating and fractionating the various cellular components prior to counting, e.g. by sequential solvent extraction of the plant material in 80% ethanol (low molecular weight solutes) then boiling water (polysaccharides), leaving a residual fraction (structural polysaccharides, proteins and nucleic acids). More sophisticated separation techniques must be used to quantify the amount of radioactivity within individual biomolecules, e.g. using column chromatography (p. 129) and autoradiography (p. 104), or 2-D thin layer chromatography (p. 128) and autoradiography. High specific activity $H^{14}CO_3^-$ and short incubation times of a few seconds are required to study the early stages in carbon photoassimilation.

Chlorophyll fluorescence This can be used to provide a non-destructive indication of photosynthetic function. When light energy is absorbed by the light harvesting apparatus of green plants, several reactions compete for the deactivation of excited chlorophyll molecules. Principally, the light energy may be trapped and used to reduce photosystem II (PS II) and drive photochemistry, it may be lost as heat, transferred to photosystem I (PS I) or re-emitted as fluorescence, measured using a fluorimeter (p. 115), at wavelengths around 685 nm at room temperature. In the dark-adapted state, a single high intensity light pulse is delivered to a leaf or chloroplast preparation and the resulting fluorescence signal is used to determine the potential maximum yield of PS II photochemistry (i.e. water splitting and O_2 evolution).

With modulated fluorimeters, the chlorophyll fluorescence signal can be separated from the daylight, allowing measurements to be made in light-adapted leaves under 'natural' conditions. Again, a single, saturating light pulse is used to reduce (close) PS II centres and, therefore, to calculate the reduction of fluorescence due to the photochemical use of absorbed light energy (photochemical quenching, Q_p). Any remaining reduction in the level of fluorescence compared to a dark-adapted standard must be due to non-photochemical thermal dissipation of energy (non-photochemical quenching, Q_n). Non-photochemical processes protect the protein components of the photosynthetic apparatus from oxidative damage due to excess light. Fluorescence can also be used to measure the quantum yield of PS II photosynthesis (ΦPS II), from which an estimate can be made of the rate of photosynthetic electron transport. The light and dark reactions of

Measuring photosynthetic quotients – in studies where both O_2 and CO_2 are measured, the photosynthetic quotient (PQ) can be determined from the relationship: PQ = O_2 evolved ÷ CO_2 consumed. In the simplest case, where fixed carbon accumulates as carbohydrate, a value of 1 should be obtained. However, the PQ may vary, depending on the amount of carbon incorporated into fats, proteins, etc. and the utilization of photosynthetic energy for other metabolic processes and growth, e.g. NO_3^- assimilation, carbon storage.

photosynthesis are co-ordinated, since electron transport generates the ATP and NADPH required for CO_2 fixation. For example, a reduction in the rate of CO_2 assimilation due to an inhibitor will be mirrored by a decrease in ΦPS II and, in many cases, by an increase in Q_n as light intensities increase to levels in excess of those required for photosynthesis (see Jones, 1992, for further details).

Studying photosynthetic electron transport

Robert Hill first showed that isolated chloroplasts can evolve oxygen in the absence of CO_2 fixation, as long as they are provided with an artificial electron acceptor that intercepts electrons from the photosynthetic electron transport chain. The net result of this 'Hill reaction' is the photolysis of water and the reduction of the Hill acceptor (A), as follows:

$$2nH_2O + 4nA \rightarrow 4nAH + nO_2 \qquad [37.2]$$

The practical significance of the Hill reaction is that it allows the photochemical reactions of the electron transport system to be studied independently of the dark reactions of photosynthesis. Many of the Hill acceptors are redox dyes that show changes in absorbance as they are reduced, e.g. ferricyanide ($Fe(CN)_6^{3-}$), which accepts electrons from the PS I complex, or 2,6-dichlorophenolindophenol (DCPIP), which intercepts electrons from the electron transport chain between PS II and PS I (Fig. 37.1). The reduction of these artificial electron acceptors can be followed spectrophotometrically, allowing the Hill reaction to be quantified. Alternatively, an oxygen electrode can be used to follow O_2 evolution from PS II in the presence of various artificial electron acceptors. Redox mediators can be used to investigate the following aspects of photosynthetic electron flow:

Measuring the Hill reaction using DCPIP – since the dye will revert to its oxidized (blue) form as soon as the chloroplasts are removed from the light, you must measure the absorbance as quickly as possible.

- the activity of PS II and PS I operating in series can be studied in whole chloroplasts using a suitable terminal acceptor, e.g. ferricyanide;
- the activity of PS II can be studied in whole chloroplasts using DCPIP (membrane-permeable), or in fragmented chloroplasts using ferricyanide;
- the activity of PS I can be measured if PS II is blocked by the herbicide diuron (DCMU) and if electron flow is maintained by the addition of an artificial electron donor, e.g. ascorbate plus DCPIP. Note that this cannot be measured by oxygen evolution, since PS II is inoperative – methyl viologen can be used as an electron acceptor from PS I and is rapidly reoxidized, consuming O_2 in a reaction which can be measured using an oxygen electrode (as O_2 *uptake*, rather than O_2 production);
- the sites of action of inhibitors (Table 37.1) can be determined by measuring their effects on the various components of the electron transport system, measured using redox dyes or the methyl viologen/oxygen electrode system.

Safety – make sure you treat *all* inhibitors with respect, observing appropriate safety precautions, and that you know the procedure to be followed in case of accident or spillage.

Table 37.1 Some inhibitors of photosynthetic electron transport[1]

Inhibitor	Target site
Hydroxylamine (NH_2OH)	photolysis of H_2O
DCMU (3(3,4-dichlorophenyl)-1,1-dimethylurea)	electron transport from PS II to plastoquinone
DBMIB (2,5-dibromo-3-methyl-6-isopropyl-*p*-benzoquinone)	electron flow from plastoquinone to cytochrome f
Methyl viologen (Paraquat)	electron flow from PS I to $NADP^+$
Atrazine (2-chloro-4-(2-propylamino)-6-ethylamine-5-triazine)	electron flow to plastoquinone
HOQNO (2-heptyl-4-hydroxyquinoline-N-oxide)	electron flow between quinones and cytochromes
DSPD (disalicylidinepropanediamine)	electron flow via ferredoxin

[1]see also Table 36.1 for uncouplers and ATPase inhibitors

Respiration

At the cellular level, respiration can be considered as the oxidation of organic compounds coupled to the production of so-called high-energy intermediates, such as ATP. The overall process is often represented in terms of the oxidation of glucose, as:

$$C_6H_{12}O_6 + 12O_2^* + 6H_2O + n\text{ADP} + n\text{P}_i \rightarrow 6CO_2 + 12H_2O^* + n\text{ATP} + nH_2O \quad [37.3]$$

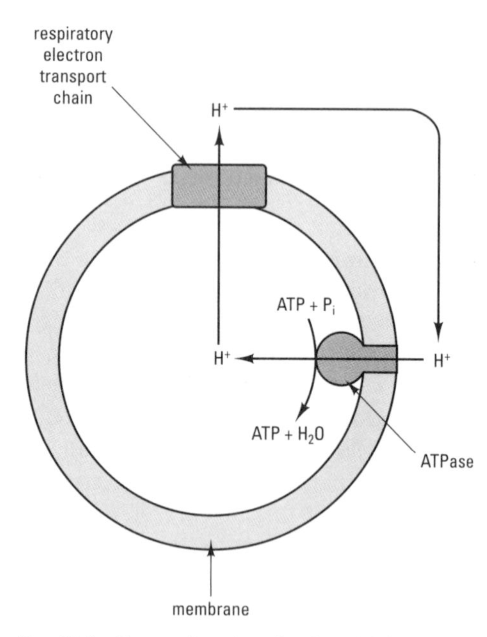

Fig. 37.2 Proton flow in mitochondrial oxidative phosphorylation: the creation of a transmembrane H^+ gradient due to the respiratory electron transport chain drives ATP synthesis via a membrane-bound ATPase.

the asterisks showing that oxygen is converted to water, and where n has a *maximum* value of 38. The process can be divided into three principal stages, namely (i) glycolysis, (ii) the tricarboxylic acid (TCA) cycle and (iii) oxidative phosphorylation. The glycolytic pathway operates in the cytosol and results in the partial breakdown of glucose (C_6) to two molecules of pyruvate (C_3) plus two molecules of NADH and two molecules of ATP. The mitochondrial TCA cycle involves the sequential dismantling of pyruvate to CO_2 (via an intermediate step involving decarboxylation to acetyl-coA), producing one molecule of $FADH_2$, four molecules of NADH and one of GTP for each pyruvate. The final stage occurs at the inner mitochondrial membrane – the movement of electrons and protons along the respiratory transport chain from reductant to oxygen as the terminal electron acceptor creates a transmembrane H^+ gradient that leads to the net synthesis of ATP, as summarized in Fig. 37.2. Thus, as in photosynthesis, respiration involves enzyme-catalysed interconversion of organic compounds plus membrane-bound electron transport reactions that are coupled to ATP synthesis.

The enzymic reactions of glycolysis and the TCA cycle can be studied using ^{14}C-labelled intermediates and pulse-chase experiments, using techniques similar to those used to investigate the dark reactions of photosynthesis (p. 206), or by purification and characterization of the individual enzymes (p. 192). Oxidative phosphorylation can be studied using techniques appropriate for electron transport reactions.

Measuring the respiratory activity of cells and tissues

The overall rate of cellular respiration is usually measured in terms of the amount of oxygen consumed by a known amount of material in a given time, e.g. as μmol O_2 mg protein^{-1} min^{-1}.

The principal techniques include:

- Manometry – this traditional approach involves measuring either the pressure change (Warburg manometer) or volume change (Gilson manometer) as gases are produced or consumed during respiration. Manometry is relatively insensitive and may be subject to large measurement errors: the principal application is in studying the relationship between the amounts of O_2 consumed and CO_2 evolved in terms of the respiratory quotient (RQ) for a particular substrate, from the relationship:

$$\text{RQ} = \frac{CO_2 \text{ evolved}}{O_2 \text{ consumed}} \quad [37.4]$$

The calculations involved in determining CO_2 production or O_2 consumption are complex and are specific to individual instruments – manufacturer's guidelines should be followed carefully.

Interpreting respiratory quotients – the complete oxidation of carbohydrates should give values close to 1.0, in agreement with Eqn. 37.4, while the oxidation of fats will give values close to 0.7 and protein oxidation will produce values of about 0.8.

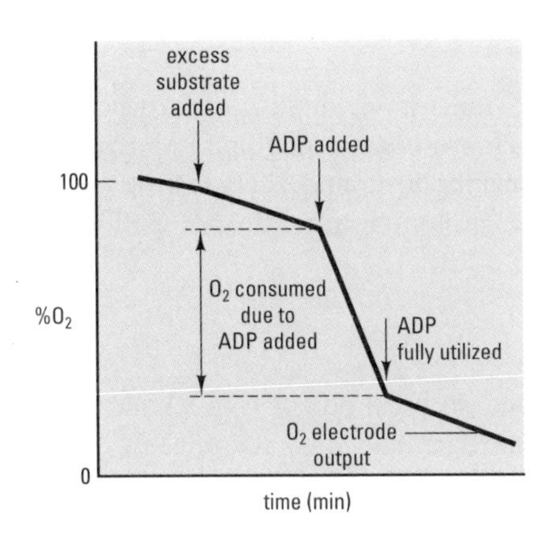

Fig. 37.3 Representative data for oxygen uptake by a suspension of mitochondria in response to additions of substrate and ADP.

- Oxygen electrode studies – this has largely replaced manometry, as the apparatus is more versatile and less complex to set up and interpret, giving a continuous readout of oxygen status via a chart recorder (p. 158). Net oxygen evolution can be measured using whole cells, mitochondrial suspensions (p. 87), or sub-mitochondrial preparations.

Using an oxygen electrode to study respiratory electron transport

Intact mitochondria suspended in an isotonic medium will show little respiratory activity unless supplied with (i) a suitable electron donor or substrate (e.g. NADH), (ii) ADP and (iii) P_i: this is termed respiratory control and the mitochondria are said to be tightly coupled, since there is a close link between electron transport, oxygen consumption and phosphorylation. The extent of this coupling can be determined using an oxygen electrode to measure the rate of oxygen consumption in the presence of substrate, P_i and ADP, with that in the absence of ADP (Fig. 37.3), as:

$$\text{respiratory control ratio} = \frac{\text{rate of } O_2 \text{ consumption with ADP}}{\text{rate of } O_2 \text{ consumption without ADP}} \quad [37.5]$$

Freshly prepared, tightly coupled mitochondria should have a respiratory control ratio of $\geqslant 4$.

To investigate the relationship between the number of ATP molecules produced per substrate molecule, substrate and P_i are supplied in excess and the oxygen uptake produced by a known amount of ADP is measured using an oxygen electrode (Fig. 37.4), allowing the P/O ($\equiv$ADP/O) ratio for a particular substrate to be calculated as:

$$\text{P/O ratio} = \frac{\mu\text{mol ADP added}}{2 \times \mu\text{mol } O_2 \text{ consumed}} \quad [37.6]$$

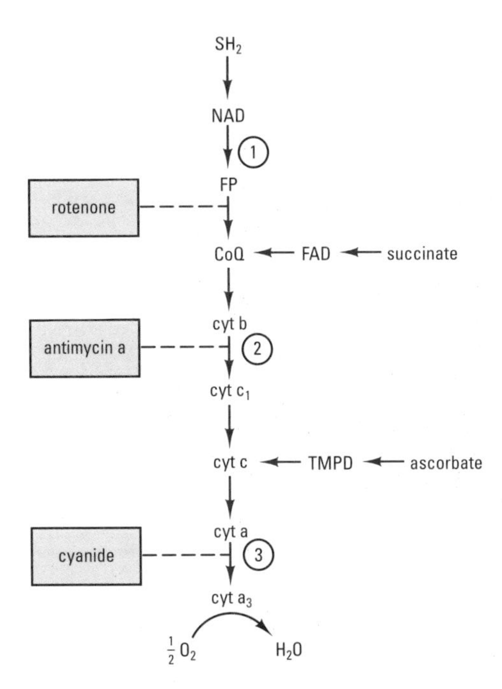

Fig. 37.4 The mitochondrial electron transport system, showing the sites of action of various inhibitors (dotted lines) and the three sites of ATP synthesis (numbered 1–3). S, substrate; FP, flavoprotein; CoQ, coenzyme Q; cyt, cytochrome; TMPD, tetramethylphenylene-diamine.

Note that the amount of O_2 is multiplied by 2, since oxygen is a diatomic molecule. The P/O ratio is determined by the site at which electrons are transferred to the respiratory chain (Fig. 37.5). Thus NADH should give a P/O ratio of 3, while succinate and $FADH_2$ should have values of 2. Artificial electron donors to cytochrome c have P/O ratios of 1, e.g. ascorbate/tetramethylphenylenediamine (TPMD). However, measured P/O ratios are always less than these values, due to partial uncoupling and the action of ATPases.

The effects of inhibitors of respiratory electron transport on the rate of oxygen consumption of mitochondria can be used to determine their site of action (Fig. 37.4). Thus, an inhibitor of electron flow will prevent oxygen uptake while an artificial substrate that donates electrons at a location beyond the site of inhibition will restore oxygen uptake, as shown in Fig. 37.5.

'Uncouplers' act by increasing the permeability of membranes to protons, causing the dissipation of the transmembrane H^+ gradient. Addition of an uncoupler increases electron flow along the respiratory chain in the absence of phosphorylation, i.e. there is loss of respiratory control. Uncouplers can reverse the effects of inhibitors such as oligomycin, whose target is the mitochondrial ATPase (Table 36.1, p. 203). Oligomycin prevents the return of H^+ to the interior of the mitochondrion, thereby inhibiting respiratory electron transport and oxygen uptake – the addition of an uncoupler (e.g. CCCP) reverses this inhibition.

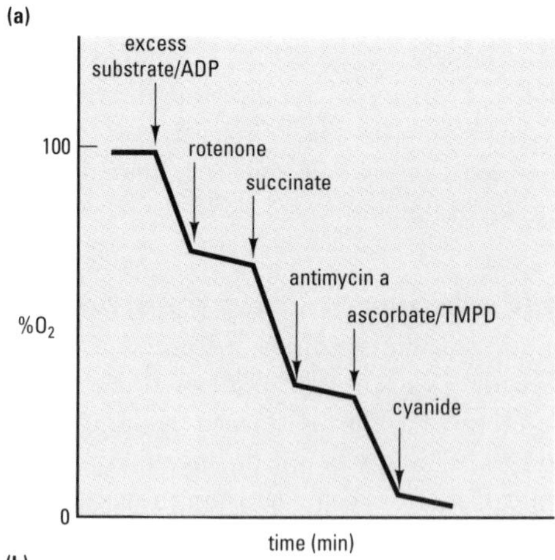

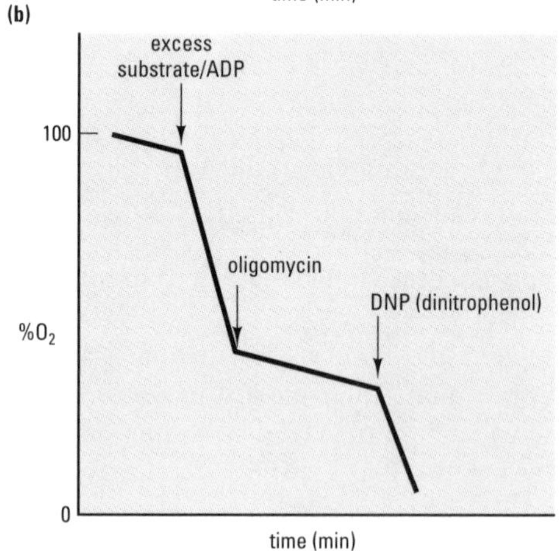

Fig. 37.5 Representative data for oxygen uptake by mitochondria in response to the addition of (a) electron transport inhibitors and substrates, or (b) inhibitors of ATP synthesis and uncouplers. Steeper slopes indicate faster rates of oxygen uptake.

Measuring the components of the proton-motive force in chloroplasts and mitochondria

According to chemiosmotic principles, the transmembrane electrochemical potential gradient of protons drives ATP synthesis in photophosphorylation and oxidative phosphorylation via a membrane-bound ATPase (Fig. 37.2). This gradient is often expressed as the proton-motive force, or pmf (Δp), expressed in mV as:

$$\Delta p = E_m - 59\,\Delta pH \qquad [37.7]$$

at 25 °C, where E_m is the transmembrane electrical potential (mV) and ΔpH is the pH gradient across the membrane. The individual components of the proton-motive force can be determined by a variety of methods:

- Transmembrane electrical potential, by the distribution of a suitable radiolabelled lipophilic cation, e.g. tetraphenylphosphonium (TPP^+), in accordance with the Nernst equation (see p. 203); by the measurement of K^+ or Rb^+ in the presence of the ionophore valinomycin (p. 203); or by the quenching of a fluorescent dye (e.g. cyanine or oxanol dyes).
- Transmembrane pH gradient from the equilibrium distribution of a radiolabelled weak permeant acid, e.g. 5,5-dimethyl-2,4-oxazolidinedione (DMO); by quenching of a fluorescent pH probe, e.g. 9-aminoacridine; or by ^{31}P-NMR (p. 120).

For radiolabelled probes, organelles or membrane vesicles can be separated from the bathing medium by silicone oil microcentrifugation (p. 125), with suitable correction for carry-over of external medium, e.g. using a membrane-impermeant solute, labelled with a second radioisotope. Care is required in such double-labelled experiments, as carry-over can represent a significant component, leading to substantial measurement errors if uncorrected.

Typical measurements for Δp across energy-transducing membranes are 180–200 mV: in mitochondria, E_m is the principal component while ΔpH represents the largest component of Δp in thylakoid membranes.

Genetics

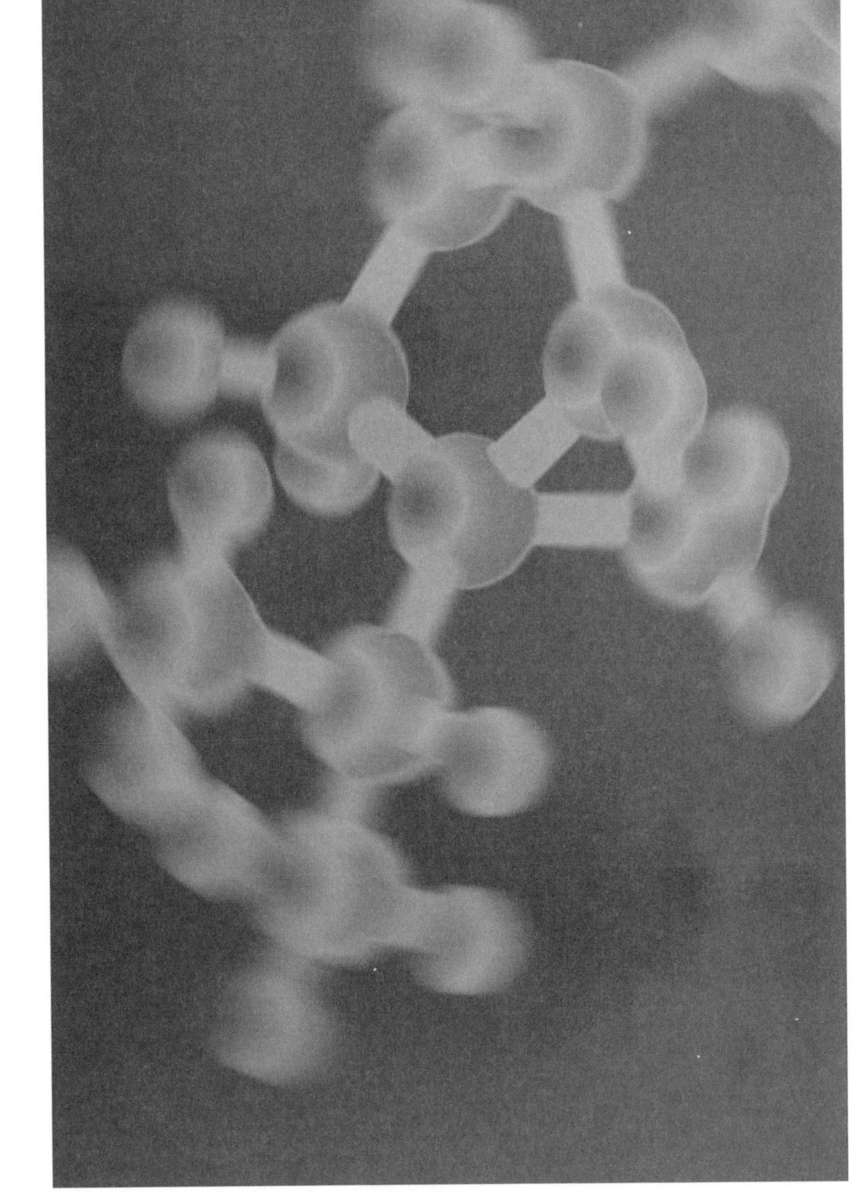

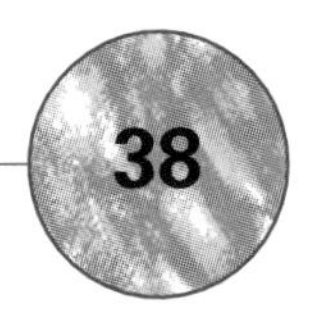

38 Mendelian genetics

Definitions

Phenotype – the observable characteristics of an individual organism; the consequence of the underlying genotype and its interaction with the environment.

Genotype – an individual's genetic make-up, i.e. the organism's genes.

Gregor Mendel, an Austrian monk, made pioneering studies of the genetics of eukaryotic organisms in the middle of the 19th century. He made crosses between different forms of flowering plants. Through careful examination and numerical analysis of the observable characteristics, or phenotype, of the parents and their progeny, Mendel was able to deduce much about their genetic characteristics, or genotype. The principles derived from these experiments explain the basis of heredity, and hence underpin our understanding of sexual reproduction, biodiversity and evolution. Mendelian genetics is concerned primarily with the transmission of genetic information, as opposed to molecular genetics which deals with the molecular details of the genome and techniques for altering genes (see Chapters 40–42). Bacterial genetics is considered separately, in Chapter 39.

KEY POINT **A common initial stumbling block in genetics is terminology. In many cases the definitions are interdependent, so your success in this subject depends on your grasp of all the definitions and underlying ideas explained below.**

Important terms and concepts

Genomes and chromosome numbers – remember that mitochondria and chloroplasts contain DNA molecules, but these are not included when calculating the chromosome number.

Each character in the phenotype is controlled by the organism's genes, the basic units of inheritance. Each gene includes the 'genetic blueprint' (DNA, see p. 180) which usually defines the amino acid sequence for a specific polypeptide or protein – often an enzyme or a structural protein. The protein gives rise to the phenotype through its activity in metabolism or its contribution to the organism's structure. The full complement of genes in an individual is known as its genome. Individual genes can exist in different forms, each of which generally leads to a different form of the protein it codes for. These different gene forms are known as alleles.

Definitions

Meiosis – division of a diploid cell which results in haploid daughter cells carrying half the original number of chromosomes. Occurs during gamete (sperm and egg) formation.

Mitosis – division of a cell into two new cells, each with the same chromosome number. Occurs in somatic cells, e.g. during growth, development, repair, replacement.

In eukaryotes, the genes are located in a particular sequence on chromosomes within the nucleus. The number of chromosomes per cell is characteristic for each organism (its chromosome number, n). For example, the chromosome number for man is 23. In cells of most 'higher' organisms, there are two of each of the chromosomes (2n). This is known as the diploid state. As a result of the process of meiosis which precedes reproduction, special haploid cells are formed (gametes) which contain only one of each chromosome (1n). In sexual reproduction, haploid gametes from two individuals fuse to form a zygote, a diploid cell with a new genome, which gives rise to a new individual through the process of mitosis. Cell numbers are increased by this process, producing genetically identical cells.

Organisms vary in the span of the diploid and haploid phases. In some 'lower' organisms, the haploid phase is the longer lasting form; in most 'higher' organisms, the diploid phase is dominant. A life-cycle diagram can be used to show how the phases are organized and what life-forms are involved in each case.

Example In the garden pea, *Pisum sativum*, studied by Mendel, the yellow seed allele Y was found to be dominant over the green seed allele y. A cross of YY x yy genotypes would give rise to Yy in all of the F_1 generation, all of which would thus have the yellow seed phenotype. If the F_1 generation were interbred, this Yy x Yy cross would lead to progeny in the next, F_2, generation with the genotypes YY:Yy:yy in the expected ratio 1 : 2 : 1. The expected phenotype ratio would be 3 : 1 for yellow : green seed.

Since each diploid individual carries two of each chromosome, it has two copies of each gene in every cell. The number of alleles of each character present depends on whether the relevant genes are the same – the homozygous state – or whether they are different – the heterozygous state. Hence, while there may be many alleles for any given gene, an individual could have two at most, and might only have one if it were homozygous. This will depend on the alleles present in the parental gametes that fused when the zygote was formed. Offspring that inherit a different combination of alleles at the two loci compared with their parents are known as recombinants.

The basis of Mendel's experiments and of many exercises in genetics are crosses, where individuals showing particular phenotypes are mated and the phenotypes of the offspring, or F_1 generation, are studied (see Box 38.1). If individuals carrying alternative alleles for a character are crossed, one of the alleles may be dominant, and all the F_1 generation will show that character in their phenotype. The character not evident is said to be recessive.

In describing crosses, geneticists denote each character with a letter of the alphabet, a capital letter being used for the dominant allele and lower case for the recessive. Taking for example a gene with dominant and recessive forms A and a respectively, there are three possibilities for each individual: it can be (a) homozygous recessive aa; (b) homozygous dominant AA; or (c) heterozygous Aa.

The reasons for dominance might relate to the activity of an enzyme coded by the relevant gene; for example, Mendel's yellow pea allele is dominant because the gene involved codes for the breakdown of chlorophyll. In the homozygous recessive case, none of the functional enzyme will be present, chlorophyll breakdown cannot occur and the seeds remain green. Not all alleles exhibit dominance in this form. In some cases, the heterozygous state results in a third phenotype (incomplete or partial dominance); in others, the heterozygous individual expresses both genotypes (codominance). Another possible situation is epistasis, in which one gene affects the expression of another gene that is independently inherited. Note also that many genes, such as those coding for human blood groups, have multiple alleles.

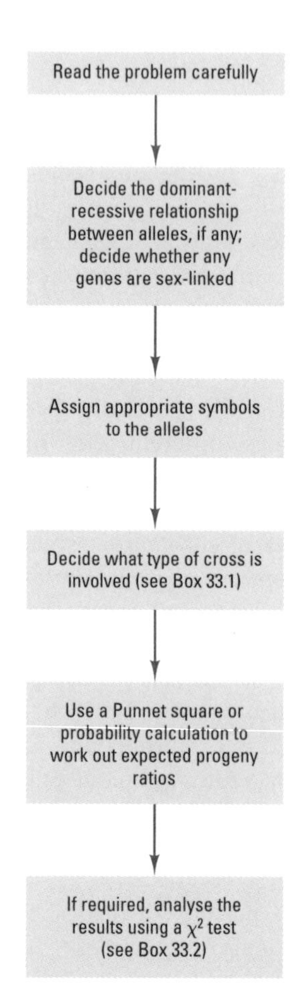

Fig. 38.1 Flowchart for tackling problems in Mendelian genetics.

KEY POINT **Genetics problems may well involve one of the 'standard' crosses shown in Box 38.1. Before tackling the problem, try to analyse the information provided to see if it fits one of these types of cross. Fig. 38.1 is a flowchart detailing the steps you should take in answering genetics problems.**

Unless otherwise stated or obvious from the evidence, you should assume that genes being considered in any given case are on separate chromosomes. This is important because it means that they will assort independently during meiosis. Thus, the fact that an allele of gene A is present in any individual will not influence the possibility of an allele of gene B being present. This allows you to apply simple probability in predicting the genetic make-up of the offspring of any cross (see below).

Where genes are present on the same chromosome, they are said to be linked genes, and thus it would appear that they would not be able to assort independently during metaphase I of meiosis, when homologous

Pedigree notation:

○ normal female

□ normal male

● ■ female or male with an inherited condition

○——□ mating

offspring

Example of simple family pedigree:

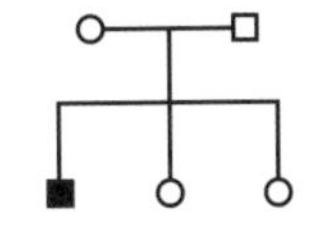

This diagram shows the offspring of a normal male and female. The two daughters are normal, but the son has the inherited condition.

Fig. 38.2 Pedigree notation and family trees.

chromosomes are independently orientated. However, although physically attached to each other, they may become separated when crossing over occurs between the homologous chromosomes at an early stage of meiosis. Exchange of genetic information between homologous chromosomes is called recombination. Linkage can be detected from a cross between individuals heterozygous and homozygous recessive for the relevant genes, e.g. AaBb × aabb. If the genes A and B are on different chromosomes, we expect the ratio of AaBb, aabb, Aabb and aaBb to be 1 : 1 : 1 : 1 in the F_1. However, if the dominant alleles of both genes occur on the same chromosome, the last two combinations will occur, but rarely. Just how rarely depends on how far apart they lie on the relevant chromosome – the further apart, the more likely it is that crossing over will occur. This is the basis of chromosome mapping (see Box 38.1).

Another complication you will come across is sex-linked genes. These occur on one of the X or Y chromosomes that control sex. Because one or other of the sexes – depending on the organism – is determined as XX and the other as XY (see Box 38.1), this means that rare recessive genes carried on the X chromosome may be expressed in XY individuals. Sex-linked genes are sometimes obvious from differences in the frequencies of phenotypes in male and female offspring. Pedigree charts (Fig. 38.2) are codified family trees that are often used to show inheritance of sex-linked characteristics through various generations.

Analysis of crosses

There are two basic ways of working out the results of crosses from known or assumed genotypes:

Denoting linked genes – these are often shown diagramatically, with a double line indicating the chromosome pair. For example, the two possible linkages for the genotype AaBb would be shown as:

$$\frac{A\ \ b}{a\ \ B} \text{ or } \frac{A\ \ B}{a\ \ b}$$

- The Punnet square method provides a good visual indication of potential combinations of gametes for a given cross. Lay out your Punnet squares consistently as shown in Fig. 38.3. Then group together the like genotypes to work out the genotype ratio and proceed to work out the corresponding phenotype ratio if required.

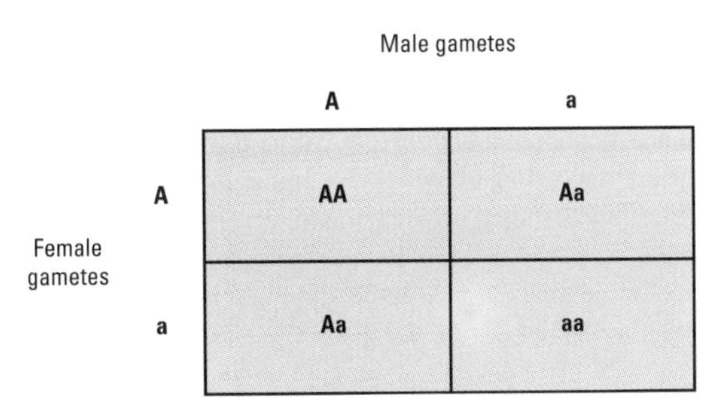

Fig. 38.3 Layout for a simple Punnett square for the cross Aa × Aa. The genotypic ratios for this cross are 1 : 2 : 1 for AA : Aa : aa, and the phenotypic ratio would be 3 : 1 for characteristic A to characteristic a. In this simple Punnet square, the allele frequencies are treated as equal ($f = 0.5$); if different from this, the probability of genotypes in each combination will be the relevant frequencies multiplied together (see p. 219).

Box 38.1 Types of cross and what you can (and can't) learn from them

Monohybrid cross – the simplest form of cross, considering two alleles of a single gene.

Example: AA × aa

If only the parental *phenotypes* are known, you can't always deduce the parental genotypes from the phenotype ratio in the F_1. An individual of dominant phenotype in the F_1 could arise from a homozygous dominant or heterozygous genotype. However, crossing the F_1 generation with themselves may provide useful information from the phenotype ratios that are found.

Dihybrid cross – a cross involving two genes, each with two alleles.

Example: AaBB × AaBb

As with a monohybrid cross, you can't always deduce the parental genotype from the phenotype ratio in the F_1.

Test cross – a cross of an unknown genotype with a homozygous recessive.

Example: AABb × aabb

A test cross is one between an individual dominant for A and B with one recessive for both genes. The progeny will all be dominant for A, revealing the homozygous nature of the parent for this gene, but the progeny phenotypes will be split approximately 1:1 dominant to recessive for gene B, revealing the heterozygous nature of the parent for this gene. This type of cross reveals the unknown parental genotype in the proportions of phenotypes in the F_1.

Sex-linked cross – a cross involving a gene carried on the X chromosome; this can be designated as dominant or recessive using appropriate superscripts (e.g. X^A and X^a).

Example: $X^AX^a \times X^AY$

In sex-linked crosses you need to know the basis of sex determination in the species concerned – which of XX or XY is male (for example, former in birds, butterflies and moths, the latter in mammals and *Drosophila*). The expected ratios of phenotypes in the offspring will depend on this. Note that the recessive genotype a will be expressed in the X^aY case.

Crosses with linked genes – genes are linked if they are on the same chromosome. This is revealed from a cross between individuals heterozygous and homozygous recessive for the relevant genes.

Example 1 (genes on separate chromosomes): expected offspring frequency from a cross AaBb × aabb is AaBb, aabb, Aabb and aaBb in the ratio 1:1:1:1.

Example 2 (linked genes): the frequencies of the last two combinations in example 1 might be skewed according to the direction of parental linkage.

Chromosome mapping uses the frequency of crossing-over of linked genes to estimate their distance apart on the chromosome on the basis that crossing over is more likely, the more distant apart are the genes. So-called 'map units' are calculated on the following basis:

$$\frac{\text{No of recombinant progeny}}{\text{Total no of progeny}} \times 100 = \%\ \text{crossing over}$$

By convention, 1% crossing over = 1 map unit (centimorgan, cM). The order of a number of genes can be worked out from their relative distances from each other. Thus, if genes A and B are 12 map units apart, while A and B are respectively 5 and 7 map units from C, the assumed order on the chromosome is ACB (Fig. 38.4).

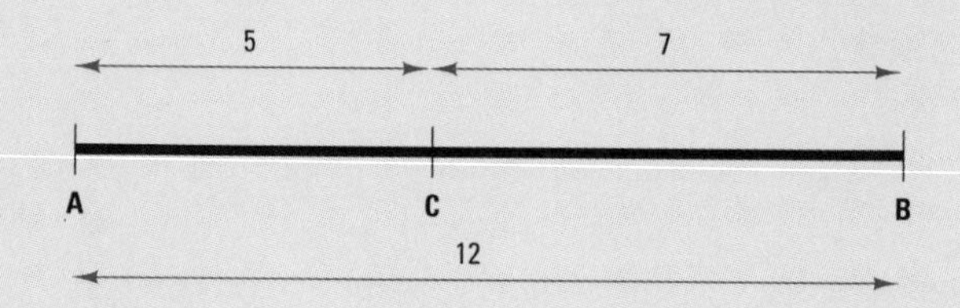

Fig. 38.4 Genetic map showing relative positions of genes A, B and C.

Using probability calculations – this can be simpler and faster than Punnet squares when two or more genes are considered.

- Probability calculations are based on the fact that the chance of a number of independent events occurring is equal to the probabilities of each event occurring multiplied together. Thus if the probability P of a child being a boy is 0.5 and the probability of the child of particular parents being blue-eyed is 0.5, then the probability of that couple having a blue-eyed son is $0.5 \times 0.5 = 0.25$, and that of having two blue-eyed boys is:

$$P = (0.5 \times 0.5) \times (0.5 \times 0.5) = 0.0625$$

How do you decide whether the results of an experimental cross fit your expectation from theory as calculated above? This isn't easy, because of the element of chance in gene crossing-over during meiosis. Thus, while you might expect to see a 3 : 1 phenotype ratio of progeny for a given cross, in 500 offspring you might actually observe a ratio of 379 : 121, which is a ratio of just over 3.13 : 1. Can you conclude that this is significantly different from 3 : 1 in the context of random error? The answer to this problem comes from statistics. However, the answer isn't certain, and your conclusion will be based on a balance of probabilities (see Box 38.2 and Table 38.1).

Box 38.2 Example of a Chi² (χ^2) test

This test allows you to assess the difference between observed (O) and expected (E) values and is extremely useful in biology. It is particularly valuable in determining whether progeny phenotype ratios fit your assumptions about their genotypes. The operation of the test is best illustrated by the use of an example. Assume that your null hypothesis (see p. 268) is that the phenotypic ratio is 3:1 and you observe that in 500 offspring the phenotype ratio is 379:121 whereas the expected ratio is 375:125.

Start the test by calculating the test statistic χ^2. The general formula for calculating χ^2 is:

$$\chi^2 = \Sigma \frac{(O - E)^2}{E}$$

In this example, this works out as:

$$\chi^2 = \frac{(379 - 375)^2}{375} + \frac{(121 - 125)^2}{125} = \frac{16}{375} + \frac{16}{125} = 0.171$$

The probability associated with this value can be obtained from χ^2 tables for $(n - 1)$ degrees of freedom (d.f.), where n = the number of categories = number of phenotypes considered. Here the d.f. value is $2 - 1 = 1$. Since the χ^2 value of 0.171 is lower than the tabulated value for 1 d.f. (3.84, Table 38.1), we therefore accept the null hypothesis and conclude that the difference between observed and expected results is not significant (since $P > 0.05$). Had χ^2 been >3.84, P would be < 0.05 and we would have rejected the null hypothesis and concluded that the difference was significant.

Table 38.1 Values of Chi² (χ^2) for which $P = 0.05$. The value for $(n - 1)$ degrees of freedom (d.f.) should be used, where n = the number of categories (= phenotypes) considered (normally fewer than 4 in genetics problems). If χ^2 is less than this value, accept the null hypothesis that the observed values arose by chance; if χ^2 is greater than this value, reject the null hypothesis and conclude that the difference between the observed and expected values is statistically significant.

Degrees of freedom	χ^2 value for which $P = 0.05$
1	3.84
2	5.99
3	7.82
4	9.49

Population genetics

Population genetics is largely concerned with the frequencies of alleles in a population and how these may change in time. The Hardy–Weinberg Principle states that the frequency of alleles f remains the same between generations, unless influenced by some outside factor(s).

To understand why this is the case, consider alleles H and h for a particular gene, which exist in the breeding population at frequencies p and q respectively. If the individuals carrying these alleles interbreed randomly, then the expected phenotype and allele ratios in the F_1 generation can be calculated simply as:

$$f(\mathrm{HH}) = p^2;$$
$$f(\mathrm{Hh}) = 2pq; \quad \text{and}$$
$$f(\mathrm{hh}) = q^2$$

If you wish to confirm this, lay out a Punnet square with appropriate frequencies for each allele. Now, by summation, the frequency of H in the $F_1 = p^2 + pq$ (A similar calculation can be made for allele h); and since in this example there are only two alleles, $p + q = 1$ and so $q = (1 - p)$. Substituting $(1 - p)$ for q, the frequency of H in the F_1 is thus:

$$p^2 + p(1 - p) = p^2 + p - p^2 = p$$

i.e. the frequency of the allele is unchanged between generations. A similar relationship exists for the other alleles.

This is known as the Hardy–Weinberg Principle after its first, independent, protagonists. It holds so long as the following criteria are satisfied:

1. random mating – so no factors influence each individual's choice of a mate;
2. large population size – so the laws of probability will apply;
3. no mutation - so no new alleles are formed;
4. no emigration, immigration or isolation – so that there is no interchange of genes with other populations nor isolation of genes within the population;
5. no natural selection – so no alleles have a reproductive advantage over others.

Population geneticists use the Hardy–Weinberg Principle to gain an idea of the rate of evolution and the influences on evolution. By ensuring that criteria 1–4 hold, if there are any changes in allele frequency between generations, then the rate of change of allele frequencies indicates the rate of evolutionary change (natural selection).

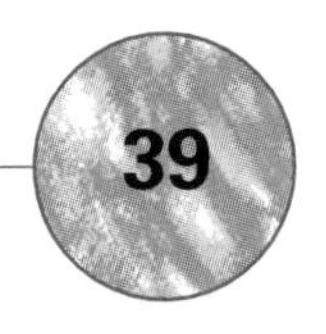

39 Bacterial and phage genetics

Definitions

Phage – a bacterial virus (bacteriophage).

Copy number – the average number of copies of a particular cellular molecule.

Transposon – a section of DNA coding for its own movement from one genomic location to another and carrying other genes in addition to those coding for transposition.

Merozygote – a cell containing two copies of a part of its genome, i.e. a partial diploid (sometimes also termed a merodiploid).

Using and interpreting standard nomenclature in bacterial genetics – different forms of three letter abbreviations are used for:

- **phenotypic features** – non-italic text, with superscripts where appropriate, e.g. Lac^+;
- **genotypic features** – lower case italic text, with individual letters to denote individual genes, e.g. *lacZ*;
- **gene products** (polypeptides and proteins) – the non-italicized equivalent of the abbreviation for the gene and with a capitalized first letter, e.g. the LacZ protein, which is a β-galactosidase and is the product of the *lacZ* gene in Lac^+ cells.

Several abbreviations may be combined, e.g. a single strain might have the phenotype Amp^r Lac^+ Trp^-.

In eukaryotic organisms, genetic reassortment usually involves the fusion of two haploid gametes to form a zygote and a new generation (p. 215).

KEY POINT **Bacterial genetics is very different from eukaryotic genetics (Chapter 38), due to the nature of the bacterial genome, which typically consists of a single chromosome plus none, one or several types of plasmid and/or phage, depending on the particular strain of bacterium.**

In most bacteria, the chromosome is usually a covalently closed circular DNA molecule, carrying genes for essential metabolic functions and structural components. As a consequence, bacteria can be regarded as haploid organisms. Plasmids are additional 'mini-chromosomes', typically coding for non-essential features, e.g. antibiotic resistance, heavy metal tolerance. They are often present at a higher copy number than the chromosome and may carry genes within mobile transposable elements (transposons). A single bacterium may contain more than one type of plasmid (though not if they are closely related plasmids, i.e. from the same incompatibility group). Plasmids can be introduced into a bacterial cell by conjugation (p. 224), or by transformation (p. 247). Bacteria can be 'cured' of their plasmids by chemical treatment, e.g. using acridine dyes that interfere with replication, or by growth under particular conditions, e.g. at high temperatures, where plasmid replication is unable to keep up with cell division.

A phage may replicate inside a bacterial cell or, in selected instances (temperate phages, p. 64), may exist within the cell in a non-replicating (latent) state, termed a prophage. As such, phages represent additional genetic elements that may be present within a bacterial cell, forming an important component of several aspects of bacterial genetics at the practical level.

The principal characteristics of experimental bacterial genetics ('crosses') are:

- the processes are completely distinct from sexual reproduction in eukaryotes;
- the processes are directional, from a donor cell (exogenote) to a recipient cell (endogenote);
- usually, only part of the donor cell's genome is transferred;
- in several instances, the recipient cell becomes a merozygote, with more than one copy of a gene, or genes. The merozygote may be a transient or a stable state, depending on circumstances;
- recombination (synapsis and 'crossing over') may or may not be involved, depending on the process and strains involved.

KEY POINT **Bacterial crosses are best described as gene *transfer* rather than gene exchange, since the latter term suggests reciprocal DNA movement.**

Working with bacterial mutants

To study bacterial genetics, it is necessary to use mutant strains, which have phenotypic characteristics that allow them to be distinguished from wild-type strains. The principal types of bacterial mutant include:

- Morphological mutants, with different structural characteristics to the wild-type, e.g. so-called 'rough' mutants of selected bacteria, such as *Streptococcus* and *Klebsiella*, are defective in their synthesis of capsular

Definitions

Minimal medium – a chemically defined medium, containing only sufficient nutrients to meet the requirements of wild-type cells, i.e. inorganic salts plus a particular carbon source.

Auxotrophs – mutants requiring an additional organic compound (e.g. an amino acid or growth factor) in order to grow in minimal medium.

Prototrophs – wild-types (from which auxotrophs are derived) and all other strains capable of growth in minimal medium.

Differential medium – usually a complex medium with additional compounds that distinguish between two types of bacteria, e.g. wild-type and mutant strains, often using pH indicator dyes (p. 27) or chromogenic/fluorogenic substrates (p. 7).

Take care when distinguishing between carbon source and nutritional mutants – note that a Lac^- mutant is unable to grow if provided with lactose as the sole carbon source, while a Trp^- mutant is unable to grow *unless* it is provided with tryptophan.

polysaccharides, giving small, dull colonies on agar-based media. This is in contrast to 'smooth' wild-type strains, where colonies are large and glistening due to the hydrophilic polysaccharide capsule. At a practical level, it is relatively straightforward to work with such mutants, since wild-type and mutant bacteria will grow on the same medium, the mutants having a feature that visibly distinguishes them from wild-types. However there are few examples in general use.

- Resistant mutants, which grow in the presence of an inhibitory substance such as an antibiotic, a toxic compound, or a particular phage. The isolation of an ampicillin-resistant mutant (Amp^r phenotype) is made possible by including ampicillin in the growth medium, since the growth of sensitive wild-type cells (Amp^s phenotype) will be inhibited.
- Carbon source mutants, which are unable to use a particular substance as a source of carbon or energy, e.g. *E. coli* mutants unable to use lactose (Lac^- phenotype), in contrast to Lac^+ wild-types. A Lac^- mutant would be unable to grow on a minimal medium containing lactose as the principal carbon source. In order to identify and characterize such mutants, a differential medium must be used, e.g. MacConkey agar (a complex medium containing lactose as a *supplementary* carbon source, plus a pH indicator dye – colonies of Lac^+ and Lac^- strains are distinguished on the basis of size and pigmentation, p. 68).
- Nutritional mutants, which have an additional requirement for a particular nutrient, compared to the wild-type. For example, a strain auxotrophic for the amino acid tryptophan (Trp^-) would grow only if the medium contained tryptophan (e.g. in a minimal medium plus tryptophan). Since it is not possible to devise a single medium which will allow an auxotroph to be distinguished from the corresponding prototrophic wild-type, the selection and identification of such mutants requires a different approach, involving replica plating from (a) nutrient-rich medium onto (b) minimal medium and (c) minimal medium supplemented with the particular nutrient (Fig. 39.1).

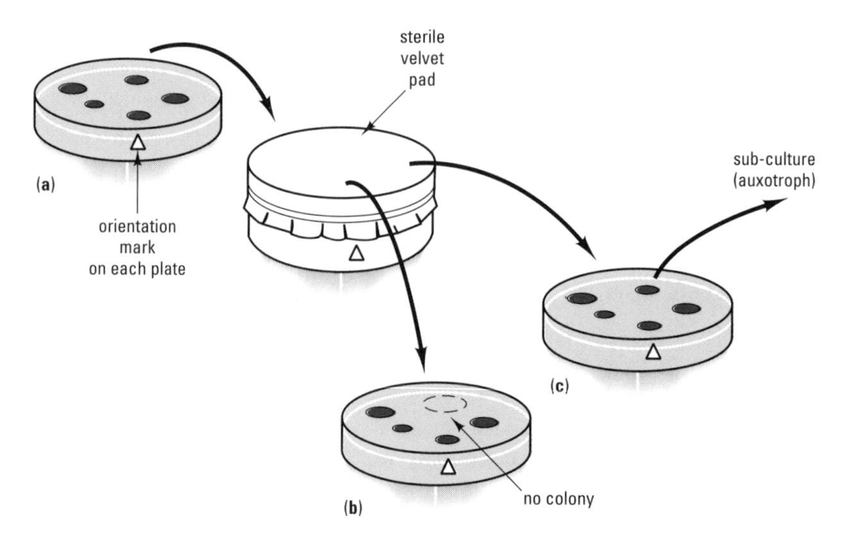

Fig. 39.1 Identification of auxotrophs using replica plating: a mixed suspension of wild-type and mutant cells is first cultured on the surface of a master plate containing nutrient-rich medium (a), then transferred to a sterile pad which is then used to inoculate minimal medium (b) and minimal medium plus a particular nutrient (c). Failure to grow on (b), coupled with growth on (c), infers auxotrophy for that particular nutrient and allows the auxotroph to be further sub-cultured and studied.

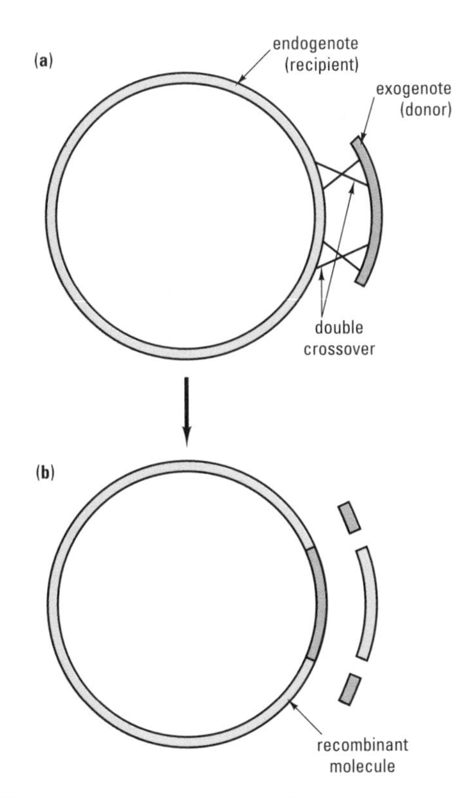

Fig. 39.2 Homologous recombination between donor DNA (from transformation, transduction or conjugation) and the recipient genome via a double crossover.

Performing co-transformation tests using auxotrophic recipients – controls must be set up without added donor DNA to determine the number of revertants (spontaneous mutants), giving the actual rate of transformation by subtraction.

- Conditional lethal mutants, which have a defect that causes death under a specific set of circumstances (the 'restrictive condition'), e.g. in *E. coli* temperature-sensitive mutants that grow at 30 °C, but not 40 °C, often as a result of the temperature-dependent inactivation of mutant enzymes.

DNA transfer in bacteria

KEY POINT The traditional approach to genetic analysis in prokaryotes involves mapping the position of individual genes using information provided from 'crosses', based on gene transfer. Typically, the recipient cell will be a mutant and the donor DNA will carry wild-type genes, enabling the transfer of wild-type characteristics to be studied in the laboratory.

The principal bacterial gene transfer processes are:

Natural transformation

In contrast to genetic engineering techniques, where DNA uptake is induced under specific laboratory conditions (p. 247), natural transformation involves the release of DNA to the external medium (e.g. death and lysis of the donor cell) and its subsequent uptake and incorporation into the genome of the recipient cell. Natural transformation is restricted to a limited number of bacterial groups and only occurs if the recipient cells are in the correct physiological state, termed *competence*, often in the early exponential growth phase (p. 80). Transformation is a relatively rare event, occurring at frequencies of $\leqslant 1$ transformant per 10^3 cells. In a competent cell, DNA will be taken up and, if homologous (i.e. from another strain of the same species), may then be incorporated into the genome of the recipient cell by homologous recombination via a double 'crossover' (two recombination events, as shown in Fig. 39.2).

The principal application of transformation has been in mapping the position of genes in those bacteria showing natural competence (e.g. *Bacillus subtilis*). The experiments are often easiest to perform with auxotrophic recipients and prototrophic (wild-type) donor DNA, since transformants can be selectively grown on media lacking one or more individual nutrients. The frequency of co-transformation of two genes is a measure of how close together they are likely to be on the donor DNA strand – a high co-transformation frequency implies that they are close together on the chromosome, reaching a recipient cell on the same fragment of DNA.

KEY POINT While co-transformation frequencies are inversely related to map distances, they are not directly equivalent to the recombination frequencies used in mapping eukaryotic genomes (p. 218), because they are also influenced by the size distribution of the fragments of donor DNA and by the likelihood of homologous recombination.

Transformation mapping has several limitations, since it requires a fairly large number of complex replica plating experiments to produce a chromosomal map, and the relative position of genes that are very far apart cannot be determined directly – the 'jigsaw' requires a large number of available pieces, before the underlying structure can be seen. It is also insensitive for small map distances – two genes that are adjacent, or very close together, will give similar high co-transformation values.

Performing generalized transduction in *E. coli* – P1 phage is often used, since it gives a high proportion of transducing particles (≈0.3% of the total), with a large fragment size (≈2.5% of the *E. coli* genome), making it easier to locate individual transductants.

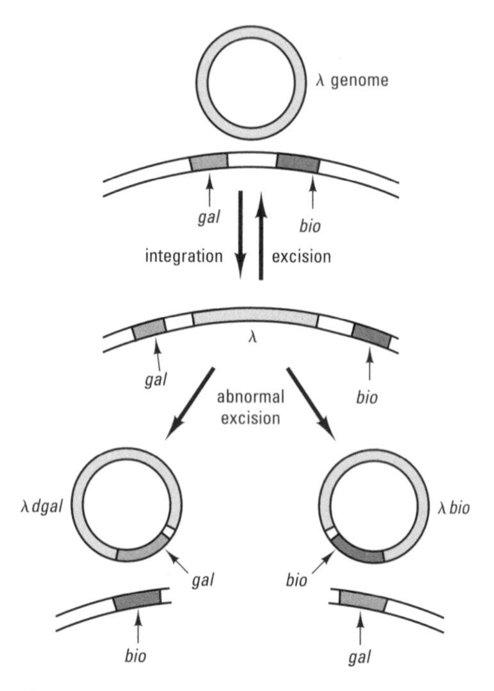

Fig. 39.3 Integration, excision and abnormal excision of λ phage, creating a modified phage genome (either λ*dgal* or λ*bio*), prior to specialized transduction.

Transduction

Here, the DNA exchange is mediated by a phage (p. 64), in one of two processes:

1. Generalized transduction: occasionally, a fragment of chromosomal or plasmid DNA within an infected bacterial cell may be packaged within the protein coat of a phage, in place of the phage genome. This fragment might be derived from any part of the host cell genome (exogenote). After release, on lysis of the donor cell, the transducing particle may introduce the DNA fragment into a new recipient cell. The introduced DNA may then be incorporated into the host genome by a double crossover, in a homologous recombination event similar to that shown in Fig. 39.2 for transformation. Generalized transduction can be used to establish gene order and for mapping purposes, using broadly similar principles to transformation, as only closely spaced genes will show co-transduction, with a co-transduction frequency that is inversely related to the distance between the two genes. Since a generalized transducing particle can carry a fairly small amount of DNA, the relative frequency of co-transduction can be used to provide finer detail of gene order over shorter distances than for transformation.
2. Specialized transduction: this is mediated only by a temperate phage, e.g. λ phage, which integrates at a specific site on the chromosome by a single crossover event. It involves a restricted number of genes – typically, a pair of genes on either side of the integration site for a particular temperate phage (Fig. 39.3). Specialized transduction will only occur if, on entry into the lytic cycle (phage replication, p. 64), there is an incorrect (abnormal) excision of the prophage and a part of the bacterial genome is incorporated, giving a specialized transducing particle (a modified phage, carrying a bacterial gene). In the case of λ phage (Fig. 39.3), the transducing particle may carry either the biotin gene (λ*bio*), or the galactose gene (λ*gal*, also termed λ*dgal*, since the modified phage is defective due to the loss of an essential part of the phage genome, and therefore is incapable of replicating without the aid of a co-infecting, non-defective 'helper' phage). On entry into the recipient cell, the modified phage may become latent, creating a partial diploid. The practical applications of specialized transduction are limited to those genes flanking the integration sites of temperate phages. The most common investigations are of complementation in merozygotes containing two copies of a particular gene. Where donor and recipients of the same mutant phenotype have mutations in different genes, the creation of a partial diploid can produce a wild-type phenotype.

Conjugation

Here, the transfer occurs as a result of cell-to-cell contact, with direct transmission of DNA from donor to recipient cell. In *E. coli*, the donor cell carries specific surface pili (protein microtubules), allowing a donor cell to attach to receptors on the surface of a recipient cell and bringing the paired cells into close contact (Fig. 39.4). In the simplest instance, the donor cell carries an additional plasmid, the F plasmid (originally termed 'F factor'), that encodes the genes responsible for conjugation, including those for the protein subunits of the specialized pili. In conventional notation, the donor is termed F^+ and the recipient F^-. During conjugation between F^+ donor and F^- recipient, a single strand of the circular F plasmid is cut and transferred (in linear form) to the recipient cell, which becomes F^+ once the entire plasmid has been transferred and a complementary strand has been synthesized – this process takes a few minutes. At a practical level, crosses involving F^+ donors give little useful information, apart from mapping the position of genes on the F plasmid. However, two other types of donor are more useful:

Example The *E. coli* strain HfrH has the F proplasmid integrated at a site close to genes for the synthesis of threonine, so these genes would be transferred early during HfrH × F^- conjugation.

1. Hfr strains, where the F plasmid DNA has become integrated into the chromosome, as a proplasmid, in an analogous manner to a temperate phage such as λ (Fig. 39.3). Such strains show a high frequency of recombination (hence Hfr), since chromosomal genes are transferred to the recipient cell at a far greater frequency than in crosses using F^+ donors. The Hfr × F^- cross is illustrated in Fig. 39.5: a part of the F proplasmid is first to be transferred, followed by chromosomal DNA and finally, the remaining fragment of the F proplasmid. After transfer, donor chromosomal DNA can be integrated into the recipient's genome by homologous recombination (Fig. 39.2). The conjugating pair usually breaks apart before the process is complete and the recipient cell remains F^-, since only the leading fragment of the F plasmid reaches the recipient cell. A number of different Hfr strains of *E. coli* are available, with the F proplasmid inserted at different chromosomal locations.
2. F′ ('F prime') donors, with a modified F plasmid incorporating one or more chromosomal genes, formed as a result of defective excision of the F proplasmid from the chromosome in a similar manner to λ*bio* (Fig. 39.3). The F′ plasmid will transfer its chromosomal genes at very high frequency to a recipient, which will also become F′ once transfer of the F′ plasmid DNA is complete. This process is sometimes referred to as F-duction (or, less appropriately, as 'sexduction').

Example F′*lac* is an F′ plasmid that incorporates wild-type chromosomal DNA coding for lactose utilization.

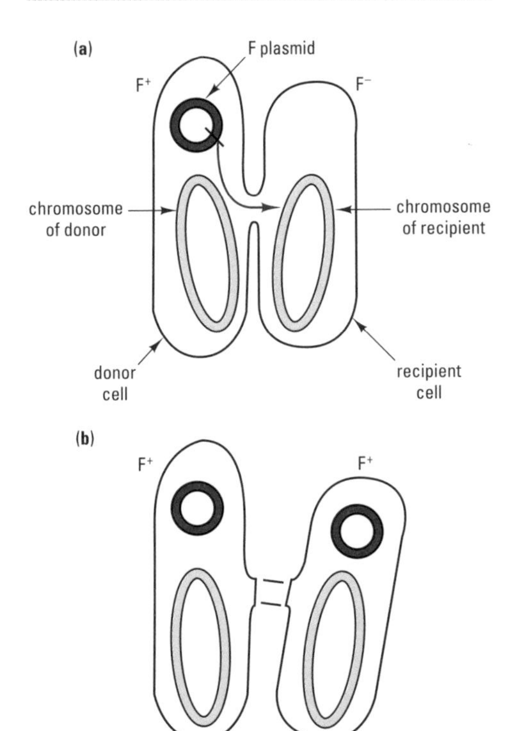

Fig. 39.4 Conjugation between F^+ donor and F^- recipient. A single strand of F plasmid DNA is cut and transferred in linear form to the recipient (a), followed by recircularization and complementary strand synthesis, converting the recipient to F^+ (b).

Mapping with Hfr donors using interrupted conjugation

Mating an antibiotic-sensitive wild-type Hfr donor with an antibiotic resistant mutant F^- recipient provides a simple means of detecting recombinants, since the donor cells will be unable to grow on a medium containing the antibiotic while unmodified recipient cells will have a different phenotype to the recombinant recipient cells. The method is even simpler for the most common crosses, using nutritional mutants, where a prototrophic Hfr donor is crossed with an auxotrophic F^- recipient and only the recombinants are able to grow on minimal medium with added antibiotic; this approach can be extended to multiple genes, using a multiple auxotroph F^- recipient. By carrying out a series of experiments where the conjugation process is terminated at different times (e.g. using vigorous mixing in a vortex mixer, p. 16), the time required to transfer a particular characteristic can be determined as the earliest time at which this interruption no longer prevents recombinants from appearing. Typical results from a cross involving several genetic markers are shown in Fig. 39.6a: by plotting the result graphically and extrapolating the curves to their intersect with the x axis, the *time of entry* of a particular marker can be determined accurately.

The different times of entry of each characteristic reflect their relative positions on the chromosome, with genes near to the origin of transfer of the F proplasmid having short times of entry and those further along the chromosome having later times of entry. The times of entry can be used to locate the genes on a chromosome map (Fig. 39.6b), representing map distances relative to the origin of transfer. This map differs from those

Genome mapping using interrupted conjugation – while this technique works well for widely spaced markers, it is less useful than transformation and transduction for closely linked genes.

produced from transformation and transduction analysis, since it is a transfer order map, rather than a linkage map. However, as transfer of the entire chromosome requires ≈100 minutes, it is difficult to map genes far from the origin of transfer, due to the decreasing number of recombinants with increasing time (Fig. 39.6a): this is overcome by combining mapping data from several Hfr strains, each with the F plasmid integrated at a different position. In fact, this approach was originally used to demonstrate the circularity of the *E. coli* chromosome and, subsequently, to determine the location of over 1500 individual genes in this organism.

Definitions

Trans (Latin 'across') – on separate DNA molecules.

Cis (Latin 'here') – on the same DNA molecule.

Genetic analysis using F′ plasmids

Since cells containing an F′ plasmid are stable merozygotes, their principal use is in studying the behaviour of genes under diploid conditions, for example:

- to determine whether a particular mutation (on the F′ plasmid) is dominant or recessive over another (on the chromosome);
- to test whether a particular gene complements another: if two mutations are in different genes, then they can complement each other in a stable merozygote, while two mutations in the same gene cannot complement each other in a stable merozygote;
- to examine whether a particular regulatory gene located on an F′ plasmid influences the expression of one or more chromosomal genes, i.e. whether it can operate in the *trans* position. The alternative test is to construct a merozygote with two genes in a chromosomal location (i.e. the *cis* position), for example, using specialized transduction (p. 224). This type of analysis has been used to investigate the molecular basis of gene expression and the function of regulatory regions of the chromosome.

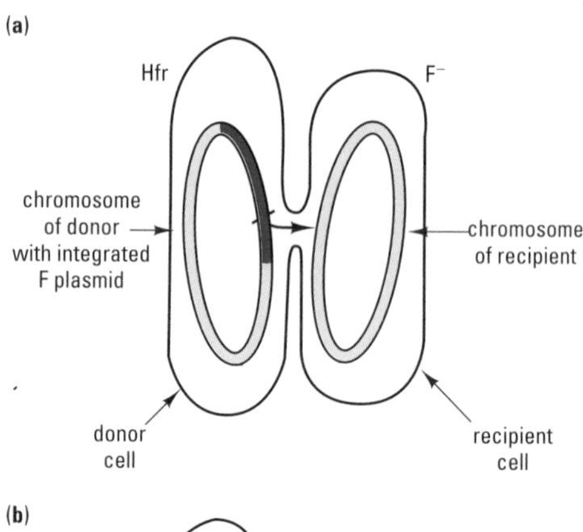

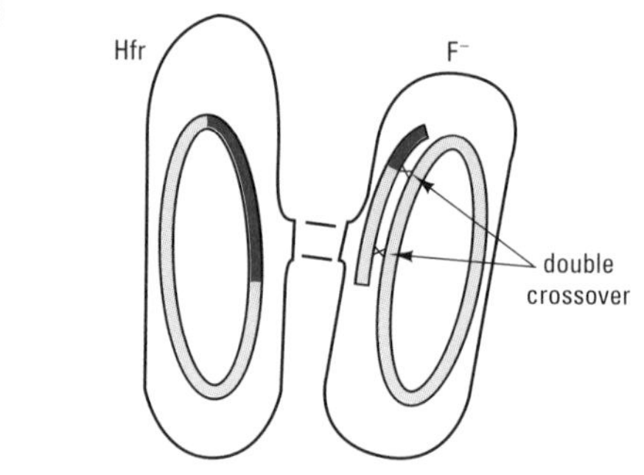

Fig. 39.5 Conjugation between Hfr donor and F⁻ recipient. The F proplasmid is cut and leads the chromosome (in single stranded form) from the donor (a), followed by homologous recombination in the recipient (b).

KEY POINT **Note that, while many of the above procedures have proved to be useful in the early stages of bacterial genome analysis, the techniques and methods of molecular biology (pp. 228–47) can be used to provide more detailed genetic information, without the need to obtain certain types of mutant, or carry out specific crosses between particular donors and recipients.**

Phage crosses

In practical classes, you may carry out experiments using phages, often using T-even phages of *E. coli* (e.g. T4 coliphage). The methods used to map the phage genome superficially resemble those used for eukaryotic organisms (Chapter 38) since the 'progeny' of two 'parental' phages with contrasting genotypes are analysed for recombinants and the frequency of recombination is used to measure the distance between linked genes.

KEY POINT **A phage cross (sometimes termed a 'mating') does not involve meiosis, gamete production or zygote formation, but is a result of a *mixed infection*, where a single bacterial cell is infected with phages of the two original genotypes.**

Performing phage crosses – large numbers of both types of phage must be used to ensure that every bacterial cell is infected with at least one of each type, i.e. there is a high multiplicity of infection.

Following a single round of replication, the 'progeny' from this mixed infection are screened for the various possible genotypes, as plaques on a 'lawn' of a susceptible strain of *E. coli*, using conventional phage culture techniques (p. 64).

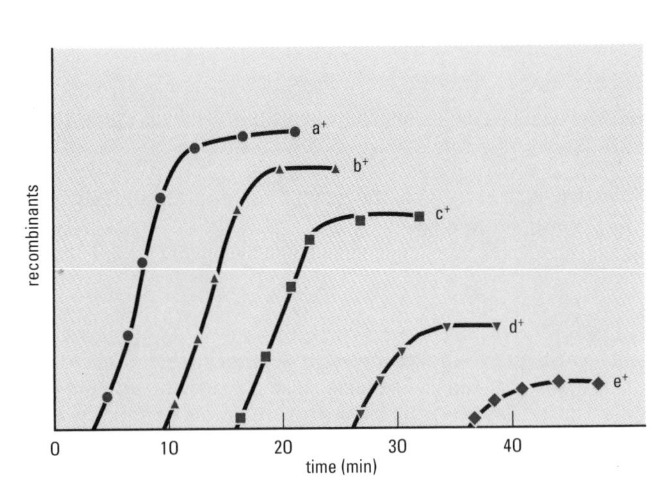

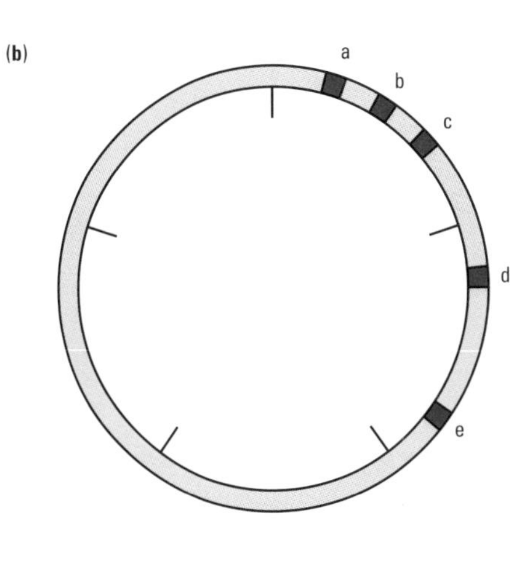

Fig. 39.6 Recombinants obtained from interrupted mating for a number of characters (a to e), plotted to show the time of entry of each character (a): the same data have been used to map the position of each character on the bacterial chromosome (b).

Phage mutants

The phage genes most widely used in crosses are those affecting:

- plaque morphology – for example, mutants may form large plaques as opposed to small plaques, or plaques with a different margin, e.g. a light turbid halo as opposed to a dark halo;
- host range – e.g. T-even mutants may infect different strains of *E. coli* to the parental types;
- conditional lethality, especially T4 phage mutants of the *rII* locus, unable to grow on *E. coli* K12(λ) in contrast to wild-type T4.

Phage mapping

By crossing a mutant for plaque morphology with a mutant for host range, the relative frequency of recombinants (i.e. progeny with double mutant and wild-type phenotypes) can be expressed in terms of the proportion of the total number of plaques, as a recombination frequency (R), where:

$$R = \frac{\text{number of double mutants and wild-type plaques}}{\text{total number of plaques}}$$

This value is sometimes multiplied by 100 and expressed as a percentage, or as map units, equivalent to 'centimorgans' (1 cM = 1% crossing over) in conventional Mendelian crosses (p. 218).

The pioneering work of Seymour Benzer established that the *rII* region of the T4 genome consisted of two distinct genes (*rIIA* and *rIIB*), each of which can give rise to the same mutant phenotype. T4 *rII* mutants are most often used for demonstrating *trans* complementation between these two genes and for demonstrating the principles of fine structure mapping of this region of the genome, since the conditional lethality of the rII phenotype simplifies the search for rare recombinant wild-type phages, as only these recombinants are able to form plaques on *E. coli* K12(λ).

Learning from Benzer's work – detailed fine structure mapping of the *rII* region of T4 coliphage established that recombination can occur within a single gene, and even between adjacent nucleotides. Before these experiments, a gene was regarded as an indivisible unit – nowadays we appreciate that the smallest unit in genetics is the base pair.

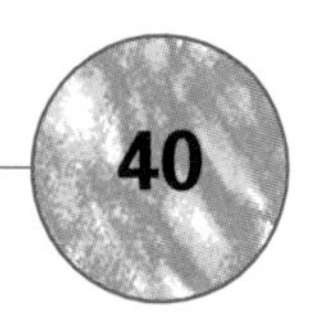

40 Molecular genetics I – fundamental principles

Deoxyribonucleic acid (DNA) is the genetic material of all cellular organisms. Its structure is outlined in Chapter 33.

KEY POINT **The sequence of the bases A, G, T and C carries the genetic information of the organism. A section of DNA that encodes the information for a single polypeptide or protein is referred to as a gene, while the entire genetic information of an organism is called the genome.**

Definitions

Units of nucleic acid size (length) –
Kilobase pair (kb) = 10^3 base pairs
Megabase pair (Mb) = 10^6 base pairs.

The amount of DNA in the genome is usually expressed in terms of base pairs (bp), rather than M_r, and its size depends on the complexity of the organism: for example, the human papilloma virus has a genome of 8×10^3 base pairs (8 kb), that of *E. coli* is 4×10^6 base pairs (4 Mb), while the human haploid genome is very large, comprising 3×10^9 base pairs (3000 Mb). The human genome contains about 100 000 genes (distributed on 23 pairs of chromosomes), which represent only about 5 per cent of the total amount of genomic DNA (i.e. 95 per cent of human DNA is non-coding). Organisms with smaller genomes have smaller amounts of non-coding DNA: some viral genomes have 'overlapping' genes, where the same base sequences carry information for more than one protein.

The size of each individual gene varies considerably: the largest ones may exceed 10 Mb. Chromosomes represent the largest organizational units of DNA: in eukaryotes, they are usually linear molecules, complexed with protein and RNA, varying in length from tens to hundreds of Mb. The unit used to denote physical distance between genes (base pairs, b) differs from that used to describe genetic distance (centimorgan, cM), which is based on recombination frequency (p. 218). In humans, 1 cM ≈ 1 Mb, though this relationship varies widely, depending on recombination frequency within particular regions of a chromosome.

Table 40.1 The Genetic Code – combinations of nucleotide bases coding for individual amino acids

	2nd Base				
1st Base	*U*	*C*	*A*	*G*	3rd Base
U	F	S	Y	C	*U*
	F	S	Y	C	*C*
	L	S	*	*	*A*
	L	S	*	W	*G*
C	L	P	H	R	*U*
	L	P	H	R	*C*
	L	P	Q	R	*A*
	L	P	Q	R	*G*
A	I	T	N	S	*U*
	I	T	N	S	*C*
	I	T	K	R	*A*
	M	T	K	R	*G*
G	V	A	D	G	*U*
	V	A	D	G	*C*
	V	A	E	G	*A*
	V	A	E	G	*G*

* = termination codons.
Standard abbreviations for the above amino acids are as given in Table 30.1.
Note that AUG (=M) is the initiation codon. The above codons are given for mRNA – the coding strand of DNA would have T in place of U, while the template strand would have complementary bases to those given above.

Each template for the synthesis of RNA (transcription) begins at a promoter site upstream of the coding sequence and terminates at a specific site at the end of the gene. The base sequence of this RNA is complementary to the 'sense strand' of the DNA. In eukaryotic cells, transcription occurs in the nucleus, where the newly synthesized RNA, or primary transcript, is also subject to processing, or 'splicing', in which non-coding regions within the gene (introns) are excised, joining the coding regions (exons) together into a continuous sequence. Further processing results in the addition of a polyadenyl 'tail' at the 3′ end and a 7-methylguanosine 'cap' at the 5′ end of what is now mature eukaryotic messenger RNA (mRNA). The mRNA then migrates from the nucleus to the cytoplasm, where it acts as a template for protein synthesis (translation) at the ribosome: the translated portion of mRNA is read in coding units, termed codons, consisting of three bases.

Each codon corresponds to a specific amino acid, including a codon for the initiation of protein synthesis (Table 40.1). Individual amino acids are brought to the ribosome by specific transfer RNA (tRNA) molecules that recognize particular codons. The amino acids are incorporated into the

Good practice in molecular genetics – this includes:

- accurate pipetting down to 1 μl, or less;
- steadiness of hand in sample loading;
- keeping enzyme solutions cold during use and frozen during storage;
- using sterile plasticware;
- using double-distilled water;
- wearing disposable gloves to avoid contamination.

growing polypeptide chain in the order dictated by the sequence of codons until a termination codon is recognized, after which the protein is released from the ribosome.

The techniques described in this and subsequent chapters are used widely in many aspects of molecular biology including the identification and characterization of genes, gene cloning and genetic engineering, medical genetics, and genetic fingerprinting.

KEY POINT **An important characteristic of nucleic acids is their ability to hybridize: two single strands with complementary base pairs will hydrogen bond (anneal) to produce a duplex, as in conventional double stranded (ds)DNA. This duplex can be converted to single stranded (ss)DNA (i.e. 'melted') by conditions that disrupt hydrogen bonding, e.g. raising the temperature or addition of salt, and then reannealed by lowering the temperature or by removal of salt.**

DNA can be easily purified and assayed (Chapter 33), and it is relatively stable in its pure form. Most of the problems of working with such a large biomolecule have been overcome by the following:

- using restriction enzymes (p. 243) to cut DNA precisely and reproducibly: mammalian DNA may yield millions of fragments, with sizes ranging from a few hundred base pairs to tens of kb, while small viral genomes may give only a few fragments;
- electrophoretic methods for the separation of DNA fragments on the basis of their sizes (p. 230);
- segments of DNA can be detected by 'probes' that specifically hybridize with the DNA of interest (p. 233);
- once separated, specific segments of DNA can be obtained in almost unlimited quantities by insertion into vectors and multiplication within suitable host cells (DNA cloning, Chapter 42);
- methods are now available for rapidly determining the base sequence of DNA segments (p. 235), together with strategies for combining information from several adjacent segments to give contiguous sequences representing large regions of a genome, from single genes to whole chromosomes (Chapter 49);
- specific target sequences of DNA within a genome can be amplified by more than a billion-fold by the polymerase chain reaction (PCR, see Chapter 41).

Producing DNA fragments

Using restriction enzymes in genetic engineering – those enzymes that cleave DNA to give single stranded 'sticky ends' are particularly useful in gene cloning (Chapter 42). Two different molecules of DNA cut with the same restriction enzyme will have complementary single stranded regions, allowing them to anneal as a result of hydrogen bonding between individual bases within these regions.

Fragments of DNA can be produced by mechanical shearing, but this is a rather haphazard process, and reproducible cleavage is better achieved by certain enzymes. Type II restriction endonucleases (commonly called restriction enzymes) recognize and cleave at a specific sequence of double stranded DNA (usually four or six base pairs), known as the restriction site. Each enzyme is given a code name derived from the name of the organism from which it is isolated, e.g. *Hin* dIII was the third restriction enzyme to be obtained from *Haemophilus influenzae* strain Rd (Fig. 40.1). Most restriction enzymes will cut each DNA strand at a slightly different position within the restriction site to produce short, single stranded regions known as cohesive ends, or 'sticky ends', as shown in Fig. 40.1. A few restriction enzymes cleave DNA to give blunt-ended fragments (e.g. *Hin* dII).

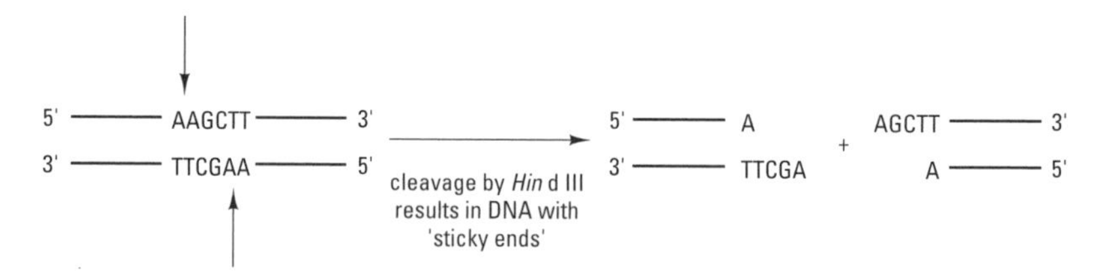

Fig. 40.1 Recognition base sequence and cleavage site for the restriction enzyme *Hin* dIII. This is the conventional representation of double-stranded DNA, showing the individual bases, where A is adenine, C cytosine, G guanine and T thymine. The cleavage site on each strand is shown by an arrow.

Separation of nucleic acids using gel electrophoresis

Separation of DNA by agarose and polyacrylamide gel electrophoresis

Maximizing recovery of DNA – these large molecules are easily damaged by mechanical forces, e.g. vigorous shaking or stirring during extraction. In addition, all glassware must be scrupulously cleaned and gloves must be worn, to prevent deoxyribonuclease contamination of solutions.

Electrophoresis is the term used to describe the movement of ions in an applied electrical field. DNA molecules are negatively charged, migrating through an agarose gel towards the anode at a rate which is dependent upon molecular size – smaller, compact DNA molecules can pass through the sieve-like agarose matrix more easily than large, extended fragments. Electrophoresis of plasmid DNA is usually carried out using a submerged agarose gel (Fig. 40.2). The amount of agarose is adjusted, depending on the size of the DNA molecules to be separated, e.g. 0.3% w/v agarose is used for large fragments (>20 000 bases) while 0.8% is used for smaller fragments. Very small fragments are best separated using a polyacrylamide gel (Table 40.2). Note the following:

- Individual samples are added to pre-formed wells using a pipettor. The volume of sample added to each well is usually less than 25 μl so a steady hand and careful dispensing are needed to pipette each sample.
- The density of the samples is usually increased by adding a small amount of sucrose, so that each sample is retained within the appropriate well.

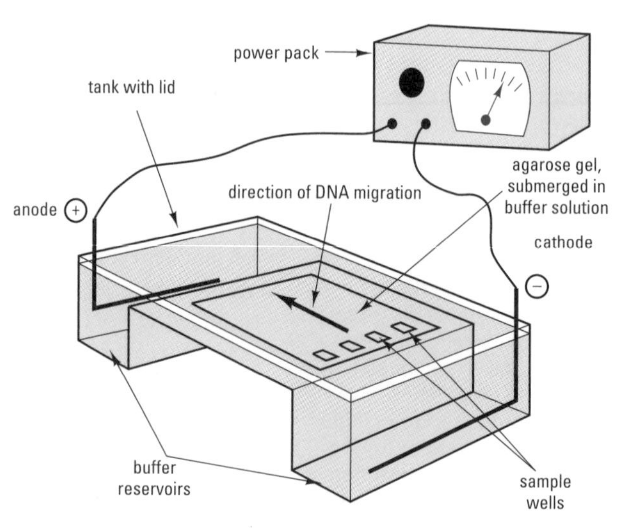

Fig. 40.2 Agarose gel electrophoresis of DNA.

Using molecular weight standards – for accurate molecular weight determination your standards must have the same conformation as the DNA in your sample, i.e. linear DNA standards for linear (restriction) fragments and closed circular standards for plasmid DNA.

Safe handling of ethidium bromide – ethidium bromide is carcinogenic so always use gloves when handling stained gels and make sure you do not spill any staining solution.

Table 40.2 Gel concentrations for the separation of DNA of various sizes.

Type of gel	% (w/v)	Range of resolution of DNA (bp)
Polyacrylamide	20.0	5–100
	15.0	20–150
	5.0	75–500
	3.5	100–1000
Agarose	2.0	100–5000
	1.2	200–8000
	0.8	400–20 000
	0.3	1000–70 000

Definition

Chromosome walking – a method for analysing areas of interest in DNA, in which the end of a segment of DNA is used as a probe to locate other segments that overlap the first segment: long stretches of DNA can be analysed by subsequent use of probes made from the ends of successive overlapping segments.

- A water-soluble anionic tracking dye (e.g. bromophenol blue) is also added to each sample, so that migration can be followed visually.
- Molecular weight standards are added to one or more wells. After electrophoresis, the relative positions of bands of known molecular weight can be used to prepare a calibration curve (usually, by plotting $\log_{10}$ of the molecular weight of each band against the distance travelled).
- The gel should be run until the tracking dye has migrated across 80% of the gel (see manufacturer's instructions for appropriate voltages/times). Note that the tank cover must be in position during electrophoresis, to prevent evaporation and to reduce the possibility of electric shock.
- After electrophoresis, the bands of DNA can be visualized by soaking the gel for around 5 min in 1 mg l^{-1} ethidium bromide, which binds to DNA by intercalation between the paired nucleotides of the double helix. An alternative approach is to incorporate the ethidium bromide into the agarose gel.
- Under UV light, bands of DNA are visible due to the intense orange–red fluorescence of the ethidium bromide. The limit of detection using this method is around 10 ng DNA per band. Suitable plastic safety glasses or goggles will protect your eyes from the UV light. The migration of each band from the well can be measured using a ruler. Alternatively, a Polaroid® photograph can be taken, using a special camera and adaptor.
- If a particular band is required for further study (e.g. a plasmid), the piece of gel containing that band is cut from the gel using a scalpel. The DNA can be separated from the gel by a variety of techniques, including phenol extraction: here the gel slice is dispersed in buffer, and on addition of phenol, the DNA remains in the upper aqueous layer, from which it can be precipitated with ethanol. If low melting point agarose is used, the gel slice can be heated to 70 °C, diluted with low salt buffer and, after cooling to 45 °C, the solution passed through an anion exchange column. The DNA is eluted at high salt concentration, and is then desalted, e.g. using a Sephadex G25 column (p. 133). Commercial kits, containing DNA-binding resins, offer an alternative approach.

Electrophoretic separation of RNA

Total cellular RNA or purified mRNA can be separated on the basis of size by electrophoretic separations similar to those used for DNA fragments. However, under the conditions used to separate dsDNA, RNA molecules tend to develop a secondary structure, and this leads to anomalous mobilities. To eliminate RNA secondary structure, samples are pre-treated by heating in dilute formamide or glyoxal, and electrophoresis is carried out in 'denaturing gels' which include buffers containing formaldehyde.

Pulsed field gel electrophoresis (PFGE)

If structural information is to be gained about large stretches of genomic DNA, then the order of the relatively short DNA segments (generated by the restriction enzymes described above) needs to be established. This is technically possible by techniques such as 'chromosome walking', but is time-consuming and potentially difficult, especially when dealing with large chromosomes such as those from yeast (a few Mb) and humans (50–100 Mb), in contrast to the smaller genomes of bacteria and viruses. The technique of PFGE allows separation of DNA fragments of up to ≈12 Mb.

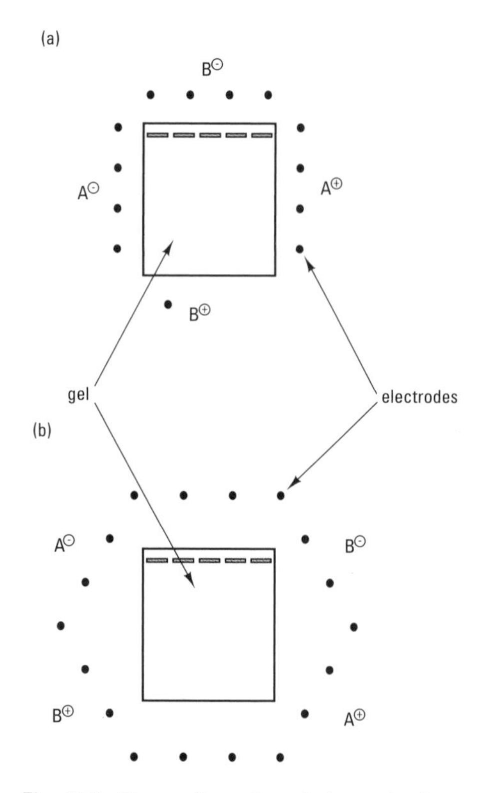

Fig. 40.3 The configuration of electrodes for conventional PFGE (a) and CHEF/PACE (b).

Very large DNA fragments (>100 kb) can be generated from chromosomal DNA by the use of certain restriction enzymes that recognize base sequences that are present at relatively low frequency, e.g. the enzyme *Not* I, which recognizes a sequence of 8 bp rather than 4–6 bp. These enzymes are sometimes called 'rare cutters'. Genomic DNA prepared in the normal way is not suitable for digestion by these enzymes, as shearing during extraction fragments the DNA. Therefore, genomic DNA for analysis by PFGE is prepared as follows:

- cells are embedded in an agarose block;
- the block is incubated in solutions containing detergent, RNase and proteinase K, lysing the cells and hydrolysing RNA and proteins. The products of RNase and proteinase digestion diffuse away, leaving behind genomic DNA molecules exceeding several thousand kb;
- the block is incubated *in situ* in a buffered solution containing an appropriate 'rare cutter': restriction fragments are produced, of up to ≈800 kb.

PFGE differs from conventional electrophoresis in that it uses two or more alternating electric fields. An explanation for the effectivess of the technique is that large DNA fragments will be distorted by the voltage gradient, tending to elongate in the direction of the electric field and 'snaking' through pores in the gel. If the original electric field is removed, and a second is applied at an angle to the first, the DNA must reorientate before it can migrate in the new direction. Larger (longer) DNA molecules will take more time to reorientate than smaller molecules, resulting in size-dependent separations.

The original configuration of electrodes used in PFGE is shown in Fig. 40.3a; this tends to produce 'bent' lanes that make lane-to-lane comparisons difficult. This can be overcome by using one of the many variants of the technique, one of which employs contour clamped homogeneous electric fields (CHEF). Here, multiple electrodes are arranged in a hexagonal array around the gel (Fig. 40.3b) and these are used to generate homogeneous electric fields with reorientation angles of up to 120°. A further development of CHEF involves programmable, autonomously controlled electrodes (PACE), which allows virtually unlimited variation of field and pulsing configurations, and can fractionate DNA molecules from 100 bp to 6 Mb.

Identification of specific nucleic acid molecules using blotting and hybridization techniques

Southern blotting

After separation by conventional agarose gel electrophoresis (p. 230), the fragments of DNA can be denatured and immobilized on a filter membrane using a technique named after its inventor, E.M. Southern. The main features of the conventional apparatus for Southern blotting are shown in Fig. 40.4. The principal stages in the procedure are:

1. The gel is soaked in alkali to denature the dsDNA to ssDNA, then neutralized.
2. A nitrocellulose membrane is then placed directly on the gel, followed by several layers of absorbent paper. The DNA is 'blotted' onto the filter as the buffer solution soaks into the paper by capillary action.
3. The filter is baked in a vacuum oven at 80 °C for 3–5 h in order to 'fix' the DNA.

Definition

Probe – a labelled DNA or RNA sequence used to detect the presence of a complementary sequence (by molecular hybridization) in a mixture of DNA or RNA fragments.

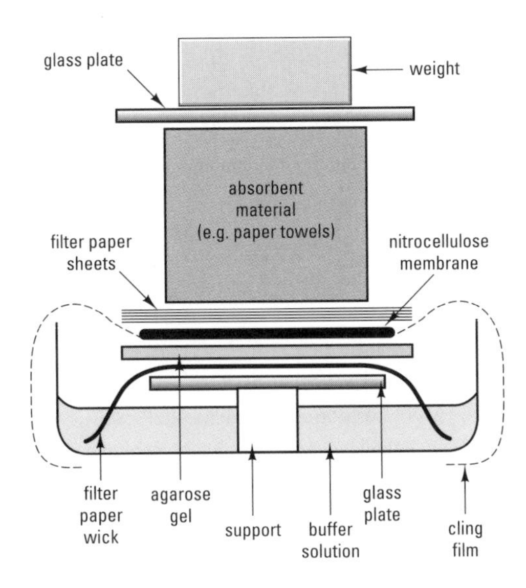

Fig. 40.4 Components of Southern blotting apparatus.

4. Specific DNA fragments are identified by incubation (6–24 h) with complementary labelled probes of ssDNA, which will hybridize with a particular sequence. If radiolabelled probes are used (see below), the desired fragments are located by autoradiography (p. 104).

More recent modifications of the method use either a vacuum apparatus or an electric field ('electroblotting') to transfer the DNA fragments from the gel to the membrane.

Northern blotting

This process is virtually identical to Southern blotting, but RNA is the molecule that is separated and probed.

Dot blotting or slot blotting

Here, samples containing denatured DNA or RNA samples are applied directly to the nitrocellulose membrane (via small individual round or slot-like templated holes) without prior digestion with restriction enzymes or electrophoretic separation. The 'blot' is then probed in a similar manner to that described for Southern blotting. This allows detection of a particular nucleic acid sequence in a sample – clinical applications include the detection of specific pathogenic microbes and the detection of particular genes.

Definition

Klenow fragment – part of the DNA polymerase I from *E. coli* that has effective polymerase activity, weak 3′ → 5′ exonuclease activity, but no 5′ → 3′ exonuclease activity.

Example For a simple tripeptide containing methionine, aspartate and phenylalanine, the synthetic oligonucleotide probes would include combinations of the following codons:
1st codon (met): ATG
2nd codon (asp): GAT, GAC
3rd codon (phe): TTT, TTC

Types of probe

The probes used in blotting and DNA hybridization can be obtained from a variety of sources including:

- cDNA (complementary or copy DNA) which is produced from isolated mRNA using reverse transcriptase. This retroviral enzyme catalyses RNA-directed DNA synthesis (rather than the normal transcription of DNA to RNA). After the mRNA has been reverse transcribed, it is degraded by alkali, leaving the ssDNA copy. This is then used as a template for a modified DNA polymerase (e.g. the 'Klenow fragment'), which directs the synthesis of the second complementary DNA strand to form dsDNA. This is denatured to ssDNA before use.
- Oligonucleotide probes (15–30 nucleotides) can be produced if the amino acid structure of the gene product is known. Since the genetic code is degenerate, i.e. some amino acids are coded for by more than one codon (see Table 40.1) it may be necessary to synthesize a mixture of oligonucleotides to detect a particular DNA sequence.
- Specific genomic DNA sequences, where the gene has been characterized.
- PCR-generated fragments (Chapter 41).
- Heterologous probes, i.e. sequences for the same gene, or its equivalent, in another organism.

Labelling of probes

Detection of very low concentrations of target DNA sequences requires probes that can be detected with high sensitivity. This is achieved by radiolabelling with ^{32}P or ^{35}S, or by using enzyme-linked methods, which are often available in kit form:

- Radiolabelled probes can be made by several methods, including the nick-translation technique. This uses DNA polymerase I from *E.coli*, which has (i) an exonuclease activity that 'nicks' dsDNA and removes a nucleotide, and (ii) a polymerase activity that can replace this with a ^{32}P-labelled deoxyribonucleotide. After several cycles of nick-translation, the labelled DNA is denatured to ssDNA for use as probes. Newer approaches include random hexamer priming of single stranded template DNA, followed by the synthesis of radiolabelled DNA fragments complementary to the template using a DNA polymerase and a radiolabelled dNTP. Probe hybridization to a target sequence is detected by autoradiography.

Using commercial kits – do not slavishly follow the protocol given by the manufacturer without making sure you understand the principles of the method and the reasons for the procedure. This will help you to recognize when things go wrong and what you might be able to do about it.

- Enzyme-linked methods involve incorporating a modified nucleotide precursor, such as biotinylated dTTP, into the DNA by nick-translation. When the probe hybridizes with the target sequence, it can be detected by addition of an enzyme (e.g. horse radish peroxidase) coupled to streptavidin. The streptavidin binds specifically to the biotin attached to the probe and the addition of a suitable chromogenic or fluorogenic substrate for the enzyme allows the probe to be located.

Hybridization of probes

The stability of the duplex formed between the probe and its target is directly proportional to the number and type of complementary base pairs that can be formed between them: stability increases with the amount of G + C, since these bases form three hydrogen bonds per base pair, rather than two (p. 180). Duplex stability is also influenced by temperature, ionic strength, and pH of the hybridization buffer, and these can be varied to suit the stringency of hybridization required:

- In 'low stringency' hybridization, duplex formation with less than perfect complementarity is promoted, either by lowering the temperature, or increasing ionic strength. This might be useful when using a heterologous probe, i.e. from another species.
- 'Stringent' hybridization conditions usually involve high temperatures, increased pH or decreased ionic strength, and will sustain only perfectly matched duplexes. An example of the use of these conditions would be for detecting a single base change in a mutant gene using a dot-blot hybridization with a specific oligonucleotide probe.

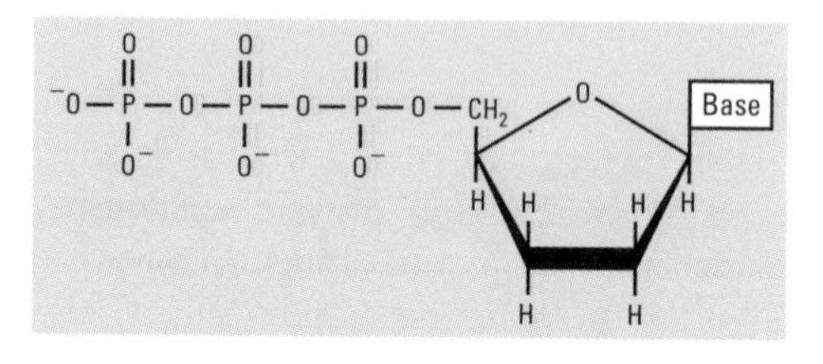

Fig. 40.5 A 2′ 3′ dideoxynucleoside triphosphate.

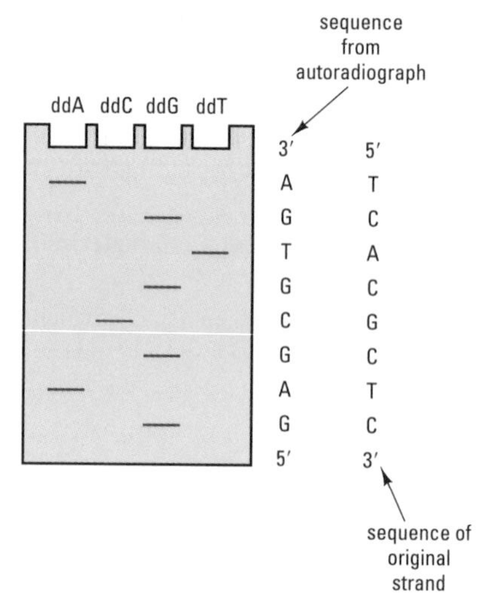

Fig. 40.6 Sanger sequencing gel, showing how the banding pattern is converted into a sequence of nucleotide bases.

DNA sequencing

By cutting a source of DNA with several restriction enzymes and then sequencing the overlapping fragments, it is possible to determine the nucleotide sequences of very large stretches of DNA, including entire genes and genomes. Sequencing methods rely on polyacrylamide gel electrophoresis (p. 145). The Sanger, or chain termination method is the most widely used in DNA sequencing. This makes use of dideoxynucleotides (Fig. 40.5), which have no –OH group at either the C-2 or C-3 of ribose. A dideoxynucleoside triphosphate (ddNTP) can be added to a growing DNA chain, but since it lacks an –OH group at the C-3 position it cannot form a phosphodiester bond with the next deoxynucleoside triphosphate (dNTP). Therefore a dideoxynucleotide acts as a terminator at the site it occupies. Details of the Sanger sequencing method are given in Box 40.1 and Fig. 40.6.

The alternative approach to sequencing is based on chemical degradation (the Maxam & Gilbert method), using different reagents to break the target DNA sequence into fragments at each point where a particular base

Box 40.1 DNA sequencing using the chain termination (Sanger) method

The DNA to be sequenced must be obtained as single stranded fragments, typically around 200 bp. This can be achieved by treating the DNA with a restriction enzyme, followed by denaturation to give single strands, which are purified by electrophoresis and cloned into a vector, e.g. M13 phage (Chapter 42). Alternatively, a denatured plasmid may be used. Each type of ssDNA fragment can be used as a template for the sequencing reaction. Sequencing is often performed using a commercial kit and the principal stages are as follows:

1. **Set up the strand synthesis reaction:** four separate tubes are required, each containing a small amount of one of the dideoxynucleoside triphosphates. Each tube contains all of the other components required for DNA synthesis, i.e. (i) the DNA selected for analysis (template DNA), (ii) an appropriate DNA polymerase (e.g. Sequenase®, or the Klenow fragment), (iii) a [^{35}S] oligonucleotide primer of known sequence, to allow synthesis of the complementary DNA strand, (iv) dGTP, dATP, dCTP, dTTP (in excess) together with (v) one type of ddNTP (i.e. ddGTP, ddATP, ddCTP or ddTTP) in limited concentration. New DNA strands are synthesized by addition of dNTPs to the primer, guided by the template DNA until a ddNTP is added. As an example of the principle of the method, consider the ddTTP tube. When T is required to pair with A on the template DNA strand, the dTTP will be competing with the ddTTP, but because the dTTP is in excess this will normally be added to the chain at the appropriate position. However on occasions a ddTTP will be inserted at a given site and this will terminate DNA synthesis on the template strand. Thus, synthesis will be halted at all possible sites where ddT has substituted for dT and several strands of different length will be formed in the reaction mixture, each ending with ddT.
2. **Terminate the strand synthesis reaction:** after 5–10 min at 37 °C add a 'stop' solution containing formamide to disrupt hydrogen bonding between complementary bases and incubate at 80 °C for 15 min, to produce ssDNA.
3. **Separate the fragments using polyacrylamide gel electrophoresis:** 4–6% (w/v) denaturing acrylamide gels are used, containing urea (46% w/v) as the denaturing agent. Thin gels (0.35 mm thickness) are used, typically 20 cm wide by 50 cm long. The products in each of the four tubes are placed in four lanes side-by-side on the same gel and are separated by electrophoresis at 35–40 W (up to 32 mA, 1.5 kV) for 2.5 h. The high voltage raises the gel temperature to ≈50 °C, helping denaturation. A single gel can separate DNA fragments that differ in length by a single nucleotide.
4. **Locate the positions of individual bands by autoradiography:** the gel is fixed for 15 min with 10% (v/v) acetic acid, covered with Whatman 3MM paper and Saran wrap, then vacuum dried, to avoid quenching the radioisotope signal (p. 103) and to prevent it from sticking to the X-ray film. The gel is unwrapped and placed next to an X-ray film for 24 h.
5. **Read the gel:** the nucleotide sequence can be read directly from the band positions that represent the newly made DNA segments of varying lengths (Fig. 40.6). The smallest segment is represented by the band at the bottom, since it travels furthest in the gel. The nucleotide sequence of the template strand can be deduced directly from that of the new strand, since the base on the template strand will be represented by an incomplete chain that terminates with the complementary dd nucleotide, i.e. template A with ddT, G with ddC, T with ddA, and C with ddG (Fig. 40.6).

Troubleshooting and other points to note: streaking can be due to damage to wells, air bubbles in wells/gel, or contamination by dirt/dust; faint and fuzzy gels may be due to insufficient template DNA, primer and/or dNTPs, due to poor annealing of primer and template, or to errors in preparing or running the gel; bands in more than one lane can indicate contamination of the template DNA, more than one primer site on the template, or secondary structure of DNA, giving 'ghost' banding.

occurs. These fragments are then separated by polyacrylamide gel electrophoresis and read in a broadly similar way to the Sanger method. The procedure is more involved than the chain termination method and the reagents are more hazardous. As a result, chemical degradation is reserved for particular, specialized applications, e.g. for sequencing DNA containing long runs of a single nucleotide, or for studying the interaction between DNA and proteins.

Automated DNA sequencing

The development of automated DNA sequencing machines (so-called 'DNA sequenators') has enabled sequencing to be performed ten times faster than with manual methods. Base-specificity is achieved by using primers labelled with fluorophores with different fluorescence characteristics in each of the four reaction tubes (Box 40.1). After the reactions are separately completed, the four sets of products can be pooled and fractionated by electrophoresis in a single lane. The fluorophore-labelled fragments are monitored as they pass the detector and the DNA sequence is determined by using both the specific wavelength emitted by the fragment (indicating the base) and by the migration time (indicating the fragment size, which corresponds to the base location in the DNA sequence). About 8000 nucleotides per day can be sequenced by current automated sequencing machines: future developments in capillary electrophoresis (p. 152) are likely to give increased efficiency.

Using PCR (p. 237) for sequencing – with the four ddNTPs tagged using different fluorophores, the products can be run in the same lane and detected by laser excitation at the end of the gel, with the sequence of fluorophores giving the base sequence.

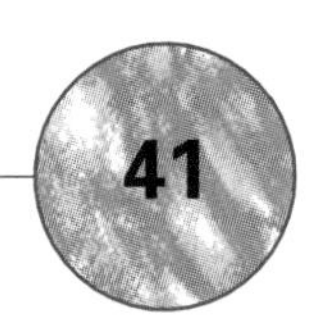

41 Molecular genetics II – PCR and related applications

Applications of PCR – these include:

- diagnosis and screening of genetic diseases and cancer;
- rapid detection of slowly growing microorganisms (e.g. mycobacteria) and viruses (e.g. HIV);
- HLA typing in transplantation;
- analysis of DNA in archival material;
- DNA fingerprinting in forensic science;
- preparation of nucleic acid probes;
- clone screening, mapping and subcloning (p. 242).

The polymerase chain reaction (PCR) is a rapid, inexpensive, and simple means of producing μg amounts of DNA from minute quantities of template.

KEY POINT **PCR offers an alternative approach to gene cloning (Chapter 42) for the production of many copies of an identical sequence of DNA. The starting material may be genomic DNA (e.g. from a single cell or a chromosome), RNA, DNA from archival specimens, cloned DNA, or forensic samples.**

The technique uses *in vitro* enzyme-catalysed DNA synthesis to create millions of identical copies of DNA. If the base sequence of the adjacent regions of the DNA to be amplified is known, this enables synthetic oligonucleotide primers to be constructed that are complementary to these so-called 'flanking regions'. Initiation of the PCR occurs when these primers are allowed to hybridize (anneal) to the component strands of the target DNA, followed by enzymatic extension of the primers (from their 3′ ends) using a thermostable DNA polymerase. A single PCR cycle consists of three distinct steps, carried out at different temperatures (Fig. 41.1), as follows:

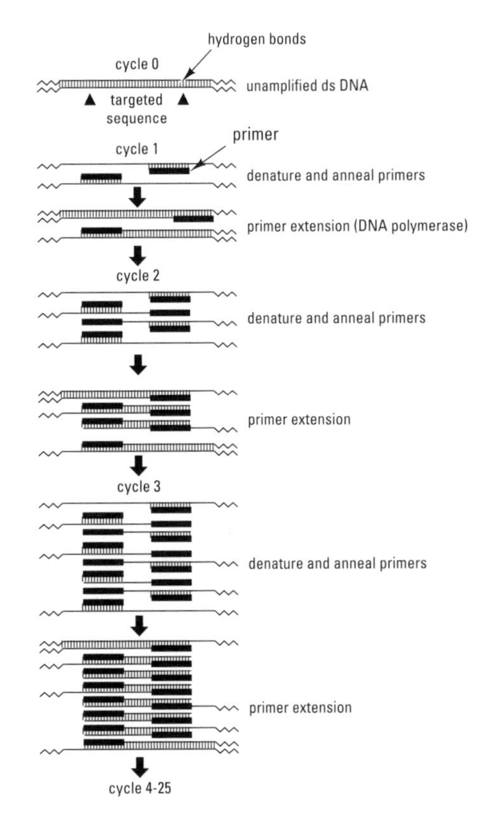

Fig. 41.1 The polymerase chain reaction (PCR).

1. Denaturation of dsDNA by heating to 94–98 °C separating the individual strands of the target DNA.
2. Annealing of the primers, which occurs when the temperature is reduced to 37–65 °C.
3. Extension of the primers by a thermostable DNA polymerase (e.g. *Taq* polymerase, isolated from *Thermus aquaticus*) at 72 °C; this step should last long enough to generate a product exceeding several hundred base pairs (e.g. 1 min).

In the first cycle, the product from one primer is extended beyond the region of complementarity of the other primer, so each newly synthesized strand can be used as a template for the primers in the second cycle (Fig. 41.1). Successive cycles will thus generate an exponentially increasing number of DNA fragments, the termini of which are bounded by the 5′ ends of the primers (length of each fragment = length of primers + length of target sequence). Since the amount of DNA produced doubles in each cycle, the amount of DNA produced = 2^n, where 'n' is the number of cycles. Up to 1 μg of amplified target DNA can be produced in 25–30 cycles from a single-copy sequence within 50 ng of genomic DNA, assuming close to 100% efficiency during the cycling process. After electrophoresis, the PCR product is normally present in sufficient quantity to be visualized directly with ethidium bromide (p. 231), rather than using labelled probes. This is a major advantage over Southern blotting.

The temperature changes in PCR are normally achieved using a thermal cycler, which is simply a purpose-built incubator block that can be programmed to vary temperatures, incubation times and cycle numbers.

Carrying out PCR with limited resources – if a thermal cycler is not available, three water baths, a stopclock and manual transfer could be used!

Storing primers for PCR – primers are best stored in ammonia solution, which remains liquid at −20 °C, avoiding the need for repeated freezing and thawing when dispensed. Before use, aliquots of stock solution should be heated in a fume cupboard, to drive off the ammonia.

Using dNTP solutions – make up stock solutions, pool in small volumes (50–100 μl of each dNTP) and store separately at −20 °C.

PCR components and conditions

These are readily available in kit form from commercial suppliers; they include *Taq* polymerase, dNTPs, buffer, non-ionic detergent, KCl, $MgCl_2$, gelatin or bovine serum albumin, primers, and target DNA. A complete PCR kit is available from Perkin-Elmer/Cetus (the Gene-Amp® kit), and this is recommended for beginners. Protocols for PCR vary considerably for particular applications – a typical procedure is given in Box 41.1. However, if you are trying to develop your own PCR procedure, the following information may be useful:

- Primers need only be 18–30 nucleotides long (primers longer than 18 nucleotides should be unique in a complex eukaryotic genome); both primers should have similar G+C content (so that they have similar annealing temperatures, p. 184), with a minimal degree of self-complementarity (to avoid formation of secondary structures), and no complementarity to each other (so that primer duplexes are not formed). For most applications, the final concentration of each primer should be 0.1 μmol l^{-1}, which gives an excess of primers of about 10^7 with respect to the template, e.g. target genomic DNA at a concentration of 50 ng per 10 μl reaction mixture.
- Annealing temperatures are based around T_m, the temperature at which 50% of the primers are annealed to their target sequence (Fig. 33.6). For primers of <20 bases, T_m can be roughly calculated in °C from the equation:

$$T_m = 4(G+C) + 2(A+T) \qquad [41.1]$$

  where G, C, A, and T are the number of bases in the primer. Using this as a starting point, the optimum annealing temperature can be determined by trial and error.
- dNTPs should be used at equal concentrations of 200 μmol l^{-1}, which should provide the initial excess required for incorporation into DNA.
- One of the key variables in PCR is the Mg^{2+} concentration; Mg^{2+} is required as a cofactor for the thermostable DNA polymerase. Excess Mg^{2+} stabilizes dsDNA and may prevent complete denaturation of product at each cycle; it also promotes spurious annealing of primers, leading to the formation of undesired products. However, very low Mg^{2+} concentration impairs polymerization. The Mg^{2+} concentration that gives optimal yield and specificity should be determined by trial and error. The purpose of gelatin and Triton X-100 is to stabilize the DNA polymerase during thermal cycling.
- The most frequently used thermostable DNA polymerase is *Taq* polymerase, which extends primers at a rate of 2–4 kb per min at 72 °C. It should be used at a concentration of $\geqslant$1 nmol l^{-1} ($\geqslant$0.1 U per 5 μl reaction mixture). A disadvantage of *Taq* polymerase is that it has a relatively high rate of misincorporation of bases (one aberrant nucleotide per 100 000 nucleotides per cycle). Other polymerases are available (e.g. VENT polymerase from *T. litoralis*, or genetically modified forms of *Taq* polymerase), which have lower misincorporation rates.

PCR variations

Nested PCR

This can be used when the target sequence is known, but the number of DNA copies is very small (e.g. a single DNA molecule from a microbial genome), or if the sample is degraded (e.g. a forensic sample). The process

Box 41.1 How to carry out the polymerase chain reaction (PCR)

The protocol given below is typical for a standard PCR. Note that temperatures, incubation times, and the number of cycles will vary with the particular application, as discussed in the text.

1. **Make sure you have the required apparatus and reagents to hand,** including: (i) a thermal cycler, or water baths at three different temperatures; (ii) genomic DNA ($\geqslant$50 ng/μl); (iii) stock solution of all dNTPs (5 mmol l^{-1} for each dNTP); (iv) *Taq* polymerase (at 5 U/μl); (v) primers at 10 μmol l^{-1}; (vi) stock buffer solution, e.g. containing 100 mmol l^{-1} TRIS (pH 8.4), 500 mmol l^{-1} KCl, 15 mmol l^{-1} $MgCl_2$, 1% (w/v) gelatin, 1% (v/v) Triton X-100 (this stock is often termed '10x PCR buffer stock').
2. **Prepare a reaction mixture:** for example, a mixture containing: 1.0 μl genomic DNA; 2.5 μl stock buffer solution; 1.0 μl primer 1; 1.0 μl primer 2; 1.0 μl of each of the stock solutions of dNTPs, 0.1 μl *Taq* polymerase; 15.4 μl distilled deionized water, to give a total volume of 25.0 μl.
3. **Use appropriate positive and negative controls:** a positive control is a PCR template that is known to work under the conditions used in the laboratory, e.g. a plasmid, with appropriate primers, buffer, etc. A commonly used negative control is the PCR mixture minus the template (genomic) DNA, though negative controls can be set up lacking any one of the reaction components.
4. **Cycle in the thermal cycler:** for example, an initial period of 5 min at 94 °C, followed by 30 cycles of 94 °C for 1 min (denaturation), then 60 °C for 1 min (primer annealing), then 72 °C for 1 min (chain extension).
5. **Assess the effectiveness of the PCR:** for example, by gel electrophoresis and ethidium bromide staining.

Troubleshooting

- If no PCR product is detected, repeat the procedure, checking carefully that all components are added to the reaction mixture. If there is still no product, check that the annealing temperature is not too high, or the denaturing temperature is not too low. The detection method must be sensitive enough (ethidium bromide detects $\approx$10 ng DNA).
- If too many bands are present, this may indicate that either (i) the primers may not be specific, (ii) the annealing temperature is too low, (iii) too many cycles have been used, or (iv) there is an excess of Mg^{2+}, dNTPs, primers or enzyme.
- Bands corresponding to primer-dimers indicate that either (i) the 3′ ends of the primers show partial complementarity, (ii) the annealing temperature is too low to encourage specific annealing to the template, or (iii) the concentration of primers is too high.

Avoiding contamination in diagnostic PCR

The sensitivity of PCR is also the major drawback, since the technique is susceptible to contamination, particularly from DNA from the skin and hair of the operator, from previous PCR products, from airborne microbes and from positive control plasmids. A number of routine precautions can be taken to avoid such contamination:

- use a laminar flow cabinet (p. 61) dedicated to PCR use and located in a separate lab from that used to store PCR products or prepare clones;
- keep separate supplies of pipettors, tips, microfuge tubes and reagents – these should be exclusive to the PCR, with separate sets for sample preparation, reagents and product analysis;
- autoclave all buffers, distilled deionized water, pipette tips and tubes;
- wear disposable gloves at all times and change them frequently: protective coverings for the face and hair are also advisable;
- avoid contamination due to carry-over by including dUTP in the PCR mixture instead of dTTP. Thus copies will contain U rather than T. Before the template denaturing step, treat the mixture with uracil-N-glycosylase (UNG, available commercially as AmpEase®): this will destroy any strands containing U, i.e. any strands carried over from a previous reaction, or any contaminating material from another PCR. The target DNA will contain T, rather than U and will not be degraded by UNG. At the first heating step, the UNG will be denatured, so any newly synthesized U-containing copies will remain intact.

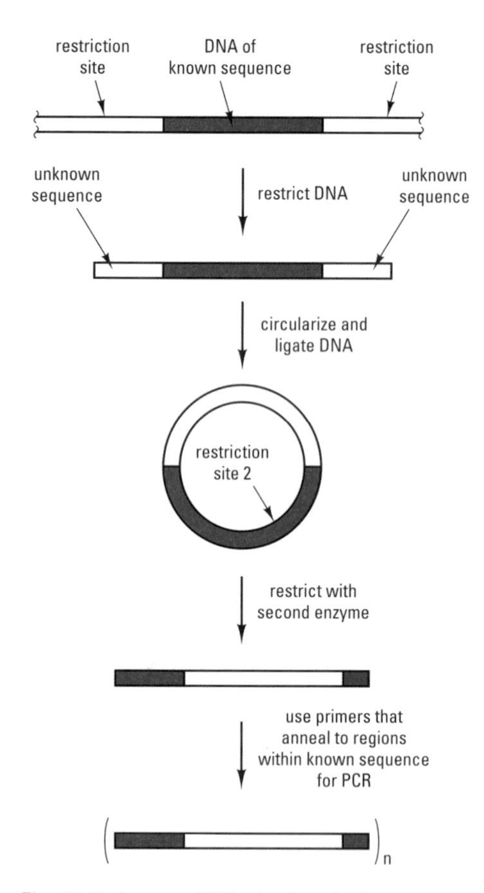

Fig. 41.2 Inverse PCR – basic principles.

involves two consecutive PCRs, each of around 25 cycles. The first PCR uses so-called 'external' primers, and the second PCR uses two 'internal' (or 'nested') primers that anneal to sequences within the product of the first PCR. This allows around 50 cycles of amplification (this would be error prone if conducted as a single PCR). Furthermore, nested PCR can be up to 1000-fold more sensitive than 50 cycles of the basic PCR, possibly because the smaller DNA fragments used in the second PCR are more effectively denatured than those in the first PCR. Nested PCR is often used to increase the specificity of the reaction, since external primers used on their own may give a reasonable yield but several bands, while the use of internal primers ensures that a unique sequence is amplified, e.g. in microbial diagnostics.

Inverse PCR

This is a useful technique for amplifying a DNA sequence flanking a region of known base sequence (Fig. 41.2), e.g. to provide material for characterizing an unknown region of DNA. The DNA is cut with a restriction enzyme so that both the region of known sequence and the flanking regions are included. This restriction fragment is then circularized and cut with a second restriction enzyme with specificity for a region in the known sequence. The now linear DNA will have part of the known sequence at each terminus, and by using primers that anneal to these parts of the known sequence, the unknown region can be amplified by conventional PCR. The product can then be sequenced and characterized (Chapter 40).

Reverse transcriptase-PCR (RT-PCR)

This technique is useful for detecting cell-specific gene expression (as evident by the presence of specific mRNA) when the amount of biological material is limited. It involves the construction of a PCR-directed cDNA library (p. 246) of mRNA purified from the tissue. Using either an oligo-dT primer to anneal to the 3′ polyadenyl 'tail' of the mRNA, or random hexamer primers, together with reverse transcriptase, cDNA is produced which is then amplified by PCR. The product is cloned, and the cDNA to the mRNA of interest is identified and quantified using an appropriate probe and blotting technique (Chapter 40). RT-PCR is often a useful method of generating a probe, the identity of which can be confirmed by sequencing (p. 235).

Overlap PCR

This technique can be used to join two PCR products together. The PCR is first set up using four primers, the central two having a small amount of overlap. After cycling, the products are added together and they will be able to anneal at the site equivalent to the overlap of the central two primers. DNA polymerase can then be used to synthesize the remainder of each complementary strand, giving a single DNA duplex for the combined region.

Amplification fragment length polymorphism (AFLP)

This term refers to several closely related techniques in which a single oligonucleotide primer of arbitrary sequence is used in a PCR reaction under conditions of low stringency, so that the primer is able to anneal to a large number of different sites within the target DNA. Some of the multiple amplification products will be polymorphic (e.g. the presence or absence of a particular annealing site will result in presence or absence of a particular band on the gel, after PCR and electrophoresis). Such polymorphisms can be used to detect differences between dissimilar DNA sequences.

Example Random amplification of polymorphic DNA (RAPD-PCR) can be used as a form of 'molecular typing' (p. 72), to identify a particular strain of an organism, e.g. in tracing the route of transmission of a pathogenic microbe.

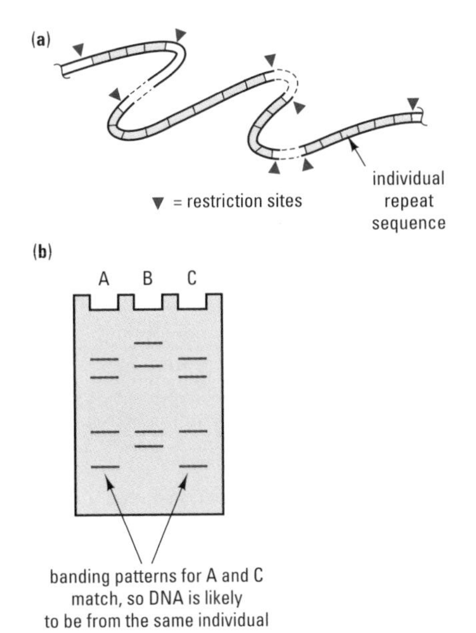

Fig. 41.3 Simplified representation of DNA fingerprinting: (a) tandem repeat sequences within DNA are cut using a restriction enzyme to yield fragments of different size; (b) these fragments are then separated by agarose gel electrophoresis on the basis of their size (M_r), giving a banding pattern that is a characteristic of the DNA used.

PCR and DNA fingerprinting techniques

Genetic mapping and sequencing studies have led to the discovery of highly variable regions in the non-coding regions of DNA between different individuals. These hypervariable regions, often termed 'minisatellite DNA', are found at many sites throughout the genome. Each minisatellite contains a defined sequence of nucleotides which is repeated a number of times in a tandem fashion (Fig. 41.3); the greater the number of repeats, the longer the minisatellite. The number of tandem repeats in any particular minisatellite varies from one person to another (i.e. they have a variable number of tandem repeats or VNTRs). This is exploited in identifying individuals on the basis of their DNA profile, by carrying out the following steps:

1. Extraction of DNA from cells of the individual (e.g. leukocytes, buccal cells, spermatozoa, etc.).
2. Digestion of the DNA with a restriction enzyme that cuts at sites other than those within the minisatellite, to produce a series of fragments of different M_r.
3. Electrophoresis and Southern blotting of the restriction fragments using probes that are specific to the particular minisatellite.

The size of the fragments identified will depend on the number of minisatellites that each fragment contains, and the pattern obtained in the Southern blot is characteristic of the individual being profiled (Fig. 41.3).

PCR is widely used in DNA profiling in circumstances where there is a limited amount of starting material. By selecting suitable primers, highly variable regions can be amplified from very small amounts of DNA, and the information from several such regions is used to decide, with a very low chance of error, whether any two samples of DNA are from the same individual or not, e.g. in forensic science (see Kirby, 1990).

KEY POINT **You should note that the PCR technique is continually being updated, with subtle new variations being produced on a regular basis – novel approaches and novel acronyms are very likely to be reported within the lifetime of this book.**

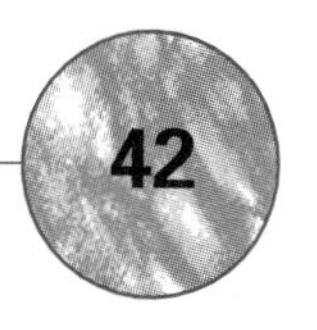

42 Molecular genetics III – genetic engineering techniques

Advances in the procedures used to manipulate nucleic acids *in vitro* have increased our understanding of the structure and function of genes at the molecular level. Additionally, these techniques can be used to alter the genome of an organism (genetic engineering), e.g. to create a bacterium capable of synthesizing a foreign protein such as a potentially useful hormone or vaccine component, or an unusual variant of a normal protein. The procedures are often termed 'recombinant DNA technology', since they involve the creation of novel combinations of DNA (i.e. recombinant DNA) under controlled laboratory conditions.

In the UK, the **Genetic Manipulation Regulations (1989)** provide the regulatory framework for all research procedures involving the genetic modification of organisms.

KEY POINT **While genetic engineering must be carried out under strict containment, in accordance with appropriate legislation, the procedures involved in the isolation, recombination and cloning of DNA are often used at undergraduate level, to illustrate the general features of the techniques.**

Basic principles

Genetic engineering involves several steps:

1. Isolation of the DNA sequence (gene) of interest from the genome of an organism, or from a gene library. This usually involves DNA purification followed by enzymic digestion or mechanical fragmentation, to liberate the target DNA sequence.
2. Creation of an artificial recombinant DNA molecule (sometimes referred to as rDNA), by inserting the (foreign) gene into a DNA molecule capable of replicating in a host cell, i.e. a 'cloning vector'. Suitable cloning vectors for bacterial cells include plasmids (p. 244) and bacteriophages (p. 245).
3. Introduction of the recombinant DNA molecule into a suitable host, e.g. *E. coli*. The process is termed transformation when a plasmid is used or transfection for a recombinant virus vector.
4. Selection and growth of the transformed (or transfected) cell, using the techniques of cell culture (p. 59). Since a single transformed host cell can be grown to give a clone of genetically identical cells, each carrying the gene of interest, the technique is often referred to as 'gene cloning', or molecular cloning.

Definition

Sub-cloning – repeated cycles of the cloning procedure, to isolate and characterize smaller portions of a particular DNA sequence, e.g. during the search for a specific gene.

Extraction and purification of plasmid DNA

Specific details of the steps involved in the isolation of DNA vary, depending upon the source material. However, the following sequence shows the principal stages in the purification of plasmid DNA from bacterial cells.

Preparing glassware – all glassware for DNA purification must be siliconized, to prevent adsorption of DNA. All glass and plastic items must be sterilized before use.

1. Cell wall digestion: incubation of bacteria in a lysozyme solution will remove the peptidoglycan cell wall. This is often carried out under isotonic conditions, to stop the cells from bursting open and releasing chromosomal DNA. Note that Gram-negative bacteria are relatively insensitive to lysozyme, requiring additional treatment to allow the

enzyme to reach the cell wall layer, e.g. osmotic shock, or incubation with a chelating agent, e.g. ethylenediaminetetra-acetate (EDTA). The latter treatment will also inactivate any bacterial deoxyribonucleases (DNases) in the solution, preventing enzymic degradation of plasmid DNA during extraction.

2. Lysis using strong alkali (NaOH) and a detergent, e.g. sodium dodecyl sulphate, to solubilize the cellular membranes and partially denature the proteins. Neutralization of this solution (e.g. using potassium acetate) causes the chromosomal DNA to aggregate as an insoluble mass, leaving the plasmid DNA in solution.
3. Removal of other macromolecules, particularly RNA and proteins by, for example, enzymic digestion using ribonuclease and proteinase. Additional chemical purification steps give further increases in purity, e.g. proteins can be removed by mixing the extract with water-saturated phenol (50% v/v), or a phenol/chloroform mixture. On centrifugation, the DNA remains in the upper aqueous layer, while the proteins partition into the lower organic layer. Repeated cycles of phenol/chloroform extraction can be used to minimize the carry-over of these macromolecules. Additional purification can be obtained using isopycnic density gradient centrifugation in CsCl (p. 124).
4. Precipitation of DNA using around 70% v/v ethanol:water (produced by adding two volumes of 95% v/v ethanol to one volume of aqueous extract), followed by centrifugation, to recover the DNA pellet. Further rinsing with 70% v/v ethanol:water will remove any salt contamination from the previous stages. The extracted DNA can then be either frozen, for future use, or redissolved in buffer solution.

Safe working practices – note that phenol is toxic and corrosive while chloroform is potentially carcinogenic (p. 12). Take appropriate safety precautions (e.g. wear gloves, extinguish all naked flames, use a fume hood, where available).

Extracting DNA from organisms other than bacteria – different homogenization techniques will be required (see Chapter 17). DNA can be extracted from the resulting homogenate by (i) complexation with the detergent cetrymethylammonium bromide (CTAB), or (ii) anion-exchange column chromatography (p. 129).

The simplest approach to quantifying the amount of nucleic acid in an aqueous solution is to measure the absorbance of the solution at 260 nm using a spectrophotometer, as detailed on p. 182. Note that the A_{260} value applies to purified DNA, whereas a plasmid DNA extract prepared using the protocol described above will contain a substantial amount of contaminating RNA, with similar absorption characteristics to DNA. Any contaminating protein would also invalidate the calculation; the protein can be detected by measuring the absorbance of the solution at 280 nm. Purified nucleic acids have a value for A_{260}/A_{280} of 1.8–2.0 and contaminating protein will give a lower ratio. If your solution gives a ratio substantially lower than this, you should repeat the phenol/chloroform extraction steps (see above).

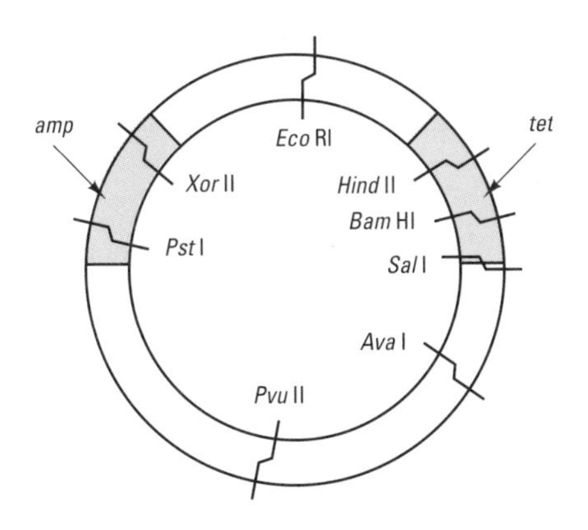

Fig. 42.1 Restriction map of the plasmid pBR 322. The position of individual restriction sites is shown together with the genes for ampicillin resistance (*amp*) and tetracycline resistance (*tet*).

Enzymatic manipulation of DNA

Restriction enzymes

Type II restriction endonucleases can be used to produce linear fragments of DNA with single stranded 'sticky ends' (Fig. 40.1): most commercial preparations of restriction enzymes will completely cleave a sample of DNA, producing a 'restriction digest', within 1 hour at 37 °C. The position of individual restriction sites can be used to create a diagnostic restriction map for a particular molecule, e.g. a plasmid cloning vector (Fig. 42.1). An important additional feature is that, under appropriate conditions, any two restriction fragments cut with the same enzyme (Fig. 42.2) can anneal (base pair), due to the formation of hydrogen bonds between individual bases within this region, allowing them to be joined together (ligated), irrespective of the sources of the two restriction fragments.

Definition

Type II restriction endonucleases – intracellular enzymes, produced by certain strains of bacteria: their function is to restrict the growth of phages within the cell by cutting phage DNA at particular sites (host DNA is protected by methylation at these sites).

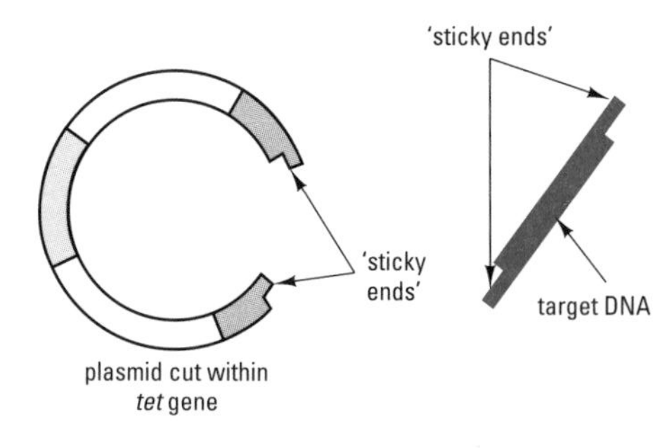

Fig. 42.2 Restriction of plasmid and foreign DNA with *Hin* dIII.

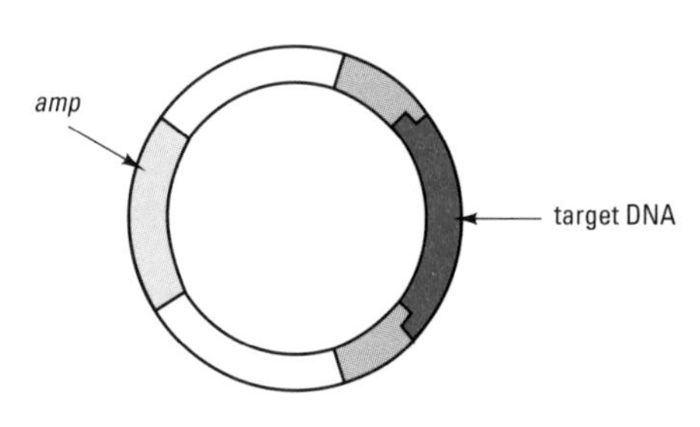

Fig. 42.3 Annealing and ligation of plasmid and target DNA to give a recombinant plasmid which confers resistance to ampicillin only. As target DNA has been inserted within the *tet* gene, the gene is now discontinuous and inactive.

Understanding the nomenclature for plasmid vectors – pBR322 is named according to standard rules: 'p' indicates the vector is a plasmid, 'BR' identifies the researchers Bolivar and Rodriguez and '322' is the specific code given to this plasmid, to distinguish it from others developed by the same workers.

DNA ligase

To construct a recombinant DNA molecule, a suitable vector must be ligated to the DNA fragment to be cloned: this is performed using another microbial enzyme, DNA ligase (usually obtained from T4-infected *E. coli*). This ATP-dependent enzyme is capable of forming covalent phosphodiester bonds between annealed DNA molecules, thereby creating recombinant DNA (Fig. 42.3). When the two molecules involved are the cloning vector and the target DNA, the size of the recombinant molecule can be predicted (e.g. a plasmid of 4.5 kb, plus a target DNA fragment of 2.5 kb will give a recombinant molecule of 7 kb), allowing separation and recovery by agarose gel electrophoresis (p. 230). Ligation is usually carried out at lower temperatures (to encourage annealing), over an extended time period of several hours (to allow the enzyme to operate), e.g. an overnight incubation at $<16\,^{\circ}\mathrm{C}$. The volume of the ligation mixture is kept as low as possible (typically, $<10\,\mu\mathrm{l}$), with approximately equimolar amounts of vector and target DNA fragments, to encourage annealing between the two different types of DNA molecule and to reduce the chance of circularization of vector or target DNA fragments.

Choosing a suitable vector

The features to be considered when selecting a suitable vector are:

- the ease of purification – e.g. reliable procedures have been developed to allow plasmid DNA to be extracted and purified from bacterial cells;
- the efficiency of insertion of the recombinant vector into a new host cell;
- the presence of single copies of suitable restriction sites;
- the presence of selectable markers (e.g. antibiotic resistance genes or 'reporter genes', p. 247);
- the size of the DNA insert to be cloned;
- the copy number of the vector in the host cell.

Plasmids

The simplest bacterial cloning vectors are those based on small plasmids – circular DNA molecules, capable of autonomous replication within a bacterial cell. The general characteristics of plasmids are described in Chapter 39. One of the first cloning vectors to be developed for *E. coli* was pBR322, a genetically engineered plasmid of 4.4 kb containing an origin of replication, two antibiotic resistance genes and single sites for a range of restriction enzymes (Fig. 42.1). Subsequently, other plasmid vectors have been developed, with additional features: for example, plasmids of the pUC series (e.g. pUC8) are now widely used for transformation of *E. coli*, having the following advantages over earlier types:

- high copy number: several thousand identical copies of the plasmid may be present in each bacterial cell, giving improved yield of plasmid DNA;
- single-step selection of recombinants using the *lacZ′* gene (p. 247);
- clustering of restriction sites within a short region of the *lacZ′* gene – restriction using two enzymes (a 'double digest') cuts a small fragment from this 'multiple cloning site' or 'polylinker', producing a cleaved plasmid with two different 'sticky ends' (e.g. *Eco* RI at one end and *Hin* dIII at the other), allowing complementary fragments of target DNA to be ligated in a particular orientation ('directional cloning').

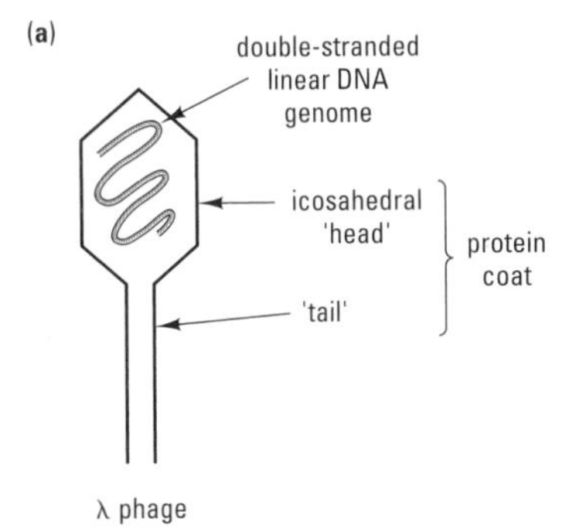

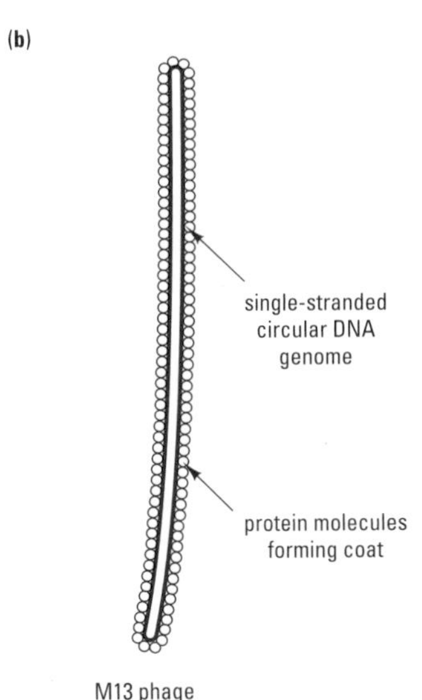

Fig. 42.4 Bacteriophages.

Phages

Bacteriophage vectors can offer several advantages over plasmid vectors:

- some phages have been engineered to carry larger fragments of foreign DNA – plasmid vectors work best with relatively short inserts of 1–2 kb, while some phage vectors can accept 10 kb or more of foreign DNA;
- a recombinant phage has a ready-made mechanism for transferring rDNA to a new host cell: the process is broadly similar to phage transduction, described on p. 224.

A number of vectors have been developed from *E. coli* phage λ (genome = 49 kb of dsDNA in linear form, Fig. 42.4a). For example, λZAPII, an insertion vector capable of accepting up to 10 kb of additional (foreign) DNA and with the *lacZ′* gene and multiple cloning site for selection, or λEMBL4, a replacement vector, where part of the vector DNA is removed and replaced by foreign DNA (up to 23 kb) during the cloning procedure.

The filamentous coliphage M13 (genome 6.4 kb of ssDNA in circular form, Fig. 42.4b) has been used to produce novel vectors that are particularly useful in providing cloned DNA in single stranded form, suitable for sequencing (Chapter 40). For example, the vector M13mp8 contains the *lacZ′* gene and a multiple cloning site equivalent to that found in pUC8. The M13 phage and its derivatives replicate within an infected cell as a double stranded 'replicative form' (RF) that can be extracted from infected cells and manipulated in the laboratory in a similar manner to a plasmid. New single stranded viral DNA is packaged within the phage coat and extruded from the infected cell without lysis. Recombinant M13 derivatives can carry up to 30 kb of additional DNA, since the filament length is variable depending on the size of the phage genome, in contrast to λ, where the size of the genome is constrained by the geometry of the polyhedral phage 'head' containing the DNA (Fig. 42.4).

Most vectors based on λ are virulent phages, released by lysis of the infected cell at the end of the replication phase: such vectors are cloned using standard techniques for phages, e.g. as plaques of lysis on a 'lawn' of bacteria growing on an agar-based medium (p. 64).In contrast, M13 derivatives do not lyse infected cells, but reduce their growth rate, and are visible as areas of limited growth (so-called 'turbid' plaques). Note that suitable strains of *E. coli* must be used for each of the various cloning vectors, e.g. M13 vectors only infect strains of *E. coli* with specific F pili (p. 224), while vectors containing *lacZ′* can only be used with those *E. coli* strains that are deficient in this part of their genome (*lacZ′* codes for the α-component of the enzyme β-galactosidase, present in wild-type *E. coli*). In some instances, a second phage (a 'helper phage') must be used along with the vector, to carry out some of the functions removed from the vector genome, e.g. phage packaging and assembly functions.

Cosmids and other vectors

Cosmid vectors contain the single stranded cohesive terminal sequences from λ phage (so-called 'cos' sites) inserted into a plasmid cloning vector. They can be used to clone up to 40 kb of foreign DNA, using the 'cos' sites to allow *in vitro* packaging into the λ phage coat: such large fragments are required for the production of gene libraries, as described below. Other hybrid vectors have been produced by the fusion of a phage and a plasmid vector – they can be manipulated in the laboratory as plasmids for ease of DNA uptake, or converted to phage-like particulate form for ease of storage.

Screening clones – a particular gene can be detected using Southern blotting (p. 232), or the expressed gene product (protein) may be detected by a suitable method, e.g. an immunoassay (p. 96). While it is relatively simple to *describe* the screening process, finding a specific gene (= clone) in a clone library is likely to be a time-consuming process.

Using cloned cDNA – information from the sequence of a particular cDNA clone may be useful in designing a suitable oligonucleotide probe to locate a particular gene in a genomic library.

Creating a gene library

For viruses with small genomes, an individual gene may be identified by hybridization of a single fragment with a suitable probe, following digestion with a particular restriction enzyme (Chapter 40). However, for most genomes, such a digest would contain a very complex mixture of fragments and a 'clone library' must be created. The isolation and identification of a particular gene is then carried out by screening individual clones from this library. Two types of clone library can be used:

1. A genomic library – prepared from the entire genome of the organism under study. The genome is fragmented to give overlapping fragments, e.g. by partial restriction (incomplete digestion, e.g. at low temperature), or by mechanical shearing. Individual fragments are then incorporated into a suitable vector, e.g. a cosmid, to create a vector library (e.g. a cosmid library). Each recombinant vector is then transferred to a separate host cell, which is cultured, giving a collection of transformants that represents the clone library. This approach is sometimes termed 'shotgun cloning'.
2. A cDNA library (cDNA = complementary, or copy DNA) – prepared by converting mRNA into DNA using the retroviral enzyme reverse transcriptase. Such a library consists only of genes expressed in those cells used to create the library, making individual genes easier to locate and identify, since the number of clones represented is usually smaller than for a genomic library. Another feature of cDNA is that it is complementary to the processed transcript (mRNA), so it contains no information on non-coding regions (introns) or transcriptional sequences (promoter regions, etc.).

Transferring rDNA to a suitable host cell

- Once a recombinant vector has been produced *in vitro*, it must be introduced into a suitable host cell. Packaged phage vectors have a built-in transfer mechanism, while naked phage DNA and plasmids must be introduced by treatments that cause a temporary increase in membrane permeability. These treatments include:

- physico-chemical shock treatment – Box 42.1 gives details of a typical procedure using $CaCl_2$ and heat shock treatment to transform *E. coli* with the plasmid pUC8;
- electroporation – cells or protoplasts are subjected to electric shock treatment (typically, $>10\,kV\,cm^{-1}$) for very short periods ($<10\,ms$);
- a range of techniques can be used for animal and plant cells, e.g. electroporation of protoplasts, or various microinjection treatments, either using a microsyringe, DNA-coated microprojectiles ('biolistics'). *Agrobacterium* can be used for rDNA transfer in certain plants.

Table 42.1 Examples of genes used to detect transformants

Gene	Product/assay
lacZ/*lacZ'*	β-galactosidase/chromogenic substrate (e.g. XGAL)
uidA	β-glucuronidase/chromogenic substrate (e.g. XGLUC)
lux	luciferase/bioluminescence in the presence of luciferin (p. 116)
amp	β-lactamase/resistance to ampicillin
cat	chloramphenicol acetyltransferase/resistance to chloramphenicol

Selection and detection of transformants

Many of the plasmid vectors used in genetic engineering carry genes coding for antibiotic resistance, e.g. pBR322 carries separate genes for ampicillin resistance, *amp*, and tetracycline resistance, *tet* (Fig. 42.1). These genes act as 'markers' for the vector. One gene (e.g. *amp*) can be used to select for transformants, which would form colonies on an agar-based medium containing the antibiotic, while non-transformed (ampicillin-sensitive) cells would be killed. The other gene (e.g. *tet*) can be used as a marker for the recombinant plasmid vector, since ligation of the target sequence into this gene causes insertional inactivation (Fig. 42.3). Thus, cells transformed with the recombinant plasmid will be resistant to ampicillin only, while cells transformed with the recircularized ('native') plasmid will be resistant to both antibiotics. These two types of transformant can be distinguished using replica plating (see p. 222). For those genes where the product is an enzyme, the presence or absence of the functional gene can be assessed using a suitable substrate. For example, the insertional inactivation of *lacZ'* can be detected by including a suitable inducer of β-galactosidase (e.g. isopropylthiogalactoside, IPTG) and the chromogenic substrate 5-bromo-4-chloro-3-indolyl-β-D-galactoside (XGAL) within the agar medium: a transformant colony derived from the native plasmid will be blue while a transformant containing the recombinant molecule will grow to produce a white colony. A number of other relevant examples are given in Table 42.1.

Recognizing transformants – after plating bacteria onto medium containing ampicillin, you may notice a few small 'feeder' colonies surrounding a single larger (transformant) colony. The feeder colonies are derived from non-transformed cells which survive due to the breakdown of antibiotic in the medium around the transformant colony, and should not be selected for sub-culture.

Box 42.1 Transformation of *E.coli* and selection of transformants

The following procedure illustrates the principal stages of the process: the efficiency of transformation will depend on the choice of *E.coli* strain and plasmid, and on the handling procedures prior to and during the process – for example, some strains do not require all of the stages listed below (for further details, see Hanahan *et al.*, 1995). Sterile equipment and appropriate technique (Chapter 13) are required at all times:

1. **Grow the cells under appropriate conditions:** the best results are often obtained using actively growing mid-log phase cells, rather than cells that have been grown to stationary phase (p. 80). Actively growing cells can be stored under appropriate conditions, e.g. as a 'frozen stock' – in suspension in concentrated glycerol at −20 °C.

2. **Induce cell competence by transfer to a suitable sterile transformation buffer:** this would normally contain divalent and monovalent salts, e.g. a salt solution containing $CaCl_2$ at 50–100 mmol l^{-1}, plus KCl and MnCl at 10–20 mmol l^{-1}. Cells are usually kept on ice in this solution for 10–15 min, to encourage the binding of plasmid DNA, added at a later stage.

3. **Add reagents to increase the permeability of cellular membranes:** typically dimethyl sulphoxide (DMSO, at up to 7% v/v) and dithiothreitol (DTT at up to 0.2 mmol l^{-1}), incubated on ice for a few minutes. DMSO is readily oxidized, reducing the effectiveness of the transformation procedure, and should be stored at −80 °C when not in use, to minimize oxidation.

4. **Add plasmid DNA:** typically at 10–1000 ng per transformation. Maintain on ice for at least 10 min in a minimal volume of solution, to allow the plasmid to become associated with the cell surface.

5. **Heat shock the cells:** briefly raise the temperature to 42–45 °C for 60–120 s, then return to ice for a few minutes. It is important that this treatment gives a *rapid* change in temperature – use a small, thin-walled (disposable) sterile plastic tube containing the minimum volume of solution to maximize the rate of temperature change.

6. **Allow the cells to recover from heat shock and to express any new genes** (e.g. to synthesize enzymes conferring antibiotic resistance in transformed cells): add sterile nutrient broth (e.g. 1 ml) and then incubate at 37 °C for up to 60 min.

7. **Plate onto an appropriate medium, to allow selection and detection of transformants:** for example, using pUC8, surface spread (p. 62) the suspension onto a medium containing ampicillin (at 25 mg l^{-1}), plus IPTG (at 15 mg l^{-1}) and XGAL (at 25 mg l^{-1}), to detect the *lacZ'* gene product, as described in the text.

Analysis and presentation of data

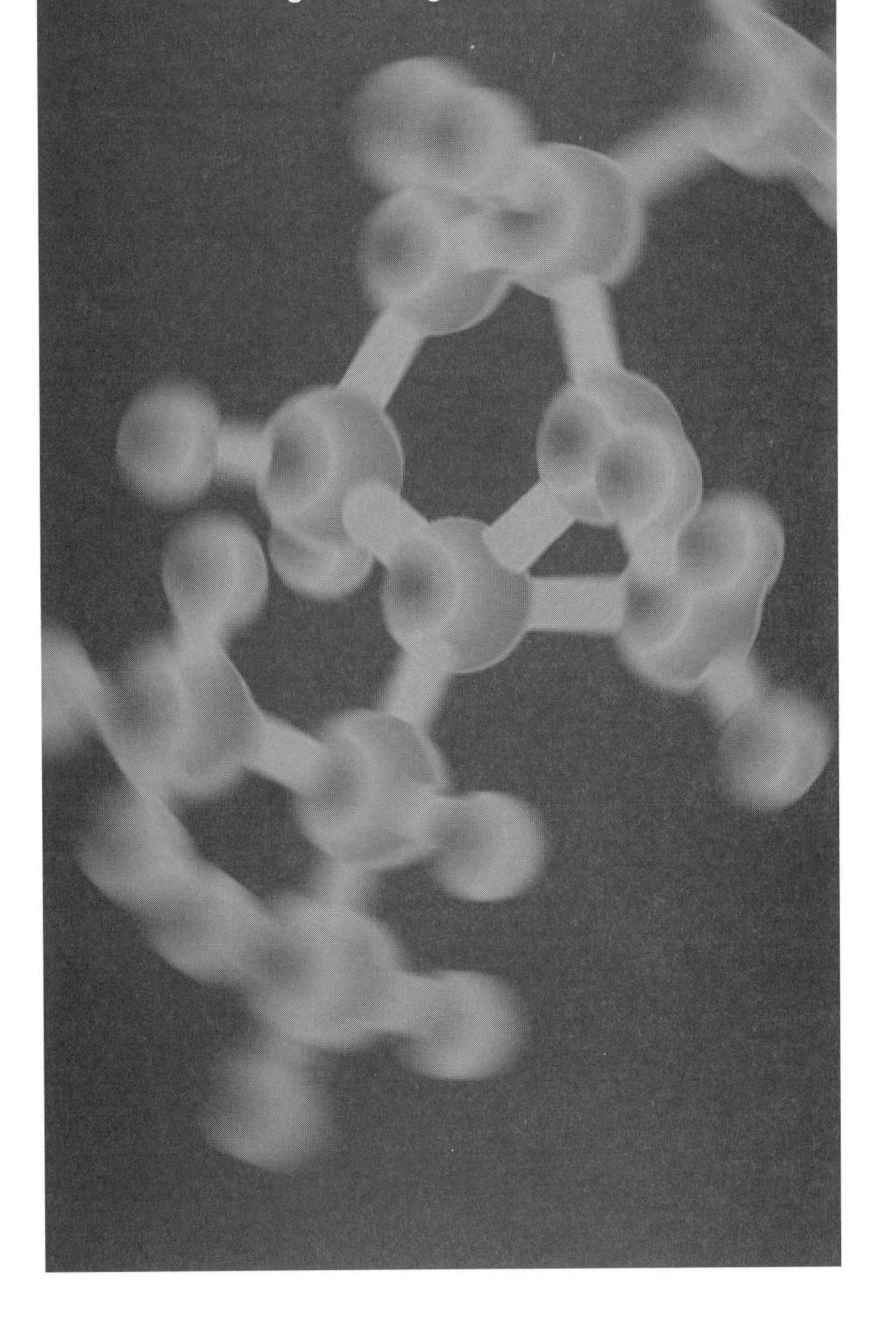

43 Using graphs

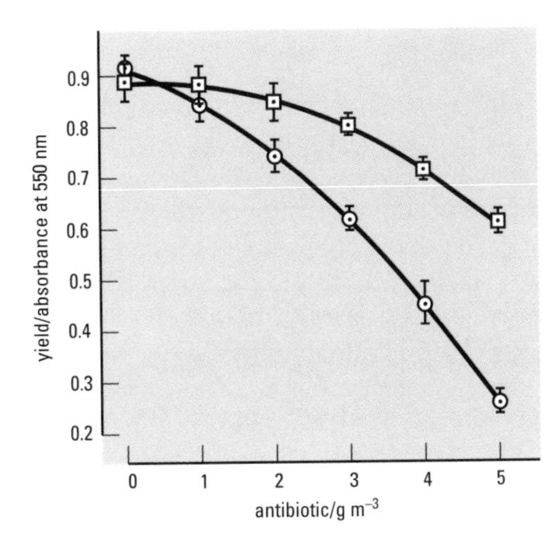

Fig. 43.1 Effect of antibiotic on yield of two bacterial isolates: ○, sensitive isolate; □, resistant isolate. Vertical bars show standard errors ($n = 6$).

Graphs can be used to show detailed results in an abbreviated form, displaying the maximum amount of information in the minimum space. Graphs and tables present findings in different ways. A graph (figure) gives a visual impression of the content and meaning of your results, while a table provides an accurate numerical record of data values. You must decide whether a graph should be used, e.g. to illustrate a pronounced trend or relationship, or whether a table (Chapter 44) is more appropriate.

A well-constructed graph will combine simplicity, accuracy and clarity. Planning of graphs is needed at the earliest stage in any write-up as your accompanying text will need to be structured so that each graph delivers the appropriate message. Therefore, it is best to decide on the final form for each of your graphs before you write your text. The text, diagrams, graphs and tables in a laboratory write-up or project report should be complementary, each contributing to the overall message. In a formal scientific communication it is rarely necessary to repeat the same data in more than one place (e.g. as a table and as a graph). However, graphical representation of data collected earlier in tabular format may be applicable in laboratory practical reports.

Practical aspects of graph drawing

The following comments apply to graphs drawn for laboratory reports. Figures for publication, or similar formal presentation are usually prepared according to specific guidelines, provided by the publisher/organizer.

Selecting a title – it is a common fault to use titles that are grammatically incorrect: a widely applicable format is to state the relationship between the independent and dependent variables within the title, e.g. 'The relationship between enzyme activity and external pH'.

KEY POINT **Graphs should be self-contained – they should include all material necessary to convey the appropriate message without reference to the text. Every graph must have a concise explanatory title to establish the content. If several graphs are used, they should be numbered, so they can be quoted in the text.**

Remembering which axis is which – a way of remembering the orientation of the *x* axis is that *x* is a 'cross', and it runs 'across' the page (horizontal axis).

- Consider the layout and scale of the axes carefully. Most graphs are used to illustrate the relationship between two variables (x and y) and have two axes at right angles (e.g. Fig. 43.1). The horizontal axis is known as the abscissa (x axis) and the vertical axis as the ordinate (y axis).
- The axis assigned to each variable must be chosen carefully. Usually the x axis is used for the independent variable (e.g. treatment) while the dependent variable (e.g. biological response) is plotted on the y axis (p. 50). When neither variable is determined by the other, or where the variables are interdependent, the axes may be plotted either way round.
- Each axis must have a descriptive label showing what is represented, together with the appropriate units of measurement, separated from the descriptive label by a solidus or 'slash' (/), as in Fig. 43.1, or brackets as in Fig. 43.2.
- Each axis must have a scale with reference marks ('tics') on the axis to show clearly the location of all numbers used.
- A figure legend should be used to provide explanatory detail, including the symbols used for each data set.

Handling very large or very small numbers

To simplify presentation when your experimental data consist of either very large or very small numbers, the plotted values may be the measured numbers

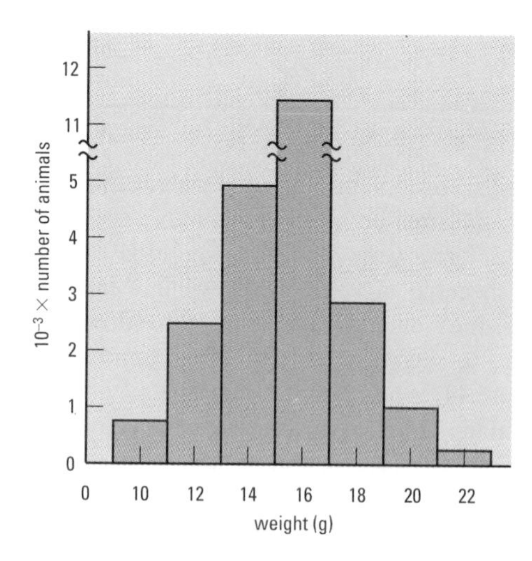

Fig. 43.2 Frequency distribution of weights for a sample of animals (sample size 24 085); the size class interval is 2 g.

multiplied by a power of 10: this multiplying power should be written immediately before the descriptive label on the appropriate axis (as in Fig. 43.2). However, it is often better to modify the primary unit with an appropriate prefix (p. 45) to avoid any confusion regarding negative powers of 10.

Size

Remember that the purpose of your graph is to communicate information. It must not be too small, so use at least half an A4 page and design your axes and labels to fill the available space without overcrowding any adjacent text. If using graph paper, remember that the white space around the grid is usually too small for effective labelling. The shape of a graph is determined by your choice of scale for the x and y axes which, in turn, is governed by your experimental data. It may be inappropriate to start the axes at zero (e.g. Fig. 43.1). In such instances, it is particularly important to show the scale clearly, with scale breaks where necessary, so the graph does not mislead. Note that Fig. 43.1 is drawn with 'floating axes' (i.e. the x and y axes do not meet in the lower left-hand corner), while Fig. 43.2 has clear scale breaks on both x and y axes.

Graph paper

In addition to conventional linear (squared) graph paper, you may need the following:

Example For a data set where the smallest number on the log axis is 12 and the largest number is 9 000, three-cycle log-linear paper would be used, covering the range 10–10 000.

- Probability graph paper. This is useful when one axis is a probability scale.
- Log-linear graph paper. This is appropriate when one of the scales shows a logarithmic progression, e.g. the exponential growth of cells in liquid culture (p. 80). Log-linear paper is defined by the number of logarithmic divisions covered (usually termed 'cycles') so make sure you use a paper with the appropriate number of cycles for your data. An alternative approach is to plot the log-transformed values on 'normal' graph paper.
- Log-log graph paper. This is appropriate when both scales show a logarithmic progression.

Types of graph

Different graphical forms may be used for different purposes, including:

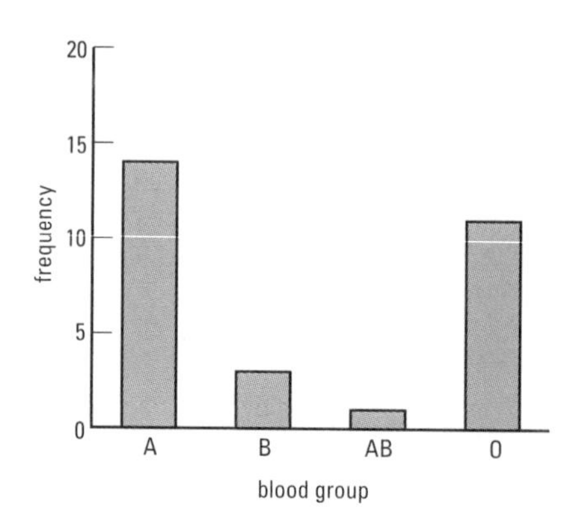

Fig. 43.3 Bar chart, showing the number of students belonging to each ABO blood group ($n = 29$).

- Plotted curves – used for data where the relationship between two variables can be represented as a continuum (e.g. Fig. 43.1).
- Scatter diagrams – used to visualize the relationship between individual data values for two interdependent variables (e.g. Fig. 47.6) often as a preliminary part of a correlation analysis (p. 275).
- Three-dimensional graphs show the interrelationships of three variables, (e.g. Fig. 25.1).
- Histograms for frequency distributions of continuous variables (e.g. Fig. 43.2).
- Frequency polygons emphasize the form of a frequency distribution by joining the co-ordinates with straight lines, in contrast to a histogram. This is particularly useful when plotting two or more sets of data values on the same graph.
- Bar charts represent frequency distributions of a discrete qualitative or quantitative variable (e.g. Fig. 43.3).
- Pie charts illustrate portions of a whole (e.g. Fig. 43.4).

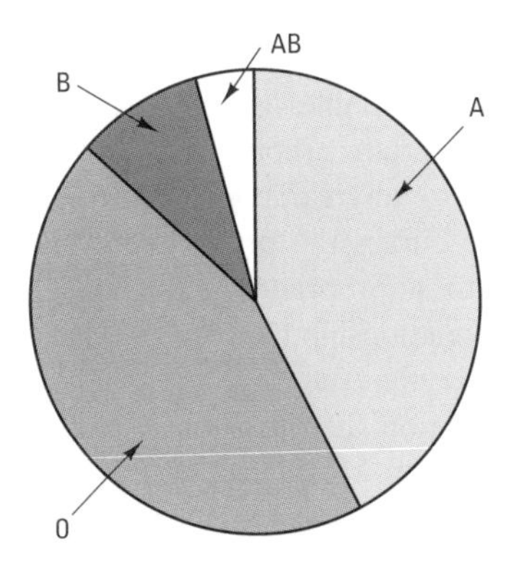

Fig. 43.4 Pie chart: relative abundance of ABO blood groups in man.

The plotted curve

This is the commonest form of graphical representation used in bioscience. The key features are outlined below and in checklist form (Box 43.1).

Data points

Each data point must be shown accurately, so that any reader can determine the exact values of x and y. In addition, the results of each treatment must be readily identifiable. A useful technique is to use a dot for each data point, surrounded by a hollow symbol for each treatment (see Fig. 43.1). An alternative is to use symbols only, though the co-ordinates of each point are defined less accurately. Use the same symbol for the same entity if it occurs in several graphs and provide a key to all symbols.

Statistical measures

If you are plotting average values for several replicates and if you have the necessary statistical knowledge, you can calculate the standard error (p. 266), or the 95% confidence limits (p. 275) for each mean value and show these on your graph as a series of vertical bars (see Fig. 43.1). Make it clear in the legend whether the bars refer to standard errors or 95% confidence limits and quote the value of n (the number of replicates per data point). Another approach is to add a least significant difference bar (p. 274) to the graph.

Interpolation

Once you have plotted each point, you must decide whether to link them by straight lines or a smoothed curve. Each of these techniques conveys a different message to your reader. Joining the points by straight lines may seem the

Box 43.1 Checklist for the stages in drawing a graph

The following sequence can be used whenever you need to construct a plotted curve: it will need to be modified for other types of graph.

1. **Collect all of the data values and statistical values** (in tabular form, where appropriate).
2. **Decide on the most suitable form of presentation**: this may include transformation, to convert the data to linear form.
3. **Choose a concise descriptive title**, together with a reference (figure) number and date, where necessary.
4. **Determine which variable is to be plotted on the x axis and which on the y axis.**
5. **Select appropriate scales for both axes** and make sure that the numbers and their location (scale marks) are clearly shown, together with any scale breaks.
6. **Decide on appropriate descriptive labels for both axes**, with SI units of measurement, where appropriate.
7. **Choose the symbols for each set of data points** and decide on the best means of representation for statistical values.
8. **Plot the points** to show the co-ordinates of each value with appropriate symbols.
9. **Draw a trend line for each set of points.**
10. **Write a figure legend**, to include a key which identifies all symbols and statistical values and any descriptive footnotes, as required.

simplest option, but may give the impression that errors are very low or non-existent and that the relationship between the variables is complex. Joining points by straight lines is appropriate only in certain graphs, e.g. recording a patient's temperature in a hospital, to emphasize any variation from one time point to the next. However, in most plotted curves the best straight line or curved line should be drawn (according to appropriate mathematical or statistical models, or by eye), to highlight the relationship between the variables – after all, your choice of a plotted curve implies that such a relationship exists! Don't worry if some of your points do not lie on the line: this is caused by errors of measurement and by biological variation. Most curves drawn by eye should have an equal number of points lying on either side of the line. You may be guided by 95% confidence limits, in which case your curve should pass within these limits wherever possible.

Conveying the correct message – the golden rule is: 'always draw the simplest line that fits the data reasonably well and is biologically reasonable'.

Curved lines can be drawn using a flexible curve, a set of French curves, or freehand. In the latter case, turn your paper so that you can draw the curve in a single, sweeping stroke by a pivoting movement at the elbow (for larger curves) or wrist (for smaller ones). Do not try to force your hand to make complex, unnatural movements, as the resulting line will not be smooth.

Extrapolation

Be wary of extrapolation beyond the upper or lower limit of your measured values. This is rarely justifiable and may lead to serious errors. Whenever extrapolation is used, a dotted line ensures that the reader is aware of the uncertainty involved. Any assumptions behind an extrapolated curve should also be stated clearly in your text.

Extrapolating plotted curves – try to avoid the need to extrapolate by better experimental design.

The histogram

While a plotted curve assumes a continuous relationship between the variables by interpolating between individual data points, a histogram involves no such assumptions and is the most appropriate representation if the number of data points is too few to allow a trend line to be drawn. Histograms are also used to represent frequency distributions (p. 262), where the y axis shows the number of times a particular value of x was obtained (e.g. Fig. 43.2). As in a plotted curve, the x axis represents a continuous variable which can take any value within a given range, so the scale must be broken down into discrete classes and the scale marks on the x axis should show either the mid-points (mid-values) of each class (Fig. 43.2), or the boundaries between the classes.

In a histogram, each datum is represented by a column with an area proportional to the magnitude of y: in most cases, you should use columns of equal width, so that the height of each column is then directly proportional to y. Shading or stippling may be used to identify individual columns, according to your needs.

The columns are adjacent to each other in a histogram, in contrast to a bar chart, where the columns are separate because the x axis of a bar chart represents discrete values.

Interpreting graphs

Whenever you look at graphs drawn by other people, make sure you understand the axes before you look at the relationship. It is all too easy to take in the shape of a graph without first considering the scale of the axes, a fact that some advertisers and politicians exploit when curves are used to misrepresent information. Such graphs are often used in newspapers and on television. Examine them critically – many would not pass the stringent requirements of scientific communication and conclusions drawn from them may be flawed.

Using computers to produce graphs – never allow a computer program to dictate size, shape and other aspects of a graph: find out how to alter scales, labels, axes, etc. and make appropriate selections. Draw curves freehand if the program only has the capacity to join the individual points by straight lines.

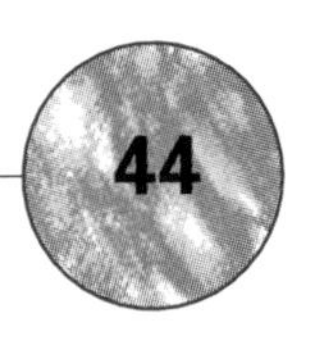

44 Presenting data in tables

A table is often the most appropriate way to present numerical data in a concise, accurate and structured form. Assignments and project reports should contain tables which have been designed to condense and display results in a meaningful way and to aid numerical comparison. The preparation of tables for recording primary data is discussed on p. 43.

Decide whether you need a table, or whether a graph is more appropriate. Histograms and plotted curves can be used to give a visual impression of the relationships within your data (p. 252). On the other hand, a table gives you the opportunity to make detailed numerical comparisons.

KEY POINT **Always remember that the primary purpose of your table is to communicate information and allow appropriate comparison, not simply to put down the results on paper!**

Preparation of tables

Title

Constructing titles – take care over titles as it is a common mistake in student practical reports to present tables without titles, or to misconstruct the title.

Every table must have a brief descriptive title. If several tables are used, number them consecutively so they can be quoted in your text. The titles within a report should be compared with one another, making sure they are logical and consistent and that they describe accurately the numerical data contained within them.

Structure

Saving space in tables – you may be able to omit a column of control data if your results can be expressed as percentages of the corresponding control values.

Display the components of each table in a way that will help the reader understand your data and grasp the significance of your results. Organize the columns so that each category of like numbers or attributes is listed vertically, while each horizontal row shows a different experimental treatment, organism, sampling site, etc. (as in Table 44.1). Where appropriate, put control values near the beginning of the table. Columns that need to be compared should be set out alongside each other. Use rulings to subdivide your table appropriately, but avoid cluttering it up with too many lines.

Table 44.1 Characteristics of selected photoautotrophic microbes

			Intracellular carbohydrate	
Division	Species	Optimum [NaCl]* ($mol\ m^{-3}$)	Identity	Quantity† (nmol (g dry wt)$^{-1}$)
Chlorophyta	*Scenedesmus quadruplicatum*	340	Sucrose	49.7
	Chlorella emersonii	780	Sucrose	102.3
	Dunaliella salina	4 700	Glycerol	910.7
Cyanobacteria	*Microcystis aeruginosa*	< 20‡	None	0.0
	Anabaena variabilis	320	Sucrose	64.2
	Rivularia atra	380	Trehalose	ND

* Determined after 28-day growth at 25 °C.
† Individual samples, analysed by gas-liquid chromatography.
‡ Poor growth in all media with added NaCl (minimum NaCl concentration 5 $mol\ m^{-3}$).
ND Sample lost: no quantitative data.

Headings and subheadings

These should identify each set of data and show the units of measurement, where necessary. Make sure that each column is wide enough for the headings and for the longest data value.

Numerical data

Within the table, do not quote values to more significant figures than necessary, as this will imply spurious accuracy (p. 42). By careful choice of appropriate units for each column you should aim to present numerical data within the range 0 to 1 000. As with graphs, it is less ambiguous to use derived SI units, with the appropriate prefixes, in the headings of columns and rows, rather than quoting multiplying factors as powers of 10. Alternatively, include exponents in the main body of the table (see Table 6.1, to avoid any possible confusion regarding the use of negative powers of 10.

Examples If you measured the width of a fungal hypha to the nearest one-tenth of a micrometre, quote the value in the form '52.6 μm'.

Quote the width of a fungal hypha as 52.6 (μm), rather than 0.000 052 6 (m) or 52.6 (10^{-6} m).

Other notations

Avoid using dashes in numerical tables, as their meaning is unclear; enter a zero reading as '0' and use 'NT' not tested or 'ND' if no data value was obtained, with a footnote to explain each abbreviation. Other footnotes, identified by asterisks, superscripts or other symbols in the table, may be used to provide relevant experimental detail (if not given in the text) and an explanation of column headings and individual results, where appropriate. Footnotes should be as condensed as possible. Table 44.1 provides examples.

Saving further space – in some instances a footnote can be used to replace a whole column of repetitive data.

Statistics

In tables where the dispersion of each data set is shown by an appropriate statistical parameter, you must state whether this is the (sample) standard deviation, the standard error (of the mean) or the 95% confidence limits and you must give the value of n (the number of replicates). Other descriptive statistics should be quoted with similar detail, and hypothesis-testing statistics should be quoted along with the value of P (the probability). Details of any test used should be given in the legend, or in a footnote.

Text

Sometimes a table can be a useful way of presenting textual information in a condensed form (see example on p. 101).

When you have finished compiling your tabulated data, carefully double-check each numerical entry against the original information, to ensure that the final version of your table is free from transcriptional errors. Box 44.1 gives a checklist for the major elements of constructing a table.

Using microcomputers and word-processing packages – these can be used to prepare high-quality versions of tables for project work (p. 29).

Box 44.1 Checklist for preparing a table

Every table should have the following components:

1. **A title**, plus a reference number and date where necessary.
2. **Headings for each column and row**, with appropriate units of measurement.
3. **Data values**, quoted to the nearest significant figure and with statistical parameters, according to your requirements.
4. **Footnotes** to explain abbreviations, modifications and individual details.
5. **Rulings to emphasize groupings** and distinguish items from each other.

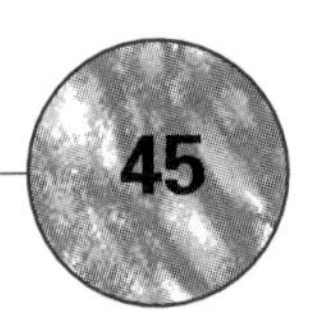

45 Hints for solving numerical problems

Life science often requires a numerical or statistical approach. Not only is mathematical modelling an important aid to understanding, but computations are often needed to turn raw data into meaningful information or to compare them with other data sets. Moreover, calculations are part of laboratory routine, perhaps required for making up solutions of known concentration (see p. 13 and below) or for the calibration of a microscope (see p. 38). In research, 'trial' calculations can reveal what input data are required and where errors in their measurement might be amplified in the final result (see p. 49).

KEY POINT **If you have a 'block' about numerical work, practice at problem-solving is especially important.**

Table 45.1 Sets of numbers and operations

Sets of numbers	
Whole numbers:	0, 1, 2, 3, ...
Natural numbers:	1, 2, 3, ...
Integers:	... −3, −2, −1, 0, 1, 2, 3, ...
Real numbers:	integers and anything between (e.g. −5, 4.376, 3/16, π, $\sqrt{5}$)
Prime numbers:	subset of natural numbers divisible by 1 and themselves only (i.e. 2, 3, 5, 7, 11, 13, ...)
Rational numbers:	p/q where p (integer) and q (natural) have no common factor (e.g. 3/4)
Fractions:	p/q where p is an integer and q is natural (e.g. −6/8)
Irrational numbers:	real numbers with no exact value (e.g. π)
Infinity:	(symbol ∞) is larger than any number (technically not a number as it does not obey the laws of algebra)
Operations and symbols	
Basic operators:	$+$, $-$, $\times$ and $\div$ will not need explanation; however, / may substitute for $\div$, $*$ may substitute for $\times$ or this operator may be omitted
Powers:	a^n, i.e. 'a to the power n', means a multiplied by itself n times (e.g. $a^2 = a \times a$ = 'a squared', $a^3 = a \times a \times a$ = 'a cubed'). n is said to be the index or exponent. Note $a^0 = 1$ and $a^1 = a$
Logarithms:	the common logarithm (log) of any number x is the power to which 10 would have to be raised to give x (i.e. the log of 100 is 2; $10^2 = 100$); the antilog of x is 10^x. Note that there is no log for 0, so take this into account when drawing log axes by breaking the axis. Natural or Napierian logarithms (ln) use the base e (= 2.71828 ...) instead of 10
Reciprocals:	the reciprocal of a real number a is $1/a$ ($a \neq 0$)
Relational operators:	$a > b$ means 'a is greater (more positive) than b', $<$ means less than, $\leqslant$ means less-than-or-equal-to and $\geqslant$ means greater-than-or-equal-to
Proportionality:	$a \propto b$ means 'a is proportional to b' (i.e. $a = kb$, where k is a constant). If $a \propto 1/b$, a is inversely proportional to b ($a = k/b$)
Sums:	Σx_i is shorthand for the sum of all x values from $i = 0$ to $i = n$ (more correctly the range of the sum is specified under the symbol)
Moduli:	$\|x\|$ signifies modulus of x, i.e. its absolute value (e.g. $\|4\| = \|-4\| = 4$)
Factorials:	$x!$ signifies factorial x, the product of all integers from 1 to x (e.g. $3! = 6$). Note $0! = 1! = 1$

Practising at problem solving:

- demystifies the procedures involved, which are normally just the elementary mathematical operations of addition, subtraction, multiplication and division (Table 45.1);
- allows you to gain confidence so that you don't become confused when confronted with an unfamiliar or apparently complex form of problem;
- helps you recognize the various forms a problem can take as, for instance, in crossing experiments in classical genetics (Box 38.1).

Steps in tackling a numerical problem

The step-by-step approach outlined below may not be the fastest method of arriving at an answer, but most mistakes occur where steps are missing, combined or not made obvious, so a logical approach is often better. Error tracing is distinctly easier when all stages in a calculation are laid out.

Have the right tools ready

A computer spreadsheet may be very useful in repetitive calculations or for 'what if?' case studies (see Chapter 50).

Scientific calculators (p. 4) greatly simplify the numerical part of problem-solving. However, the seeming infallibility of the calculator may lead you to accept an absurd result which could have arisen because of faulty key-pressing or faulty logic. Make sure you know how to use all the features on your calculator, especially how the memory works; how to introduce a constant multiplier or divider; and how to obtain an exponent (note that the 'exp' button on most calculators gives you 10^x, not 1^x or y^x; so 1×10^6 would be entered as [1] [exp] [6], *not* [10] [exp] [6]).

Approach the problem thoughtfully

If the individual steps have been laid out on a worksheet, the 'tactics' will already have been decided. It is more difficult when you have to adopt a strategy on your own, especially if the problem is presented in a descriptive style and it isn't obvious which equations or rules need to be applied.

Table 45.2 Simple algebra – rules for manipulating equations

If $a = b + c$, then $b = a - c$ and $c = a - b$
If $a = b \times c$, then $b = a/c$ and $c = a/b$
If $a = b^c$, then $b = a^{1/c}$ and $c = \log a/\log b$
$a^{1/n} = \sqrt[n]{a}$
$a^{-n} = 1/a^n$
$a^b \times a^c = a^{(b+c)}$ and $a^b/a^c = a^{(b-c)}$
$(a^b)^c = a^{(b \times c)}$
$a \times b = \text{antilog}(\log a + \log b)$

- Read the problem carefully as the text may give clues as to how it should be tackled. Be certain of what is required as an answer before starting.
- Analyse what kind of problem it is, which effectively means deciding which equation(s) or approach will be applicable. If this is not obvious, consider the dimensions/units of the information available and think how they could be fitted to a relevant formula. In examinations, a favourite ploy of examiners is to present a problem such that the familiar form of an equation must be rearranged (see Table 45.2 and Box 45.1). Another is to make you use two or more equations in series. If you are unsure whether a recalled formula is correct, a dimensional analysis can help: write in all the units for the variables and make sure that they cancel out to give the expected answer.
- Check that you have, or can derive, all of the information required to use your chosen equation(s). It is unusual but not unknown for examiners to supply redundant information. So, if you decide not to use some of the information given, be sure why you do not require it.
- Decide on what format and units the answer should be presented in. This is sometimes suggested to you. If the problem requires many changes in the prefixes to units, it is a good idea to convert all data to base SI units (multiplied by a power of 10) at the outset.
- If a problem appears complex, break it down into component parts.

Box 45.1 Example of using the rules of Table 45.2

Problem: if $a = (b - c)/(d + e^n)$, find e

1. Multiply both sides by $(d + e^n)$; formula becomes:

   $$a(d + e^n) = (b - c)$$

2. Divide both sides by a; formula becomes: $d + e^n = \frac{b - c}{a}$

3. Subtract d from both sides; formula becomes: $e^n = \frac{b - c}{a} - d$

4. Raise each side to the power $1/n$; formula becomes:

   $$e = \left\{\frac{b - c}{a} - d\right\}^{1/n}$$

Present your answer clearly

The way you present your answer obviously needs to fit the individual problem. Guidelines for presenting an answer include:

Show the steps in your calculations – most markers will only penalize a mistake once and part marks will be given if the remaining operations are performed correctly. This can only be done if those operations are visible!

(a) Make your assumptions explicit. Most mathematical models of biological phenomena require that certain criteria are met before they can be legitimately applied (e.g. 'assuming the tissue is homogeneous . . .'), while some approaches involve approximations which should be clearly stated (e.g. 'to estimate the mouse's skin area, its body was approximated to a cylinder with radius x and height y . . .').
(b) Explain your strategy for answering, perhaps giving the applicable formula or definitions which suit the approach to be taken. Give details of what the symbols mean (and their units) at this point.
(c) Rearrange the formula to the required form with the desired unknown on the left-hand side (see Table 45.2).
(d) Substitute the relevant values into the right-hand side of the formula, using the units and prefixes as given (it may be convenient to convert values to SI beforehand). Convert prefixes to appropriate powers of 10 as soon as possible.
(e) Convert to the desired units step-by-step, i.e. taking each variable in turn.
(f) When you have the answer in the desired units, rewrite the left-hand side and underline the answer. Make sure that the result is presented with an appropriate number of significant figures (see p. 42).

Units – never write any answer without its unit(s) unless it is truly dimensionless.

Rounding off – do not round off numbers until you arrive at the final answer.

Check your answer

Having written out your answer, you should check it methodically, answering the following questions:

- Is the answer of a realistic magnitude? You should be alerted to an error if an answer is absurdly large or small. In repeated calculations, a result standing out from others in the same series should be double-checked.
- Do the units make sense and match up with the answer required? Don't, for example, present a volume in units of m^2.
- Do you get the same answer if you recalculate in a different way? If you have time, recalculate the answer using a different 'route', entering the numbers into your calculator in a different form and/or carrying out the operations in a different order.

Some reminders of basic mathematics

Errors in calculations sometimes appear because of faults in mathematics rather than computational errors. For reference purposes, Tables 45.1 and 45.2 give some basic mathematical principles that may be useful. Eason *et al.* (1992) should be consulted for more advanced needs.

Exponents

Example $2^3 = 2 \times 2 \times 2 = 8$

Exponential notation is an alternative way of expressing numbers in the form a^n ('a to the power n'), where a is multiplied by itself n times. The number a is called the base and the number n the exponent (or power or index). The exponent need not be a whole number, and it can be negative if the number being expressed is less than 1. See Table 45.2 for other mathematical relationships involving exponents.

Scientific notation

Example Avogadro's number, $\approx$ 602 352 000 000 000 000 000 000, is more conveniently expressed as 6.02352×10^{23}.

In scientific notation, also known as 'standard form', the base is 10 and the exponent a whole number. To express numbers that are not whole powers of 10, the form $c \times 10^n$ is used, where the coefficient c is normally between 1 and 10. Scientific notation is valuable when you are using very large numbers and wish to avoid suggesting spurious accuracy. Thus if you write 123 000, this suggests that you know the number to ± 0.5, whereas 1.23×10^5 might give a truer indication of measurement accuracy (i.e. implied to be ± 500 in this case). Engineering notation is similar, but treats numbers as powers of ten in groups of 3, i.e. $c \times 10^0$, 10^3, 10^6, 10^9, etc. This corresponds to the SI system of prefixes (p. 45).

A useful property of powers when expressed to the same base is that when multiplying two numbers together, you simply add the powers, while if dividing, you subtract the powers. Thus, suppose you counted 8 bacteria in a 10^{-7} dilution (see p. 84), there would be 8×10^7 in the same volume of undiluted solution; if you now dilute this 500-fold (5×10^2), then the number present in the same volume would be $8/5 \times 10^{(7-2)} = 1.6 \times 10^5 = 160\,000$.

Logarithms

Example (use to check the correct use of your own calculator)
102963 as a log = 5.012681 (to 6 decimal places)
$10^{5.012681} = 102962.96$
(Note loss of accuracy due to loss of decimal places).

When a number is expressed as a logarithm, this refers to the power n that the base number a must be raised to give that number, e.g. $\log_{10}(1000) = 3$, since $10^3 = 1000$. Any base could be used, but the two most common are 10, when the power is referred to as $\log_{10}$ or simply log, and the constant e (2.718282), used for mathematical convenience in certain situations, when the power is referred to as $\log_e$ or ln. Where a coefficient would be used in scientific notation, then the log is not a whole number.

To obtain logs, you will need to use the log key on your calculator, or special log tables (now largely redundant). To convert back, use:

- the $\boxed{10^x}$ key, with x = log value;
- the $\boxed{\text{inverse}}$ then the log key; or
- the $\boxed{y^x}$ key, with $y = 10$ and x = log value.

With log tables, you will find complementary antilogarithm tables to do this.

There are many uses of logarithms in biology, including pH ($= -\log[H^+]$), where $[H^+]$ is expressed in $\text{mol}\,l^{-1}$ (p. 26), and the exponential growth of microorganisms, where if log(cell number) is plotted against time, a straight-line relationship is obtained (p. 80).

Hints for some typical problems

Example A lab schedule states that 5 g of a compound of M_r 220 g mol^{-1} are dissolved in 400 ml of solvent. For writing up your Materials and Methods, you wish to express this as mol l^{-1}.

1. If there are 5 g in 400 ml, then there are 5/400 g in 1 ml.
2. Hence, 1000 ml will contain $5/400 \times 1000 \quad 400$ g = 12.5 g.
3. 12.5 g = 12.5/220 mol = 0.0568 mol, so [solution] = 56.8 mmol l^{-1} (= 56.8 mol m^{-3}).

Calculations involving proportions or ratios

The 'unitary method' is a useful way of approaching calculations involving proportions or ratios, such as those required when making up solutions from stocks (see also Chapter 5) or as a subsidiary part of longer calculations.

1. If given a value for a multiple, work out the corresponding value for a single item or 'unit'.
2. Use this 'unitary value' to calculate the required new value.

Calculations involving series

Series (used in e.g. dilutions, see also p. 15) can be of two main forms:

- arithmetic, where the *difference* between two successive numbers in the series is a constant, e.g. 2, 4, 6, 8, 10, ...
- geometric, where the *ratio* between two successive numbers in the series is a constant, e.g. 1, 10, 100, 1000, 10000, ...

Note that the logs of the numbers in a geometric series will form an arithmetic series (e.g. 0, 1, 2, 3, 4, ... in the above case). Thus, if a quantity y varies with a quantity x such that the rate of change in y is proportional to the value of y (i.e. it varies in an exponential manner), a semi-log plot of such data will form a straight line. This form of relationship is relevant for exponentially growing cell cultures (p. 80) and radioactive decay (p. 101).

Statistical calculations

The need for long, complex calculations in statistics has largely been removed because of the widespread use of spreadsheets with statistical functions (Chapter 50) and specialized programs such as Minitab™ (p. 289). It is, however, important to understand the principles behind what you are trying to do (see Chapters 46 and 47) and interpret the program's output correctly, either using the 'help' function or a reference manual.

Problems in Mendelian genetics

These cause difficulties for many students. The key is to recognize the different types of problem and to practise so you are familiar with the techniques for solving them. Chapter 38 deals with the different types of cross you will come across and methods of analysing them, including the use of the χ^2 (Chi^2) test.

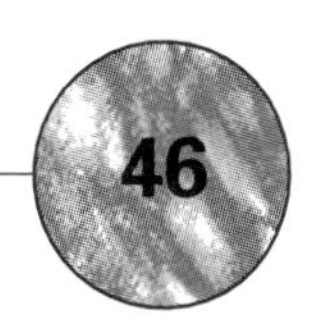

46 Descriptive statistics

The purpose of most practical work is to observe and measure a particular characteristic of a biomolecule or biological system. However, it would be extremely rare if the same value was obtained every time the characteristic was measured, or with every experimental subject. More commonly, such measurements will show variability, due to measurement error, sampling variation and/or biological variability (p. 42). Such variability can be displayed as a frequency distribution (e.g. Fig. 43.2), where the y axis shows the number of times (frequency, f) each particular value of the measured variable (Y) has been obtained. Descriptive (or summary) statistics quantify aspects of the frequency distribution of a sample. You can use them to condense a large data set, for presentation in figures or tables. An additional application of descriptive statistics is to provide estimates of the true values of the underlying frequency distribution of the population being sampled, allowing the significance and precision of the experimental observations to be assessed (p. 269).

Fig. 46.1 Two distributions with different locations but the same dispersion. The data set labelled B could have been obtained by adding a constant to each datum in the data set labelled A.

KEY POINT The appropriate descriptive statistics to choose will depend on both the type of data, i.e. whether quantitative, ranked or qualitative (see p. 41) and the nature of the underlying frequency distribution.

In many instances, the normal (Gaussian) distribution best describes the observed pattern, giving a symmetrical, bell-shaped frequency distribution (p. 271), for example, measurements of a quantitative continuous variable in a number of individuals (e.g. blood plasma pH of several adult males), or replicate measurements of a particular characteristic (e.g. repeated measurements of the blood plasma pH of an individual adult male).

Three important features of a frequency distribution that can be summarized by descriptive statistics are:

- the sample's location, i.e. its position along a given dimension representing the dependent (measured) variable (Fig. 46.1);
- the dispersion of the data, i.e. how spread out the values are (Fig. 46.2);
- the shape of the distribution, i.e. whether symmetrical, skewed, U-shaped, etc. (Fig. 46.3).

Fig. 46.2 Two distributions with different dispersions but the same location. The data set labelled A covers a relatively narrow range of values of the dependent (measured) variable while that labelled B covers a wider range.

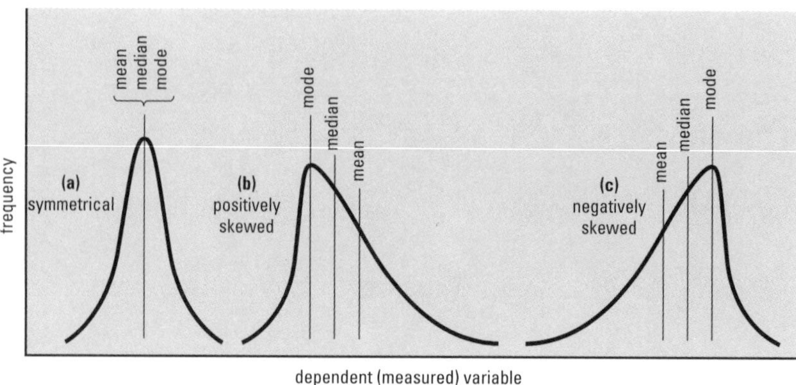

Fig. 46.3 Symmetrical and skewed frequency distributions, showing relative positions of mean, median and mode.

Box 46.1 Calculation of descriptive statistics for a sample of grouped data

Value (Y)	Frequency (f)	Cumulative frequency	fY	fY^2
1	0	0	0	0
2	1	1	2	4
3	2	3	6	18
4	3	6	12	48
5	8	14	40	200
6	5	19	30	180
7	2	21	14	98
8	0	21	0	0
Totals	$21 = \Sigma f (= n)$		$104 = \Sigma fY$	$548 = \Sigma fY^2$

In this example, for simplicity and ease of calculation, integer values of Y are used. In many practical exercises, where continuous variables are measured to several significant figures and where the number of data values is small, giving frequencies of 1 for most of the values of Y, it may be simpler to omit the column dealing with frequency and list all the individual values of Y and Y^2 in the appropriate columns. To gauge the underlying frequency distribution of such data sets, you would need to group individual data into broader classes (e.g. all values between 1.0 and 1.9, all values between 2.0 and 2.9, etc.) and then draw a histogram (p. 252).

Statistic	Value*	How calculated
Mean	4.95	$\Sigma fY/n$, i.e. 104/21
Median	5	Value of the $(n + 10/2$ variate, i.e. the value ranked $(21 + 1)/2 = 11$th (obtained from the cumulative frequency column)
Mode	5	The most common value (Y value with highest frequency)
Upper extreme	7	Highest Y value in data set
Lower extreme	2	Lowest Y value in data set
Range	5	Difference between upper and lower extremes
Variance (s^2)	1.65	$s^2 = \dfrac{\Sigma fY^2 - (\Sigma fY)^2/n}{n-1} = \dfrac{548 - (104)^2/21}{20}$
Standard deviation (s)	1.28	$\sqrt{s^2}$
Standard error (SE)	0.280	$s/\sqrt{n}$
Coefficient of variation (cov)	25.9%	$100s/\bar{Y}$

*Rounded to appropriate signfiicant figures

Use of symbols – Y is used in Chapters 46 and 47 to signify the dependent variable in statistical calculations (following the example of Sokal and Rohlf, 1994, Wardlaw, 1985 and Heath, 1995). Note, however, that some authors use X or x in analogous formulae and many calculators refer to e.g. $\bar{x}$, Σx^2, etc. for their statistical functions.

Measuring location

Here, the objective is to pinpoint the 'centre' of the frequency distribution, i.e. the value about which most of the data are grouped. The chief measures of location are the mean, median and mode.

Mean

The mean (denoted $\bar{Y}$ and also referred to as the arithmetic mean) is the average value of the data. It is obtained from the sum of all the individual data values divided by the number of data values (in symbolic terms, $\Sigma Y/n$). The mean is a good measure of the centre of symmetrical frequency distributions of qualitative variables. It uses all of the numerical values of the sample and therefore incorporates all of the information content of the data. However, the value of a mean is greatly affected by the presence of outliers. The arithmetic mean is a widely used statistic in life science, but there are situations when you should be careful about using it (see Box 46.2 for examples).

Definitions

An outlier – any datum which has a value much smaller or bigger than most of the data.

Rank – the position of a data value when all the data are placed in order of ascending magnitude. If ties occur, an average rank of the tied variates is used. Thus, the rank of the datum 6 in the sequence 1,3,5,6,8,8,10 is 4; the rank of each datum with value 8 is 5.5.

Median

The median is the mid-point of the observations when ranked in increasing order. For odd-sized samples, the median is the middle observation; for even-sized samples it is the mean of the middle pair of observations. For a quantitative variable, the median may represent the location of the main body of data better than the mean when the distribution is asymmetric or when there are outliers.

Box 46.2 Three examples where simple arithmetic means are inappropriate

Mean	*n*
6	4
7	7
8	1

1. If means of samples are themselves meaned, an error can arise if the samples are of different size. For example, the arithmetic mean of the means in the table shown left is 7, but this does not take account of the different 'reliabilities' of each mean due to their sample sizes. The correct weighted mean is obtained by multiplying each mean by its sample size (*n*) (a 'weight') and dividing the sum of these values by the total number of observations, i.e. in the case shown, $(24 + 49 + 8)/12 = 6.75$.
2. When making a mean of ratios (e.g. percentages) for several groups of different sizes, the ratio for the combined total of all the groups is not the mean of the proportions for the individual groups. For example, if 20 rats from a batch of 50 are male, this implies 40% are male. If 60 rats from a batch of 120 are male, this implies 50% are male. The mean percentage of males $(50 + 40)/2 = 45\%$ is *not* the percentage of males in the two groups combined, because there are $20 + 60 = 80$ males in a total of 170 rats = 47.1% approx.
3. If the measurement scale is not linear, arithmetic means may give a false value. For example, if three media had pH values 6, 7 and 8, the appropriate mean pH is not 7 because the pH scale is logarithmic. The definition of pH is $-\log_{10}[H]$, where [H] is expressed in $mol\,l^{-1}$ ('molar'); therefore, to obtain the true mean, convert data into [H] values (i.e. put them on a linear scale) by calculating $10^{(-\text{pH value})}$ as shown. Now calculate the mean of these values and convert the answer back into pH units. Thus, the appropriate answer is pH 6.43 rather than 7. Note that a similar procedure is necessary when calculating statistics of dispersion in such cases, so you will find these almost certainly asymmetric about the mean.

pH value	[H] ($mol\,l^{-1}$)
6	1×10^{-6}
7	1×10^{-7}
8	1×10^{-8}
mean	3.7×10^{-7}
$-\log_{10}$ mean	6.43

Mean values of log-transformed data are often termed geometric means – they are sometimes used in microbiology and in cell culture studies, where log-transformed values for cell density counts are averaged and plotted (p. 80), rather than using the raw data values. The use of geometric means in such circumstances serves to reduce the effects of outliers on the mean.

Mode

The mode is the most common value in the sample. The mode is easily found from a tabulated frequency distribution as the most frequent value. The mode provides a rapidly and easily found estimate of sample location and is unaffected by outliers. However, the mode is affected by chance variation in the shape of a sample's distribution and it may lie distant from the obvious centre of the distribution. Note that the mode is the only statistic to make sense of qualitative data, e.g. 'the modal (most frequent) eye colour was blue'.

The mean, median and mode have the same units as the variable under discussion. However, whether these statistics of location have the same or similar values for a given frequency distribution depends on the symmetry and shape of the distribution. If it is near-symmetrical with a single peak, all three will be very similar; if it is skewed or has more than one peak, their values will differ to a greater degree (see Fig. 46.3).

Measuring dispersion

Here, the objective is to quantify the spread of the data about the centre of the distribution. The principal measures of dispersion are the range, variance, standard deviation and coefficient of variation.

Range

The range is the difference between the largest and smallest data values in the sample (the extremes) and has the same units as the measured variable. The range is easy to determine, but is greatly affected by outliers. Its value may also depend on sample size: in general, the larger this is, the greater will be the range. These features make the range a poor measure of dispersion for many practical purposes.

Variance and standard deviation

For symmetrical frequency distributions of quantitative data, an ideal measure of dispersion would take into account each value's deviation from the mean and provide a measure of the average deviation from the mean. Two such statistics are the sample variance, which is the sum of squared deviations from the mean ($\Sigma(Y - \bar{Y})^2$) divided by $n - 1$ (where n is the number of data values), and the sample standard deviation, which is the positive square root of the sample variance.

The sample variance (s^2) has units which are the square of the original units, while the sample standard deviation (s) is expressed in the original units, one reason s is often preferred as a measure of dispersion. Calculating s or s^2 longhand is a tedious job and is best done with the help of a calculator or computer. If you don't have a calculator that calculates s for you, an alternative formula that simplifies calculations is:

Using a calculator for statistics – make sure you understand how to enter individual data values and which buttons will give the sample mean (usually shown as $\bar{X}$ or $\bar{x}$) and sample standard deviation (often shown as σ_{n-1}). In general, you should not use the population standard deviation (usually shown as σ_n).

$$s = +\sqrt{\frac{\Sigma Y^2 - (\Sigma Y)^2/n}{n-1}} \qquad [46.1]$$

To calculate s using a calculator:

1. Obtain ΣY, square it, divide by n and store in memory.
2. Square Y values, obtain ΣY^2, subtract memory value from this.
3. Divide this answer by $n - 1$.
4. Take the positive square root of this value.

Take care to retain significant figures, or errors in the final value of s will result. If continuous data have been grouped into classes, the class mid-values or their squares must be multiplied by the appropriate frequencies before summation. When data values are large, longhand calculations can be simplified by coding the data, e.g. by subtracting a constant from each datum, and decoding when the simplified calculations are complete (see Sokal and Rohlf, 1994).

Coefficient of variation

The coefficient of variation (CoV) is a dimensionless measure of variability relative to location which expresses the sample standard deviation as a percentage of the sample mean, i.e.

$$\mathrm{CoV} = 100s/\bar{Y}\,(\%) \qquad [46.2]$$

This statistic is useful when comparing the relative dispersion of data sets with widely differing means or where different units have been used for the same or similar quantities.

A useful application of the CoV is to compare different analytical methods or procedures, so that you can decide which involves the least proportional error – create a standard stock solution, then compare the results from several sub-samples analysed by each method. You may find it useful to use the CoV to compare the precision of your own results with those of a manufacturer, e.g. for an autopipettor (p. 9). The smaller the CoV, the more precise (repeatable) is the apparatus or technique (note: this does not mean that it is necessarily more *accurate*, see p. 41).

Measuring the precision of the sample mean as an estimate of the true value

Understanding statistical symbols – note that sample statistics are given Roman character symbols while population statistics are given Greek symbols.

Most practical exercises are based on a limited number of individual data values (a sample) which are used to make inferences about the population from which they were drawn. For example, the haemoglobin content might be measured in blood samples from 100 adult females and used as an estimate of the adult female haemoglobin content, with the sample mean ($\bar{Y}$) and sample standard deviation (s) providing estimates of the true values of the underlying population mean (μ) and the population standard deviation (σ). The reliability of the sample mean as an estimate of the true (population) mean can be assessed by calculating the standard error of the sample mean (often abbreviated to standard error or SE), from:

$$\mathrm{SE} = s/\sqrt{n} \qquad [46.3]$$

Strictly, the standard error is an estimate of the standard deviation of the means of n-sized samples from the population. At a practical level, it is clear from Eqn 46.3 that the SE is directly affected by sample dispersion and inversely related to sample size. This means that the SE will decrease as the number of data values in the sample increases, giving increased precision.

Summary descriptive statistics for the sample mean are often quoted as $\bar{Y} \pm \mathrm{SE}\,(n)$, with the SE being given to one significant figure more than the mean. You can use such information to carry out a t-test between two samples (Box 47.1); the SE is also useful because it allows calculation of confidence limits for the sample mean (p. 275).

Example Summary statistics for the sample mean and standard error for the data shown in Box 46.1 would be quoted as 4.95 ± 0.280 ($n = 21$).

Describing the 'shape' of frequency distributions

Frequency distributions may differ in the following characteristics:

- number of peaks;
- skewness or asymmetry;
- kurtosis or pointedness.

The shape of a frequency distribution of a small sample is affected by chance variation and may not be a fair reflection of the underlying population frequency distribution: check this by comparing repeated samples from the same population or by increasing the sample size. If the original shape were due to random events, it should not appear consistently in repeated samples and should become less obvious as sample size increases.

Genuinely bimodal or polymodal distributions may result from the combination of two or more unimodal distributions, indicating that more than one underlying population is being sampled (Fig. 46.5). An example of a bimodal distribution is the height of adult humans (females and males combined).

A distribution is skewed if it is not symmetrical, a symptom being that the mean, median and mode are not equal (Fig. 46.3). Positive skewness is where the longer 'tail' of the distribution occurs for higher values of the measured variable; negative skewness where the longer tail occurs for lower values. Some biological examples of characteristics distributed in a skewed fashion are volumes of plant protoplasts, insulin levels in human plasma and bacterial colony counts.

Kurtosis is the name given to the 'pointedness' of a frequency distribution. A platykurtic frequency distribution is one with a flattened peak, while a leptokurtic frequency distribution is one with a pointed peak (Fig. 46.4). While descriptive terms can be used, based on visual observation of the shape and direction of skew, the degree of skewness and kurtosis can be quantified and statistical tests exist to test the 'significance' of observed values (see Sokal and Rohlf, 1994), but the calculations required are complex and best done with the aid of a computer.

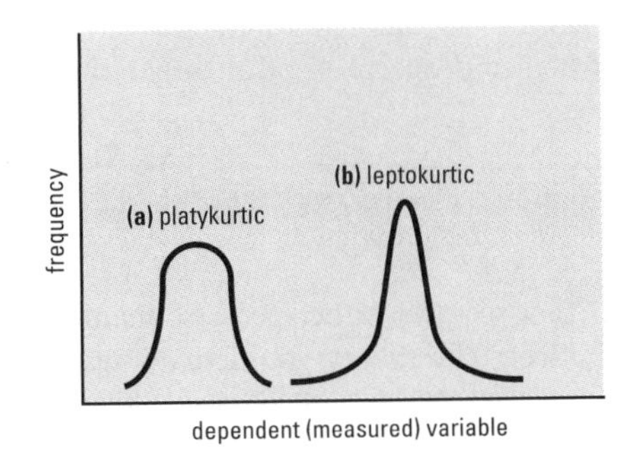

Fig. 46.4 Examples of the two types of kurtosis.

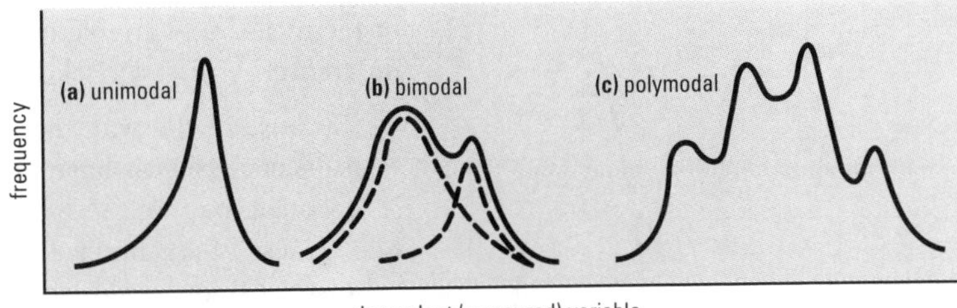

Fig. 46.5 Frequency distributions with different numbers of peaks. A unimodal distribution (a) may be symmetrical or asymmetrical. The dotted lines in (b) indicate how a bimodal distribution could arise from a combination of two underlying unimodal distributions. Note here how the term 'bimodal' is applied to any distribution with two major peaks – their frequencies do not have to be exactly the same.

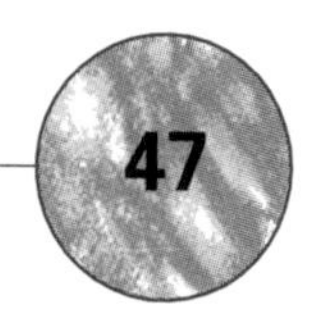

Choosing and using statistical tests

This chapter outlines the philosophy of hypothesis-testing statistics, indicates the steps to be taken when choosing a test, and discusses features and assumptions of some important tests. For details of the mechanics of tests, consult appropriate texts (e.g. Wardlaw, 1985; Sokal and Rohlf, 1994; Heath, 1995). Most tests are now available in statistical packages for computers (see p. 295).

To carry out a statistical test:

1. Decide what it is you wish to test (create a null hypothesis and its alternative).
2. Determine whether your data fit a standard distribution pattern.
3. Select a test and apply it to your data.

Setting up a null hypothesis

Hypothesis-testing statistics are used to compare the properties of samples either with other samples or with some theory about them. For instance, you may be interested in whether two samples can be regarded as having different means, whether the counts of an organism in different quadrats can be regarded as randomly distributed, or whether property A of an organism is linearly related to property B.

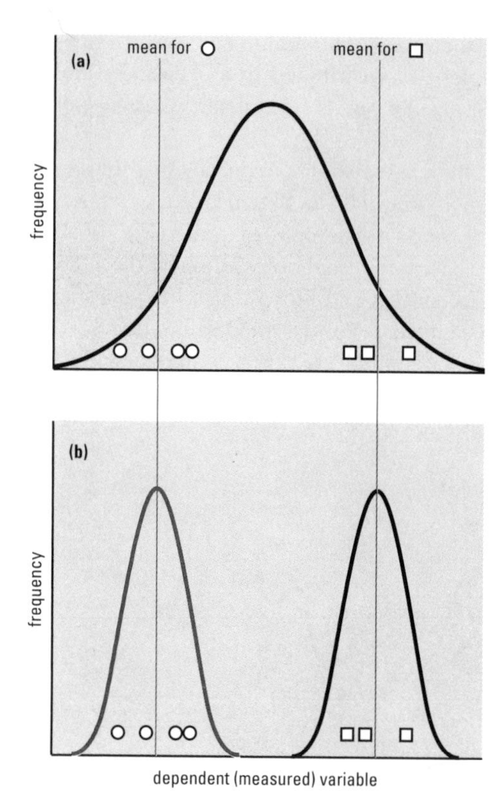

Fig. 47.1 Two explanations for the difference between two means. In case (a) the two samples happen by chance to have come from opposite ends of the same frequency distribution, i.e. there is no true difference between the samples. In case (b) the two samples come from different frequency distributions, i.e. there is a true difference between the samples. In both cases, the means of the two samples are the same.

KEY POINT **You can't use statistics to *prove* any hypothesis, but they can be used to assess *how likely* it is to be wrong.**

Statistical testing operates in what at first seems a rather perverse manner. Suppose you think a treatment has an effect. The theory you actually test is that it has no effect; the test tells you how improbable your data would be if this theory were true. This 'no effect' theory is the null hypothesis (NH). If your data are very improbable under the NH, then you may suppose it to be wrong, and this would support your original idea (the 'alternative hypothesis'). The concept can be illustrated by an example. Suppose two groups of subjects were treated in different ways, and you observed a difference in the mean value of the measured variable for the two groups. Can this be regarded as a 'true' difference? As Fig. 47.1 shows, it could have arisen in two ways:

- Because of the way the subjects were allocated to treatments, i.e. all the subjects liable to have high values might, by chance, have been assigned to one group and those with low values to the other (Fig. 47.1a).
- Because of a genuine effect of the treatments, i.e. each group came from a distinct frequency distribution (Fig. 47.1b).

A statistical test will indicate the probabilities of these options. The NH states that the two groups come from the same population (i.e. the treatment effects are negligible in the context of random variation). To test this, you calculate a test statistic from the data, and compare it with tabulated critical values giving the probability of obtaining the observed or a more extreme result by chance (see Boxes 47.1 and 47.2). This probability is sometimes called the significance of the test.

Note that you must take into account the degrees of freedom (d.f.) when looking up critical values of most test statistics. The d.f. is related to the size(s) of the samples studied; formulae for calculating it depend on the test being used. Life scientists normally use two-tailed tests, i.e. we have no certainty beforehand that the treatment will have a positive or negative effect compared to the control (in a one-tailed test we expect one particular treatment to be bigger than the other). Be sure to use critical values for the correct type of test.

Definition

Modulus – the absolute value of a number, e.g. modulus −3.385 = 3.385.

By convention, the critical probability for rejecting the NH is 5% (i.e. $P = 0.05$). This means we reject the NH if the observed result would have come up less than one time in twenty by chance. If the modulus of the test statistic is less than the tabulated critical value for $P = 0.05$, then we accept the NH and the result is said to be 'not significant' (NS for short). If the modulus of the test statistic is greater than the tabulated value for $P = 0.05$, then we reject the NH in favour of the alternative hypothesis that the treatments had different effects and the result is 'statistically significant'.

Two types of error are possible when making a conclusion on the basis of a statistical test. The first occurs if you reject the NH when it is true and the second if you accept the NH when it is false. To limit the chance of the first type of error, choose a lower probability, e.g. $P = 0.01$, but note that the critical value of the test statistic increases when you do this and results in the probability of the second error increasing. The conventional significance levels given in statistical tables (usually 0.05, 0.01, 0.001) are arbitrary. Increasing use of statistical computer programs is likely to lead to the actual probability of obtaining the calculated value of the test statistic being quoted (e.g. $P = 0.037$).

Choosing between parametric and non-parametric tests – always plot your data graphically when determining whether they are suitable for parametric tests as this may save a lot of unnecessary effort later.

Note that if the NH is rejected, this does not tell you which of many alternative hypotheses is true. Also, it is important to distinguish between statistical and practical significance: identifying a statistically significant difference between two samples doesn't mean that this will carry any biological importance.

Comparing data with parametric distributions

A parametric test is one which makes particular assumptions about the mathematical nature of the population distribution from which the samples were taken. If these assumptions are not true, then the test is obviously invalid, even though it might give the answer we expect! A non-parametric test does not assume that the data fit a particular pattern, but it may assume some things about the distributions. Used in appropriate circumstances, parametric tests are better able to distinguish between true but marginal differences between samples than their non-parametric equivalents (i.e. they have greater 'power').

The distribution pattern of a set of data values may be biologically relevant, but it is also of practical importance because it defines the type of statistical tests that can be used. The properties of the main distribution types found in biology are given below with both rules-of-thumb and more rigorous tests for deciding whether data fit these distributions.

Binomial distributions

These apply to samples of any size from populations when data values occur independently in only two mutually exclusive classes (e.g. type A or type B). They describe the probability of finding the different possible combinations of the attribute for a specified sample size k (e.g. out of 10 specimens, what is the chance of 8 being type A). If p is the probability of the attribute being of type A and q the probability of it being type B, then the expected mean sample number of type A is kp and the standard deviation is $\sqrt{kpq}$. Expected frequencies can be calculated using mathematical expressions (see Sokal and

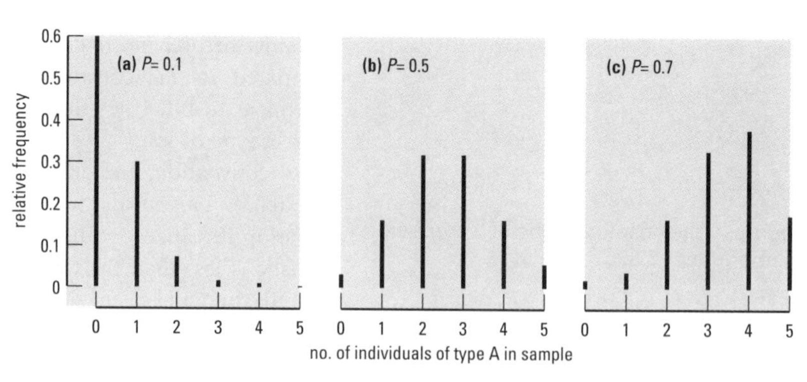

Fig. 47.2 Examples of binomial frequency distributions with different probabilities. The distributions show the expected frequency of obtaining *n* individuals of type A in a sample of 5. Here *P* is the probability of an individual being type A rather than type B.

Rohlf, 1994). Examples of the shapes of some binomial distributions are shown in Fig. 47.2. Note that they are symmetrical in shape for the special case $p = q = 0.5$ and the greater the disparity between p and q, the more skewed the distribution.

Some biological examples of data likely to be distributed in binomial fashion are: possession of two alleles for seed coat morphology (e.g. smooth and wrinkly); whether an organism is infected with a microbe or not; whether an animal is male or female. Binomial distributions are particularly useful for predicting gene segregation in Mendelian genetics (Chapter 38) and can be used for testing whether combinations of events have occurred more frequently than predicted (e.g. more siblings being of the same sex than expected). To establish whether a set of data is distributed in binomial fashion: calculate expected frequencies from probability values obtained from theory or observation, then test against observed frequencies using a χ^2-test or a G-test.

Tendency towards the normal distribution – under certain conditions, binomial and Poisson distributions can be treated as normally distributed:

- where samples from a binomial distribution are large (i.e. > 15) and p and q are close to 0.5;
- for Poisson distributions, if the number of counts recorded in each outcome is greater than about 15.

Poisson distributions

These apply to discrete characteristics which can assume low whole number values, such as counts of events occurring in area, volume or time. The events should be 'rare' in that the mean number observed should be a small proportion of the total that could possibly be found. Also, finding one count should not influence the probability of finding another. The shape of Poisson distributions is described by only one parameter, the mean number of events observed, and has the special characteristic that the variance is equal to the mean. The shape has a pronounced positive skewness at low mean counts, but becomes more and more symmetrical as the mean number of counts increases (Fig. 47.3).

Some examples of characteristics distributed in a Poisson fashion are: number of microbes per unit volume of medium; number of radioactive disintegrations per unit time. One of the main uses for the Poisson distribution is to quantify errors in count data such as estimates of cell densities in dilute suspensions (p. 82). To decide whether data are Poisson distributed:

- Use the rule-of-thumb that if the coefficient of dispersion ≈ 1, the distribution is likely to be Poisson.
- Calculate 'expected' frequencies from the equation for the Poisson distribution and compare with actual values using a χ^2-test or a G-test.

Definition

Coefficient of dispersion = $s^2/\bar{Y}$. This is an alternative measure of dispersion to the coefficient of variation (p. 266)

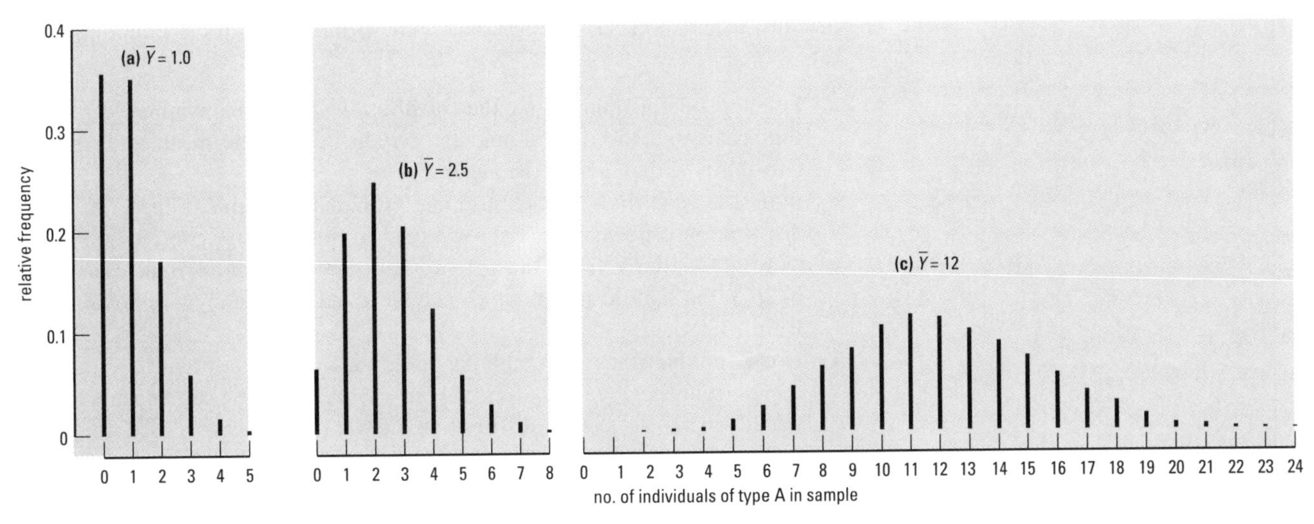

Fig. 47.3 Examples of Poisson frequency distributions differing in mean. The distributions are shown as line charts because the independent variable (events per sample) is discrete.

It is sometimes of interest to show that data are *not* distributed in a Poisson fashion. If $s^2/\bar{Y} > 1$, the data are 'clumped' and occur together more than would be expected by chance; if $s^2/\bar{Y} < 1$, the data are 'repulsed' and occur together less frequently than would be expected by chance.

Normal distributions (Gaussian distributions)

These occur when random events act to produce variability in a continuous characteristic (quantitative variable). This situation occurs frequently in biology, so normal distributions are very useful and much used. The bell-like shape of normal distributions is specified by the population mean and standard deviation (Fig. 47.4): it is symmetrical and configured such that 68.27% of the data will lie within ± 1 standard deviation of the mean, 95.45% within ± 2 standard deviations of the mean, and 99.73% within ± 3 standard deviations of the mean.

Some biological examples of data likely to be distributed in a normal fashion are: linear dimensions of bacterial cells; height of either adult women

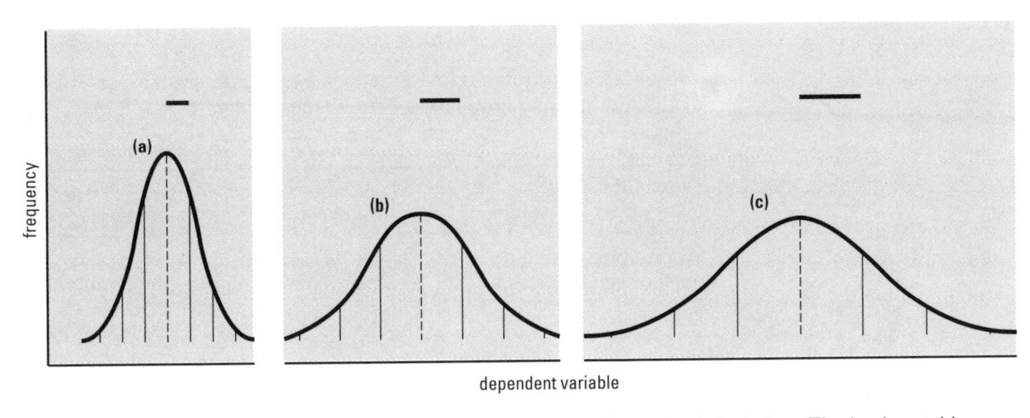

Fig. 47.4 Examples of normal frequency distributions differing in mean and standard deviation. The horizontal bars represent population standard deviations for the curves, increasing from (a) to (c). Vertical dashed lines are population means, while vertical solid lines show positions of values ± 1, 2 and 3 standard deviations from the means.

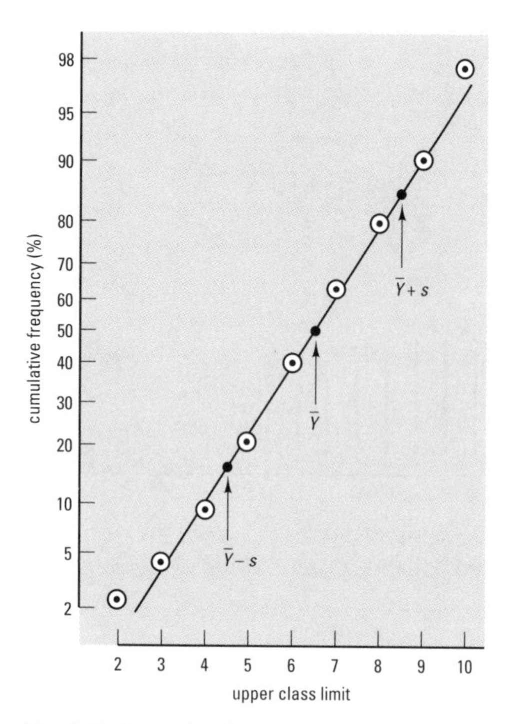

Fig. 47.5 Example of a normal probability plot. The plotted points are from a small data set where the mean $\bar{Y} = 6.93$ and the standard deviation $s = 1.895$. Note that values corresponding to 0% and 100% cumulative frequency cannot be used. The straight line is that predicted for a normal distribution with $\bar{Y} = 6.93$ and $s = 1.895$. This is plotted by calculating the expected positions of points for $\bar{Y} \pm s$. Since 68.3% of the distribution falls within these bounds, the relevant points on the cumulative frequency scale are $50 \pm 34.15\%$; thus this line was drawn using the points (4.495, 15.85) and (8.285, 84.15) as indicated on the plot.

or adult men. To check whether data come from a normal distribution, you can:

- Use the rule-of-thumb that the distribution should be symmetrical and that nearly all the data should fall within $\pm 3s$ of the mean and about two-thirds within $\pm 1s$ of the mean.
- Plot the distribution on normal probability graph paper. If the distribution is normal, the data will tend to follow a straight line (see Fig. 47.5). Deviations from linearity reveal skewness and/or kurtosis (see p. 267), the significance of which can be tested statistically (see Sokal and Rohlf, 1994).
- Use a suitable statistical computer program to generate predicted normal curves from the $\bar{Y}$ and s values of your sample(s). These can be compared visually with the actual distribution of data and can be used to give 'expected' values for a χ^2-test or a G-test.

The wide availability of tests based on the normal distribution and their relative simplicity means you may wish to transform your data to make them more like a normal distribution. Table 47.1 provides transformations that can be applied. The transformed data should be tested for normality as described above before proceeding - don't forget that you may need to check that transformed variances are homogeneous for certain tests (see below).

A very important theorem in statistics, the Central Limit Theorem, states that as sample size increases, the distribution of a series of means from any frequency distribution will become normally distributed. This fact can be used to devise an experimental or sampling strategy that ensures that data are normally distributed, i.e. using means of samples as if they were primary data.

Choosing a suitable statistical test

Comparing location (e.g. means)

If you can assume that your data are normally distributed, the main test for comparing two means from independent samples is Student's t-test (see Boxes 47.1 and 47.2, and Table 47.2). This assumes that the variances of the data sets are homogeneous. Tests based on the t-distribution are also available for comparing paired data or for comparing a sample mean with a chosen value.

When comparing means of two or more samples, analysis of variance (ANOVA) is a very useful technique. This method also assumes data are

Table 47.1 Suggested transformations altering different types of frequency distribution to the normal type. To use, modify data by the formula shown; then examine effects with the tests described on p. 269–71.

Type of data; distribution suspected	Suggested transformation(s)
Proportions (including percentages); binomial	arcsine $\sqrt{x}$ (also called the angular transformation)
Scores; Poisson	$\sqrt{x}$ or $\sqrt{(x + 1/2)}$ if zero values present
Measurements; negatively skewed	x^2, x^3, x^4, etc. (in order of increasing strength)
Measurements; positively skewed	$1/\sqrt{x}$, $\sqrt{x}$, $\ln x$, $1/x$ (in order of increasing strength)

Box 47.1 How to carry out a *t*-test

The *t*-test was devised by a statistician who used the pen-name 'Student', so you may see it referred to as Student's *t*-test. It is used when you wish to decide whether two samples come from the same population or from different ones (Fig. 47.1). The samples might have been obtained from two different sources or by applying two different treatments to an originally homogeneous population (Chapter 1).

The null hypothesis (NH) is that the two groups can be represented as samples from the same overlying population (Fig.47.1a). If, as a result of the test, you accept this hypothesis, you can say that there is no significant difference between the groups.

The alternative hypothesis is that the two groups come from different populations (Fig.46.1b). By rejecting the NH as a result of the test, you can accept the alternative hypothesis and say that there is a significant difference between the samples, or, where an experiment has been carried out, that the two treatments affected the samples differently.

How can you decide between these two hypotheses? On the basis of certain assumptions (see below), and some relatively simple calculations, you can work out the probability that the samples came from the same population. If this probability is very low, then you can reasonably reject the NH in favour of the alternative hypothesis, and if it is high, you will accept the NH.

To find out the probability that the observed difference between sample means arose by chance, you must first calculate a '*t* value' for the two samples in question. Some computer programs (e.g. Minitab®) provide this probability as part of the output, otherwise you can look up statistical tables (e.g. Table 47.2). These tables show 'critical values' – the borders between probability levels. If your value of *t* exceeds the critical value for probability *P*, you can reject the NH at this probability ('level of significance').

Note that:

- for a given difference in the means of the two samples, the value of *t* will get larger the smaller the scatter within each data set; and
- for a given scatter of the data, the value of *t* will get larger, the greater the difference between the means.

So, at what probability should you reject the NH? Normally, the threshold is arbitrarily set at 5% – in scientific literature you often see descriptions like 'the sample means were significantly different ($P < 0.05$)'. At this 'significance level' there is still up to a 5% chance of the *t* value arising by chance, so about 1 in 20 times, on average, the conclusion will be wrong. If *P* turns out to be lower, then this kind of error is much less likely.

Tabulated probability levels are generally given for 5%, 1% and 0.1% significance levels (see Table 47.2). Note that this table is designed for 'two-tailed' tests, i.e. where the treatment or sampling strategy could have resulted in either an increase *or* a decrease in the measured values. These are the most likely situations you will deal with in bioscience.

Examine Table 47.2 and note the following:

- The larger the size of the samples (i.e. the greater the 'degrees of freedom'), the smaller *t* needs to be to exceed the critical value at a given significance level.
- The lower the probability, the greater *t* needs to be to exceed the critical value.

The mechanics of the test

A calculator that can work out means and standard deviations is helpful.

1. **Work out the sample means $\bar{Y}_1$ and $\bar{Y}_2$ and calculate the difference between them ($\bar{Y}_1 - \bar{Y}_2$).**
2. **Work out the sample standard deviations s_1 and s_2.** (NB if your calculator offers a choice, chose the 'n-1' option for calculating s – see p. 265).
3. **Work out the sample standard errors $SE_1 = s_1/\sqrt{n_1}$ and $SE_2 = s_2/\sqrt{n_2}$; now square each, add the squares together, then take the square root of this** (n_1 and n_2 are the respective sample sizes, which may, or may not, not be equal) $\sqrt{(SE_1)^2 + (SE_2)^2)}$.
4. **Calculate *t*** from the formula:

$$t = \frac{\bar{Y}_1 - \bar{Y}_2}{\sqrt{((SE_1)^2) + (SE_2)^2)}} \qquad [47.1]$$

   The value of *t* can be negative or positive, depending on the values of the means; this does not matter and you should compare the modulus (absolute value) of *t* with the values in tables.
5. **Work out the degrees of freedom** $= (n_1 - 1) + (n_2 - 1)$.
6. **Compare the *t* value with the appropriate critical value in a table**, e.g. Table 47.2.

Box 47.2 provides a worked example – use this to check that you understand the above procedures.

Assumptions that must be met before using the test

The most important assumptions are:

- The two samples are independent and randomly drawn (or if not, drawn in a way that does not create bias). The test assumes that the samples are quite large.
- The underlying distribution of each sample is normal. This can be tested with a special statistical test, but a rule of thumb is that a frequency distribution of the data should be (a) symmetrical about the mean and (b) nearly all of the data should be within 3 standard deviations of the mean and about two-thirds within 1 standard deviation of the mean (see p. 271).
- The two samples should have equal variances. This again can be tested (by an F-test), but may be gauged from inspection of the two standard deviations.

Box 47.2 Worked example of a *t*-test

Suppose the following data were obtained in an experiment (the units are not relevant):

Control: 6.6, 5.5, 6.8, 5.8, 6.1, 5.9
Treatment: 6.3, 7.2, 6.5, 7.1, 7.5, 7.3

Using the steps outlined in Box 47.1, the following values are obtained (denoting control with subscript 1, treatment with subscript 2):

1. $\bar{Y}_1 = 6.1167$; $\bar{Y}_2 = 6.9833$: difference between means $= \bar{Y}_1 - \bar{Y}_2 = -0.8666$
2. $s_1 = 0.49565$; $s_2 = 0.47504$
3. $SE_1 = 0.49565/2.44949 = 0.202348$
   $SE_2 = 0.47504/2.44949 = 0.193934$
4. $t = \dfrac{-0.8666}{\sqrt{(0.202348^2 + 0.193934^2)}} = \dfrac{-0.8666}{0.280277} = -3.09$
5. d.f. $= (5 + 5) = 10$
6. Looking at Table 47.2, we see that the modulus of this t value exceeds the tabulated value for $P = 0.05$ at 10 degrees of freedom $(= 2.23)$. We therefore reject the NH, and conclude that the means are different at the 5% level of significance. If the modulus of t had been < 2.23, we would have accepted the NH. If modulus of t had been > 3.17, we could have concluded that the means are different at the 1% level of significance.

Table 47.2 Critical values of Student's t statistic (for two-tailed tests). Reject the Null Hypothesis at probability P if your calculated t value exceeds the value shown for the appropriate degrees of freedom $= (n_1 - 1) + (n_2 - 1)$

Degrees of freedom	Critical values for $P = 0.05$	Critical values for $P = 0.01$	Critical values for $P = 0.001$
1	12.71	63.66	636.62
2	4.30	9.92	31.60
3	3.18	5.84	12.94
4	2.78	4.60	8.61
5	2.57	4.03	6.86
6	2.45	3.71	5.96
7	2.36	3.50	5.40
8	2.31	3.36	5.04
9	2.26	3.25	4.78
10	2.23	3.17	4.59
12	2.18	3.06	4.32
14	2.14	2.98	4.14
16	2.12	2.92	4.02
20	2.09	2.85	3.85
25	2.06	2.79	3.72
30	2.04	2.75	3.65
40	2.02	2.70	3.55
60	2.00	2.66	3.46
120	1.98	2.62	3.37
∞	1.96	2.58	3.29

normally distributed and that the variances of the samples are homogeneous. The samples must also be independent (e.g. not sub-samples). The nested types of ANOVA are useful for letting you know the relative importance of different sources of variability in your data. Two-way and multi-way ANOVAs are useful for studying interactions between treatments.

For data satisfying the ANOVA requirements, the least significant difference (LSD) is useful for making planned comparisons among several means. Any two means that differ by more than the LSD will be significantly different. The LSD is useful for showing on graphs.

Checking the assumptions of a test – always acquaint yourself with the assumptions of a test. If necessary, test them before using the test.

The chief non-parametric tests for comparing locations are the Mann–Whitney U-test and the Kolmogorov–Smirnov test. The former assumes that the frequency distributions of the data sets are similar, whereas the latter makes no such assumption. In the Kolmogorov–Smirnov test, significant differences found with the test may be due to differences in location or shape of the distribution, or both.

Suitable non-parametric comparisons of location for paired quantitative data (sample size $\geqslant 6$) include Wilcoxon's signed rank test, which assumes that the distributions have similar shape.

Non-parametric comparisons of location for three or more samples include the Kruskal–Wallis H-test. Here, the two data sets can be unequal in size, but again the underlying distributions are assumed to be similar.

Comparing dispersions (e.g. variances)

If you wish to compare the variances of two sets of data that are normally distributed, use the F-test. For comparing more than two samples, it may be sufficient to use the F_{max}-test, on the highest and lowest variances. The Scheffé–Box (log-ANOVA) test is recommended for testing the significance of differences between several variances. Non-parametric tests exist but are not widely available: you may need to transform the data and use a test based on the normal distribution.

Determining whether frequency observations fit theoretical expectation

The χ^2-test (Box 38.2) is useful for tests of 'goodness of fit', e.g. comparing expected and observed progeny frequencies in genetical experiments or comparing observed frequency distributions with some theoretical function. One limitation is that simple formulae for calculating χ^2 assume that no expected number is less than 5. The G-test ($2I$ test) is used in similar circumstances.

Comparing proportion data

When comparing proportions between two small groups (e.g. whether 3/10 is significantly different from 5/10), you can use probability tables such as those of Finney *et al.*, (1963) or calculate probabilities from formulae; however, this can be tedious for large sample sizes. Certain proportions can be transformed so that their distribution becomes normal.

Placing confidence limits on an estimate of a population parameter

Confidence limits for statistics other than the mean – consult an advanced statistical text (e.g. Sokal and Rohlf, 1994) if you wish to indicate the reliability of estimates of e.g. population variances.

On many occasions, sample statistics are used to provide an estimate of the population parameters. It is extremely useful to indicate the reliability of such estimates. This can be done by putting a confidence limit on the sample statistic. The most common application is to place confidence limits on the mean of a sample from a normally distributed population. This is done by working out the limits as $\bar{Y} - (t_{P[n-1]} \times \text{SE})$ and $\bar{Y} + (t_{P[n-1]} \times \text{SE})$ where $t_{P[n-1]}$ is the tabulated critical value of Student's t statistic for a two-tailed test with $n - 1$ degrees of freedom and SE is the standard error of the mean (p. 266). A 95% confidence limit (i.e. $P = 0.05$) tells you that on average, 95 times out of 100, this limit will contain the population mean.

Regression and correlation

These methods are used when testing relationships between samples of two variables. If one variable is assumed to be dependent on the other then regression techniques are used to find the line of best fit for your data. This

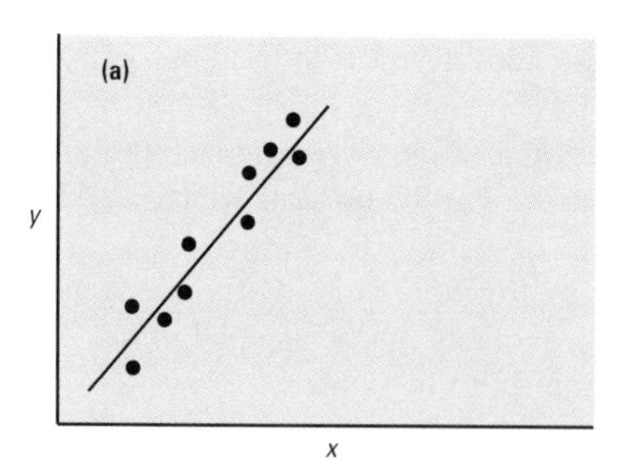

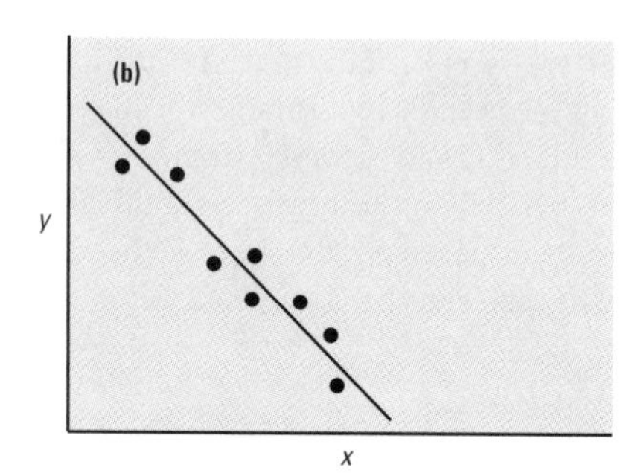

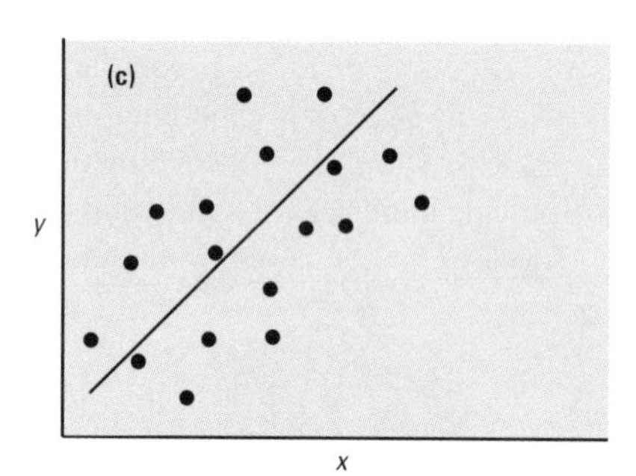

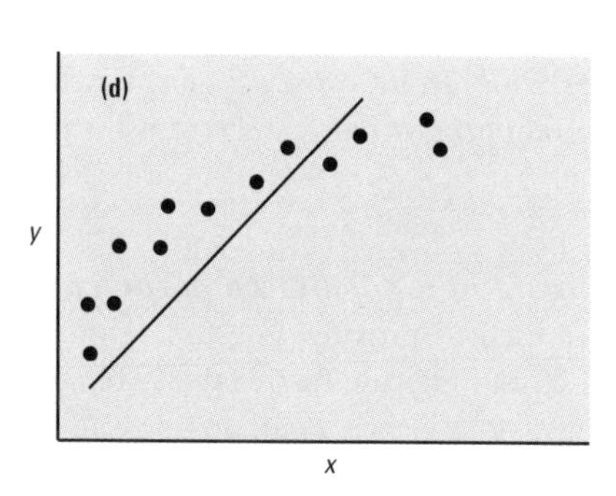

Fig. 47.6 Examples of correlation. The linear regression line is shown. In (a) and (b), the correlation between *x* and *y* is good: for (a) there is a positive correlation and the correlation coefficient would be close to 1; for (b) there is a negative correlation and the correlation coefficient would be close to −1. In (c) there is a weak positive correlation and *r* would be close to 0. In (d) the correlation coefficient may be quite large, but the choice of linear regression is clearly inappropriate.

does not tell you how well the data fit the line: for this, a correlation coefficient must be calculated. If there is no *a priori* reason to assume dependency between variables, correlation methods alone are appropriate.

If graphs or theory indicate a linear relationship between a dependent and an independent variable, linear regression can be used to estimate the equation that links them. If the relationship is not linear, a transformation may give a linear relationship. For example, this is sometimes used in analysis of enzyme kinetics (see Fig. 35.3). However, 'linearizations' can lead to errors when carrying out regression analysis: take care to ensure (a) that the data are evenly distributed throughout the range of the independent variable and (b) that the variances of the dependent variable are homogeneous. If these criteria cannot be met, weighting methods may reduce errors. In this situation, it may be better to use non-linear regression using a suitable computer program.

Model I linear regression is suitable for experiments where a dependent variable Y varies with an *error-free* independent variable X and the mean (expected) value of Y is given by $a + bX$. This might occur where you have carefully controlled the independent variable and it can therefore be assumed to have zero error (e.g. a calibration curve). Errors can be calculated for estimates of a and b and predicted values of Y. The Y values should be normally distributed and the variance of Y constant at all values of X.

Model II linear regression is suitable for experiments where a dependent variable Y varies with an independent variable X which has an error associated with it and the mean (expected) value of Y is given by $a + bX$. This might occur where the experimenter is measuring two variables and believes there to be a causal relationship between them; both variables will be subject to errors in this case. The exact method to use depends on whether your aim is to estimate the functional relationship or to estimate one variable from the other.

A correlation coefficient measures the strength of relationships but does not describe the relationship. These coefficients are expressed as a number between −1 and 1. A positive coefficient indicates a positive relationship while a negative coefficient indicates a negative relationship (Fig. 47.6). The nearer the coefficient is to −1 or 1, the stronger the relationship between the variables, i.e. the less scatter there would be about a line of best fit (note that this does not imply that one variable is dependent on the other!). A coefficient of 0 implies that there is no relationship between the variables. The importance of graphing data is shown by the case illustrated in Fig. 47.6d.

Pearson's product-moment correlation coefficient (r) is the most commonly used correlation coefficient. If both variables are normally distributed, then r can be used in statistical tests to test whether the degree of correlation is significant. If one or both variables are not normally distributed you can use Kendall's coefficient of rank correlation (τ) or Spearman's coefficient of rank correlation (r_s). They require that data are ranked separately and calculation can be complex if there are tied ranks. Spearman's coefficient is said to be better if there is uncertainty about the reliability of closely ranked data values.

Information technology and library resources

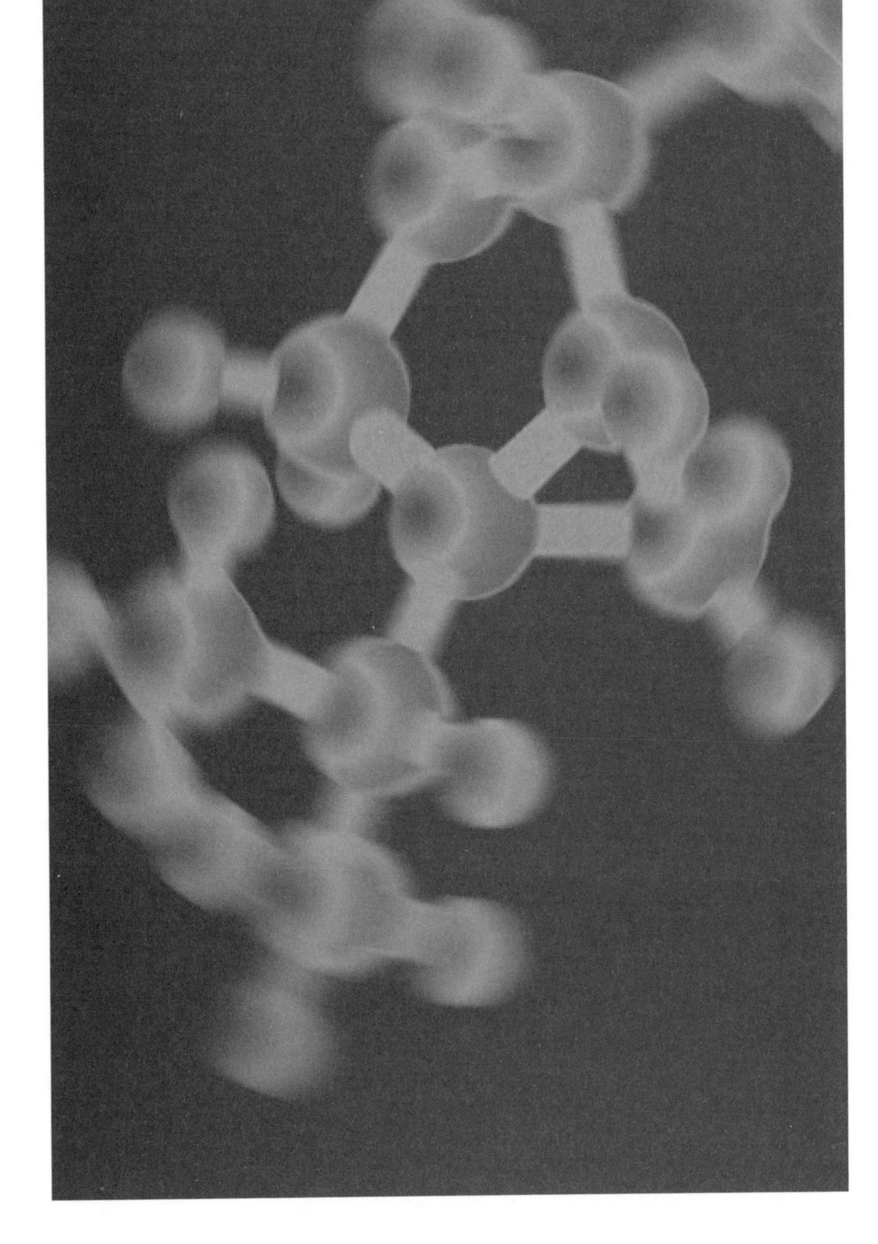

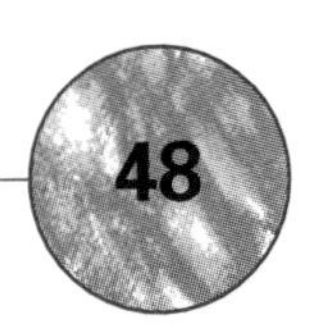

48 Introduction to the Internet and World Wide Web

Definitions

Browser – a program to display Web pages and other Internet resources.

FAQ – Frequently Asked Question; sometimes used as a file extension (.faq) for down-loadable files.

FTP – File Transfer Protocol; means of down-loading files.

URL – Uniform Resource Locator: the 'address' for WWW resources.

Making the most of information technology ('IT') requires skills related to the use of computers for finding, retrieving, recording, analysing and communicating information, especially via the developing global Internet environment. This includes using programs for:

- searching databases and Internet resources, e.g. using Web 'browsers' such as Netscape™;
- retrieving network resources, e.g. applying FTP to obtain copies of files;
- storing, modifying and analysing information, e.g. using databases, spreadsheets and statistical packages (see Chapters 50 and 51);
- communicating information, e.g. by e-mail, word processors, desk-top publishing packages and programs for making presentations.

The Internet as a global resource

The key to the rapid development of the Internet was the evolution and expansion of networks – collections of computers which can communicate with each other. They operate at various scales such as Local Area Networks (LANs) and Wide Area Networks (WANs) and these can be connected to the Internet, which is a complex network of networks (Fig. 48.1). The Internet is loosely organized; no one group runs it or owns it. Instead, many private organizations, universities and government organizations pay for and run discrete parts of it. Private organizations include commercial on-line service providers such as America On-line™ and CompuServe™.

You can gain access the Internet either through a LAN at your place of work or from home via a modem connected to a dial-in service provider over the telephone line. You do not need to understand the technology of the network to use it – most of it is invisible to the user. However, if you do wish to understand more, sources such as Gralla (1996) and Winship and McNab (1996) are recommended. What you do need to know are the options available to you in terms of using its facilities and their relative merits or disadvantages. You also need to understand a little about the nature of Internet addresses.

Fig. 48.1 Diagram of a network system.

KEY POINT **Most material on the Internet has not been subject to peer review, vetting or editing. Information obtained from the WWW or posted on newsgroups may be inaccurate, biased or spoof; do not assume that everything you read is true or even legal.**

Communicating on the Internet: e-mail and newsgroups

Using e-mail, you are able to send messages to anyone who is connected to the Internet, directly or indirectly. You can attach text files, data files, pictures, video clips, sounds and executable files to your messages. The messages themselves are usually only very simple in format but sufficient for most purposes: formatted material can be attached as a file if necessary. The uses that can be made of the system vary from personal and business-related to the submission of work to a tutor in an educational system. Note,

Junk and chain mail on the Internet – do not get involved in distributing junk or chain mail (the electronic equivalent of chain letters). This activity is likely to be against your institute's rules for the use of its computer facilities because it can foul up the operation of networks.

Examples Common domains and sub-domains include:

.com	commercial (USA mainly)
.edu	education (USA mainly)
.gov	government (USA only)
.mil	military (USA only)
.net	network companies
.org	organization
.uk	United Kingdom

however, that the system is not secure (confidential) and the transmission of sensitive information should be done with caution. Another downside of e-mail is the junk-mail that you may receive once your e-mail address is distributed, e.g. in newsletters.

Specific address information is required to exchange information and mail between computers. Although the computer actually uses a complex series of numbers for this purpose (the 'I.P.' address), the Domain Name System/Service (DNS) was developed to make this easier for users. Each computer on the Internet is given a domain name (= Internet address) which is a hierarchy of lists and addresses. Thus, 'unn.ac.uk' is the DNS-registered name of the University of Northumbria at Newcastle: the top (root) level domain is 'uk', identifying its country as United Kingdom, and the next is 'ac', identifying the academic community sub-domain. The final sub-domain is 'unn' identifying the specific academic institute. For the purposes of e-mail, the names of individuals at that site may be added before the domain name and the @ sign is used to separate them. Thus, the e-mail address of the first author of this book is 'rob.reed@unn.ac.uk'.

Internet tools

There are various facilities available for use on the Internet depending on your own system and method of accessing the system. The best way to learn how to use them is simply to try them out.

Locating information on the WWW – useful searching systems are located at the following URLs (some may be directly accessible from your browser):

http://www.altavista.com

http://www.goto.com/

http://www.lycos.com/

http://www.webcrawler.com/

http://www.yahoo.com

The World Wide Web (WWW) The 'Web' is the most popular Internet application. It allows easy links to information and files which may be located on computers anywhere in the world. The WWW allows access to millions of 'home pages' or 'Web sites', the initial point of reference with companies, institutes and individuals. Besides their own text and images, these contain 'hypertext links', highlighted words or phrases that you click on to take you to another page on the same Web site or to a completely different site with related subject material. Certain sites specialize in such links, acting like indices to other Web sites: these are particularly useful.

When using a Web browser program to get to a particular page of information on the Web all you require is the name and location of that page. The page location is commonly referred to as a URL (Uniform Resource Locator). The URL always takes the same basic format, beginning with 'http://' and followed by the various terms which direct the system to the resource pages. If you don't have a specific URL in mind but wish to find appropriate sites, use a 'search engine' within the browser: enter appropriate and limited keywords on which to search and note the site(s) which may be of interest. There is so much available on the Web, from company products through library resources to detailed information on specialist biological and environmental topics, that much time can be spent searching and reading information: try to stay focused!

Wide Area Information Systems – the WWW is an example of a WAIS, but there are others. Use Telnet to access other public WAIS systems on the Internet. 'Gophers' can be used to search one WAIS database at a time. Gophers are predecessors of Web browsers but use text only. They are menu-based and allow you to navigate through 'Gopher-space' using searching systems such as *Veronica*. This is much more powerful than using Telnet or Gopher alone.

Telnet This is a simple system allowing connection between computers on the net so that they work as if you were directly connected to each. It is most frequently used to access structured collections such as a library catalogue or bibliographic database. You normally follow a pre-arranged connection routine involving log-in passwords, but some sites will allow limited access to users who log in as 'guest'. An easy way to discover the uses of Telnet is to log into *http://access.usask.ca*.

FTP (File Transfer Protocol) and file transmission FTP is a method of transferring files across the Internet. In may cases the files are made available for 'anonymous' FTP access, i.e. you do not need previously arranged passwords. Log in as 'anonymous' and give your e-mail address as the password. Use your Web browser to locate the file you want and then use its FTP software to transfer it to your computer.

KEY POINT **The transfer of files can result in the transfer of associated viruses. Always check your files for viruses before running them.**

Archie This searching system keeps track of the main FTP servers on the Internet. Use it for locating a file by specifying the name of the file or search keywords, and it will locate the file or files for you. You can use Archie either by e-mailing your request to *archie@archie.doc.ic.ac.uk* with the message: 'find keyword/s' or by Telnet to an Archie site and typing the *find* command at the prompt. Use *quit* to finish.

When using information technology, including the Internet, always remember the basic rules of using computers and networks (Box 48.1).

Box 48.1 Important guidelines for using computers and networks

Hardware

- Don't drink or smoke around the computer.
- Try not to turn the computer off more than is necessary.
- Never turn off the electricity supply to the machine while in use.
- Rest your eyes at frequent intervals if working for extended periods at a computer monitor.
- Never try to re-format the hard disk except in special circumstances.

Floppy disks

- Protect floppy disks when not in use by keeping them in holders or boxes.
- Never touch a floppy disk's recording surface.
- Keep disks away from dampness, excess heat or cold.
- Keep disks well away from magnets; remember these are present in e.g. loudspeakers, TVs, etc.
- Don't use disks from others unless first checked for viruses.
- Don't insert or remove a disk from the drive when it is operating (drive light on).
- Try not to leave a disk in the drive when you switch the computer off.

File management

- Always use virus-checking programs on imported files before running them.
- Make backups of all important files and at frequent intervals (say every half hour) during the production of your own work e.g. when using a word processor or spreadsheet.
- Periodically clear out and reorganize your file storage areas.

Network rules

- Be polite when sending messages.
- Never attempt to 'hack' into other people's files.
- Do not play games without approval – they can hinder the operation of the system.
- Periodically reorganize your e-mail folder(s). These rapidly become filled with acknowledgements and redundant messages which reduce server efficiency.
- Remember to log out of the network when finished: others can access your files if you forget to log out.

The golden rule: Always make backup copies of important disks/files and store them well away from your working copies. Be sure that the same accident cannot happen to both copies.

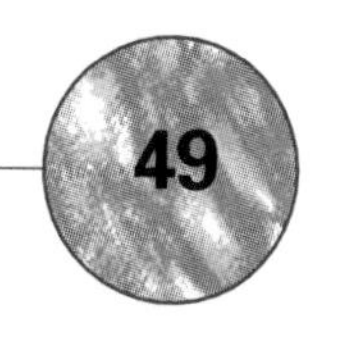

49 Internet resources for molecular biology

Revisiting Web sites – 'bookmark' sites of interest, so that you can return to them at a later date. Make a copy of your bookmark file occasionally, to avoid loss of relevant information.

A common way to find information on the Internet is by browsing ('surfing') the Internet, and especially the World Wide Web (WWW). However, as this can be time-consuming and wasteful, browsing should be focused on relevant sites. Many of the most useful Web sites are those providing detailed lists and hypertext links to other locations. A useful place to find out what the Internet offers the life scientist is *The Biologist's Guide to Internet Resources*, available from the URL http://www.csc.fi/molbio/una/unapost.index.html, a Yale University site. Alternatively, you could try the molecular biology 'gateway' at http://www.almac.net/a/gateway.html, or the OMNI gateway (Organizing Medical Networked Information) at http://omni.ac.uk/.

KEY POINT **Remember that the Internet should not be viewed as a substitute for your University library and other local resources, but should complement, rather than replace, more traditional printed texts and CD-ROM material.**

General information

Some of the principal resources you can utilize via the WWW are:

- Libraries, publishers and companies. These organizations recognize the significance of the Internet as a means of communication – for example, the Addison Wesley Longman higher education Web site at http://www.awl-he.com/ allows information on specific catalogues and books to be requested on-line. A number of conventional scientific journals are also available in electronic format, e.g. the Journal of Molecular Biology (at http://www.hbuk.co.uk/jmb/), though some journals require a fee for access – check whether your institution subscribes. A broad range of scientific journals published by Academic Press is available via the site http://www.apnet.com/, and students at UK colleges and universities should have full access without additional subscription (at http://www.janet.idealibrary.com/). You can keep up to date using New Scientist pages (at http://www.newscientist.com/), Scientific American (at http://sciam.com/) or Nature (at http:/www.nature.com/). A few of the newer titles are available only in electronic format, e.g. the e-journals Biochemistry On-line (at http://www.arach-net.com/~jlyon/biochem/index.html), or Molecular Vision (at http://www1.cc.emory.edu/MOLECULAR_VISION/), and electronic publishing is likely to develop further in future years.
- Institutions. Many research organizations, societies and educational institutions around the world are now on-line, with their own Web pages. There is a detailed list of scholarly biological societies at http://www.lib.uwaterloo.ca/society/biol_soc.html. You can use the Web sites of these organizations to obtain specific information – they frequently provide hypertext links to other relevant sites for particular groups or topics: for example, the Society for General Microbiology Web page (http://www.socgenmicrobiol.org.uk/links.htm) has links to various sites of interest to microbiologists and molecular biologists. Use the WWW to obtain details of the activities of research organizations (e.g. the Imperial Cancer Research Fund, at http://www.icnet.uk/) or individual laboratories and researchers at specific universities. Some other relevant Web sites are given in Table 49.1.

Table 49.1 Selected examples of useful Web sites

Biochemical Society: http://www.biochemsoc.org.uk
American Society for Microbiology: http://www.asmusa.org/
Physiology Society: http://physiology.cup.com.ac.uk
Sanger Centre, Cambridge, UK: http://www.sanger.ac.uk
E. coli index: http://sun1.bham.ac.uk/bcm4ght6/res.html
Molecular biology protocols: http://research.nwfsc.noaa.gov/protocols.html
Cells alive (animated graphics site): http://www.comet.net/quill/

Using Internet addresses – note that the locations given in the Chapter may change as sites are updated: you can make a keyword search using a searching system or 'search engine' (p. 280) to find a particular Web site if necessary.

- Data and pictures. Archives of text material, photographs and video clips can be accessed and downloaded using Telnet and anonymous FTP (p. 280). However, downloading graphical images may take quite a while, especially for remote links or at busy times, leading to high costs.
- Newsgroups (Usenet). News articles on a wide range of topics are 'posted' at appropriate sites, where they are placed into subject groups (newsgroups). Any user can contribute to the discussion by posting his/her own message, to be read by other users via appropriate news software. While the number of newsgroups is very large, some of them may include relevant information – further information on topics of biological interest is available at http://www.bio.net/ (BIOSCI/bionet newsgroups). Student access to newsgroups may be limited at some universities and colleges.

KEY POINT Remember that the information from Internet newsgroups and similar Web sites may be unedited and represents the personal opinion of the author of the article.

- Mailing lists. Messages sent to the list address are distributed automatically to all members of the mailing list, via their personal e-mailbox, keeping them up to date on the particular topic of the mailing list. To receive such messages, you will need to join the mailing list. Relevant mailing lists for biological sciences can be found at http://www.n2h2.com/kovacs/s0028s.html. Take care not to join too many lists, as you will receive a large number of messages, and many are likely to be of only marginal interest. A number of mailing lists also have archived files, offering a more selective means of locating relevant material.
- Databases. Sites such as Bath Information and Data Services (BIDS) provide access to abstracts of recent publications – use this to find relevant literature for specific topics. Access is via the Web site at http://www.bids.ac.uk/, or via a Telnet link (p. 280). You will need a username and password – check with your Department or library. Other databases include MEDLINE, which is primarily a subscription database, although free versions with a restricted range of facilities are available via the Web (e.g. at http://www.healthgate.com/), and OMIM (online mendelian inheritance in man, dealing with aspects of human inherited disease, at http://www3.ncbi.nlm.nih.gov/omim/).

Analysis of biological macromolecules

Example The chromosome of *E. coli* contains around 4 million base pairs, coding for several thousand genes, while the human genome (haploid) contains ≈3 billion base pairs, p. 228.

This is one of the most useful features of the Internet and World Wide Web. Complex biological macromolecules such as nucleic acids and proteins contain a large amount of information within their structures, and analysis of this information by visual examination is impractical for all but the smallest polynucleotide or polypeptide fragments. The only practical approach is to use a computer – often, by comparing information with a database (Table 49.2). A range of programs and packages is available – some can be used on-line, via the Internet, while others can be downloaded for local operation. Alternatively, commercial packages can be used, with database information downloaded via the Internet or supplied on CD-ROM, e.g. for use with a class of students. The newcomer may find that commercial packages are easier to use than Internet programs: consult suppliers' handbooks or make use of help/tutorials to familiarize yourself with the details for particular

Table 49.2 Some useful databases and other locations relevant to nucleic acid analysis

Primer on molecular genetics	http://www.gdb.org/Dan/DOE/intro.html
EMBL (European Bioinformatics Institute nucleic acid database)	http://www.ebi.ac.uk/services/services.html
GenBank (NCBI, USA nucleic acid database)	http://www.ncbi.nlm.nih.gov/
GDB (UK Medical Research Council human genome database)	http://www.hgmp.mrc.ac.uk//gdb/gdbtop.html
SEQNET (Daresbury laboratory, UK)	http://www.dl.ac.uk/seqnet/
Saccharomyces genomic information resource	http://genome-www.standford.edu/
E. coli genetic stock centre	http://cgsc.biology.yale.edu/cgsc.html
Drosophila database (flybase)	http://morgan.harvard.edu/
Plant genome data and information centre	http://www.nalusda.gov/answers/info_centers/pgdic/
Caenorhabditis elegans genome project	http://www.sanger.ac.uk/
Integrated human gene map	http://www.ncbi.nlm.nih.gov/SCIENCE96/

programs or packages. The number of different programs available is a reflection of the various algorithms used for analysis, and such programs are likely to be updated in parallel with advances in computer hardware.

Nucleic acid analysis

A common problem is to investigate a nucleic acid sequence obtained as part of a cloning project (p. 242), to see whether the sequence, or anything similar, has been reported previously. The principal databases for DNA sequences are Genbank® (National Center for Biotechnology Information, USA), the EMBL database (European Molecular Biology Laboratory outstation, European Biotechnology Institute, UK) and the DNA database of Japan (DDBJ). These hold comprehensive information submitted by researchers and sequencing groups, with collaborative data exchange, while others provide data for a particular organism (Table 49.2). The information is often available via anonymous FTP (p. 281); or on CD-ROM, providing local access without waiting for online Internet connections; or via e-mail to the database.

Nucleic acid databases can be searched using programs that allow you to:

- enter a particular nucleotide sequence and edit the sequence if required; some packages have automated data entry, e.g. using a digitizer to convert the banding pattern on a gel into a sequence of bases, though ambiguous bands may give problems with automated systems;
- retrieve a particular sequence from the database using a keyword, author name or accession number;
- search for homology between a particular sequence and those of the database – such sequence comparisons can be carried out with various degrees of sensitivity and stringency, e.g. the program can be set to search only for perfectly matched sequences, or for a particular percentage match, etc.;
- carry out phylogenetic analysis, constructing 'ancestry trees' to show the most likely evolutionary relationships between sequences from various organisms obtained from the database.

Example A rapid search of the EMBL nucleic acid database for closely related sequences can be obtained using the program QUICKSEARCH, while more distantly related DNA sequences can be obtained via BLAST.

Other programs allow you to manipulate nucleic acid sequences:

- to assemble/align sequence fragments into contiguous sequences, also known as 'contigs'. This type of alignment is particularly useful for sets of cloned fragments from a clone library (p. 246) – some programs will highlight individual mismatched bases within overlapping sequences, pointing to areas that need to be checked, e.g. by repeating the sequencing procedure;

Example The Cadgene® program allows users to simulate gene cloning and mapping exercises using operator-defined nucleotide sequences, or sequences from a database, e.g. Genbank®.

- to simulate *in vitro* manipulation and cloning procedures, including the preparation of restriction digests, ligation of a DNA fragment to a particular vector (p. 244), designing oligonucleotide probes and primers, and simulating PCR (p. 237);
- to search for structural and functional 'motifs' within the nucleic acid sequence, including palindromes, restriction sites, likely promoters or control sites, and regions that might form secondary structures, e.g. hairpin loops;
- to identify likely coding regions (open reading frames, or ORFs, and exons) and non-coding regions (e.g. repeat sequences) in eukaryotic DNA, start and stop codons, etc.;
- to translate a nucleotide sequence into an amino acid sequence and *vice versa*.

Analysis of proteins and polypeptides

Example The BLAST program can be used to search the EBI databases for either nucleotide or amino acid sequences, while the BLITZ program is specific to amino acid sequences (SwissProt database).

The principal protein sequence databases are PIR (Protein Identification Resource, USA), and SwissProt (EMBL, UK outstation, EBI), while PDB (Protein Database, Brookhaven Laboratory, USA) holds extensive information on the three-dimensional structure of proteins. A number of other relevant sites are listed in Table 49.3. The protein sequence databases may be searched via a range of programs equivalent to those described above for nucleic acid analysis, allowing the operator to:

- locate and retrieve a particular protein or amino acid sequence from the database using a keyword, author name or accession number;
- enter and edit (cut and paste) a particular amino acid sequence, or convert a nucleotide sequence into an amino acid sequence;
- search the database (locally, or on-line) for sequences with a perfect match, or for similarities in primary or secondary structure at a specified level of homology: such searches may be more useful than those based on DNA homology, since protein sequence information (based on 20 different amino acid residues) is more complex than nucleic acid sequence information (based on four different nucleotides);
- compare and align the amino acid sequences of several proteins, to identify particular regions of homology and non-homology.

Table 49.3 Some useful databases and other locations relevant to protein analysis

Swiss-prot (EBI protein sequence database)	http://expasy.hcuge.ch/sprot/sprottop.html
PIR (NBRF/MIPS protein database)	http://www.gdb.org/Dan/proteins/pir.html
PDB (database of 3-D protein structures)	http://pb1.pdb.bnl.gov/
Swiss 3-D (3-D protein database)	http://expasy.hcuge.ch/sw3d/sw3d-top.html
LiMB (Listing of Molecular Biology databases)	(e-mail)limb@life.lanl.gov
FTP archives for molecular biology	gopher://genome-gopher.stanford.edu/11/ftp
Prosite (protein motifs, etc.)	http://expasy.hcuge.ch/sprot/prosite.html
Motif server	(email)motif@genome.ad.jp
Principles of protein structure (interactive course)	http://www.cryst.bbk.ac.uk/pps/index/html
Biomolecules in CHIME and RASMOL formats	http://www.umass.edu/microbio/rasmol/tutbymol.html
Enzyme nomenclature database	http://expasy.hcuge.ch/sprot/enzyme.html
NIH center for molecular modeling	http://cmm.info.nih.gov/modeling/
Weblab™ viewer for molecular simulation	http://www.msi.com/
ChemWeb	http://www.softshell.com/FREE/chemweb/cw_pages.html

Example Prosite (Table 49.3) is a database of biologically significant sites, patterns and profiles for known proteins that can be used to investigate the likely function of a particular sequence in a novel protein.

Example RasMol is a molecular graphics program enabling the user to visualize the structure of proteins, nucleic acids and small biomolecules – it can be obtained by anonymous FTP from ftp.dcs.ed.ac.uk (directory pub/rasmol).

For further analysis, other programs and packages allow you to:

- assemble overlapping peptide fragments, based on their amino acid sequences, e.g. during sequence analysis of an individual protein (p. 170);
- search for motifs and patterns within the sequence that might provide information on the possible function of the protein;
- predict secondary structure from a primary sequence, identifying regions most likely to exist as α-helices or β-sheets, etc.
- investigate properties such as hydropathy, solvent accessibility and antigenicity;
- model the likely three-dimensional structure of the protein, e.g. via Hyperchem^{TM}, and to identify likely membrane-spanning regions, interactive sites and other features, or to undertake molecular mechanics and quantum mechanics calculations;
- predict and model the effects of changes in primary or secondary structure on protein characteristics and three-dimensional organization.

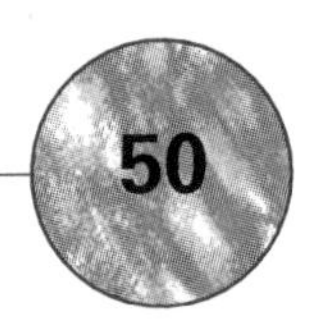

50 Using spreadsheets

Definitions

Spreadsheet – a display of a grid of cells into which numbers, text or formulae can be typed to form a worksheet. Each cell is uniquely identifiable by its column and row number combination (i.e. its 2-D coordinates) and can contain a formula which makes it possible for an entry to one cell to alter the contents of one or more other cells.

Template – a pre-designed spreadsheet without data but including all formulae necessary for (repeated) data analysis.

Macro – a sequence of user-defined instruction carried out by the spreadsheet, allowing complex repeated tasks to be 'automated'.

KEY POINT The spreadsheet is one of the most powerful and wide-ranging of all microcomputer applications. It is the electronic equivalent of a huge sheet of paper with calculating powers and provides a dynamic method of storing and manipulating data sets.

Statistical calculations and graphical presentations are available in many versions and most have scientific functions. Spreadsheets can be used to:

- manipulate raw data by removing the drudgery of repeated calculations, allowing easy transformation of data and calculation of statistics;
- graph out your data rapidly to get an instant evaluation of results. Printout can be used in practical and project reports;
- carry out limited statistical analysis by built-in procedures or by allowing construction of formulae for specific tasks;
- model 'what if' situations where the consequences of changes in data can be seen and evaluated;
- store data sets with or without statistical and graphical analysis.

The spreadsheet (Fig. 50.1) is divided into rows (identified by numbers) and columns (identified by alphabetic characters). Each individual combination of column and row forms a cell which can contain either a data item, a formula, or a piece of text called a label. Formulae can include scientific and/or statistical functions and/or a reference to other cells or groups of cells (often called a range). Complex systems of data input and analysis can be constructed (models). The analysis, in part or complete, can be printed out. New data can be added at any time and the sheet recalculated. You can construct templates, pre-designed spreadsheets containing the formulae required for repeated data analyses.

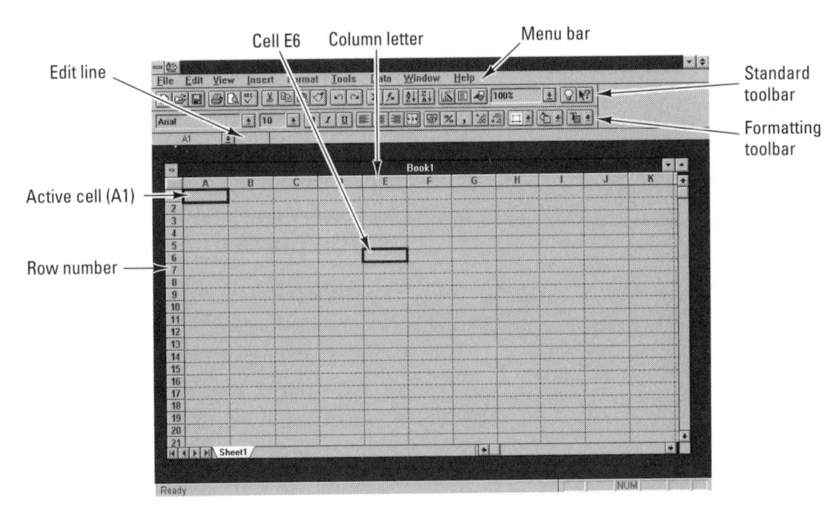

Fig. 50.1 A screen dump of a typical spreadsheet, showing cells, rows and columns; toolbars etc. Screen shot reprinted by permission from Microsoft Corporation.

Templates – these should contain:

- a data input section,
- data transformation and/or calculation sections,
- a results section, which can include graphics,
- text in the form of headings and annotations,
- a summary section.

Constructing a spreadsheet – start with a simple model and extend it gradually, checking for correct operation as you go.

The power a spreadsheet offers is directly related to your ability to create models that are accurate and templates that are easy to use. The sequence of operations required is:

1. Determine what information/statistics you want to produce.
2. Identify the variables you will need to use, both for original data that will be entered and for any intermediate calculations that might be required.
3. Set up areas of the spreadsheet for data entry, calculation of intermediate values (statistical values such as sums of squares, etc.), calculation of final parameters/statistics and, if necessary, a summary area.
4. Establish the format of the numeric data if it is different from the default values. This can be done globally (affecting the entire spreadsheet) or locally (affecting only a specified part of the spreadsheet).
5. Establish the column widths required for the various activities.
6. Enter labels: use extensively for annotation.
7. Enter a test set of values to use during formula entry: use a fully worked example to check that formulae are working correctly.
8. Enter the formulae required to make all the calculations, both intermediate and final. Check that results are correct using the test data.

The spreadsheet is then ready for use. Delete all the test data values and you have created your template. Save the template to a disk and it is then available for repeated operations.

Data entry

Spreadsheets have built-in commands which allow you to control the layout of data in the cells. These include number format, the number of decimal places to be shown (the spreadsheet always calculates using eight or more places), the cell width and the location of the entry within the cell (left, right or centre). An auto-entry facility assists greatly in entering large amounts of data by moving the entry cursor either vertically or horizontally as data is entered. Recalculation default is usually automatic so that when a new data value is entered the entire sheet is recalculated immediately. This can dramatically slow down data entry so select manual recalculation mode before entering new data sets if the spreadsheet is large with many calculations.

The parts of a spreadsheet

Labels

These identify the contents of rows and columns. They are text characters, and cannot be used in calculations. Separate them from the data cells by drawing lines, if this feature is available. Programs make assumptions about the nature of the entry being made: most assume that if the first character is a number, then the entry is a number or formula. If it is a letter, then it will be a label. If you want to start a label with a number, you must override this assumption by typing a designated character before the number to tell the program that this is a label; check your program manual for details.

Using hidden (or zero-width) columns – these are useful for storing intermediate calculations which you do not wish to be displayed on the screen or printout.

Numbers

You can also enter numbers (values) in cells for use in calculations. Many programs let you enter numbers in more than one way and you must decide which method you prefer. The way you enter the number does not affect the way it is displayed on the screen as this is controlled by the cell format at the point of entry. There are usually special ways to enter data for percentages, currency and scientific notation for very large and small numbers.

Formulae

These are the 'power tools' of the spreadsheet because they do the calculations. A cell can be referred to by its alphanumeric code, e.g. A5 (column A, row 5) and the value contained in that cell manipulated within a formula, e.g. (A5 + 10) or (A5 + B22) in another cell. Formulae can include a diverse array of pre-programmed functions which can refer to a cell, so that if the value of that cell is changed, so is the result of the formula calculation. They may also include limited branching options through the use of logical operators.

Definition

Function – a pre-programmed code for the transformation of values (mathematical or statistical functions) or selection of text characters (string functions).

Functions

A variety of functions is usually offered, but only mathematical and statistical functions will be considered here.

Example = sin(A5) is an example of a function in Microsoft® Excel®. If you write this in a cell, the spreadsheet will calculate the sine of the number in cell A5 (assuming it to be an angle in radians) and write it in the cell. Different programs may use a slightly different syntax.

Mathematical functions

Spreadsheets have program-specific sets of predetermined functions but they almost all include trigonometrical functions, angle functions, logarithms (p. 260) and random number functions. Functions are invaluable for transforming sets of data rapidly and can be used in formulae required for more complex analyses. Spreadsheets work with an order of preference of the operators in much the same way as a standard calculator and this must always be taken into account when operators are used in formulae. They also require a very precise syntax – the program should warn you if you break this!

Empty cells – note that these may be given the value 0 by the spreadsheet for certain functions. This may cause errors e.g. by rendering a minimum value inappropriate. Also, an error return may result for certain functions if the cell content is zero.

Statistical functions

Modern spreadsheets incorporate many sophisticated statistical functions, and if these are not appropriate, the spreadsheet can be used to facilitate the calculations required for most of the statistical tests found in textbooks. The descriptive statistics normally available include:

- sums of all data present in a column, row or block;
- minima and maxima of a defined range of cells;
- counts of cells – a useful operation if you have an unknown or variable number of data values;
- averages and other statistics describing location;
- standard deviations and other statistics describing dispersion.

A useful function where you have large numbers of data allows you to create frequency distributions using pre-defined class intervals.

Statistical calculations – make sure you understand whether any functions you employ are for populations or samples (see p. 265).

The hypothesis-testing statistical functions may be reasonably powerful (e.g. *t*-test, ANOVA, regressions) and they often return the *probability* P of obtaining the test statistic (where $0 < P < 1$), so there may be no need to refer to statistical tables. Again, check on the effects of including empty cells.

Example In Microsoft® Excel®, copying is normally *relative*, and if you wish a cell reference to be *absolute* when copied, this is done by putting a dollar ($) sign before and after the column reference letter e.g. C56.

Copying

All programs provide a means of copying (replicating) formulae or cell contents when required and this is a very useful feature. When copying, references to cells may be either relative, changing with the row/column as they are copied or absolute, remaining a fixed cell reference and not changing as the formulae are copied. This distinction between cell references is very important and must be understood; it provides one of the most common forms of error when copying formulae. Be sure to understand how your spreadsheet performs these operations.

Naming blocks

When a group of cells (a block) is carrying out a particular function, it is often easier to give the block a name which can then be used in all formulae referring to that block. This powerful feature also allows the spreadsheet to be more readable.

Graphics display

Most spreadsheets now offer a wide range of graphics facilities which are easy to use and this represents an ideal way to examine your data sets rapidly and comprehensively. The quality of the final graphics output (to a printer) is variable but is usually perfectly sufficient for initial investigation of your data. Many of the options are business graphics styles but there are usually histogram, bar chart, *X-Y* plotting, line and area graphics options available. Note that some spreadsheet graphics may not come up to the standards expected for the formal presentation of scientific data (p. 251).

Printing spreadsheets

This is usually a straightforward menu-controlled procedure, made difficult only by the fact that your spreadsheet may be too big to fit on one piece of paper. Try to develop an area of the sheet which contains only the data that you will be printing, i.e. perhaps a summary area. Remember that columns can usually be hidden for printing purposes and you can control whether the printout is in portrait or landscape mode, and for continuous paper or single sheets (depending on printer capabilities). Use a screen preview option, if available, to check your layout before printing. Most spreadsheets are now WYSIWYG (What You See Is What You Get) so that the appearance on the screen is a realistic impression of the printout. A 'print to fit' option is also available in some programs, making the output fit the page dimensions.

Using string functions – these allow you to manipulate text within your spreadsheet and include functions such as 'search and replace' and alphabetical or numerical 'sort'.

Use as a database

Many spreadsheets can be used as databases, using rows and columns to represent the fields and records (see Chapter 51). For many bioscience applications, the spreadsheet form of database is perfectly adequate and should be seriously considered before using a full-feature database program.

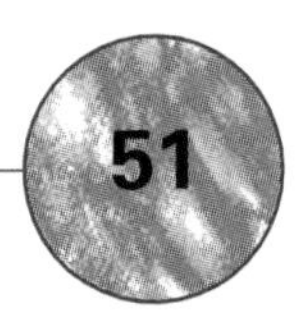

51 Word processors, databases and other packages

Word processors

The word processor has facilitated writing because of the ease of revising text. Word processing is a transferable skill valuable beyond the immediate requirements of your biology course. Using a word processor should improve your writing skills and speed because you can create, check and change your text on the screen before printing it as 'hard copy' on paper. Once entered and saved, multiple uses can be made of a piece of text with little effort.

When using a word processor you can:

- refine material many times before submission;
- insert material easily, allowing writing to take place in any sequence;
- use a spell-checker to check your text;
- use a thesaurus when composing your text;
- produce high quality final copies;
- reuse part or all of the text in other documents.

The potential disadvantages of using a word processor include:

- lack of ready access to a computer, software and/or a printer;
- time taken to learn the operational details of the program;
- the temptation to make 'trivial' revisions;
- loss of files due to computer breakdown or disk loss or failure.

The computerized office – many word processors are now sold as part of an integrated suite, e.g. PerfectOffice® and Microsoft® Office, with the advantage that they share a common interface in the different components (word processor, spreadsheet, database, etc.) and allow ready exchange of information (e.g. text, graphics) between component programs.

Word processors come as 'packages' comprising the program and a manual, often with a tutorial program. Examples are Microsoft® Word and WordPerfect®. Most word processors have similar general features but differ in operational detail; it is best to pick one and stick to it as far as possible so that you become familiar with it. Learning to use the package is like learning to drive a car – you need only to know how to drive the computer and its program, not to understand how the engine (program) and transmission (data transfer) work, although a little background knowledge is often helpful and will allow you to get the most from the program.

In most word processors, the appearance of the screen realistically represents what the printout on paper will look like (WYSIWYG). Word processing files actually contain large amounts of code relating to text format, etc., but these clutter the screen if visible, as in non-WYSIWYG programs. Some word processors are menu-driven, others require keyboard entry of codes: menus are easier to start with and the more sophisticated programs allow you to choose between these options.

Using textbooks, manuals and tutorials – the manuals that come with some programs may not be very user-friendly and it is often worth investing in one of the textbooks that are available for most word processing programs. Alternatively, use an on-line 'help' tutorial, available with the more sophisticated packages.

Because of variation in operational details, only general and strategic information is provided in this chapter: you must learn the details of your word processor through use of the appropriate manual and 'help' facilities.

Before starting you will need:

- the program (ideally on a hard disk);
- a floppy disk for storage, retrieval and back-up of your own files when created;
- the appropriate manual or textbook giving operational details;

- a draft page layout design: in particular you should have decided on page size, page margins, typeface (font) and size, type of text justification, and format of page numbering;
- an outline of the text content;
- access to a suitable printer: this need not be attached to the computer you are using since your file can be taken to an office where such a printer is available, providing that it has the same word processing program.

Using a word processor – take full advantage of the differences between word processing and 'normal' writing (which necessarily follows a linear sequence and requires more planning):

- Simply jot down your initial ideas for a plan, preferably to paragraph topic level. The order can be altered easily and if a paragraph grows too much it can easily be split.
- Start writing wherever you wish and fill in the rest later.
- Just put down your ideas as you think, confident in the knowledge that it is the concepts that are important to note; their order and the way you express them can be adjusted later.
- Don't worry about spelling and use of synonyms – these can (and should) be checked during a special revision run through your text, using the spelling checker first to correct obvious mistakes, then the thesaurus to change words for style or to find the exact word.
- Don't forget that a draft printout may be required to check (a) for pace and spacing – difficult to correct for on a screen; and (b) to ensure that words checked for spelling fit the required sense.

Laying out (formatting) your document

Although you can format your text at any time, it is good practice to enter the basic commands at the start of your document: entering them later can lead to considerable problems due to reorganization of the text layout. If you use a particular set of layout criteria regularly, e.g. an A4 page with space for a letterhead, make a template containing the appropriate codes that can be called up whenever you start a new document. Note that various printers may respond differently to particular codes, resulting in a different spacing and layout.

Typing the text

Think of the screen as a piece of typing paper. The cursor marks the position where your text/data will be entered and can be moved around the screen by use of the cursor-control keys. When you type, don't worry about running out of space on the line because the text will wrap around to the next line automatically. Do not use a carriage return (usually the [ENTER] or [↵] key) unless you wish to force a new line, e.g. when a new paragraph is wanted. If you make a mistake when typing, correction is easy. You can usually delete characters or words or lines and the space is closed automatically. You can also insert new text in the middle of a line or word. You can insert special codes to carry out a variety of tasks, including changing text appearance such as underlining, **bolding** and *italics*. Paragraph indentations can be automated using [TAB] or [⇆] as on a typewriter but you can also indent or bullet whole blocks of text using special menu options. The function keys are usually pre-programmed to assist in many of these operations.

Deleting and restoring text – because deletion can sometimes be made in error, there is usually an 'undelete' or 'restore' feature which allows the last deletion to be recovered.

Editing features

Word processors usually have an array of features designed to make editing documents easy. In addition to the simple editing procedures described above, the program usually allows blocks of text to be moved ('cut and paste'), copied or deleted.

An extremely valuable editing facility is the search procedure: this can rapidly scan through a document looking for a specified word, phrase or punctuation. This is particularly valuable when combined with a replace facility so that, for example, you could replace the word 'test' with 'trial' throughout your document simply and rapidly.

Most WYSIWYG word processors have a command (e.g. Show/Hide in Microsoft® Word) which reveals the non-printing characters, including paragraph and space markers. This can be useful when editing text, as spacing may alter in apparently mysterious ways if the precise position of the cursor *vis a vis* these markers is not taken into account.

Fonts and line spacing

Most word processors offer a variety of fonts depending upon the printer being used. Fonts come in a wide variety of types and sizes, but they are defined in particular ways as follows:

Presenting your documents – it is good practice not to mix typefaces too much in a formal document; also the font size should not differ greatly for different headings, subheadings and the text.

- Typeface: the term for a family of characters of a particular design, each of which is given a particular name. The most commonly used for normal text is Times Roman as used here for the main text but many others are widely available, particularly for the better quality printers. They fall into three broad groups: serif fonts with curves and flourishes at the ends of the characters (e.g. Times Roman); sans serif fonts without such flourishes, providing a clean, modern appearance (e.g. Helvetica, also known as Swiss); and decorative fonts used for special purposes only, such as the production of newsletters and notices (e.g. *Freestyle Script*).
- Size: measured in points. A point is the smallest typographical unit of measurement, there being 72 points to the inch (about 28 points per cm). The standard sizes for text are 10, 11 and 12 point, but typefaces are often available up to 72 point or more.
- Appearance: many typefaces are available in a variety of styles and weights. Many of these are not designed for use in scientific literature but for desk-top publishing.
- Spacing: can be either fixed, where every character is the same width, or proportional, where the width of every character, including spaces, is varied. Typewriter fonts such as Elite and Prestige use fixed spacing and are useful for filling in forms or tables, but proportional fonts make the overall appearance of text more pleasing and readable.
- Pitch: specifies the number of characters per horizontal inch of text. Typewriter fonts are usually 10 or 12 pitch, but proportional fonts are never given a pitch value since it is inherently variable.
- Justification is the term describing the way in which text is aligned vertically. Left justification is normal, but for formal documents, both left and right justification may be used (as here).

Preparing draft documents – use double spacing to allow room for your editing comments on the printed page.

You should also consider the vertical spacing of lines in your document. Drafts and manuscripts are frequently double-spaced to allow room for editing. If your document has unusual font sizes, this may well affect line spacing, although most word processors will cope with this automatically.

Preparing final documents – for most work, use a 12 point proportional serif typeface with spacing dependent upon the specifications for the work.

Table construction

Tables can be produced by a variety of methods:

- Using the tab key [⇆] as on a typewriter: this moves the cursor to predetermined positions on the page, equivalent to the start of each tabular column. You can define the positions of these tabs as required at the start of each table.
- Using special table-constructing procedures. Here the table construction is largely done for you and it is much easier than using tabs, providing you enter the correct information when you set up the table.
- Using a spreadsheet to construct the table and then copying it to the word processor. This procedure requires considerably more manipulation than using the word processor directly and is best reserved for special circumstances, such as the presentation of a very large or complex table of data, especially if the data are already stored as a spreadsheet.

Graphics and special characters

Many word processors can incorporate graphics from other programs into the text of a document. Files must be compatible (see your manual) but if this is so, it is a relatively straightforward procedure. For highly professional documents this is a valuable facility, but for most undergraduate work it is probably better to produce and use graphics as a separate operation, e.g. employing a spreadsheet.

You can draw lines and other graphical features directly within most word processors and special characters may be available dependent upon your printer's capabilities. It is a good idea to print out a full set of characters from your printer so that you know what it is capable of. These may include symbols and Greek characters, often useful in biological work.

Tools

Many word processors also offer you special tools, the most important of which are:

- Macros: special sets of files you can create when you have a frequently repeated set of keystrokes to make. You can record these keystrokes as a 'macro' so that it can provide a short-cut for repeated operations.
- Thesaurus: used to look up alternative words of similar or opposite meaning while composing text at the keyboard.
- Spell-check: a very useful facility which will check your spellings against a dictionary provided by the program. This dictionary is often expandable to include specialist words which you use in your work. The danger lies in becoming too dependent upon this facility as they all have limitations: in particular, they will not pick up incorrect words which happen to be correct in a different context (i.e. 'was' typed as 'saw' or 'meter' rather than 'metre'). Beware of American spellings in programs from the USA, e.g. 'color' instead of 'colour'. The rule, therefore, is to use the spell-check first and then carefully read the text for errors which have slipped through.
- Word count: useful when you are writing to a prescribed limit.

Using a spell-check facility – do not rely on this to spot all errors. Remember that spell-check programs do not correct grammatical errors.

Printing from your program

Word processors require you to specify precisely the type of printer and/or other style details you wish to use. Most printers also offer choices as to text and graphics quality, so choose draft (low) quality for all but your final copy since this will save both time and materials.

Use a print preview option to show the page layout if it is available. Assuming that you have entered appropriate layout and font commands, printing is a straightforward operation carried out by the word processor at your command. Problems usually arise because of some incompatibility between the criteria you have entered and the printer's own capabilities. Make sure that you know what your printer offers before starting to type: although parameters are modifiable at any time, changing the page size, margin size, font size, etc., all cause your text to be re-arranged, and this can be frustrating if you have spent hours carefully laying out the pages!

Using the print preview mode – this can reveal errors of several types, e.g. spacing between pages, that can prevent you wasting paper and printer ink unnecessarily.

KEY POINT **It is vital to save your work frequently to a hard or floppy disk (or both). This should be done every 10 min or so. If you do not save regularly, you may lose hours or days of work. Many programs can be set to 'autosave' every few minutes.**

Databases

A database is an electronic filing system whose structure is similar to a manual record card collection. Its collection of records is termed a file. The individual items of information on each record are termed fields. Once the database is constructed, search criteria can be used to view files through various filters according to your requirements. The computerized catalogues in your library are just such a system; you enter the filter requirements in the form of author or subject keywords.

You can use a database to catalogue, search, sort, and relate collections of information. The benefits of a computerized database over a manual card-file system are:

- The information content is easily amended/updated.
- Printout of relevant items can be obtained.
- It is quick and easy to organize through sorting and searching/selection criteria, to produce sub-groups of relevant records.
- Record displays can easily be redesigned, allowing flexible methods of presenting records according to interest.
- Relational databases can be combined, giving the whole system immense flexibility. The older 'flat-file' databases store information in files which can be searched and sorted, but cannot be linked to other databases.

Choosing between a database and a spreadsheet – use a database only after careful consideration. Can the task be done better within a spreadsheet? A database program can be complex to set up and usually needs to be updated regularly.

Relatively simple database files can be constructed within the more advanced spreadsheets using the columns and rows as fields and records respectively. These are capable of limited sorting and searching operations and are probably sufficient for the types of databases you are likely to require as an undergraduate. You may also make use of a bibliography database especially constructed for that purpose.

Statistical analysis packages

Statistical packages vary from small programs designed to carry out very specific statistical tasks to large sophisticated packages (Statgraphics®, Minitab®, etc.) intended to provide statistical assistance, from experimental design to the analysis of results. Consider the following features when selecting a package:

- The data entry and editing section should be user-friendly, with options for transforming data.
- Options should include descriptive statistics and exploratory data analysis techniques.
- Hypothesis testing techniques should include ANOVA, regression analysis, multivariate techniques and parametric and non-parametric statistics.
- Output facilities should be suitable for graphical and tabular formats.

Some programs have complex data entry systems, limiting ease of use. The data entry and storage system should be based upon a spreadsheet system, so that subsequent editing and transformation operations are straightforward.

KEY POINT **Make sure that you understand the statistical basis for your test and the computational techniques involved *before* using a particular program.**

Graphics/presentation packages

Many of these packages are specifically designed for business graphics rather than science. They do, however, have considerable value in the preparation of materials for posters and talks where visual quality is an important factor. There are several packages available for microcomputers such as Freelance Graphics®, Harvard Graphics® and Microsoft® PowerPoint®, which provide templates for the preparation of overhead transparencies, slide transparencies and paper copy, both black and white and in colour. They usually incorporate a 'freehand' drawing option, allowing you to make your own designs.

Although the facilities offered are often attractive, the learning time required for some of the more complex operations is considerable and they should be considered only for specific purposes: routine graphical presentation of data sets is best done from within a spreadsheet or statistical package. There may be a service provided by your institution for the preparation of such material and this should be seriously considered before trying to learn to use these programs.

The most important points regarding the use of graphics packages are:

- Graphics quality: the built-in graphics are sometimes of only moderate quality. Use of annotation facilities can improve graphics considerably. Do not use inappropriate graphics for scientific presentation.
- The production of colour graphics: this requires a good quality colour printer/plotter.
- Importing of graphics files: graphs produced by spreadsheets or other statistical programs can usually be imported into graphics programs – this is useful for adding legends, annotations, etc., when the facilities offered by the original programs are inadequate. Check that the format of files produced by your statistics/spreadsheet program can be recognized by your graphics program. The different types of file are distinguished by the three-character filename extension.

KEY POINT **Computer graphics are not always satisfactory for scientific presentation. While they may be useful for exploratory procedures, they may need to be re-drawn by hand for your final report. It may be helpful to use a computer-generated graph as a template for the final version.**

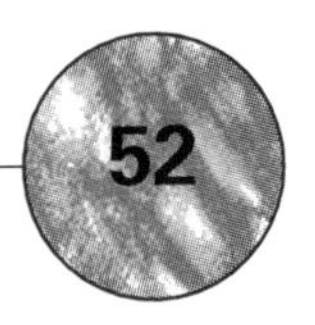

52 Finding and citing published information

The ability to find scientific information is a skill required for many exercises in your degree programme. You will need to research facts and published findings as part of writing essays, literature reviews and project introductions, and when amplifying your lecture notes and revising for exams. You must also learn how to follow scientific convention in citing source material as the authority for each statement you make.

Browsing in a library – this *may* turn up interesting material, but remember the books on the shelves are those *not* currently out on loan. Almost by definition, the latter may be more up-to-date and useful. To find out a library's full holding of books in any subject area, you need to search its catalogue (normally available as a computerized database).

Example The book *The Selfish Gene* by Richard Dawkins (1976; Oxford University Press) is likely to be classified as follows:

Dewey decimal system: 591.51

where 591	refers to zoology
591.5	refers to ecology of animals
591.51	refers to habits and behaviour patterns

Library of Congress system: QL751

where Q	refers to science
QL	refers to zoology
QL75	refers to animal behaviour
QL751	refers to general works and treatises

Computer databases – several databases are now produced on CD-ROM for open use in libraries (e.g. Medline, Applied Science and Technology Index). Especially useful to life scientists is the *Biosis Previews Database* based on *Biological Abstracts*, accessed via Dialog Information Services, Inc. (Palo Alto, California, USA). Some databases can be accessed via the Internet, such as BIDS (Bath Information and Data Service), a UK service providing access to information from over 7 000 periodicals (username and password required via your library). Each of these databases usually has its own easy-to-follow menu instructions. It is worthwhile to consider keywords for your search beforehand to focus your search and save time.

Sources of information

For essays and revision

You are unlikely to delve into the primary literature for these purposes – books and reviews are much more readable! If a lecturer or tutor specifies a particular book, then it should not be difficult to find out where it is shelved in your library, as most now have a computerized index system and their staff will be happy to assist with any queries. If you want to find out which books your library holds on a specified topic, use the system's subject index. You will also be able to search by author or by key words.

There are two main systems used by libraries to classify books: the Dewey Decimal system and the Library of Congress system. Libraries differ in the way they employ these systems, especially by adding further numbers and letters after the standard classification marks to signify e.g. shelving position or edition number. Enquire at your library for a full explanation of local usage.

The World Wide Web is an expanding resource for gathering both general and specific information (see Chapter 48). Sites fall into analogous categories to those in the printed literature: there are sites with original information, sites that review information and bibliographic sites. One considerable problem is that Web sites may be frequently updated, so information present when you first looked may be altered or even absent when the site is next consulted. Further, very little of the information on the WWW has been monitored or refereed. Another disadvantage is that the site information may not state the origin of the material, who wrote it or when it was written.

For literature surveys and project work

You will probably need to consult the primary literature. If you are starting a new research project or writing a report from scratch, you can build up a core of relevant papers by using the following methods:

- Asking around: supervisors or their postgraduate students will almost certainly be able to supply you with a reference or two that will start you off.
- Searching a computer database: these cover very wide areas and are a convenient way to start a reference collection, although a charge is often made for access and sending out a listing of the papers selected (your library may or may not pass this on to you).
- Consulting the bibliography of other papers in your collection – an important way of finding the key papers in your field. In effect, you are taking advantage of the fact that another researcher has already done all the hard work!

Definitions

Journal/periodical/serial – any publication issued at regular intervals. In biosciences, usually containing papers describing original research findings and reviews of literature.
The primary literature – this comprises original research papers, published in specialist scientific periodicals. Certain prestigious general journals (e.g. *Nature*) contain important new advances from a wide subject area.
Monograph – a specialized book covering a single topic.
Review – an article in which recent advances in a specific area are outlined and discussed.
Proceedings – volume compiling written versions of papers read at a scientific meeting on a specific topic.
Abstracts – shortened versions of papers, often those read at scientific meetings. These may later appear in the literature as full papers.
Bibliography – a summary of the published work in a defined subject area.

- Referring to 'current awareness' journals or computer databases: these are useful for keeping you up to date with current research; they usually provide a monthly listing of article details (title, authors, source, author address) arranged by subject and cross-referenced by subject and author. Current awareness journals cover a wider range of primary journals than could ever be available in any one library. Examples include:
  (a) *Current Contents*, published by the Institute of Scientific Information, Philadelphia, USA, which reproduces the contents pages of journals of a particular subject area and presents an analysis by author and subject.
  (b) *Current Advances*, published by Pergamon Press, Oxford, UK, which subdivides papers by subject within research areas and cross-references by subject and author.
  (c) *Biological Abstracts*, in which each paper's abstract is also reproduced. Papers may be cross-referenced according to various taxa, which is useful in allowing you to find out what work has been done on a particular organism.
- Using the *Science Citation Index* (SCI): this is a very valuable source of new references, because it lets you see who has cited a given paper; in effect, SCI allows you to move forward through the literature from an existing reference. The Index is published regularly during the year and issues are collated annually. Some libraries have copies on CD-ROM: this allows rapid access and output of selected information.

For specialized information

You may need to consult reference works, such as encyclopaedias, maps and books providing specialized information. Much of this is now available on CD-ROM (consult your library's information service). Three books worth noting are:

- The *Handbook of Chemistry and Physics* (Lide and Frederikse, 1996): the Chemical Rubber Company's publication (affectionately known as the 'Rubber Bible') giving all manner of physical constants, radioisotope half-lives, etc.
- *The Merck Index* (Budavari, 1996), which gives useful information about organic chemicals, e.g. solubility, whether poisonous, etc.
- The *Geigy Scientific Tables* (8th edition), a series of 6 volumes (Lentner, 1981; 1984; 1988; 1990; 1992; Lentner *et al.*, 1982) provides a wide range of information centred on biochemistry, e.g. buffer formulae, properties of constituents of living matter.

Copyright law – In Europe, copyright regulations were harmonized in 1993 (Directive 93/98/EEC) to allow literary copyright for 70 years after the death of an author and typographical copyright for 25 years after publication. This was implemented in the UK in 1996, where, in addition, the Copyright, Designs and Patents Act (1988) allows the Copyright Licensing Agency to license institutions so that lecturers, students and researchers may take copies for teaching and personal research purposes – no more than a single article per journal issue, one chapter of a book, or extracts to a total of 10% of a book.

Obtaining and organizing research papers

Obtaining a copy

It is usually more convenient to have personal copies of key research articles for direct consultation when working in a laboratory or writing. The simplest way of obtaining these is to photocopy the originals. For academic purposes, this is normally acceptable within copyright law. If your library does not take the journal, it may be possible for them to borrow it from a nearby institute or obtain a copy via a national borrowing centre (an 'inter-library loan'). If the latter, you will have to fill in a form giving full bibliographic details of the paper and where it was cited, as well as signing a copyright clearance statement concerning your use of the copy.

Storing research papers – these can easily be kept in alphabetical order within filing boxes or drawers, but if your collection is likely to grow large, it will need to be refiled as it outgrows the storage space. An alternative is to keep an alphabetical card index system (useful when typing out lists of references) and file the papers by 'accession number' as they accumulate. New filing space is only required at one 'end' and you can use the accession numbers to form the basis of a simple cross-referencing system.

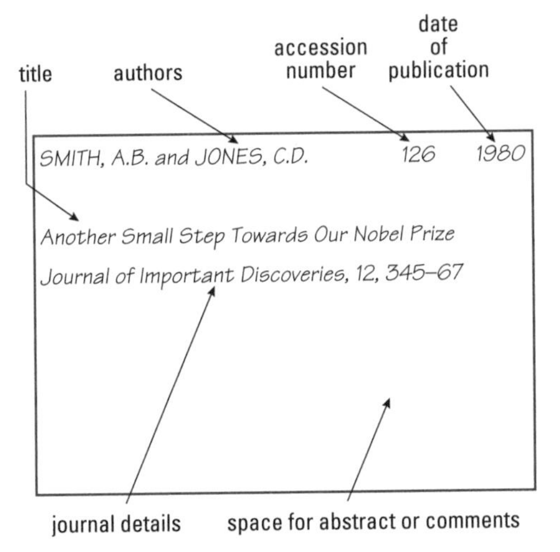

Fig. 52.1 A typical reference card. Make sure the index card carries all the bibliographic information of potential relevance.

Your department might be able to supply 'reprint request' postcards to be sent to the designated author of a paper. This is an unreliable method of obtaining a copy because it may take some time (allow at least a 1–3 months!) and some requests will not receive a reply. Taking into account the waste involved in postage and printing, it is probably best simply to photocopy or send for a copy via inter-library loan.

Organizing papers

Although the numbers of papers you accumulate may be small to start with, it is worth putting some thought into their storage and indexing before your collection becomes disorganized and unmanageable. Few things are more frustrating than not being able to lay your hands on a vital piece of information, and this can seriously disrupt your flow when writing or revising.

Card index systems

Index cards (Fig. 52.1) are a useful adjunct to any filing system. Firstly, you may not have a copy of the paper to file yet may still wish the reference information to be recorded somewhere for later use. Secondly, a selected pile of cards can be used when typing out different bibliographies. Thirdly, the cards can help when organizing a review (see p. 314). Fourthly, the card can be used to record key points and comments on the paper. The priority rule for storage in card boxes is again first author name, subsequent author name(s), date. Computerized card index systems simplify cross-referencing and can provide computer files for direct insertion into word-processed documents; however, they are very time-consuming to set up and maintain, so you should only consider using one if the time invested will prove worthwhile.

Making citations in text

There are two main ways of citing articles and creating a Bibliography (also referred to as 'References' or 'Literature Cited').

The Harvard system

For each citation, the author name(s) and the date of publication are given at the relevant point in the text. The Bibliography is organized alphabetically and by date of publication for papers with the same authors. Formats normally adopted are, for example, 'Smith and Jones (1983) stated that ...' or 'it has been shown that ... (Smith and Jones, 1983)'. Lists of references within parentheses are separated by semi-colons, e.g. '(Smith and Jones, 1983; Jones and Smith, 1985)', normally in order of date of publication. To avoid repetition within the same paragraph, a formula such as 'the investigations of Smith and Jones indicated that' could be used following an initial citation of the paper. Where there are more than two authors it is usual to write 'et al.' (or *et al.* if an italic font is available); this stands for the Latin *et alia* meaning 'and others'. If citing more than one paper with the same authors, put, for example, 'Smith and Jones (1987; 1990)' and if papers by a given set of authors appeared in the same year, letter them (e.g. Smith and Jones, 1989a; 1989b).

The numerical or Vancouver system

Papers are cited via a superscript or bracketed reference number inserted at the appropriate point. Normal format would be, for example: 'DNA sequences[4,5] have shown that '... or 'Jones [55,82] has claimed that ...'.

Repeated citations use the number from the first citation. In the true numerical method (e.g. as in *Nature*), numbers are allocated by order of citation in the text, but in the alpha-numerical method (e.g. the *Annual Review* series), the references are first ordered alphabetically in the Bibliography, then numbered, and it is this number which is used in the text. Note that with this latter method, adding or removing references is tedious, so the numbering should be done only when the text is finalized.

KEY POINT The main advantages of the Harvard system are that the reader might recognize the paper being referred to and that it is easily expanded if extra references are added. The main advantages of the Vancouver system are that it aids text flow and reduces length.

How to list your citations in a bibliography

Whichever citation method is used in the text, comprehensive details are required for the bibliography so that the reader has enough information to find the reference easily. Citations should be listed in alphabetical order with the priority: first author, subsequent author(s), date. Unfortunately, in terms of punctuation and layout, there are almost as many ways of citing papers as there are journals!

Examples

Paper in journal:
Smith, A. B., Jones, C.D. and Professor, A. (1998). Innovative results concerning our research interest. Journal of New Results, 11, 234–5.

Book:
Smith, A. B. (1998). Summary of my life's work. Megadosh Publishing Corp., Bigcity. ISBN 0-123-45678-9.

Chapter in edited book:
Jones, C. D. and Smith, A. B. (1998). Earth-shattering research from our laboratory. In: Research Compendium 1998 (ed. A. Professor), pp 123–456. Bigbucks Press, Booktown.

Thesis:
Smith, A. B. (1995). Investigations on my favourite topic. PhD thesis, University of Life, Fulchester.

Note that underlining used here specifies italics in print: use an italic font if working with a word processor.

KEY POINT Your Department may specify an exact format for project work; if not, decide on a style and be consistent – if you do not pay attention to the details of citation you may lose marks.

Take special care with the following aspects:

- Authors and editors: give details of *all* authors and editors in your bibliography, even if given as *et al.* in the text.
- Abbreviations for journals: while there are standard abbreviations for the titles of journals (consult library staff), it is a good idea to give the whole title, if possible.
- Books: the edition should always be specified as contents may change between editions. Add, for example, '(5th edition)' after the title of the book. You may be asked to give the International Standard Book Number (ISBN), a unique reference number for each book published.
- Unsigned articles, e.g. unattributed newspaper articles and instruction manuals – refer to the author(s) in text and bibliography as 'Anon.'.
- Unread articles: you may be forced to refer to a paper via another without having seen it. If possible, refer to another authority who has cited the paper, e.g. '... Jones (1980), cited in Smith (1990), claimed that ...'. Alternatively, you could denote such references in the bibliography by an asterisk and add a short note to explain at the start of the reference list.
- Personal communications: information received in a letter, seminar or conversation can be referred to in the text as, for example, '... (Smith, pers. comm.)'. These citations are not generally listed in the bibliography of papers, though in a thesis you could give a list of personal communicants and their addresses.

Citing Web sites – there is no widely accepted format at present. We suggest providing author name(s) and date in the text if using the Harvard system, while in the bibliography giving the above, plus site title and full URL reference (e.g. Hacker, A. (1998) University of Cybertown homepage on aardvarks. http://www.myserver.ac.uk/homepage).

Communicating information

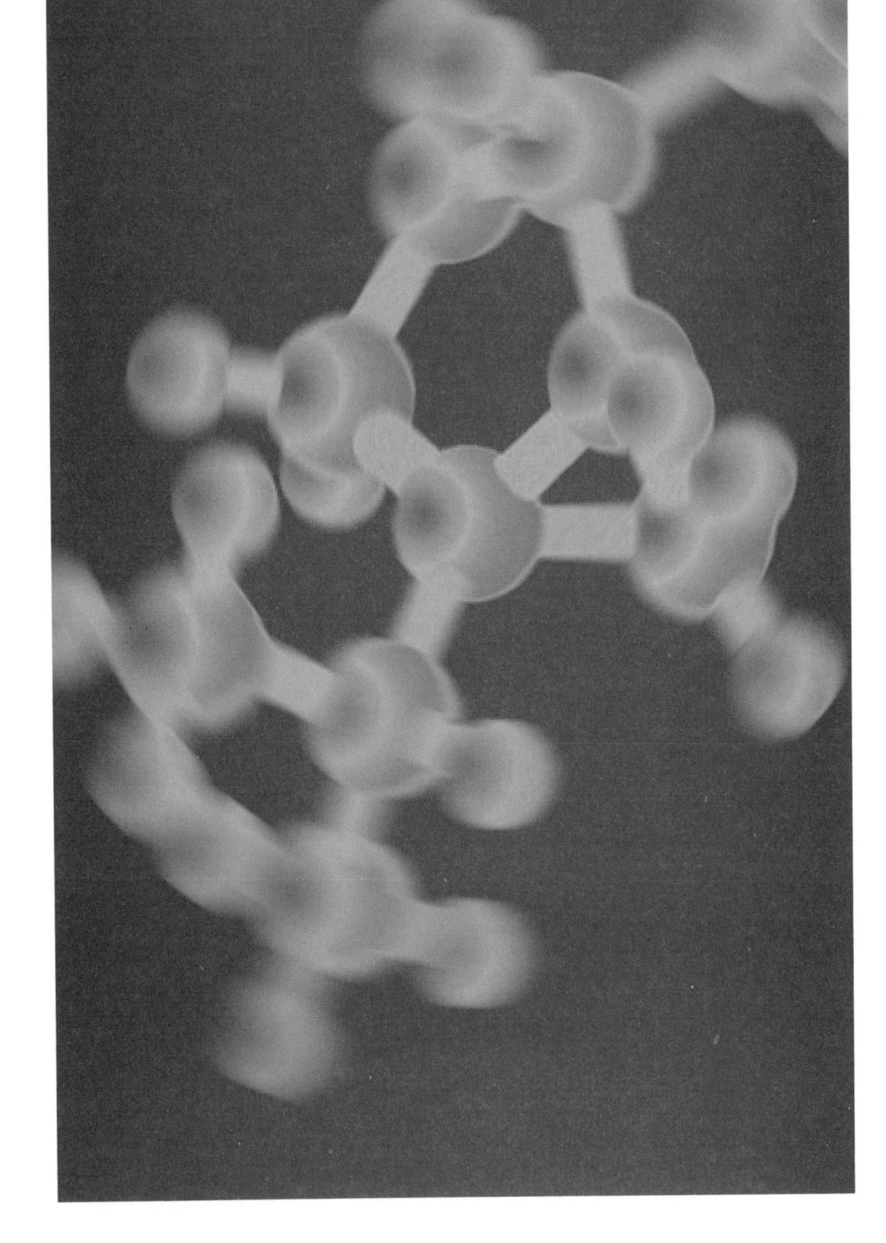

53 General aspects of scientific writing

Written communication is an essential component of all sciences. Most courses include writing exercises in which you will learn to describe ideas and results accurately, succinctly and in an appropriate style and format. The following are features common to all forms of scientific writing.

Organizing time

Making a timetable at the outset helps ensure that you give each stage adequate attention and complete the work on time. To create and use a timetable:

1. Break down the task into stages.
2. Decide on the proportion of the total time each stage should take.
3. Set realistic deadlines for completing each stage, allowing some time for slippage.
4. Refer to your timetable frequently as you work: if you fail to meet one of your deadlines, make a serious effort to catch up as soon as possible.

KEY POINT **The appropriate allocation of your time to reading, planning, writing and revising will differ according to the task in hand (see Chapters 54 and 56).**

Organizing information and ideas

Before you write, you need to gather and/or think about relevant material (Chapter 52). You must then decide:

- what needs to be included and what doesn't;
- in what order it should appear.

Start by jotting down headings for everything of potential relevance to the topic (this is sometimes called 'brainstorming'). A spider diagram (Fig. 53.1) will help you organize these ideas. The next stage is to create an outline of your text (Fig. 53.2). Outlines are valuable because they:

Creating an outline – an informal outline can be made simply by indicating the order of sections on a spider diagram (as in Fig. 53.1).

- force you to think about and plan the structure;
- provide a checklist so nothing is missed out;
- ensure the material is balanced in content and length;
- help you organize figures and tables by showing where they will be used.

In an essay or review, the structure of your writing should help the reader to assimilate and understand your main points. Sub-divisions of the topic could simply be related to the nature of the subject matter (e.g. levels of organization of a protein) and should proceed logically (e.g. primary structures, then secondary, etc.).

A chronological approach is good for evaluation of past work (e.g. the development of the concept of DNA as the hereditary material), whereas a step-by-step comparison might be best for certain exam questions (e.g. 'Discuss the differences between prokaryotes and eukaryotes'). There is little choice about structure for practical and project reports (see p. 310).

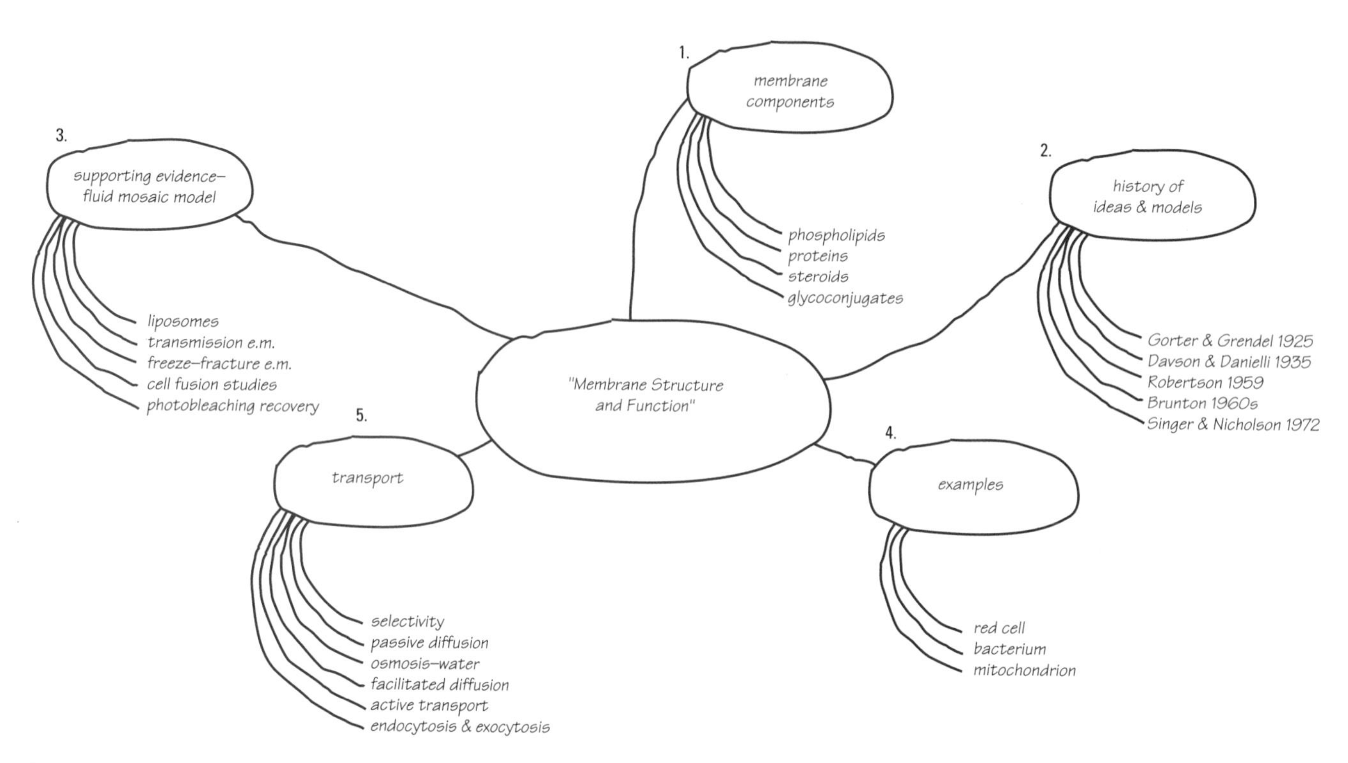

Fig. 53.1 Spider diagram showing how you might 'brainstorm' an essay with the title 'Membrane Structure and Function'. Write out the essay title in full to form the spider's body, and as you think of possible content, place headings around this to form its legs. Decide which headings are relevant and which are not and use arrows to note connections between subjects, if required. This may influence your choice of order and may help to make your writing flow because the links between paragraphs will be natural. You can make an informal outline directly on a spider diagram by adding numbers indicating a sequence of paragraphs (as shown). This method is best when you must work quickly, as with an essay written under exam conditions.

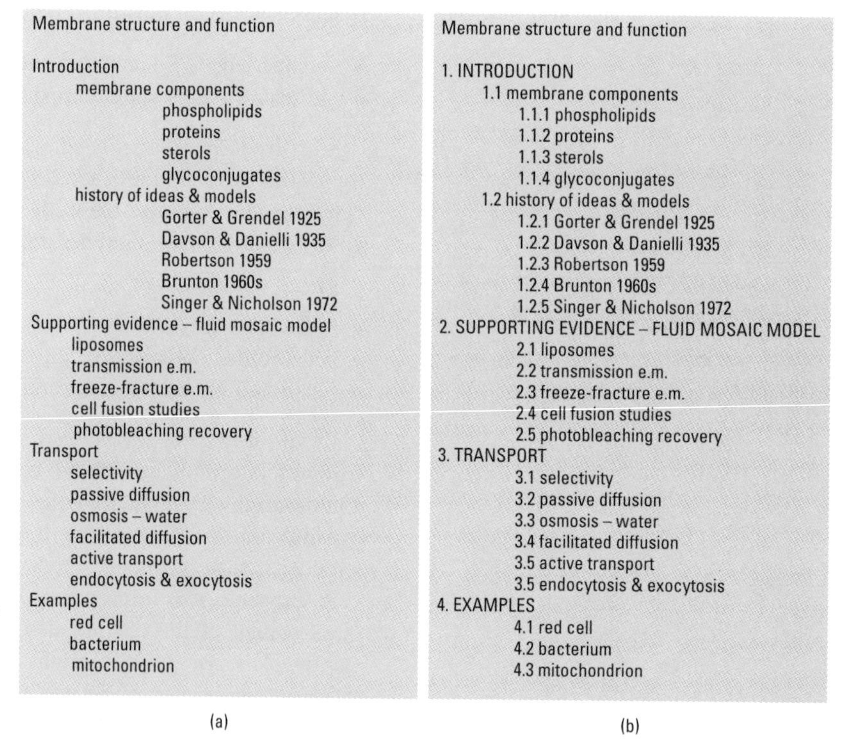

(a)

Membrane structure and function

Introduction
- membrane components
  - phospholipids
  - proteins
  - sterols
  - glycoconjugates
- history of ideas & models
  - Gorter & Grendel 1925
  - Davson & Danielli 1935
  - Robertson 1959
  - Brunton 1960s
  - Singer & Nicholson 1972

Supporting evidence – fluid mosaic model
- liposomes
- transmission e.m.
- freeze-fracture e.m.
- cell fusion studies
- photobleaching recovery

Transport
- selectivity
- passive diffusion
- osmosis – water
- facilitated diffusion
- active transport
- endocytosis & exocytosis

Examples
- red cell
- bacterium
- mitochondrion

(b)

Membrane structure and function

1. INTRODUCTION
   - 1.1 membrane components
     - 1.1.1 phospholipids
     - 1.1.2 proteins
     - 1.1.3 sterols
     - 1.1.4 glycoconjugates
   - 1.2 history of ideas & models
     - 1.2.1 Gorter & Grendel 1925
     - 1.2.2 Davson & Danielli 1935
     - 1.2.3 Robertson 1959
     - 1.2.4 Brunton 1960s
     - 1.2.5 Singer & Nicholson 1972
2. SUPPORTING EVIDENCE – FLUID MOSAIC MODEL
   - 2.1 liposomes
   - 2.2 transmission e.m.
   - 2.3 freeze-fracture e.m.
   - 2.4 cell fusion studies
   - 2.5 photobleaching recovery
3. TRANSPORT
   - 3.1 selectivity
   - 3.2 passive diffusion
   - 3.3 osmosis – water
   - 3.4 facilitated diffusion
   - 3.5 active transport
   - 3.5 endocytosis & exocytosis
4. EXAMPLES
   - 4.1 red cell
   - 4.2 bacterium
   - 4.3 mitochondrion

Fig. 53.2 Formal outlines. These are useful for a long piece of work where you or the reader might otherwise lose track of the structure. The headings for sections and paragraphs are simply written in sequence with the type of lettering and level of indentation indicating their hierarchy. Two different forms of formal outline are shown, a minimal form (a) and a numbered form (b). Note that the headings used in an outline are often repeated within the essay to emphasize its structure. The content of an outline will depend on the time you have available and the nature of the work, but the most detailed hierarchy you should reasonably include is the subject of each paragraph.

Writing

Adopting a scientific style

Your main aim in developing a scientific style should be to get your message across directly and unambiguously. While you can try to achieve this through a set of 'rules' (see Box 53.1), you may find other requirements driving your writing in a contradictory direction. For instance, the need to be accurate and complete may result in text littered with technical terms, and the flow may be continually interrupted by references to the literature. The need to be succinct also affects style and readability through the use of, for example, stacked noun-adjectives (e.g. 'restriction fragment length polymorphism') and acronyms (e.g. 'RFLP'). Finally, style is very much a matter of taste and each tutor, examiner, supervisor or editor will have pet loves and hates which you may have to accommodate.

Developing technique

Writing is a skill that can be improved, but not instantly. You should analyse your deficiencies with the help of feedback from your tutors, be prepared to change work habits (e.g. start planning your work more carefully), and willing to learn from some of the excellent texts that are available on scientific writing.

KEY POINT **You need to take a long-term view if you wish to improve your writing skills. An essential preliminary is to invest in and *make full use of* a personal reference library (see Box 53.2).**

Getting started

A common problem is 'writer's block' – inactivity or stalling brought on by a variety of causes. If blocked, ask yourself these questions:

- Are you comfortable with your surroundings? Make sure you are seated comfortably at a reasonably clear desk and have minimized the possibility of interruptions and distractions.
- Are you trying to write too soon? Have you clarified your thoughts on the subject? Have you done enough preliminary reading? Talking to a friend about your topic might bring out ideas or reveal deficiencies in your knowledge.
- Are you happy with the underlying structure of your work? If you haven't made an outline, try this. If you are unhappy because you can't think of a particular detail at the planning stage, just start writing – it is more likely to come to you while you are thinking of something else.
- Are you trying to be too clever? Your first sentence doesn't have to be earth-shattering in content or particularly smart in style. A short statement of fact or a definition is fine. If there will be time for revision, get your ideas down on paper and revise grammar, content and order later.
- Do you really need to start writing at the beginning? Try writing the opening remarks after a more straightforward part. With reports of

Box 53.1 How to achieve a clear, readable style

Words and phrases

- Choose short clear words and phrases rather than long ones: e.g. use 'build' rather than 'fabricate'; 'now' rather than 'at the present time'. At certain times, technical terms must be used for precision, but don't use jargon if you don't have to.
- Don't worry too much about repeating words, especially when to introduce an alternative might subtly alter your meaning.
- Where appropriate, use the first person to describe your actions ('We decided to '; 'I conclude that '), but not if this is specifically discouraged by your supervisor.
- Favour active forms of speech ('the solution was placed in a beaker') rather than the passive voice ('the beaker was filled with the solution').
- Use tenses consistently. Past tense is always used for materials and methods ('samples were taken from...') and for reviewing past work ('Smith (1990) concluded that...'). The present tense is used when describing data ('Fig. 1 shows...'), for generalizations ('Most authorities agree that...') and conclusions ('I conclude that...').
- Use statements in parentheses sparingly – they disrupt the reader's attention to your central theme (see above section for examples).
- Avoid clichés and colloquialisms – they are usually inappropriate in a scientific context.

Sentences

- Don't make them over-long or complicated.
- Introduce variety in structure and length.
- Make sure you understand how and when to use punctuation.
- If unhappy with the structure of a sentence, try chopping it into a series of shorter sentences.

Paragraphs

- Keep short and restrict them to a distinct theme.
- Use repeated key words (same subject or verb) or appropriate linking phrases (e.g. 'On the other hand...') to connect sentences and emphasize the flow of text.
- The first sentence should introduce the topic of a paragraph and the following sentences explain, illustrate or give examples.

Note: If you're not sure what is meant by any of the terms used here, consult a guide on writing (see Box 53.2).

experimental work, the Materials and Methods section may be the easiest to start at.

- Are you too tired to work? Don't try to 'sweat it out' by writing for long periods at a stretch: stop frequently for a rest.

Revising your text

Wholesale revision of your first draft is strongly advised for all writing apart from in exams. If a word processor is available, this can be a simple process. Where possible, schedule your writing so you can leave the first draft to 'settle' for at least a couple of days. When you return to it fresh, you will see more easily where improvements can be made. Try the following structured revision process, each stage being covered in a separate scan of your text:

Revising your text – to improve clarity and shorten your text, 'distil' each sentence by taking away unnecessary words and 'condense' words or phrases by choosing a shorter alternative.

1. Examine content. Have you included everything you need to? Is all the material relevant?
2. Check the grammar and spelling. Can you spot any 'howlers'?
3. Focus on clarity. Is the text clear and unambiguous? Does each sentence really say what you want it to say?
4. Try to achieve brevity. What could be missed out without spoiling the essence of your work? It might help to imagine an editor has set you the target of reducing the text by 15%.
5. Improve style. Could the text read better? Consider the sentence and paragraph structure and the way your text develops to its conclusion.

Box 53.2 Improve your writing ability by consulting a personal reference library

Using dictionaries

We all know that a dictionary helps with spelling and definitions, but how many of us use one effectively? You should:

- Keep a dictionary beside you when writing and always use it if in any doubt about spelling or definitions.
- Use it to prepare a list of words which you have difficulty in spelling: apart from speeding up the checking process, the act of writing out the words helps commit them to memory.
- Use it to write out a personal glossary of terms. This can help you memorize definitions. From time to time, test yourself.

Not all dictionaries are the same! Ask your tutor or supervisor whether he/she has a preference and why. Try out the *Oxford Advanced Learner's Dictionary*, which is particularly useful because it gives examples of use of all words and helps with grammar, e.g. by indicating which prepositions to use with verbs. Dictionaries of biology tend to be variable in quality, possibly because the subject is so wide and new terms are continually being coined. *Henderson's Dictionary of Biological Terms* (Addison Wesley Longman) is a useful example.

Using a thesaurus

A thesaurus contains lists of words of similar meaning grouped thematically; words of opposite meaning always appear nearby.

- Use a thesaurus to find a more precise and appropriate word to fit your meaning, but check definitions of unfamiliar words with a dictionary.
- Use it to find a word or phrase 'on the tip of your tongue' by looking up a word of similar meaning.
- Use it to increase your vocabulary.

Roget's Thesaurus is the standard. Collins publish a combined dictionary and thesaurus.

Using guides for written English

These provide help with the use of words.

- Use a guide to solve grammatical problems such as when to use 'shall' or 'will', 'which' or 'that', 'effect' or 'affect', etc.
- Use it for help with the paragraph concept and the correct use of punctuation.
- Use it to learn how to structure writing for different tasks.

Recommended guides include the following:

Kane, T.S. (1983) *The Oxford Guide to Writing*. Oxford University Press, New York. This is excellent for the basics of English – it covers grammar, usage and the construction of sentences and paragraphs.

Partridge, E. (1953) *You Have a Point There*. Routledge and Kegan Paul, London. This covers punctuation in a very readable manner.

Tichy, H.J. (1988) *Effective Writing for Engineers, Managers and Scientists*. John Wiley and Sons, New York. This is strong on scientific style and clarity in writing.

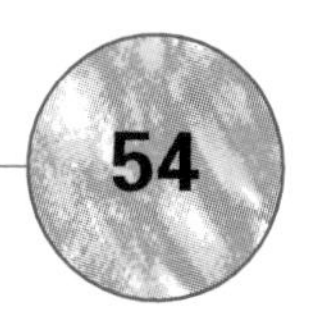

54 Writing essays

The function of an essay is to show how much you understand about a topic and how well you can organize and express your knowledge.

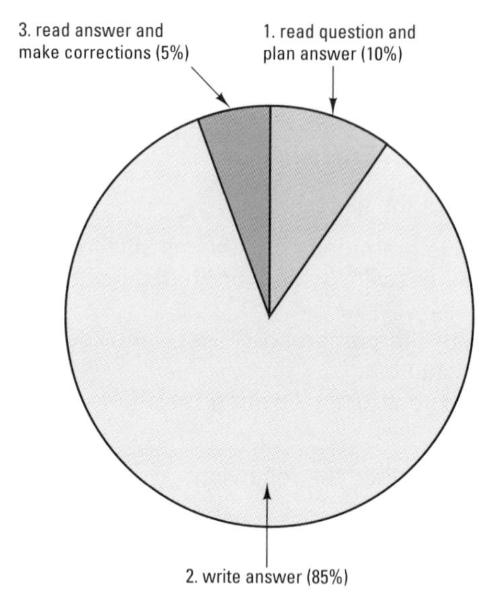

Fig. 54.1 Pie chart showing a typical division of time for an essay.

Organizing your time

Most essays have a relatively straightforward structure and it is best to divide your time into three main parts (Fig. 54.1). For exam strategies, see Chapter 59.

Making a plan for your essay

Dissect the meaning of the essay question or title

Read the title very carefully and think about the topic before starting to write. Consider the definitions of each of the important nouns (this can help in approaching the introductory section). Also think about the meaning of the verb(s) used and try to follow each instruction precisely (see Table 54.1). Don't get side-tracked because you know something about one word or phrase in the title: consider the whole title and all its ramifications. If there are two or more parts to the question, make sure you give adequate attention to each part.

Table 54.1 Instructions often used in essay questions and their meanings. When more than one instruction is given (e.g. compare and contrast; describe and explain), make sure you carry out *both* or your may lose a large proportion of the available marks

Account for:	give the reasons for
Analyse:	examine in depth and describe the main characteristics of
Assess:	weigh up the elements of and arrive at a conclusion about
Comment:	give an opinion on and provide evidence for your views
Compare:	bring out the similarities between
Contrast:	bring out dissimilarities between
Criticize:	judge the worth of (give both positive and negative aspects)
Define:	explain the exact meaning of
Describe:	use words and diagrams to illustrate
Discuss:	provide evidence or opinions about, arriving at a balanced conclusion
Enumerate:	list in outline form
Evaluate:	weigh up or appraise; find a numerical value for
Explain:	make the meaning of something clear
Illustrate:	use diagrams or examples to make clear
Interpret:	express in simple terms, providing a judgement
Justify:	show that an idea or statement is correct
List:	provide an itemized series of statements about
Outline:	describe the essential parts only, stressing the classification
Prove:	establish the truth of
Relate:	show the connection between
Review:	examine critically, perhaps concentrating on the stages in the development of an idea or method
State:	express clearly
Summarize:	without illustrations, provide a brief account of
Trace:	describe a sequence of events from a defined point of origin

Consider possible content and examples

The spider diagram technique (p. 304) is a speedy way of doing this. If you have time to read several sources, consider their content in relation to the essay title. Can you spot different approaches to the same subject? Which do you prefer as a means of treating the topic in relation to your title? Which examples are most relevant to your case, and why?

Construct an outline

Every essay should have a structure related to its title. Most marks for essays are lost because the written material is badly organized or is irrelevant. An essay plan, by definition, creates order and, if thought about carefully, can ensure relevance. Your plan should be written down (but scored through later if written in an exam book). Think about an essay's content in three parts:

Essay content – it is rarely enough simply to lay down facts for the reader – you must analyse them and comment on their significance.

1. The introductory section, in which you should include definitions and some background information on the context of the topic being considered. You should also tell your reader how you plan to approach the subject.
2. The middle of the essay, where you develop your answer and provide relevant examples. Decide whether a broad analytical approach is appropriate or whether the essay should contain more factual detail.
3. The conclusion, which you can make quite short. You should use this part to summarize and draw together the components of the essay, without merely repeating previous phrases. You might mention such things as: the broader significance of the topic; its future; its relevance to other important areas of biology. Always try to mention both sides of any debate you have touched on, but beware of 'sitting on the fence'.

KEY POINT Use paragraphs (p. 306) to make the essay's structure obvious. Emphasize them with headings and sub-headings unless the material beneath the headings would be too short or trivial.

Now start writing!

Using diagrams – give a title and legend for each diagram so that it makes sense in isolation and point out in the text when the reader should consult it (e.g. 'as shown in Fig. 1 . . .' or 'as can be seen in the accompanying diagram, . . .').

- Never lose track of the importance of content and its relevance. Repeatedly ask yourself: 'Am I really answering this question?' Never waffle just to increase the length of an essay. Quality rather than quantity is important.
- Illustrate your answer appropriately. Use examples to make your points clear, but remember that too many similar examples can stifle the flow of an essay. Use diagrams where a written description would be difficult or take too long. Use tables to condense information.
- Take care with your handwriting. You can't get marks if your writing is illegible! Try to cultivate an open form of handwriting, making the individual letters large and distinct. If there is time, make out a rough draft from which a tidy version can be copied.

Reviewing your answer

Learning from lecturers' and tutors' comments – ask for further explanations if you don't understand a comment or why an essay was less successful than you thought it should have been.

Don't stop yet!

- Re-read the question to check that you have answered all points.
- Re-read your essay to check for errors in punctuation, spelling and content. Make any corrections obvious. Don't panic if you suddenly realize you've missed a large chunk out as the reader can be redirected to a supplementary paragraph if necessary.

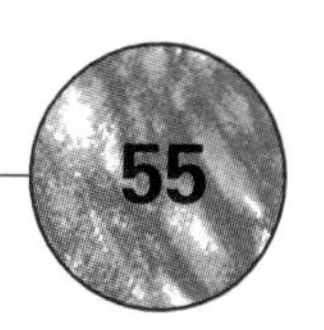

55 Reporting practical and project work

Practical reports, project reports, theses and scientific papers differ greatly in depth, scope and size, but they all have the same basic structure (Box 55.1). Some variation is permitted, however (see Box 55.1), and you should always follow the advice or rules provided by your department.

Additional parts may be specified: for theses, a title page is often required and a List of Figures and Tables as part of the Contents section. When work is submitted for certain degrees, you may need to include certain declarations and statements made by the student and supervisor. In scientific papers, a list of Key Words is often added following the Abstract: this information may be combined with words in the title for computer cross-referencing systems.

KEY POINT **Department or Faculty regulations may specify an exact format for producing your report or thesis. Obtain a copy of these rules at an early stage and follow them closely.**

Options for discussing data – the main optional variants of the general structure include combining Results and Discussion into a single section and adding a separate Conclusions section.

- The main advantage of a joint Results and Discussion section is that you can link together different experiments, perhaps explaining why a particular result led to a new hypothesis and the next experiment.
- The main advantage of having a separate Conclusions section is to draw together and emphasize the chief points arising from your work, when these may have been 'buried' in an extensive Discussion section.

Practical and project reports

These are exercises designed to make you think more deeply about your experiments and to practise and test the skills necessary for writing up research work. Special features are:

- Introductory material is generally short and unless otherwise specified should outline the aims of the experiment(s) with a minimum of background material.
- Materials and methods may be provided by your supervisor for practical reports. With project work, your lab notebook (see p. 43) should provide the basis for writing this section.
- Great attention in assessment will be paid to presentation and analysis of data. Take special care over graphs (see p. 254). Make sure your conclusions are justified by the evidence.

Theses

Theses are submitted as part of the examination for a degree following an extended period of research. They act to place on record full details about your experimental work and will normally only be read by those with a direct interest in it – your examiners or colleagues. Note the following:

- You are allowed scope to expand on your findings and to include detail that might otherwise be omitted in a scientific paper.
- You may have problems with the volume of information that has to be organized. One method of coping with this is to divide your thesis into chapters, each having the standard format (as in Box 55.1). A General Introduction can be given at the start and a General Discussion at the end. Discuss this option with your supervisor as it is not universally favoured.
- There may be an oral exam ('viva') associated with the submission of the thesis. The primary aim of the examiners will be to ensure that you understand what you did and why you did it.

Box 55.1 The structure of reports of experimental work

Undergraduate practical and project reports are generally modelled on this structure or a close variant of it, because this is the structure used for nearly all research papers and theses. The more common variations include Results and Discussion combined into a single section for convenience and Conclusions appearing separately as a series of points arising from the work. In scientific papers, a list of Key Words (for computer cross-referencing systems) may be included following the Abstract. Regarding variations in positioning, Acknowledgements may appear after the Contents section, rather than near the end. Department or faculty regulations for producing theses and reports may specify a precise format; they often require a title page to be inserted at the start and a list of figures and tables as part of the Contents section. These regulations may also specify declarations and statements to be made by the student and supervisor.

Part (in order)	Contents/purpose	Checklist for reviewing content
Title	Explains what the project was about	Does it explain what the text is about succinctly?
Authors plus their institutions	Explains who did the work and where; also where they can be contacted now	Are all the details correct?
Abstract/Summary	Synopsis of methods, results and conclusion of work described. Allows the reader to grasp quickly the essence of the work	Does it explain why the work was done? Does it outline the whole of your work and your findings?
List of Contents	Shows the organization of the text (not required for short papers)	Are all the sections covered? Are the page numbers correct?
Abbreviations	Lists all the abbreviations used (but not those of SI, chemical elements, or standard biochemical terms)	Have they all been explained? Are they all in the accepted form? Are they in alphabetical order?
Introduction	Orientates the reader, explains why the work has been done and its context in the literature, why the methods used were chosen, why the experimental organisms were chosen. Indicates the central hypothesis behind the experiments	Does it provide enough background information and cite all the relevant references? Is it of the correct depth for the readership? Have all the technical terms been defined? Have you explained why you investigated the problem? Have you explained your methodological approach to the problem?
Materials and Methods	Explains how the work was done. Should contain sufficient detail to allow another competent worker to repeat the work	Is each experiment covered and have you avoided unnecessary duplication? Is there sufficient detail to allow repetition of the work? Are proper scientific names and authorities given for all organisms? Have you explained where you got them from? Are the correct names, sources and grades given for all chemicals?
Results	Displays and describes the data obtained. Should be presented in a form which is easily assimilated (graphs rather than tables, small tables rather than large ones)	Is the sequence of experiments logical? Are the parts adequately linked? Are the data presented in the clearest possible way? Have SI units been used properly throughout? Has adequate statistical analysis been carried out? Is all the material relevant? Are the figures and tables all numbered in the order of their appearance? Are their titles appropriate? Do the figure and table legends provide all the information necessary to interpret the data without reference to the text? Have you presented the same data more than once?
Discussion/ Conclusions	Discusses the results: their meaning, their importance; compares the results with those of others; suggests what to do next	Have you explained the significance of the results? have you compared your data with other published work? Are your conclusions justified by the data presented?
Acknowledgements	Gives credit to those who helped carry out the work	Have you listed everyone that helped, including any grant-awarding bodies?
Literature Cited (Bibliography)	Lists all references cited in appropriate format: provides enough information to allow the reader to find the reference in a library	Do all the references in the text appear on the list? Do all the listed references appear in the text? Do the years of publications and authors match? Are the journal details complete and in the correct format? Is the list in alphabetical order, or correct numerical order?

Steps in the production of a practical report or thesis

Choose the experiments you wish to describe and decide how best to present them

Try to start this process before your lab work ends, because at the stage of reviewing your experiments, a gap may become apparent (e.g. a missing control) and you might still have time to rectify the deficiency. Irrelevant material should be ruthlessly eliminated, at the same time bearing in mind that negative results can be extremely important (see p. 49). Use as many different forms of data presentation as are appropriate, but avoid presenting the same data in more than one form. Graphs are generally easier for the reader to assimilate, while tables can be used to condense a lot of data into a small space. Relegate large tables of data to an appendix and summarize the important points. Make sure that the experiments you describe are representative: always state the number of times they were repeated and how consistent your findings were.

Repeating your experiments – remember, if you do an experiment twice, you have repeated it only once!

Make up plans or outlines for the component parts

The overall plan is well defined (see Box 55.1), but individual parts will need to be organized as with any other form of writing (see Chapter 53).

Write!

Presenting your results – remember that the order of results presented in a report need not correspond with the order in which you carried out the experiments: you are expected to rearrange them to provide a logical sequence of findings.

The Materials and Methods section is often the easiest to write once you have decided what to report. Remember to use the past tense and do not allow results or discussion to creep in. The Results section is the next easiest as it should only involve description. At this stage, you may benefit from jotting down ideas for the Discussion – this may be the hardest part to compose as you need an overview both of your own work and of the relevant literature. It is also liable to become wordy, so try hard to make it succinct. The Introduction shouldn't be too difficult if you have fully understood the aims of the experiments. Write the Abstract and complete the list of references at the end. To assist with the latter, it is a good idea as you write to jot down the references you use or to pull out their cards from your index system.

Revise the text

Once your first draft is complete, try to answer all the questions given in Box 55.1. Show your work to your supervisors and learn from their comments. Let a friend or colleague who is unfamiliar with your subject read your text; they may be able to pinpoint obscure wording and show where information or explanation is missing. If writing a thesis, double-check that you are adhering to your institution's thesis regulations.

Prepare the final version

Markers appreciate neatly produced work but a well-presented document will not disguise poor science! If using a word processor, print the final version with the best printer available. Make sure figures are clear and in the correct size and format.

Submit your work

Your department will specify when to submit a thesis or project report, so plan your work carefully to meet this deadline or you may lose marks. Tell your supervisor early of any circumstances that may cause delay.

Producing a scientific paper

Scientific papers are the means by which research findings are communicated to others. They are published in journals with a wide circulation among academics and are 'peer reviewed' by one or more referees before being accepted. Each journal covers a well-defined subject area and publishes details of the format they expect. It would be very unusual for an undergraduate to submit a paper on his or her own – this would normally be done in collaboration with your project supervisor and only then if your research has satisfied appropriate criteria. However, it is important to understand the process whereby a paper comes into being (Box 55.2), as this can help you when interpreting the primary literature.

Box 55.2 Steps in producing a scientific paper

Scientific papers are the lifeblood of any science and it is a major landmark in your scientific career to publish your first paper. The major steps in doing this should include the following:

Assessing potential content
The work must be of an appropriate standard to be published and should be 'new, true and meaningful'. Therefore, before starting, the authors need to review their work critically under these headings. The material included in a scientific paper will generally be a subset of the total work done during a project, so it must be carefully selected for relevance to a clear central hypothesis – if the authors won't prune, the referees and editors of the journal certainly will!

Choosing a journal
There are thousands of journals covering biology and each covers a specific area (which may change through time). The main factors in deciding on an appropriate journal are the range of subjects it covers, the quality of its content and the number and geographical distribution of its readers. The choice of journal always dictates the format of a paper since authors must follow to the letter the journal's 'Instructions to Authors'.

Deciding on authorship
In multi-author papers, a contentious issue is often who should appear as an author and in what order they should be cited. Where authors make an equal contribution, an alphabetical order of names may be used. Otherwise, each author should have made a substantial contribution to the paper and should be prepared to defend it in public. Ideally, the order of appearance will reflect the amount of work done rather than seniority. This may not happen in practice!

Writing
The paper's format will be similar to that shown in Box 55.1 and the process of writing will include outlining, reviewing, etc., as discussed elsewhere in this chapter. Figures must be finished to an appropriate standard and this may involve preparing photographs of them.

Submitting
When completed, copies of the paper are submitted to the editor of the chosen journal with a simple covering letter. A delay of one to two months usually follows while the manuscript is sent to one or more anonymous referees who will be asked by the editor to check that the paper is novel, scientifically correct and that its length is fully justified.

Responding to referees' comments
The editor will send on the referees' comments and the authors then have a chance to respond. The editor will decide on the basis of the comments and replies to them whether the paper should be published. Sometimes quite heated correspondence can result if the authors and referees disagree!

Checking proofs and waiting for publication
If a paper is accepted, it will be sent off to the typesetters. The next the authors see of it is the proofs (first printed version in style of journal), which have to be corrected carefully for errors and returned. Eventually, the paper will appear in print, but a delay of six months following acceptance is not unusual. Most journals offer the authors reprints, which can be sent to other researchers in the field or to those who send in reprint request cards.

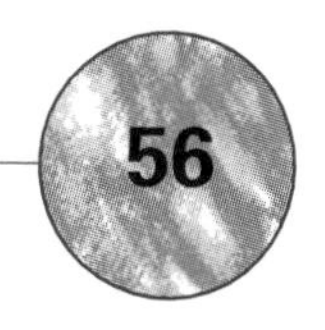

56 Writing literature surveys and reviews

8 (5%) 1 (5%)
7 (10%)
6 (5%)
5 (5%)
2 (45%)
4 (10%)
3 (15%)

Fig. 56.1 Pie chart showing how you might allocate time for a literature survey:
1. select a topic;
2. scan the literature;
3. plan the review;
4. write first draft;
5. leave to settle;
6. structured review of text;
7. write final draft;
8. produce top copy.

The literature survey or review is a specialized form of essay which summarizes and reviews the evidence and concepts concerning a particular area of research.

KEY POINT **A literature review should *not* be a recitation of facts. The best reviews are those which analyse information rather than simply describe it.**

Making up a timetable

Figure 56.1 illustrates how you might divide up your time for writing a literature survey. There are many sub-divisions in this chart because of the size of the task: in general, for lengthy tasks, it is best to divide up the work into manageable chunks. Note also that proportionately less time is allocated to writing itself than with an essay. In a literature survey, make sure that you spend adequate time on research and revision.

Selecting a topic

You may have no choice in the topic to be covered, but if you do, carry out your selection as a three-stage process:

1. Identify a broad subject area that interests you.
2. Find and read relevant literature in that area. Try to gain a broad impression of the field from books and general review articles. Discuss your ideas with your supervisor.
3. Select a relevant and concise title. The wording should be considered very carefully as it will define the content expected by the reader. A narrow subject area will cut down on the amount of literature you will be expected to review, but will also restrict the scope of the conclusions you can make (and vice versa for a wide subject area).

Scanning the literature and organizing your references

You will need to carry out a thorough investigation of the literature before you start to write. The key problems are as follows:

- Getting an initial toe-hold in the literature. Seek help from your supervisor, who may be willing to supply a few key papers to get you started. Hints on expanding your collection of references are given on p. 297.
- Assessing the relevance and value of each article. This is the essence of writing a review, but it is difficult unless you already have a good understanding of the field (Catch 22!). Try reading earlier reviews in your area.
- Clarifying your thoughts. Sometimes you can't see the wood for the trees! Sub-dividing the main topic and assigning your references to these smaller subject areas may help you gain a better overview of the literature.

Using index cards (see p. 299) – these can help you organize large numbers of references. Write key points on each card – this helps when considering where the reference fits into the literature. Arrange the cards in subject piles, eliminating irrelevant ones. Order the cards in the sequence you wish to write in.

Deciding on structure and content

The general structure and content of a literature survey is described below.

Introduction

The introduction should give the general background to the research area, concentrating on its development and importance. You should also make a statement about the scope of your survey; as well as defining the subject matter to be discussed, you may wish to restrict the period being considered.

Main body of text

The review itself should discuss the published work in the selected field and maybe subdivided into appropriate sections. Within each portion of a review, the approach is usually chronological, with appropriate linking phrases (e.g. 'Following on from this, ...'; 'Meanwhile, Bloggs (1980) tackled the problem from a different angle ...'). However, a good review is much more than a chronological list of work done. It should:

- allow the reader to obtain an overall view of the current state of the research area, identifying the key areas where knowledge is advancing;
- show how techniques are developing and discuss the benefits and disadvantages of using particular organisms or experimental systems;
- assess the relative worth of different types of evidence – this is the most important aspect. Do not be intimidated from taking a critical approach as the conclusions you may read in the primary literature aren't always correct;
- indicate where there is conflict in findings or theories, suggesting if possible which side has the stronger case;
- indicate gaps in current knowledge.

Balancing opposing views – even if you favour one side of a disagreement in the literature, your review should provide a fair description of all the published views of the topic. Having done this, if you do wish to state a preference, give reasons for your opinion.

Conclusions

The conclusions should draw together the threads of the preceding parts and point the way forward, perhaps listing areas of ignorance or where the application of new techniques may lead to advances.

References, etc.

The References or Literature Cited section should provide full details of all papers referred to in the text (see p. 300). The regulations for your department may also specify a format and position for the title page, list of contents, acknowledgements, etc.

Making citations – a review of literature poses stylistic problems because of the need to cite large numbers of papers; in the *Annual Review* series this is overcome by using numbered references (see p. 299).

Style of literature surveys

The Annual Review series (available in most university libraries) provides good examples of appropriate style for reviews of the biosciences.

57 Organizing a poster display

A scientific poster is a visual display of the results of an investigation, usually mounted on a rectangular board. Posters are used at scientific meetings, to communicate research findings, and in undergraduate courses, to display project results or assignment work.

In a written report you can include a reasonable amount of specific detail and the reader can go back and re-read difficult passages. However, if a poster is long-winded or contains too much detail, your reader is likely to lose interest.

KEY POINT **A poster session is like a competition – you are competing for the attention of people in a room. Because you need to attract and hold the attention of your audience, make your poster as interesting as possible. Think of it as an advertisement for your work and you will not go far wrong.**

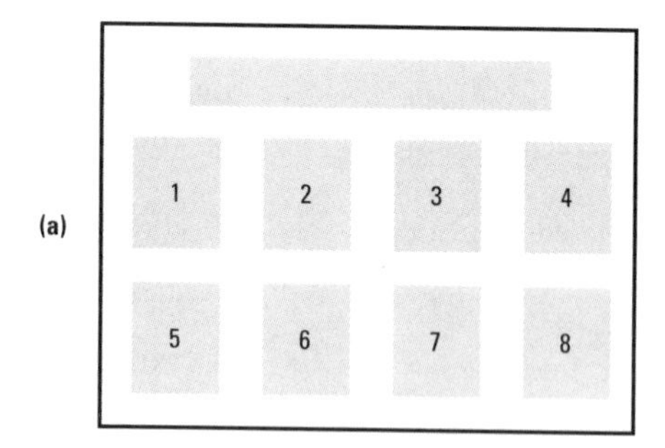

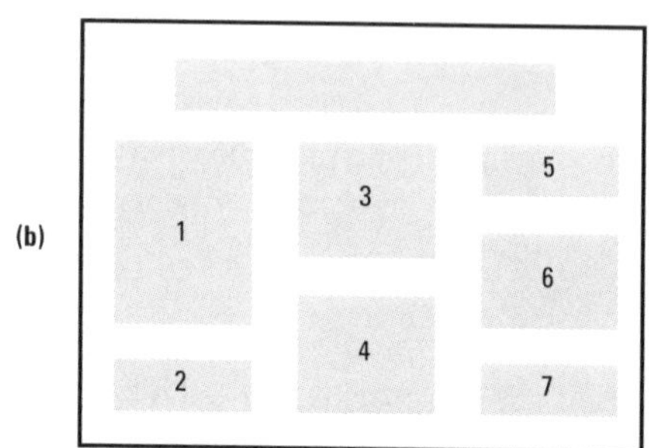

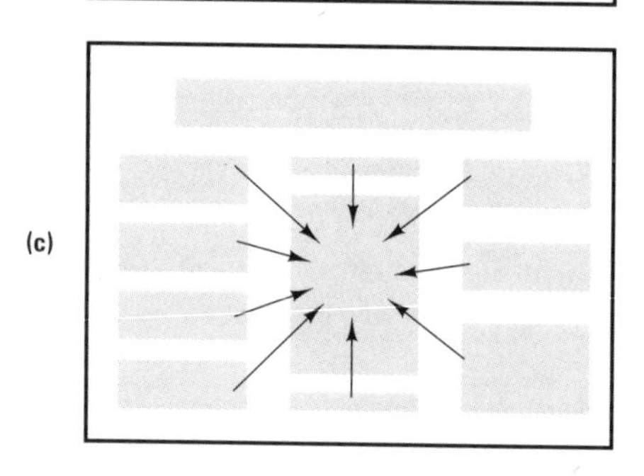

Fig. 57.1 Poster design. (a) An uninspiring design: sub-units of equal area, reading left to right, are not recommended. (b) This design is more interesting and the text will be easier to read (column format). (c) An alternative approach, with a central focus and arrows/tapes to guide the reader.

Preliminaries

Before considering the content of your poster, you should find out:

- the linear dimensions of your poster area, typically up to 1.5 m wide by 1.0 m high;
- the composition of the poster board and the method of attachment, whether drawing pins, Velcro® tape, or some other form of adhesive; and whether these will be provided – in any case, it's safer to bring your own;
- the time(s) when the poster should be set up and when you should attend;
- the room where the poster session will be held.

Design

Plan your poster with your audience in mind, as this will dictate the appropriate level for your presentation. Aim to make your poster as accessible as possible to a broad audience. Since a poster is a visual display, you must pay particular attention to the presentation of information: work that may have taken hours to prepare can be ruined in a few minutes by the ill-considered arrangement of items (Fig. 57.1). Begin by making a draft sketch of the major elements of your poster. It is worth discussing your intended design with someone else, as constructive advice at the draft stage will save a lot of time and effort when you prepare the final version (or consult Simmonds and Reynolds, 1994).

Layout

Usually the best approach is to divide the poster into several smaller areas, perhaps six or eight in all, and prepare each as a separate item on a piece of thick card. Some people prefer to produce a single large poster on one sheet of paper or card and store it inside a protective cardboard tube. However, a single large poster will bend and crease, making it difficult to flatten out. In addition, photographs and text attached to the backing sheet often work loose.

Sub-dividing your poster means that each smaller area can be prepared on a separate piece of paper or card, of A4 size or slightly larger, making transport and storage easier. It also breaks the reading matter up into smaller pieces, looking less formidable to a potential reader. By using pieces of card of different colours you can provide emphasis for key aspects, or link text with figures or photographs.

You will need to guide your reader through the poster. It is often appropriate to use either a numbering system, with large, clear numbers at the top of each piece of card, or a system of arrows (or thin tapes), to show the relationship of sections within the poster (see Fig. 57.1). Make sure that the relationship is clear and that the arrows or tapes do not cross.

Title

Your chosen title should be concise (no more than eight words), specific and interesting, to encourage people to read the poster. Make the title large and bold – it should run across the top of your poster, in letters at least 4 cm high, so that it can be read from the other side of the room. Coloured spirit-based marker and block capitals drawn with a ruler work well, as long as your writing is readable and neat (the colour can be used to add emphasis). Alternatively, you can use Letraset®, or similar lettering. Details of authors, together with their addresses (if appropriate), should be given, usually in the top right-hand corner in somewhat smaller lettering than the title. At conferences, a passport-sized photograph of the contributor is sometimes useful for identification.

Making up your poster – text and graphics printed on good quality paper can be glued directly onto a contrasting mounting card: use photographic spray mountant or Pritt® rather than liquid glue. Trim carefully using a guillotine to give equal margins, parallel with the paper. Photographs should be placed in a window mount to avoid the tendency for their corners to curl. Another approach is to trim pages or photographs to their correct size, then encapsulate in plastic film: this gives a highly professional finish and is less weighty to transport.

Text

Keep text to a minimum – aim to have a maximum of 500 words in your poster. Write in short sentences and avoid verbosity. Keep your poster as visual as possible and make effective use of the spaces between the blocks of text. Your final text should be double-spaced and should have a minimum capital letter height of 5–10 mm, preferably greater, so that the poster can be read at a distance of 1 m. One method of obtaining text of the required size is to photo-enlarge standard typescript (using a good-quality photocopier), or use a high-quality (laser) printer. It is best to avoid continuous use of text in capitals, since it slows reading and makes the text less interesting to the reader. Also avoid italic, 'balloon' or decorative styles of lettering.

Sub-titles and headings

These should have a capital letter height of 12–20 mm, and should be restricted to two or three words. They can be produced by photo-enlargement, by stencilling, Letraset® or by hand, using pencilled guidelines (but make sure that no pencil marks are visible on your finished poster).

Colour

Consider the overall visual effect of your chosen display, including the relationship between text, diagrams and the backing board. Colour can be used to highlight key aspects of your poster. However, it is very easy to ruin a poster by the inappropriate choice and application of colour. Careful use of two, or at most three, complementary colours will be easier on the eye and may aid comprehension. Colour can be used to link the text with the visual images (e.g. by picking out a colour in a photograph and using the same colour on the mounting board for the accompanying text). Use coloured inks or water-based paints to provide colour in diagrams and figures, as felt pens rarely give satisfactory results.

Content

The typical format is that of a scientific report (see Box 55.1), i.e. with the same headings, but with a considerably reduced content. Never be tempted to spend the minimum amount of time converting a piece of scientific writing into poster format. At scientific meetings, the least interesting posters are those where the author simply displays pages from a written communication (e.g. a journal article) on the poster board! Keep references within the text to a minimum – interested parties can always ask you for further information.

Introduction

This should give the reader background information on the broad field of study and the aims of your own work. It is vital that this section is as interesting as possible, to capture the interest of your audience. It is often worth listing your objectives as a series of numbered points.

Designing the materials and methods section – photographs or diagrams of apparatus can help to break up the text of the Materials and Methods section and provide visual interest. It is sometimes worth preparing this section in a smaller typeface.

Materials and Methods

Keep this short, and describe only the principal techniques used. You might mention any special techniques, or problems of general interest.

Results

Don't present your raw data: use data reduction wherever possible, i.e. figures and simple statistical comparisons. Graphs, diagrams, histograms and pie charts give clear visual images of trends and relationships and should be used in place of tabulated data (see p. 251). Final copies of all figures should be produced so that the numbers can be read from a distance of 1 m. Each should have a concise title and legend, so that it is self-contained: if appropriate, a series of numbered points can be used to link a diagram with the accompanying text. Where symbols are used, provide a key on each graph (symbol size should be at least 5 mm). Avoid using graphs straight from a written version, e.g. a project report, textbook, or a paper, without considering whether they need modification to meet your requirements.

Keeping graphs and diagrams simple – avoid composite graphs with different scales for the same axis, or with several trend lines (use a maximum of three trend lines per graph).

Conclusions

This is where many readers will begin, and they may go no further unless you make this section sufficiently interesting. This section needs to be the strongest part of your poster. Refer to your figures here to draw the reader into the main part of your poster. A slightly larger or bolder typeface may add emphasis, though too many different typefaces can look messy.

Listing your conclusions – a series of numbered points is a useful approach, if your findings fit this pattern.

The poster session

If you stand at the side of your poster throughout the session, you are likely to discourage some readers, who may not wish to become involved in a detailed conversation about the poster. Stand nearby. Find something to do – talk to someone else, or browse among the other posters, but remain aware of people reading your poster and be ready to answer any queries they may raise. Do not be too discouraged if you aren't asked lots of questions: remember, the poster is meant to be a self-contained, visual story, without need for further explanation.

Consider providing a handout – this is a useful way to summarize the main points of your poster, so that your readers have a permanent record of the information you have presented.

A poster display will never feel like an oral presentation, where the nervousness beforehand is replaced by a combination of satisfaction and relief as you unwind after the event. However, it can be a very satisfying means of communication, particularly if you follow these guidelines.

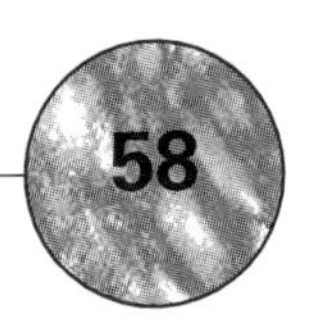

58 Giving an oral presentation

Most students feel very nervous about giving talks. This is natural, since very few people are sufficiently confident and outgoing that they look forward to speaking in public. Additionally, the technical nature of your subject matter may give you cause for concern, especially if you feel that some members of the audience have a greater knowledge than you have. However, this is a fundamental method of scientific communication and it therefore forms an important component of many courses.

The comments in this chapter apply equally to informal talks, e.g. those based on assignments and project work, and to more formal conference presentations. It is hoped that the advice and guidance given below will encourage you to make the most of your opportunities for public speaking, but there is no substitute for practice. Do not expect to find all of the answers from this, or any other, book. Rehearse, and learn from your own experience.

KEY POINT **The three 'Rs' of successful public speaking are: Reflect – give sufficient thought to all aspects of your presentation, particularly at the planning stage. Rehearse – to improve your delivery. Rewrite – modify the content and style of your material in response to your own ideas and to the comments of others.**

Preparation

Preliminary information

Begin by marshalling all of the details you need to plan your presentation, including:

- the duration of the talk;
- whether time for questions is included;
- the size and location of the room;
- the projection/lighting facilities provided, and whether pointers or similar aids are available.

It is especially important to find out whether the room has the necessary equipment for slide projection (slide projector and screen, black-out curtains or blinds, appropriate lighting) or overhead projection before you prepare your audio-visual aids. If you concentrate only on the spoken part of your presentation at this stage, you are inviting trouble later on. Have a look around the room and try out the equipment at the earliest opportunity, so that you are able to use the lights, projector, etc., with confidence.

Learning from experience – use your own experience of good and bad lecturers to shape your performance. Some of the more common errors include:

- speaking too quickly
- reading to notes and ignoring the audience
- unexpressive, impersonal or indistinct speech
- distracting mannerisms
- poorly structured material with little emphasis on key information
- factual information too complex and detailed
- too few visual aids

Audio-visual aids

Find out whether your department has facilities for preparing overhead transparencies and slides, whether these facilities are available for your use and the cost of materials. Adopt the following guidelines:

- Keep text to a minimum: present only the key points, with up to 20 words per slide/transparency.
- Make sure the text is readable: try out your material beforehand.
- Use several simpler figures rather than a single complex graph.
- Avoid too much colour on overhead transparencies: blue and black are easier to read than red or green.
- Don't mix slides and transparencies as this is often distracting.

- Use spirit-based pens for transparencies: use alcohol for corrections.
- Transparencies can be produced from typewritten or printed text using a photocopier, often giving a better product than pens. Note that you must use special heat-resistant acetate sheets for photocopying.

Audience

You should consider your audience at the earliest stage, since they will determine the appropriate level for your presentation. If you are talking to fellow students you may be able to assume a common level of background knowledge. In contrast, a research lecture given to your department, or a paper at a meeting of a scientific society, will be presented to an audience from a broader range of backgrounds. An oral presentation is not the place for a complex discussion of specialized information: build up your talk from a low level. The speed at which this can be done will vary according to your audience. As long as you are not boring or patronizing, you can cover basic information without losing the attention of the more knowledgeable members in your audience. The general rule should be: 'do not overestimate the background knowledge of your audience'. This sometimes happens in student presentations, where fears about the presence of 'experts' can encourage the speaker to include too much detail, overloading the audience with facts.

Content

While the specific details in your talk will be for you to decide, most oral presentations share some common features of structure, as described below.

Introductory remarks

It is vital to capture the interest of your audience at the outset. Consequently, you must make sure your opening comments are strong, otherwise your audience will lose interest before you reach the main message. Remember it takes a sentence or two for an audience to establish a relationship with a new speaker. Your opening sentence should be some form of preamble and should not contain any key information. For a formal lecture, you might begin with 'Mr Chairman, ladies and gentlemen, my talk today is about . . .' then re-state the title and acknowledge other contributors, etc. You might show a transparency or slide with the title printed on it, or an introductory photograph, if appropriate. This should provide the necessary settling-in period.

After these preliminaries, you should introduce your topic. Begin your story on a strong note – this is no place for timid or apologetic phrases. You should:

- explain the structure of your talk;
- set out the aims and objectives of your work;
- explain your approach to the topic.

Opening remarks are unlikely to occupy more than 10% of the talk. However, because of their significance, you might reasonably spend up to 25% of your preparation time on them. Make sure you have practised this section, so that you can deliver the material in a flowing style, with less chance of mistakes.

The main message

This section should include the bulk of your experimental results or literature findings, depending on the type of presentation. Keep details of methods to the minimum needed to explain your data. This is *not* the place for a detailed description of equipment and experimental protocol (unless it is a talk about methodology!). Results should be presented in an easily digested format.

Do not expect your audience to cope with large amounts of data; use a maximum of six numbers per slide. Present summary statistics rather than individual results. Show the final results of any analyses in terms of the statistics calculated, and their significance (p. 269), rather than dwelling on details of the procedures used. Remember that graphs and diagrams are usually better than tables of raw data, since the audience will be able to see the trends and relationships in your data (p. 251). However, figures should not be crowded with unnecessary detail. Every diagram should have a concise title and the symbols and trend lines should be clearly labelled, with an explanatory key where necessary. When presenting graphical data always 'introduce' each graph by stating the units for each axis and describing the relationship for each trend line or data set. Summary slides can be used at regular intervals, to maintain the flow of the presentation and to emphasize the key points.

Allowing time for slides – as a rough guide you should allow at least two minutes per illustration, although some diagrams may need longer, depending on content.

Take the audience through your story step-by-step at a reasonable pace. Try not to rush the delivery of your main message due to nervousness. Avoid complex, convoluted story-lines – one of the most distracting things you can do is to fumble backwards through slides or overhead transparencies. If you need to use the same diagram or graph more than once then you should make two (or more) copies. In a presentation of experimental results, you should discuss each point as it is raised, in contrast to written text, where the results and discussion may be in separate sections. The main message typically occupies approximately 80% of the time allocated to an oral presentation.

Concluding remarks

Having captured the interest of your audience in the introduction and given them the details of your story in the middle section, you must now bring your talk to a conclusion. At all costs, do not end weakly, e.g. by running out of steam on the last slide. Provide your audience with a clear 'take-home message', by returning to the key points in your presentation. It is often appropriate to prepare a slide or overhead transparency listing your main conclusions as a numbered series.

Final remarks – make sure you give the audience sufficient time to assimilate your final slide: some of them may wish to write down the key points. Alternatively, you might provide a handout, with a brief outline of the aims of your study and the major conclusions.

Signal the end of your talk by saying 'finally ...', 'in conclusion ...', or a similar comment and then finish speaking after that sentence. Your audience will lose interest if you extend your closing remarks beyond this point. You may add a simple end phrase (for example, 'thank you') as you put your notes into your folder, but do not say 'that's all folks!', or make any similar offhand remark. Finish as strongly and as clearly as you started.

Hints on presentation

Notes

Many accomplished speakers use abbreviated notes for guidance, rather than reading from a prepared script. When writing your talk:

- Prepare a first draft as a full script: write in spoken English, keeping the text simple and avoiding an impersonal style. Aim to *talk* to your audience, not read to them.
- Use note cards with key phrases and words: it is best to avoid using a full script at the final presentation. As you rehearse and your confidence improves, a set of cards may be a more appropriate format for your notes.
- Consider the structure of your talk: keep it as simple as possible and announce each sub-division, so your audience is aware of the structure.

- Mark the position of slides/key points, etc.: each note card should contain details of structure, as well as content.
- Memorize your introductory/closing remarks: you may prefer to rely on a full written version for these sections, in case your memory fails.
- Use notes: write on only one side of the card/paper, in handwriting large enough to be read easily during the presentation. Each card or sheet must be clearly numbered, so that you do not lose your place.
- Rehearse your presentation: ask a friend to listen and to comment constructively on parts that were difficult to follow.
- Use 'split times' to pace yourself: following rehearsal, note the time at which you should arrive at key points of your talk. These timing marks will help you keep to time during the 'real thing'.

Using slides – check that the lecture theatre has a lectern light, otherwise you may have problems reading your notes when the lights are dimmed.

Image

Ensure that the image you project is appropriate for the occasion:

- Consider what to wear: aim to be respectable without 'dressing up', otherwise your message may be diminished.
- Develop a good posture: it will help your voice projection if you stand upright, rather than slouching or leaning over the lectern.
- Project your voice: speak towards the back of the room.
- Make eye contact: look at members of the audience in all parts of the room. Avoid talking to your notes, or to only one section of the audience.
- Deliver your material with expression: arm movements and subdued body language will help maintain the interest of your audience. However, you should avoid extreme gestures (it may work for some TV personalities, but it isn't recommended for the beginner!).
- Manage your time: avoid looking at your watch as it gives a negative signal to the audience. Use a wall clock, if one is provided, or take off your watch and put it beside your notes, so you can glance at it without distracting your audience.
- Try to identify and control any distracting repetitive mannerisms, e.g. repeated empty phrases, fidgeting with pens, keys, etc., as this will distract your audience. Practising in front of a mirror may help.
- Practise your delivery: use the comments of your friends to improve your performance.

Questions

Many speakers are worried by the prospect of questions after their oral presentation. Once again, the best approach is to prepare beforehand:

- Consider what questions you may be asked: prepare brief answers.
- Do not be afraid to say 'I don't know': your audience will appreciate honesty, rather than vacillation, if you don't have an answer for a particular question.
- Avoid arguing with a questioner: suggest a discussion afterwards rather than becoming involved in a debate about specific details.
- If no questions are forthcoming you may pose a question yourself, and then ask for opinions from the audience: if you use this approach you should be prepared to comment briefly if your audience has no suggestions. This will prevent the presentation from ending in an embarrassing silence.

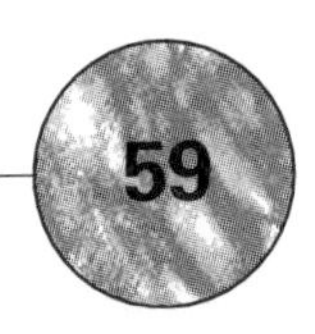

Examinations

You are unlikely to have reached this stage in your education without being exposed to the examination process. However, the following comments should help you to identify and improve on the skills required for exam success.

Information gathering

KEY POINT To do well in an examination, you need to put in effective work long before you go into the examination hall and even before you start to revise. You need to base your revision on accurate, tidy notes with an appropriate amount of subject detail and depth.

Examples Commonly used abbreviations include:

Symbol	Meaning
$\exists$	there are, there exist(s)
$\therefore$	therefore
$\because$	because
$\propto$	is proportional to
$\rightarrow$	leads to, into
$\leftarrow$	comes from, from
$\rightarrow\rightarrow$	involves several processes in a sequence
1°, 2°	primary, secondary (etc.)
$\approx$, $\cong$	approximately, roughly equal to
$=$, $\neq$	equals, not equal to
$\equiv$, $\not\equiv$	equivalent, not equivalent to
$<$, $>$	smaller than, bigger than
$\gg$	much bigger than
[X]	concentration of X
$\sum$	sum
f	function
#	number
∞	infinity, infinite

You should also make up your own abbreviations relevant to the context, e.g. if a lecturer is talking about photosynthesis, you could write 'PS' instead.

Taking notes from lectures

Taking good lecture notes is essential if you are to make sense of them later. Start by noting the date, course, topic and lecturer. Number every page in case they get mixed up later. The most popular way of taking notes is to write in a linear sequence down the page; however, the alternative 'pattern' method (Fig. 59.1) has its advocates: experiment to see which you prefer.

Whatever method you use to take notes, you shouldn't try to take down all the lecturer's words, except when an important definition or example is being given, or when the lecturer has made it clear that he/she is dictating. Listen first, then write. Your goal should be to abstract the structure and reasoning behind the lecturer's approach. Use headings and leave plenty of space, but don't worry too much about being tidy at this stage – it is more important that you get down the appropriate information in a form that you at least can read. Use abbreviations to save time. Make sure you note down references to texts and take special care to ensure accuracy of definitions and numerical examples. If the lecturer repeats or otherwise emphasises a point, make a margin note of this – it could come in useful when revising. If there is something you don't understand, ask at the end of the lecture, and make an appointment to discuss the matter if there isn't time to deal with it then. Tutorials may provide an additional forum for discussing course topics.

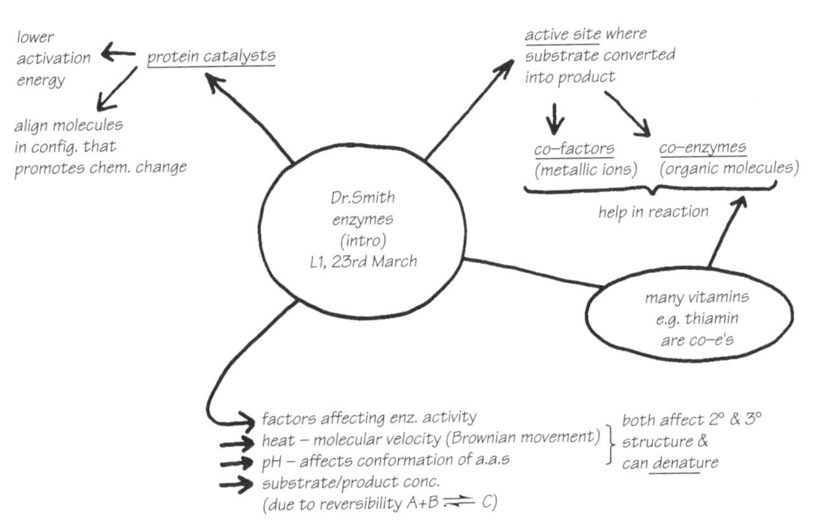

Fig. 59.1 An example of 'pattern' notes, an alternative to the more commonly used 'linear' format.

'Making up' your notes

As soon as possible after the lecture, work through your notes, tidying them up and adding detail where necessary. Add emphasis to any headings you have made, so that the structure is clearer. Compare your notes with material in a textbook and correct any inconsistencies. Make notes from, or photocopy, any useful material you see in textbooks, ready for revision.

Skimming texts

This is a valuable means of gaining the maximum amount of information in the minimum amount of time, by reading as little of a text as is required. It can be used to decide what parts to read in detail and to make notes of, perhaps when you have already read the text in detail at some point in the past.

Essentially, the technique requires you to look at the structure of the text, rather than the detail. In a sense, you are trying to see the writer's original plan and the purpose behind each part of the text. Look through the whole of the piece first, to gain an overview of its scope and structure. Headings provide an obvious clue to structure, if present. Next, look for the 'topic sentence' in each paragraph (p. 306), which is often the first. You might then decide that the paragraph contains a definition that is important to note, or it may contain examples, so may not be worth reading for your purpose.

Seeing sequences – writers often number their points (firstly, secondly, thirdly, etc.) and looking for these words in the text can help you skim it quickly.

Preparing for an exam

Begin by finding out as much as you can about the exam, including:

- its format and duration;
- the date and location;
- the types of question;
- whether any questions/sections are compulsory;
- whether the questions are internally or externally set or assessed;
- whether calculators are required.

Your course tutor is likely to give you details of exam structure and timing well beforehand, so that you can plan your revision: the course handbook and past papers (if available) can provide further useful details. Check with your tutor that the nature of the exam has not changed before you consult past papers.

Organizing and using lecture notes, assignments and practical reports

Given their importance as a source of material for revision, you should have sorted out any deficiencies or omissions in lecture notes/practical reports at an early stage. For example, you may have missed a lecture or practical due to illness, etc., but the exam is likely to assume attendance throughout the year. Make sure you attend classes whenever possible and keep your notes up to date.

Your practical reports and any assignment work will contain specific comments from the teaching staff, indicating where marks were lost, corrections, mistakes, inadequacies, etc. It is always worth reading these comments as soon as your work is returned, to improve the standard of your subsequent reports. If you are unsure about why you lost marks in an assignment, or about some particular aspects of a topic, ask the appropriate member of staff for further explanation. Most lecturers are quite happy to discuss such details with students on a one-to-one basis and this information may provide you with 'clues' to the expectations of individual lecturers that may be useful in exams set by the same members of staff. However, you should *never* 'fish' for specific information on possible exam questions, as this is likely to be counter-productive.

Revision

Begin your revision early, to avoid last-minute panic. Start in earnest about 6 weeks beforehand:

- Prepare a revision timetable – an 'action plan' that gives details of specific topics to be covered. Find out at an early stage when (and where) your examinations are to be held, and plan your revision around this. Try to keep to your timetable. Time management during this period is as important as keeping to time during the exam itself.
- Remember, your concentration span is limited to 15–20 min: make sure you have two or three short (5 min) breaks during each hour of revision.
- Make your revision as active and interesting as possible: the least productive approach is simply to read and re-read your notes.
- Include recreation within your schedule: there is little point in tiring yourself with too much revision, as this is unlikely to be profitable.
- Ease back on the revision near the exam: plan your revision, to avoid last-minute cramming and overload fatigue.

Active revision

The following techniques may prove useful in devising an active revision strategy:

- Prepare revision sheets with details for a particular topic on a single sheet of paper, arranged as a numbered checklist. Wall posters are another useful revision aid.
- Memorize definitions and key phrases: definitions can be a useful starting point for many exam answers.
- Use mnemonics and acronyms to commit specific factual information to memory. The dafter they are, the better they work!
- Prepare answers to past or hypothetical questions, e.g. write essays or work through calculations and problems, within appropriate time limits. However, you should not rely on 'question spotting': this is a risky practice!
- Use spider diagrams as a means of testing your powers of recall on a particular topic (p. 304).
- Try recitation as an alternative to written recall.
- Draw diagrams from memory: make sure you can label them fully.
- Form a revision group to share ideas and discuss topics with other students.
- Use a variety of different approaches to avoid boredom during revision (e.g. record information on audio tape, use cartoons, or *any* other method, as long as it's not just reading notes!).

Preparing for an exam – make a checklist of the items you'll need (e.g. pens, pencils, sharpener and eraser, ruler, calculator, paper tissues, watch).

The evening before your exam should be spent in consolidating your material, and checking through summary lists and plans. Avoid introducing new material at this late stage: your aim should be to boost your confidence, putting yourself in the right frame of mind for the exam itself.

Final preparations – try to get a good night's sleep before an exam. Last minute cramming will be counter-productive if you are too tired during the exam.

The examination

On the day of the exam, give yourself sufficient time to arrive at the correct room, without the risk of being late (e.g. what if your bus breaks down?).

The exam paper

Begin by reading the instructions at the top of the exam paper carefully, so that you do not make any errors based on lack of understanding. Make sure that you know:

- how many questions are set;
- how many must be answered;
- whether the paper is divided into sections;
- whether any parts are compulsory;
- what each question/section is worth, as a proportion of the total mark;
- whether different questions should be answered in different books.

If you are unsure about anything, ask! – the easiest way to lose marks in an exam is to answer the wrong number of questions, or to answer a different question from the one set by the examiner. Underline the key phrases in the instructions, to reinforce their message.

Using the exam paper – unless this is specifically forbidden, you *should* write on the question paper to plan your strategy, keep to time and organize your answers.

Next, read through the set of questions. If there is a choice, decide on those questions to be answered and decide on the order in which you will tackle them. Prepare a timetable which takes into account the amount of time required to complete each question and which reflects the allocation of marks – there is little point in spending one-quarter of the exam period on a question worth only 5% of the total marks! Use the exam paper to mark the sequence in which the questions will be answered and write the finishing times alongside: refer to this timetable during the exam to keep yourself on course.

Do not be tempted to spend too long on any one question: the return in terms of marks will not justify the loss of time from other questions (see Fig. 59.2). Take the first 10 min or so to read the paper and plan your strategy, before you begin writing. Do not be put off by those who begin immediately; it is almost certain they are producing unplanned work of a poor standard.

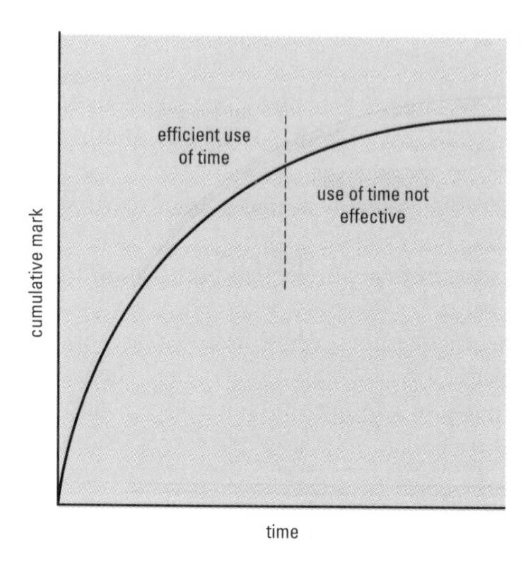

Fig. 59.2 Exam marks as a function of time. The marks awarded in a single answer will follow the law of diminishing returns – it will be far more difficult to achieve the final 25% of the available marks than the initial 25%. Do not spend too long on any one question.

Providing answers

Before you tackle a particular question, you must be sure of what is required in your answer. Ask yourself 'What is the examiner looking for in this particular question?' and then set about providing a *relevant* answer. Consider each individual word in the question and highlight, underline or circle the key words. Make sure you know the meaning of the terms given in Table 54.1 (p. 308) so that you can provide the appropriate information, where necessary. Refer back to the question as you write, to confirm that you are keeping to the subject matter. Box 59.1 gives advice on writing essays under exam conditions.

It is usually a good idea to begin with the question that you are most confident about. This will reassure you before tackling more difficult parts of the paper. If you run out of time, write in note form. Examiners are usually understanding, as long as the main components of the question have been addressed and the intended structure of the answer is clear.

The final stage

At the end of the exam, you should allow at least 10 min to read through your script, to check for:

- grammatical and spelling errors;
- mathematical errors.

Box 59.1 Writing under exam conditions

Never go into an exam without a strategy for managing the available time.

- **Allocate some time (say 5% of the total) to consider which questions to answer and in which order.**
- **Share the rest of the time among the questions.** Aim to optimize the marks obtained. A potentially good answer should be allocated *slightly* more time than one you don't feel so happy about. However, don't concentrate on any one answer (see Fig. 59.2).
- **For each question divide the time into planning, writing and revision phases** (see p. 308).

Employ time-saving techniques as much as possible.

- **Use spider diagrams** (p. 304) to organize and plan your answer.
- **Use diagrams and tables** to save time in making difficult and lengthy explanations.
- **Use abbreviations** to save time repeating text but *always* explain them at the first point of use.
- **Consider speed of writing and neatness** especially when selecting the type of pen to use – ball-point pens are fastest, but they can smudge. You can only gain marks if the examiner can read your script!
- **Keep your answer simple and to the point**, with clear explanations of your reasoning.

Make sure your answer is relevant.

- **Don't include irrelevant facts** just because you memorized them during revision as this may do you more harm than good. You must answer the specific question that has been set.
- **Time taken to write irrelevant material is time lost from another question.**

After the exam – try to avoid becoming involved in prolonged analyses with other students over the 'ideal' answers to the questions; after all, it is too late to change anything at this stage. Go for a walk, watch TV for a while, or do something else that helps you relax, so that you are ready to face the next exam with confidence.

Make sure your name is on each exam book and on all other sheets of paper, including graph paper, even if securely attached to your script, as it is in your interest to ensure that your work does not go astray.

Never leave any exam early. Most exams assess work carried out over several months in a time period of 2–3 h and there is always something constructive you can do with the remaining time to improve your script.

Practical exams: special considerations

The prospect of a practical examination may cause you more concern than a theory exam. This may be due to a limited experience of practical examinations, or the fact that practical and observational skills are tested, as well as recall, description and analysis of factual information. Your first thoughts may be that it is not possible to prepare for a practical exam but, in fact, you can improve your performance by mastering the various practical techniques described in this book. The principal types of question you are likely to encounter include:

- Manipulative exercises, often based on work carried out as part of your practical course (e.g. sterile technique, p. 59).
- 'Spot' tests: short answer questions requiring identification, or brief descriptive notes on a specific item (e.g. a prepared slide).
- Numerical exercises, including the preparation of aqueous solutions at particular concentrations (p. 13) and statistical exercises (p. 268). General advice is given in Chapter 45.
- Data analysis, including the preparation and interpretation of graphs (p. 254) and numerical information, from data either obtained during the exam or provided by the examiner.
- Drawing a specimen, with accurate representation and labelling.
- Preparation of a specimen for examination with a microscope: this may test staining technique and skills in light microscopy (Chapter 8).

- Interpretation of photographic material: sometimes used when it is not possible to provide living specimens, e.g. electron micrographs or photographs of DNA sequencing gels.

Practical reports

You may be allowed to take your laboratory reports and other texts into the practical exam. Don't assume that this is a soft option, or that revision is unnecessary: you will not have time to read large sections of your reports or to familiarize yourself with basic principles, etc. The main advantage of 'open book' exams is that you can check specific details of methodology, reducing your reliance on memory, provided you know your way around your practical manual. In all other respects, your revision and preparation for such exams should be similar to theory exams. Make sure you are familiar with all of the practical exercises, including any work carried out in class by your partner (since exams are assessed on individual performance). Check with the teaching staff to see whether you can be given access to the laboratory, to complete any exercises that you have missed.

The practical exam

At the outset, determine or decide on the order in which you will tackle the questions. A question in the latter half of the paper may need to be started early on in the exam period (e.g. an enzyme assay requiring 2-h incubation in a 3-h exam). Such questions are included to test your forward-planning and time-management skills. You may need to make additional decisions on the allocation of material, e.g. if you are given 30 sterile test tubes, there is little value in designing an experiment that uses 25 of these to answer question 1, only to find that you need at least 15 tubes for subsequent questions!

Make sure you explain your choice of apparatus and experimental design. Calculations should be set out in a stepwise manner, so that credit can be given, even if the final answer is incorrect (see p. 259). If there are any questions that rely on recall of factual information and you are unable to remember specific details, e.g. you cannot identify a particular specimen, or slide, make sure that you describe the item fully, so that you gain credit for observational skills. Alternatively, leave a gap and return to the question at a later stage.

References

Anon. (1963) Tables of Spectrophotometric Absorption Data for Compounds used for the Colorimetric Detection of Elements, *International Union of Pure and Applied Chemistry*, Butterworth-Heinemann, London.

Billington, D., Jayson, G.G. and Maltby, P.J. (1992) *Radioisotopes*. Bios, Oxford.

Briscoe, M.H. (1990) *A Researcher's Guide to Scientific and Medical Illustrations*. Springer Verlag, Berlin.

Budavari, S., *et al.* (1996) *The Merck Index: An Encyclopedia of Chemicals, Drugs and Biologicals*, 12th edn. Merck & Co., Inc., Rahway, New Jersey.

Causton, D.R. and Venus, J.C. (1981) *The Biometry of Plant Growth*. Edward Arnold, London.

Clausen, J. (1989) *Immunochemical Techniques for the Identification and Estimation of Macromolecules*, 3rd edn. Elsevier, Amsterdam.

Collins, C.H., Lyne, P.M. and Grange, J.M. (1995) *Microbiological Methods*, 7th edn. Butterworth-Heinemann, London.

Dodds, J.H. and Robert, L.W. (1995) *Experiments in Plant Tissue Culture*, 3rd edn. Cambridge University Press, Cambridge.

Eason, G., Coles, C.W. and Gettinby, G. (1992) *Mathematics and Statistics for the Bio-Sciences*. Ellis Horwood, Chichester.

Finney, D.J., Latscha, R., Bennett, B.M. and Hsu, P. (1963) *Tables for Testing Significance in a 2 × 2 Table*. Cambridge University Press, Cambridge.

Flowers, T.J. and Yeo, A.R. (1992) *Solute Transport in Plants*. Blackie, London.

Ford, T.C. and Graham, J.M. (1991) *An Introduction to Centrifugation*. Bios, Oxford.

Freshney, R.I. (1994) *Culture of Animal Cells: A Manual of Basic Techniques*, 3rd edn. A.R. Liss, New York.

Gadian, D.G. (1995) *NMR and its Applications to Living Systems*, 2nd edn. Oxford Science Publications, Oxford.

Geider, R.J. and Osborne, B.A. (1992) *Algal photosynthesis*. Chapman and Hall, New York.

Gerhardt, P. (ed.) (1994) *Manual of Methods for General and Molecular Bacteriology*. American Society for Microbiology, Washington DC.

Gersten, D.M. (1996) *Gel Electrophoresis of Proteins*. Essential Techniques Series. Bios, Oxford.

Golterman, H.L., Clymo, R.S. and Ohnstad, M.A.M. (1978) *Methods for Physical and Chemical Analysis of Fresh Waters*. Blackwell Scientific Publications, Oxford.

Gralla, P. (1996) *How the Internet Works*. Ziff-Davis Press, Emeryville, California.

Green, E.J. and Carritt, D.E. (1967) New Tables for Oxygen Saturation of Seawater, *Journal of Marine Research*, 25, pp. 140–7.

Gunstone, F.D., Harwood, J.L. and Padley, F.B. (1986) *The Lipid Handbook*. Chapman & Hall, London.

Hanahan, D., Jessee, J. and Bloom, F.R. (1995) Techniques for transformation of *E. coli*. In: *DNA Cloning I* (ed D.M. Glover & B.D. Hames) Practical Approach Series, IRL Press, Oxford.

Heath, D. (1995) *An Introduction to Experimental Design and Statistics for Biology*. UCL Press, London.

Johnstone, R.A.W. and Rose, M.E. (1996) *Mass Spectrometry for Chemists and Biochemists*, 2nd edn. Cambridge University Press, Cambridge.

Jones, H.G. (1992) *Plants and Microclimate: a Quantitative Approach to Environmental Plant Physiology*, 2nd edn. Cambridge University Press, Cambridge.

Kates, M. (1986) *Techniques of Lipidology*, 2nd edn. Elsevier, Amsterdam.

Kirby, L.T. (1990) *DNA Fingerprinting: an Introduction*. Stockton Press, New York.

Lentner, C. (ed.) (1981) *Geigy Scientific Tables, 8th edn, vol. 1: Units of Measurement, Body Fluids, Composition of the Body, Nutrition*. Ciba-Geigy, Basel.

Lentner, C. (ed). (1981) *Geigy Scientific Tables, 8th edn, vol. 3: Physical Chemistry, Composition of Blood, Hematology, Somatometric Data*. Ciba-Geigy, Basel.

Lentner, C. (ed. (1988) *Geigy Scientific Tables, 8th edn, vol. 4: Biochemistry, Metabolism of Xenobiotics, Inborn Errors of Metabolism, Pharmacogenetics, Ecogenetics*. Ciba-Geigy, Basel.

Lentner, C. (ed.) (1981) *Geigy Scientific Tables, 8th edn, vol. 5: Heart and Circulation*. Ciba-Geigy, Basel.

Lentner, C. (ed.) (1992) *Geigy Scientific Tables, 8th edn, vol. 6: Bacteria, Fungi, Protozoa, Helminths*. Ciba-Geigy, Basel.

Lentner, C, Diem, K. and Seldrup, J. (eds) (1982) *Geigy Scientific Tables, 8th edn, vol. 2: Introduction to Statistics, Statistical Tables, Mathematical Formulae*. Ciba-Geigy, Basel.

Lide, D.R. and Frederikse, H.P.R. (eds) (1996) *CRC Handbook of Chemistry and Physics*, 7th edn. CRC Press, Boca Raton, Florida.

Lüning, K.J. (1981) 'Light', in C.S. Lobban and M.J. Wynne (eds), *The Biology of Seaweeds*, pp. 326–55, Blackwell Scientific, Oxford.

Milazzo, G., Caroli, S. and Sharma, V.K. (1978) *Tables of Standard Electrode Potentials*. Wiley, London.

Nicholls, D.G. and Ferguson, S.J. (1992) *Bioenergetics 2*. Academic Press, London.

Nobel, P.S. (1991) *Physicochemical and Environmental Plant Physiology*. Academic Press, New York.

Perrin, D.D. and Dempsey, B. (1974) *Buffers for pH and Metal Ion Control*. Chapman and Hall, London.

Robinson, R.A. and Stokes, R.H. (1970) *Electrolyte Solutions*. Butterworth-Heinemann, London.

Roitt, I. (1997) *Essential Immunology*, 9th edn. Blackwell Scientific Publications, Oxford.

Simmonds, D. and Reynolds, L. (1994) *Data Presentation and Visual Literacy in Medicine and Science*. Butterworth-Heineman, London.

Skoog, D.A. and Leary, J.J. (1992) *Principles of Instrumental Analysis*, 4th edn. Harcourt Brace, Fort Worth.

Sokal, R.R. and Rohlf, F.J. (1994) *Biometry*, 3rd edn. W.H. Freeman and Co., San Francisco.

Stace, C. (1997) *New Flora of the British Isles*, 2nd edn. Cambridge University Press, Cambridge.

Stahl, E. (1965) *Thin Layer Chromatography – a Laboratory Handbook*. Springer Verlag, Berlin.

Stein, W.D. (1990) *Channels, Carriers and Pumps: an Introduction to Membrane Transport*. Academic Press, New York.

Warburg, O. and Christian, W. (1942) Isolierung und Kristallisation des Garungferments Enolase, *Biochemische Zeitschrift*, **310**, pp. 384–421.

Wardlaw, A.C. (1985) *Practical Statistics for Experimental Biologists*. John Wiley and Sons Inc., New York.

Westermeier, R. (1993) *Electrophoresis in Practice*. VCH, Weinheim.

White, B. (1991) *Studying for Science*. E. & F.N. Spon, Chapman and Hall, London.

Winship, I. and McNab, A. (1996) *The Student's Guide to the Internet*. Library Association, London.

Index

Index

Index